Contemporary Research Methods in Neuroanatomy

Edited by

Walle J. H. Nauta
Sven O. E. Ebbesson

With 190 Figures

Springer-Verlag New York Heidelberg Berlin 1970

This volume contains the Proceedings of an International Conference on

Contemporary Research Methods in Neuroanatomy

held at the Laboratory of Perinatal Physiology, San Juan, Puerto Rico, in January 1969, under the auspices of the National Institute of Neurological Diseases and Stroke and the University of Puerto Rico.

ISBN 978-3-642-85988-5 ISBN 978-3-642-85986-1 (eBook)
DOI 10.1007/978-3-642-85986-1

Preface

As a source of information on neuroanatomical research methods this Volume is not without precedent. In 1957, at the initiative of Dr. W. F. Windle, a conference was held at the National Institutes of Health, the proceedings of which, edited by Dr. Windle and published by C. C. Thomas under the title "New Research Techniques of Neuroanatomy", rapidly became something like a standard reference in the field of Neuromorphology. The present editors were emboldened to seek support for a second exposé of contemporary research methods in Neuroanatomy by the success of this earlier publication, as well as by the consideration that the years elapsed since its appearance have been, perhaps, more productive of new research methods and strategies in Neuroanatomy than were any dozen consecutive years since the golden decades of the 1870's and 1880's.

The decision, which methods to include in this conference, has been a difficult one. For reasons of space alone it would have been impossible to do equal justice to techniques approaching the brain as a neuronal system, the brain as a tissue, or the neuron as a cell. As a brief inspection of the contents of this volume will show, the weight of choice fell upon the first of these alternatives. The reader will find, furthermore, that not all of the book is devoted to new methods in the strict sense. In fact, a majority of the chapters deal with such matters as: re-interpretation of the results that are obtained by traditional methods, new applications for long-established techniques, and promising combinations of old and newer procedures.

The conference reported in this volume could not have been held without the generous sponsorship of the National Institute of Neurological Diseases and Stroke, and the administrative and organizational support provided by the University of Puerto Rico. At the latter Institute, special thanks are due Dr. Adán Nigaglioni, Chancellor of the Faculty of Medicine, and Professor Walter L. Stiehl, Chairman of the Department of Anatomy, for their thoughtful guidance in planning the conference.

But credits are owed a wider circle of supporters. Dr. Ronald E. Myers, himself a participant, generously placed the facilities of the Laboratory of Perinatal Physiology at the disposal of the conference. Mrs. Paulina Papp, Miss Lolyn Lopez, Miss Emelie Ortiz, and Messrs. C. Torres, G. Hammond and O. Nieves deserve credit for making the necessary domestic and administrative arrangements, Messrs. P. Sola and J. Camunas for their efficient handling of projection and recording, Dr. Kalman Rubinson and Mrs. Sarah Powers for transcribing the records of the discussions, and Miss Elizabeth B. Jones and Mrs. Lauvonda Young for secretarial assistance in making the manuscripts ready for the Press.

January 1970

W. J. H. Nauta
S. O. E. Ebbesson

Contents

Participants in the Conference

THEODORE BLACKSTAD, Department of Anatomy, University of Aarhus, Denmark

C. B. G. CAMPBELL, Center for Neural Sciences, The University of Indiana, Bloomington, Indiana 47401

W. MAXWELL COWAN, Department of Anatomy, Washington University School of Medicine, St. Louis, Missouri 63130

SVEN O. E. EBBESSON, Department of Neurosurgery, University of Virginia Medical School, Charlottesville, Virginia 22901

CLEMENT A. FOX, Department of Anatomy, Wayne State University, Detroit, Michigan 48207

JOSÉ G. FRONTERA, Department of Anatomy, University of Puerto Rico, San Juan, Puerto Rico

KJELL FUXE, Department of Histology, Karolinska Institute, Stockholm, Sweden

GUNNAR GRANT, Department of Human Anatomy, University of Uppsala, Uppsala, Sweden

R. W. GUILLERY, Department of Anatomy, University of Wisconsin, Madison, Wisconsin 53706

LENNART HEIMER, Department of Psychology, Massachusetts Institute of Technology, Cambridge, Massachusetts 02139

HARVEY J. KARTEN, Department of Psychology, Massachusetts Institute of Technology, Cambridge, Massachusetts 02139

BRUCE W. KONIGSMARK, Department of Pathology, Johns Hopkins University, School of Medicine, Baltimore, Maryland 21218

RONALD E. MYERS, Laboratory of Perinatal Physiology, National Institute of Neurological Diseases and Stroke, San Juan, Puerto Rico

WALLE J. H. NAUTA, Department of Psychology, Massachusetts Institute of Technology, Cambridge, Massachusetts 02139

ALAN PETERS, Department of Anatomy, Boston University, School of Medicine, Boston, Massachusetts 02118

J. M. PETRAS, Department of Neurophysiology, Walter Reed Army Institute of Research, Washington, D.C. 20012

ENRIQUE RAMÓN-MOLINER, Département d'Anatomie, Université de Sherbrooke, Faculté de Médecine, Sherbrooke, P. Q., Canada

A. RUIZ-MARCOS, Instituto Cajal, Madrid, Spain

Arnold and Mila Scheibel, Department of Anatomy, University of California School of Medicine, Los Angeles, California 90024

Richard L. Sidman, Department of Neuropathology, Harvard Medical School, Boston, Massachusetts 02115

Facundo Valverde, Instituto Cajal, Madrid, Spain

John Van Buren, National Institute of Neurological Diseases and Stroke, Bethesda, Maryland 20014

Theodore J. Voneida, Department of Anatomy, Case Western Reserve University, Cleveland, Ohio 44106

The Rapid Golgi Method. Indian Summer or Renaissance?

MADGE E. SCHEIBEL and ARNOLD B. SCHEIBEL[1, 2]

It is not unreasonably maintained that the golden days of Golgi morphology were already numbered by the turn of the century. The great flowering of descriptive histology in the hands of Golgi, Cajal, Van Gehuchten, Lenhossek, Held, Retzius, P. Ramon, and others was the kind of phenomenon which can only occur at one stage in the life of a powerful investigative technique. Once the maximal exploitation of its powers had occurred and "the cream was skimmed", a more realistic assessment of its role in neurologic research was possible. It could then become part of the technological armamentarium available to all investigators and might expect the continued, if less heralded, utilization it had earned.

It is a curious thing that the method of Golgi did not actually follow this course. For after the heroic age of chrome-silver exploration, the number of its practitioners rapidly fell away and by the late thirties and early forties, its last distinguished advocates, Lorente de Nó and O'Leary, had turned to other techniques. The uncanny silencing of Golgi studies in the literature for more than a decade deserves consideration.

We believe that the phenomenon is a direct outgrowth of the power of the method itself. In the quarter century of its maximal use, a wealth of structural data had been accumulated. Such data were peculiarly the product of this method, data that stressed organization of neuropil fields, the intimate morphology of synapses, the position and pattern of axons and their collaterals, etc. As elaborate as the circuitry was revealed to be, there was still no adequate means of exploring the functional implications of these patterns except through notion and hypothesis. The Golgi method had literally supplied a detailed map of freeway systems before automobiles had been developed to use them.

The entire field of the neurosciences, especially neurophysiology, behavior, and information theory had to develop to the point where this reservoir of data could become functionally meaningful before the Golgi was once more pressed into service. Now, at the end of the sixties, the Golgi study has again become fashionable at a time when techniques of vastly augmented resolution would seem ready to supersede it. It is our contention, on the other hand, that its relevance is greater now than ever, and that it must become the indispensable companion of electron microscope and microelectrode by defining the matrix within which data from these more modern techniques can be interpreted.

[1] Supported by U.S.P.H.S. Grant 01063, N.I.N.D.S.

[2] We wish to thank Miss Lore Liepmann for her help in preparation of the histological material.

The electron microscope has increased the resolution of neurohistological structure by at least 2 orders of magnitude. Its ultra thin sections and tremendous magnification combine to provide an image of the nervous system consisting essentially of collections of membranes. Intensive correlation studies with other techniques combined with some degree of educated guessing have enabled identification of various membranes as presynaptic or postsynaptic, as portions of intracytoplasmic or intranuclear organelles, etc. The presence or absence of extracellular space has been argued depending on the technical handling of the tissue (Harreveld, 1965), and there are even hints as to some degree of organization in portions of the extracellular compartment (Robertson, 1966). The use of the electron microscope following differential centrifugation of homogenized brain tissue through density gradients has proven a powerful tool for derivation and study of reasonably pure samples of certain endocellular structures such as mitochondria, synaptic vesicles, etc. (Gray and Whittaker, 1960; De Robertis, 1966). The goal of eventual visualization of individual molecules is already within grasp as resolution, under special circumstances has been pushed to the level of 6 to 8 Å (Perkin-Elmer Corp., 1968).

It is obvious that the raw information obtainable from these increasingly elegant techniques becomes useful only as it can be related to known structural entities with definable physiological properties. We are in possession of one type of information when we can identify a profile complex as that of a small dendrite complete with spine and spine apparatus, microtubules and granular material. Equally important is the ability to specify that it is the external segment of a basilar dendrite generated by a small pyramid of layer III in the second somatic (S2) receptive area. This kind of translational information is best provided by the Golgi methods[3]. It is obvious that powerful advantages can accrue to those willing to use electron microscopy secondarily on fragments of tissue or even on individual neural elements stained and identified by initial Golgi procedures. Such combined methodologies are described in the chapter by Blackstad.

Microelectrophysiological techniques can reveal a great deal of information about spike discharge characteristics of individual or small groups of cells. In the former, firing patterns may increase, decrease, or be quenched entirely; or they may be structured in clusters with regular or irregular parameters. When the stimulus is under experimental control, a number of inferences may be drawn as to the nature and position of presynaptic inputs responsible for the behavior. The rapid Golgi method with its capacity for staining fine axonal as well as soma-dendrite components is clearly the method of choice in establishing the relevance (or irrelevance) of each inference. In some cases, especially where cortical structures are involved, impregnation may be attempted directly on the tissue block containing the mark of the microelectrode tip. Success may also follow where brain stem or spinal cord recording has been achieved in the early postnatal stage before myelinization is substantially advanced. For adult axial core tissue, the rapid Golgi may be attempted following perfusion with formalin or glutaraldehyde; or one of a number of other modifications using substances other than osmic acid may be substituted. An obvious alternative, especially in poor Golgi-risk tissue such as adult brain stem, is to establish location of the microelectrode tip

[3] The Ehrlich methylene blue technique remains too capricious to be used regularly as a research tool in the vertebrate nervous system.

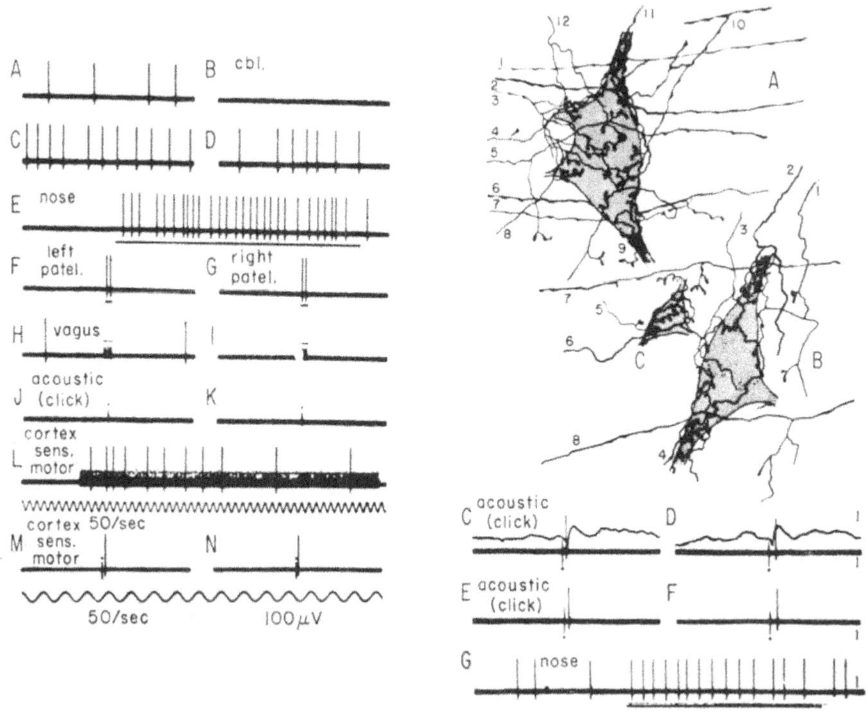

Fig. 1 A—N. Convergence of heterogenous afferents upon individual neurons of the medullary reticular core demonstrated by microphysiological and Golgi histological techniques. Strips A through N and C through G illustrate patterns of extracellularly derived spike discharges from two elements of the bulbar reticular formation. A: spontaneous discharge; B: inhibition by cerebellar polarization (anterior lobe, surface positive); C: rebound following cessation of polarization; D: return toward normal spontaneous discharge pattern; E: driven by nose pressure; F and G: driven by patellar tendon taps bilaterally; H and I: unaffected by short bursts of vagal nerve stimulation; J and K: unaffected by auditory stimuli (clicks); L: driven by repetitive cortical stimulation; M and N: driven after brief latency by single shocks to cortex (note expanded time base under these two records). Strips C, D, E, F, show another unit that is sensitive to nose pressure and can also be driven by auditory clicks. Cells A, B, and C are bulbar reticular neurons in a ten day old kitten, whose synaptic scale of terminating afferents is partly shown. Horizontal-running fibers such as A, 1 through 7 and B, 6 through 8 may belong to long spino-reticular and reticulo-reticular components while B, 1 through 4 represent sensory collaterals and cerebello-reticular terminals which could be traced back far enough to permit identification of their cells of origin. (From Scheibel and Scheibel, in: Reticular Formation of the Brain, by permission of Little Brown & Co., Boston 1958)

in routine Nissl or Klüver preparations, and then study the same region in tissue blocks from younger animals where the power of the Golgi method can be expressed more effectively. Clearly, a trade-off must be made between the precision with which the recording locus is identified, and the likelihood of obtaining a usable impregnation in the area of concern (Fig. 1).

Developmental neurology has become an active subspecialty of its own over the past 15 years. In part, perhaps, because of its rather late start among neurobiological disciplines, its practice has triggered an unusual degree of correlative research,

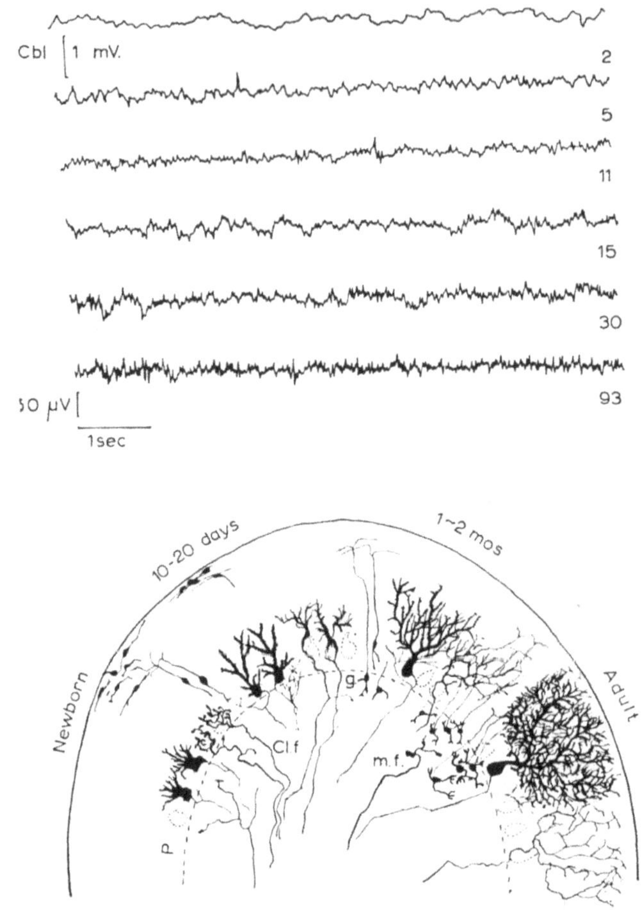

Fig. 2. Structuro-functional correlates of cerebellar cortical maturation in the young cat as revealed by electrophysiological and Golgi histological methods. Upper series of traces shows progressive maturation of the electro-cerebellogram from the amorphous pattern of the neonatal period toward the adult rapid (circa 300/sec) pattern reached between the second and third postnatal month. Drawing at bottom summarizes in clockwise fashion the growth sequence of several histological components of the cerebellar cortex including Purkinje cells, *P*; granule cells, *g*; climbing fibers, *Cl.f.*; and mossy fibers, *m.f.* (From Scheibel and Scheibel, in: The Developing Brain, by courtesy of Elsevier Publishing Co., Amsterdam 1964)

spreading over the areas of neurobiochemistry, histochemistry, electrophysiology, neuroanatomy, and behavior. It is highly desirable that progressive developmental changes be related with reasonable rigor to the changing histological substrate. Most of the usual histological, and histochemical stains are capable of showing changes in certain neural parameters such as cortical thickness, size of neuronal somata and nuclei, number and size of glial cells, cell packing density, degree of chromatophilia, etc. But it is also clear that the vast majority of glial and neuronal surfaces remain invisible. And it is becoming equally clear that many of the most significant developmental changes, especially in the postnatal period, occur in the dendrites and axons of nerve cells, and the peripheral processes of glia. Once again, the rapid Golgi

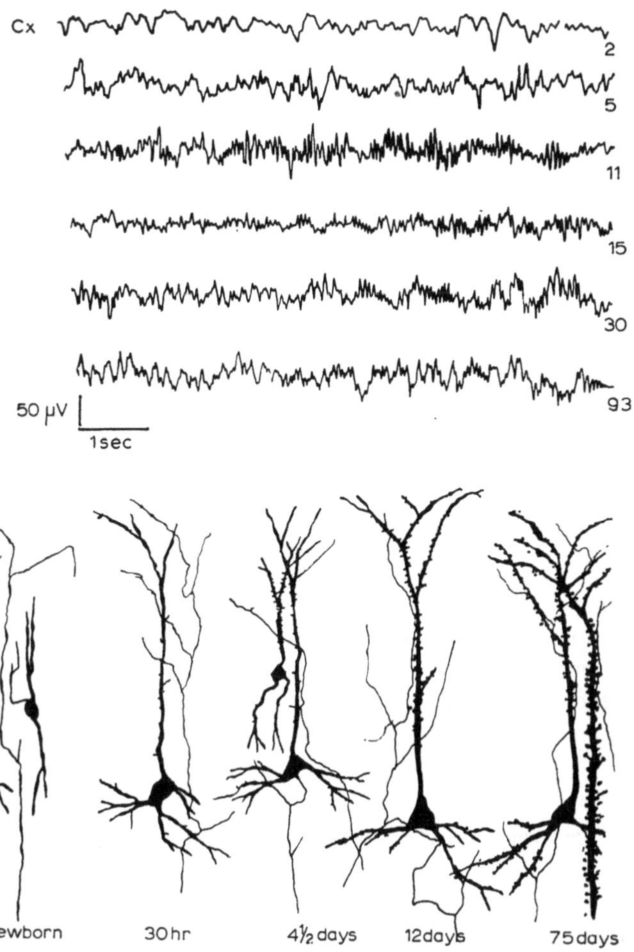

Fig. 3. Structuro-functional correlates of cortical maturation in the young cat as revealed by electrophysiological and Golgi histological methods. Upper series of traces shows development of cortical electrical patterns from the second to the ninety-third postnatal day. Spindle bursts appeapr by the eleventh day and the record is rich in alpha rhythms by the thirtieth day. By the ninety-third day, the record is indistinguishable from the adult tracing. The lower series of drawings summarize the development of the dendritic apparatus of cortical pyramids and their increasingly intensive synaptic relations with presumed nonspecific corticipetal fibers. (From Scheibel, in: Recent Advances in Biological Psychiatry, Vol. 4, by permission of Plenum Press, New York 1962)

method fulfills a unique role in the area of optical microscopy by enabling visualization of these systems as they become increasingly complex. Fig. 2 epitomizes a correlative structural and electrophysiological study of the developing feline cerebellum. No one-to-one correspondence can yet be drawn between any component of the electrocerebellogram and the underlying structural components. On the other hand, even cursory examination shows that massive dendritic arborization and complex terminal axonal branching go along with increasingly rapid cerebellar rhythms. The high-dry and/or oil immersion objectives reveal further details of pre and postsynaptic surfaces, including the progressive development of dendritic spines, elaborating axonal terminal fields and growing glial complexity. Similar changes of an equally dramatic nature can be followed in postnatal cerebral cortex (Fig. 3) where the establishment of contact between long ascending corticipetal fibers of presumed reticular origin (the nonspecific system of the brain stem) with dendrite (apical shaft) spines correlates closely in time with the development of EEG spindling, consistent EEG arousal to reticular stimulation and the appearance of 'activated sleep' patterns (Scheibel and Scheibel, 1964 and unpublished). Once the nature and location of these presumably significant structural elements have been established with the rapid Golgi method, electron microscopy can then, logically, be employed for follow-up, higher resolution study.

Model building has become an increasingly significant part of activity in the brain sciences. Two varieties of this pursuit have special relevance to our area of concern. The first devolves around development of models from electrophysiological (usually intracellular microphysiological) data and is exemplified by the interneuronal (short axon cell) theory of inhibition. The second depends on analytic, conceptual techniques ranging from "elementary quantitative and logical models to differential and field equations, including 'black box' models of sensory-motor systems, digital computer simulation programs, and electronic and gas-ion analogs" (Reiss, 1964). Currently popular models of dendrite systems emphasizing spatio-temporal patterns of synaptic inputs are included in this category. Ultimately, most of the conceptual systems depend on, or imply, a particular circuit paradigm considered capable of generating the subsumed properties. The relevance of the proposed circuit can best be tested by studying the natural system impregnated by the rapid Golgi method or one of its variants.

The use of intracellular recording as a routine technique in neurophysiology led to the discovery of the hyperpolarizing potential (I.P.S.P.) as an apparently invariable concomitant of the inhibitory process (Eccles, 1953). This suggested, in turn, the necessity for chemical mediation of synaptic processes. The short axoned cell hypothesis of inhibition, originally intended to provide electrotonic biasing of the postsynaptic membrane (Brooks and Eccles, 1947) now was reconverted to supply appropriate inhibitory mediator substances. In this manner, Dale's thesis (1935) was conserved and simultaneously, an extra synapse was provided to account for the slightly greater latency to initiation of the I.P.S.P. (Eccles, 1953). The spinal cord was the testing ground for development of these models of postsynaptic inhibition, the best examples of which were taken to be the flexor side of the myotatic reflex arc (also known as afferent collateral inhibition), and the antidromic inhibitory effect of Renshaw, also known as recurrent inhibition (Eccles, Fatt, and Koketsu, 1954; Eccles, 1964).

The circuits proposed for each of these systems have been investigated by means of rapid Golgi methods in young and adult cat spinal cord. Careful study of the relevant histology in thousands of preparations from several hundred animals has led us to conclude that in each case, the postulated circuitry is not supported by the synaptic patterns that can be visualized.

1. Large caliber (presumably 1 A) primary afferent collateral fibers can be followed directly to their synaptic terminals on motoneurons. Identification of the nature of

these cell groups is based on comparison with previous retrograde degeneration studies such as those of Romanes (1951). We have found that approximately 40% of large primary afferents project to one motoneuron pool alone (simple fibers) whereas 30% project to two or more motor nuclei of similar function (compound fibers) and the remaining 30% to motor pools of antagonistic function (complex fibers) (Scheibel and Scheibel, 1969a). If only a few percent of such fibers simultaneously innervating flexor and extensor pools were seen, they might be dismissed as displaced or variant elements which were statistically insignificant. However, approximately one third of all the large primary afferent collaterals we have studied appear to innervate both extensor and flexor pools without an intercalated interneuron on either leg of the circuit, forcing us to conclude that this represents a functionally meaningful pattern. Apparently the intercalated inhibitory interneuron is not a necessary feature on the flexor (inhibitory) side of the myotatic reflex arc.

2. In the case of recurrent inhibition, a group of physiological data has been interpreted as indicating that the inhibition of motoneurons following antidromic stimulation of the ventral root fibers depends on axon recurrent collaterals playing back cholinergically upon a special group of ventro-medially situated short axoned internuncials called Renshaw cells (Eccles *et al.*, 1954). Having reversed the sign of the synaptic process, these projected in turn to adjacent motoneurons whose inhibition was presumed to enhance the precision of the ongoing motor performance.

Golgi analysis of the anterior horn revealed that there were no short axoned cells in the expected ventro-medial locus, nor in fact anywhere else in the ventral two thirds of the spinal grey matter. All internuncials in this area were long axoned cells which, furthermore, projected to the contralateral side. Since the Renshaw effect had been found to be exclusively ipsilateral, we were forced to conclude that Renshaw interneurons, if they existed, had to be long axoned projecting cells (proprioneurons) located either centrally or laterally in the anterior horn (Scheibel and Scheibel, 1966).

Further difficulties were discovered with the initial leg of the proposed arc, as it became clear that 30 to 50% of large motoneuron axons were not provided with collaterals. In general, those cells which were the farthest from the ventral root exit zone appeared to be the ones most likely to develop recurrent branches. But even here, the demands of the model were not easily met since almost one half of the collaterals which we could follow were traced back to synaptic loci along the dendrites of the parent motoneuron. Another appreciable fraction (approximately one third) could be followed to terminal sites on other motoneurons (identified by their own axon trajectories) without intervention of intercalated interneurons. Quite clearly, anatomical analysis of spinal motoneuron systems using rapid Golgi methods did not support the Renshaw cell model of postsynaptic inhibition (Scheibel and Scheibel, 1966, 1969a).

Conceptual models of the type recently popularized by Rall (1962) bear a somewhat different relationship to the methods of Golgi. Here, the anatomical substrate, expressed in terms of the dendritic domain, branching pattern of individual dendrite shafts and the site and density of presynaptic terminals, comes first. With the structural characteristics of the system in hand, idealized models can then be generated to facilitate prediction of the properties inherent in such a system (Fig. 4). Finally the validity of such models can be subjected to electrophysiological analysis (Rall *et al.*, 1967). Farley (1964) reminds us that "... an unlimited number of models exist for any set of experimental data ..." In one context or another, the rapid Golgi method remains crucial in establishing the most appropriate fit.

Experimental methods. The power of the Golgi as a descriptive anatomical technique, and the quasi-experimental nature of each attempted impregnation has tended to obscure its usefulness as an experimental method in the modern sense. However, over the past few years it has been found peculiarly effective in answering certain

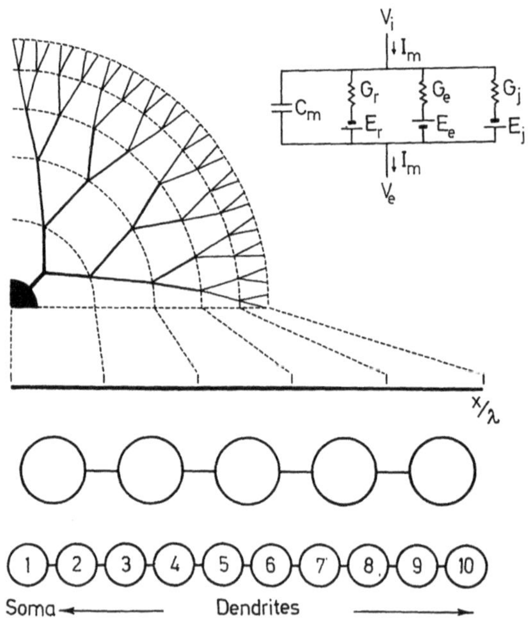

Fig. 4. Schematic diagrams representing simple models of nerve membrane (upper right), portion of dendritic tree (left), and and the soma-dendrite surface (bottom). The major portion of the figure indicates the mathematical transformation of a class of dendritic trees into an equivalent cylinder and into approximately equivalent chains of equal compartments. [From Rall: J. Neurophysiol. **30**, 1138—1168 (1967), by permission of author and Publisher]

kinds of questions put to the nervous system. This arises in part from its ability, unique among silver stains, for revealing the entire postsynaptic (dendritic) as well as the presynaptic topography. This capability has proven especially useful in determining the precise site of termination of a group of axonal terminals along an extended dendritic system such as the apical shafts of cortical pyramids. In this case where the individual dendrites may extend as far as 2 mm from the cell body, position of the terminals along the shaft can significantly affect their impact upon final neuronal output (Rall, 1962). Since it seems quite clear that dendritic spines disappear following their deafferentation (Colonnier, 1964; Walberg, 1963), it follows that a selected group of lesions placed in relevant afferent systems should lead via appropriate spine density determinations to post hoc mapping of the entire terminal axonal array originally investing the dendrite system. This has already been accomplished in part, in the cerebral cortex of young rabbits (Globus and Scheibel, 1967 a, b) with unexpected results.

Lesions in a corticipetal projection (geniculo-calcarine) led to spine loss limited to the central three fifths of the apical shafts of cortical pyramids. There was no convincing evidence that stellate cells of the fourth layer were directly involved in this projection despite the inferences of a number of previous investigators (Cajal, 1908; Lorente de Nó 1949, etc.). Commissural afferents crossing in the corpus callosum were found to terminate only on the oblique branches of the apical shafts (Globus and Scheibel, 1967 b) (Fig. 5) while it seems likely on the basis of partial data that recurrent collaterals of pyramidal cell axons terminate mainly on basilar dendrites and perhaps on apical arches (Globus and Scheibel, unpublished). As a result of such findings, we were led to the working hypothesis that extra-cortical afferents terminate selectively along vertical dendrite components (those perpendicular to the cortical surface) while afferents of intra-cortical origin are localized on the horizontal elements. Further support for this notion seems to come from observations suggesting that the non-specific corticipetal projection from the intralaminar system of the thalamus terminates along the entire length of the apical shafts, though in a low density pattern (Scheibel, Globus, and Scheibel, unpublished). Further cortical studies employing this technique seem desirable. The method is applicable to any other part of the neuraxis where spiny dendrites exist.

Neuropathology. The rapid Golgi methods are not ordinarily thought of as lending themselves to the study of pathological tissue. Indeed, one may thumb through the index of a standard neuropathological text such as Greenfields' (1967) without finding a single reference to Golgi use. It has become increasingly obvious, however, that in those conditions where there may be involvement of nerve cell processes, especially dendrites, the technique can prove one of the most powerful available to optical microscopy. Over 15 years age we began to look at temporal lobe biopsy tissue from epileptic patients (Scheibel and Scheibel, unpublished), and recently have been able to continue the program, on both fresh and formalin fixed material. In addition to gliosis of the hippocampus, the rapid Golgi has clearly shown beaded, spine-poor, and distorted dendrite profiles associated with shrunken neurons, a picture also recently described in experimental material by Ward's group (Westrum *et al.*, 1964). It seems reasonable to predict more extensive use of Golgi methods to augment present conceptions of the histological appearance of abnormal brain tissue and to establish a frame of reference within which the findings of histochemist and electron microscopist can better be interpretated.

Despite the large number of Golgi modifications which can be found in any textbook of technique and which are periodically augmented in contemporary literature, few fundamental improvements have appeared. Though victimized by its own mystique, it has never been a difficult method to master. Unlike almost any other histological technique in current use, however, it is at best a partial method. Depending on a number of variables, some of which can be controlled, it demonstrates 1 to 5% of the elements in a field, often with startling clarity and completeness. Here lies its strength, and for the unwary, its weakness. For the rapid Golgi may be considered a chemical dissection wherein the nature of the chemistry is still uncertain, and the laws of the dissection, unknown.

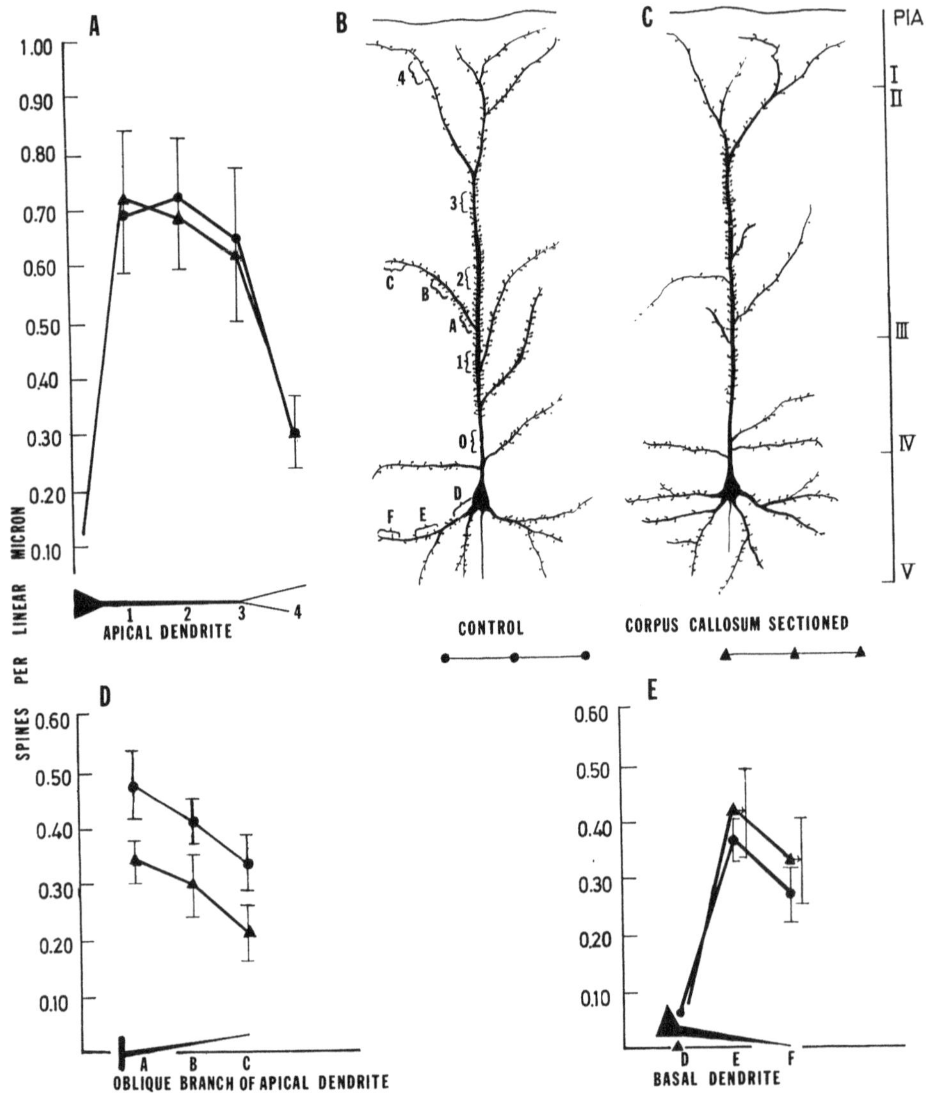

Fig. 5A—E. The use of the rapid Golgi method as an experimental technique is epitomized in this semi-diagrammatic and graphic representation of the effect of section of the corpus callosum on pyramidal cells of parietal cortex. Graphs A, D, and E represent spine counts on apical dendrites, oblique branches, and basilar dendrites respectively. Values for callosum-sectioned animals are shown by filled triangles, those for control animals by filled circles. There is significant loss of spines along the oblique branches. Brackets indicate range of ± 1 standard deviation. B and C represent cells from control and callosum-sectioned cortex and indicate the loss of spines along oblique branches of the apical shaft. Standard counting stations for each class of dendrite branches are shown in B. [From Globus and Scheibel: Science **156**, 1127—1129 (1967), by permission of the Publisher]

But the secret to its use is a simple one as the Sage of Pavia taught us — "provando e riprovando" (Golgi, 1878, 1886), and in the repetition, the development of a total image growing, mosaic-like, from innumerable glimpses of the brain. The picture which is gradually reconstructed is more than just the sum of its parts. For it also represents a summation of the observers subjective impressions of the neural complex, seen by him in so many phases of partial impregnation. It is in fact this subjective component, called 'functional intuition' for want of a better name, which, in the last analysis gives life to the structural idea. It is, perhaps, the ability to fuse rigorously objective synthesis with imaginative integration which distinguishes the work of the masters of this powerful method from so many others who must remain at best adequate Golgi technicians. And it may well be this dynamic breathing quality of the construct as it takes shape, the combination of explicit delineation, tipped with aesthetic judgment, which ensures its continued relevance in a technologically changing setting.

The Golgi Method. A Tool for Comparative Structural Analyses

F. Valverde

For almost a century the rapid Golgi method has been recognized as one of the best, and probably the most elegant procedure of inquiry into the general morphology of the brain. The technique provides in unsurpassed clarity and faithful delineation the shape and spatial arrangement of isolated nerve units as no other method can.

In the following presentation I will give a description of the technique currently employed in this laboratory, to be followed by a discussion of its capacities and limitations, with some comments on what is known about the chemistry of the reaction. A review of its application as a tool in the analysis of certain structural patterns and the modification of these patterns as a consequence of sensory deprivation will conclude this presentation.

The Method

I am not going to deal with any one of the multiple modifications of the procedure used and developed since many years; instead I will limit myself to the *rapid Golgi method*, the very same procedure which was used so profitably by Cajal.

The Golgi staining of nerve cells and their processes is achieved by the immersion of pieces of nervous tissue in an osmium-dichromate solution, followed by their impregnation in dilute solutions of silver nitrate. The treatment may be repeated for a total of two or three times. After this cycle has been completed the pieces are adequately attached and oriented on the microtome, sectioned and mounted serially.

Tissue specimens may be taken directly from brain removed immediately after death following a lethal dose of an anesthetic or, alternatively, from animals perfused with the osmium-dichromate solution. In either case the procedure for staining the pieces is as follows:

1. Pieces of nervous tissue no more than 4—5 mm in thickness are immersed in an aqueous solution containing 2.33% w/w potassium dichromate and 0.19% w/w osmium tetroxide. The solution is made up by dissolving 12 g of potassium dichromate and 1 g of osmium tetroxide in 500 ml of distilled water (Valverde, 1965); it can be stored in the refrigerator (2° C) for several weeks. A minimum of 20 cc of this solution should be used for each piece. Usually two or three pieces can be immersed in an adequate volume of the solution in the same glass container. The material is kept at room temperature (18—25° C) for *seven* days. It is not necessary to keep the pieces in the dark.

2. The pieces are rinsed briefly in a small volume of 0.75% aqueous silver nitrate, and stored for *24 hours* in a fresh volume of 0.75% silver nitrate. For each piece 50 cm^3 of the solution should be used.

3. The pieces are blotted and transferred without rinsing to the same osmium-dichromate solution used in 1. They are left in this solution for *six* days.

4. Step 2 is repeated, using a new silver solution in which the pieces are left for *48 hours*.

5. The pieces are blotted and transferred to a new osmium-dichromate solution (same volume) for *five* days.

6. Step 2 is repeated, using a new silver solution and increasing the time in this solution to *three days*.

7. The pieces are dehydrated in absolute alcohol for 5 minutes and embedded in a soft paraffin matrix by pressing each piece gently into the melted paraffin. During this operation the piece should be kept wet with absolute alcohol applied by an eye-dropper.

8. Sections 100—300 μ are cut wet with absolute alcohol on a sliding microtome. The pieces may be cleared before sectioning in oil of cloves or terpineol for 10 minutes. The sections are transferred individually from the surface of the knife to absolute alcohol for 15 minutes. They can be kept in sequence by flattening them against the bottom of the container. Subsequently, the sections are transferred individually into oil of cloves or terpineol. Each section is allowed to float and is then pushed down with a brush onto the bottom of the dish in serial alignment. The sections should be left in the oil for 15 minutes.

9. The sections are mounted on glass slides, washed in xylene to remove the oil, and covered thinly with any usual neutral mounting medium. No cover slip is used.

Results and Interpretations

The method can be used to study the nervous system of any animal species. It is applicable equally to central nervous structures, peripheral sense organs, and free nerve endings. Although it still remains to be applied for the first time to many groups of animals, it can already be said that the Golgi method is universally applicable.

The nerve cells, dendrites, non-myelinated axons and fine terminal arborizations are stained dark brown against a yellow background. Neuroglial cells are only rarely impregnated. Blood capillaries are stained light brown.

The value of the Golgi technique rests upon the fact that only about 10% of the nerve cells appear stained, a circumstance allowing complete studies of individual cells and their processes. Provided an appropriate orientation of the section, the axonal relations of neighboring cells can be readily traced. The technique outlined above yields impregnation of very fine axonal ramifications and their terminal structures, whether complex axonal endings such as those of the olfactory filaments, the grape-like terminals of retinal fibers in the lateral geniculate body, the mossy hippocampal and cerebellar fibers, the laminated-type of endings in the ventral thalamus and trapezoid nucleus, the perisomal cerebellar baskets and the delicate baskets in the septal nuclei and the spinal gelatinous substance, or simpler forms such as the finger-like short collateral twigs in the cerebral cortex and reticular formation of the brain stem. Several examples illustrate the results of the above procedure in some terminal structures (Figs. 1—3, 4A, 5B) and in various types of nerve cells (Figs. 4B, 5A, 6). Nerve cells and dendrites appear clearly outlined and the dendritic spines are clearly and consistently stained.

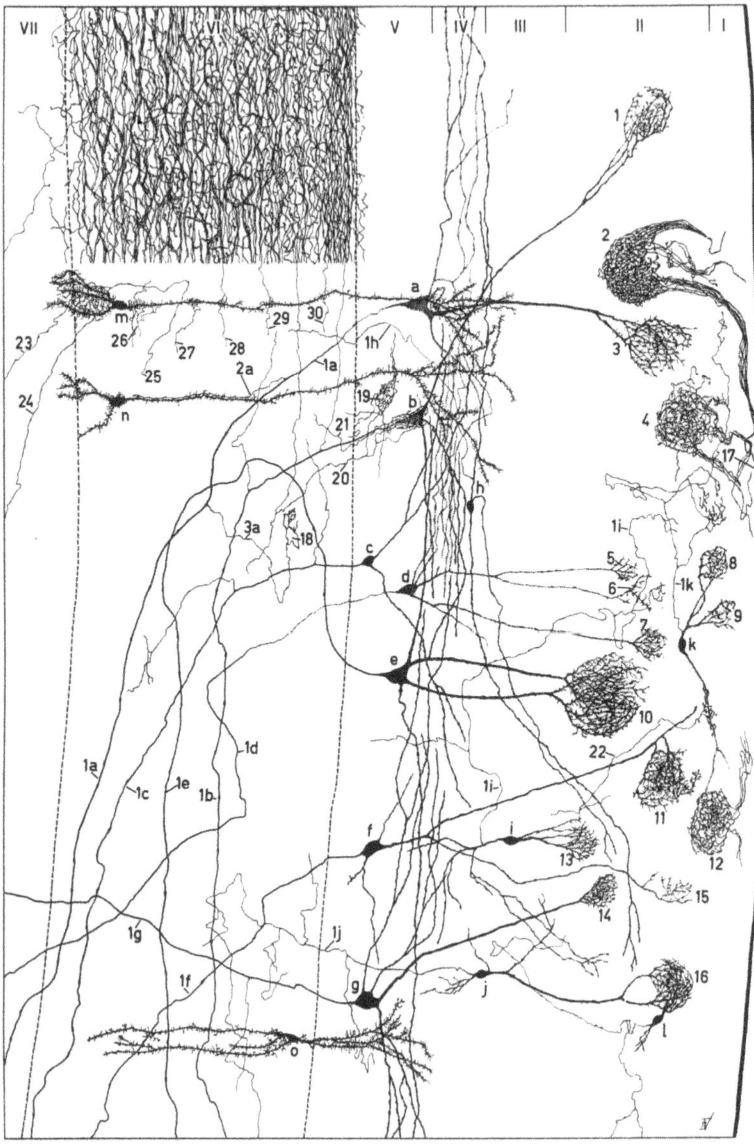

Fig. 1. Horizontal section through the olfactory bulb. Albino rat 9 days old. The illustration integrates several aspects of cellular and fibrillar arrangement and details of synaptic contacts. The terminal ramifications of olfactory nerve fibers (*17*) contact the dendritic ramifications of mitral (a, b, d, e, f and g), tufted (i and j) and short-axoned periglomerular (k and l) cells at the olfactory glomeruli, an interlacement of fibers and dendrites whose different components have been drawn separately in the figure (*1* to *16*). The olfactory glomerulus appears as one of the most interesting places of synaptic integration; it is of historical interest to recall that Cajal's observations on the olfactory glomerulus played a major role in the formulation of the neuron doctrine. Another part of the drawing shows contacts of the descending processes of the internal granular cells (m, n and o) with horizontal dendrites of mitral cells. The nature of these internal granular cells and the mode of their contacts with mitral cells have been a matter of considerable recent interest. Golgi method. (From F. Valverde, 1965; reproduced with permission of Harvard University Press)

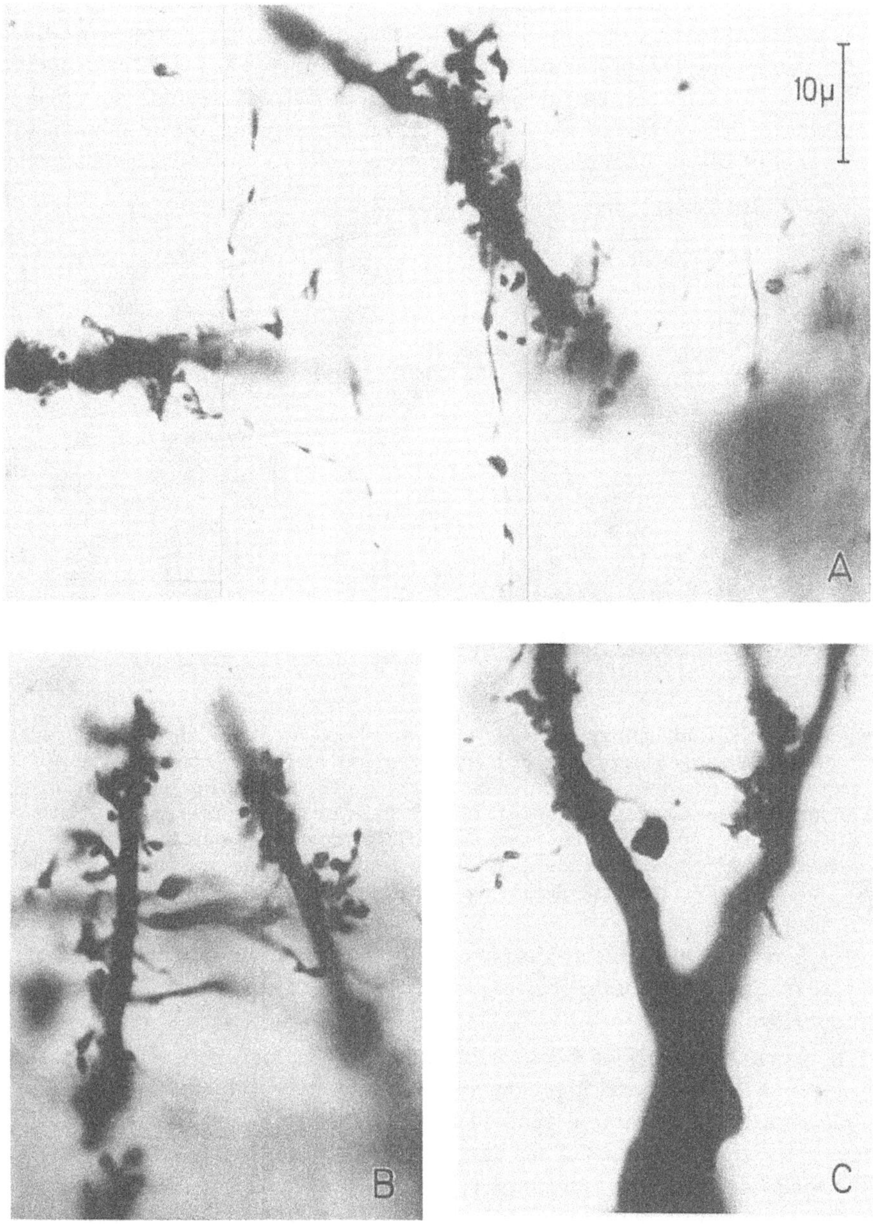

Fig. 2A—C. Mossy fiber synapses in the hippocampus. A. Mouse 19 days old. Mossy fiber synapses on dendritic portions of pyramidal cells. Mosaic photomicrographic reconstruction. B. Mouse 19 days old. Another detail of synaptic contacts between mossy fibers and the dendritic spines of two large dendrites of one pyramidal cell. C. Mouse 21 days old. A third example of synaptic contacts between mossy fibers and the dendrites of one pyramidal cell. Note the large axon terminal on one of the dendrites. Same magnification for the three pictures. Golgi method

Successful impregnation of mice brains was obtained in this laboratory by the use of a perfusion-fixation technique similar to that described by Morest and Morest (1966). Following this procedure, a saline solution buffered at pH 7.2 with sodium barbital at 37° C is perfused through the heart for 10—20 seconds, followed by perfusion of 30—50 cm³ of the above-described osmium-dichromate solution (unbuffered, pH 4.1) at room temperature. The preparatory steps and the construction

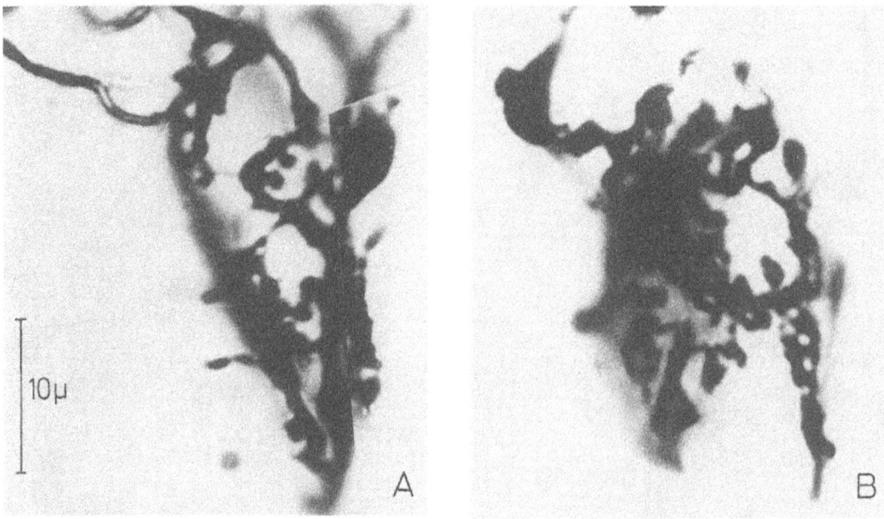

Fig. 3A and B. Rudimentary cerebellar climbing fibers. A. Mouse 10 days old. Mosaic photomicrographic reconstruction. Plan 100/1.25 objective. B. Cat 7 days old. Plan 40/0.65 objective. Same final magnification for both pictures. The afferent Purkinje fibers are still in an immature stage, the so-called stage of pericellular nest. Thick fibers arriving from the white substance develop a complex of interlacing variçosities which approach and then embrace the Purkinje bodies. Later these fibers will climb along the main Purkinje dendritic stem, establishing the adult stage. Golgi method

of the perfusion apparatus are in accordance with the usual procedures and techniques of perfusion-fixation (Koenig *et al.*, 1945; Palay *et al.*, 1962; Morest and Morest, 1966).

It is impossible to give rules for obtaining the impregnation of any particular region or structure. Local impregnation tendencies vary with the age and species of the animal, and appear to depend on many unknown and therefore uncontrollable factors. Particularly successful staining has, for example, been obtained in myelin-deficient *quaking* and *jimpy* mutant mice (Valverde and Sidman, 1965). However, even though local impregnation requirements are quite unpredictable, it is often possible to promote particular results by pragmatic means. When a particular region is to be studied, a detailed trial schedule should be prepared in advance, starting with several pieces subjected individually to different times of immersion in the osmium-dichromate solution. Variation of the time of immersion in the silver baths seems to have no effect on the degree and distribution of the impregnation, but the optimal schedule of immersion of the tissue in the osmium-dichromate solution (for instance 7—6—5 in the above schedule) varies from one brain region to the other, and can be established by preliminary experimentation. Once an optimum time schedule has been determined, a reasonably consistent Golgi impregnation of the region in question can be obtained routinely.

The following example will illustrate the importance of the time sequences employed in the recurring osmium-dichromate treatments. In several cases, successful staining of nerve cells and their processes was obtained after a second impregnation (7—3). In these cases the pieces had been immersed for the second time in a new

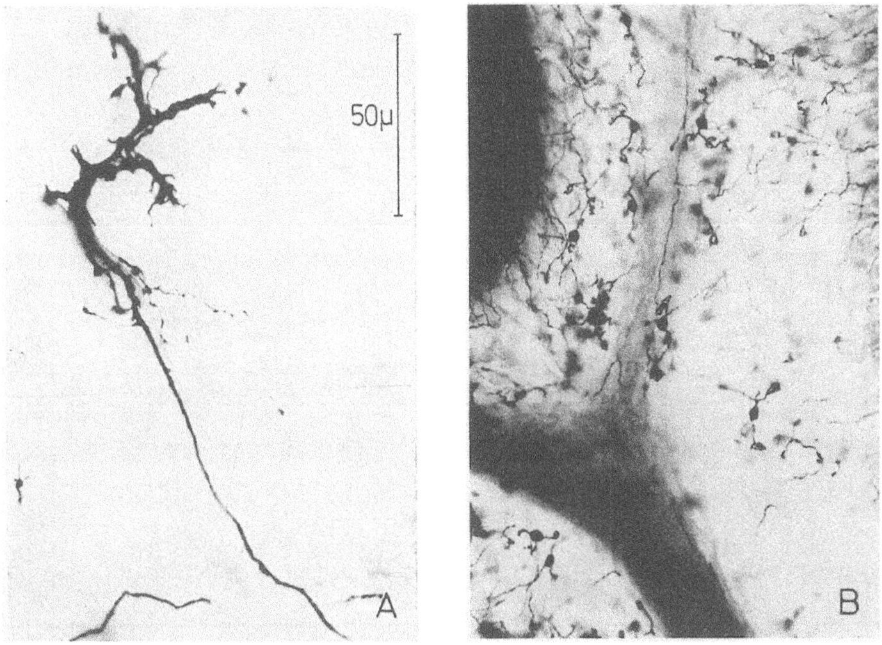

Fig. 4A and B. Cerebellum. A. Mouse 12 days old. A climbing fiber in a developmental stage more advanced than that shown in Fig. 3. Mosaic photomicrographic reconstruction. The afferent fiber climbs along the main dendrites of a (unstained) Purkinje cell. B. Hamster 21 days old. Granule cells showing characteristic claw-shaped dendritic ramifications, special dendritic formations destined to synapse with mossy excrescences and other types of axons forming the cerebellar glomeruli. This later formation had represented another example of the fundaments of the neuron doctrine. (B by courtesy of M. Marin-Padilla.) Golgi method

osmium-dichromate solution for *three* days, rinsed and silvered for two days, and sectioned. This procedure caused the staining of a large number of cells, but failed to impregnate terminal arborizations and delicate end structures in the neuropil which would appear completely stained only after completing the triple impregnation.

An important safety factor in the evaluation of comparative structural analyses lies in the use of subjects of common genetic lineage. In any planned work this precaution would help to avoid the presence of possible morphological inter-group differences derived from random mating. In our current studies on neurons in the mouse visual system and the structural modifications which such neurons undergo as a consequence of sensory deprivation, a line of mice derived from an inbred stock was therefore used. Comparisons between closely related subjects can thus be made as to dendritic spine distribution, dendritic orientation and axonal organization under a variety of experimental conditions such as dark rearing or eye enucleation.

The structures impregnated by the Golgi method can be adequately reproduced only by means of camera lucida drawings. Microphotography does not allow to follow in single pictures the complete trajectory of long dendritic or axonal processes oriented in diverse directions. In exceptionally sharp and clear preparations, nervous structures can be photographed with low numerical aperture objectives

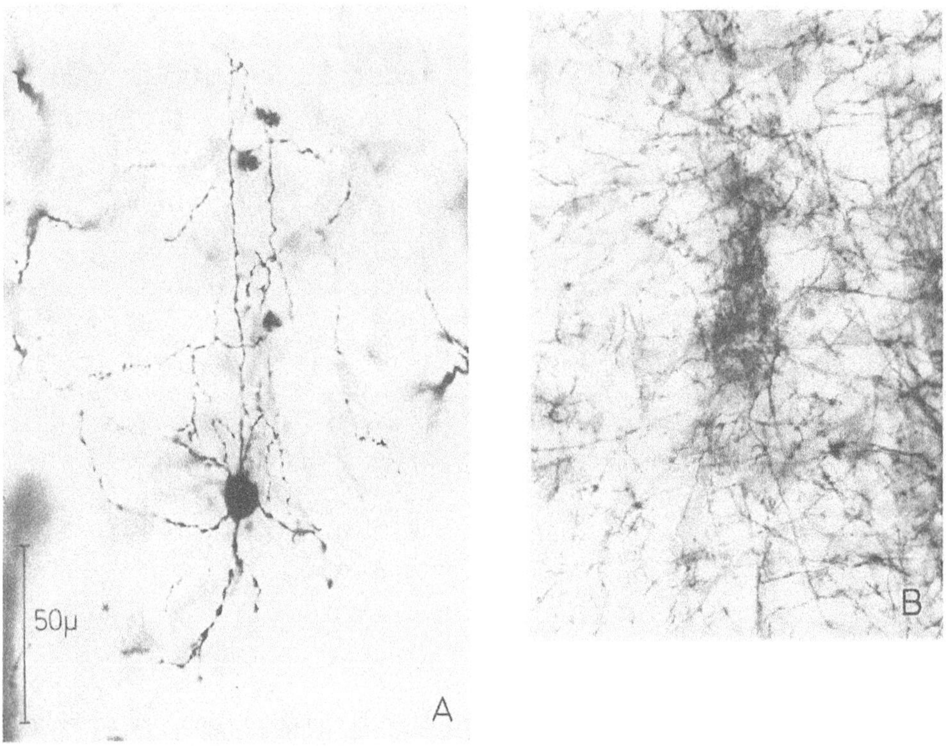

Fig. 5A and B. Human cerebral cortex. A. Man 16 years old. A Golgi type II cell in layer IV of the supramarginal gyrus. Mosaic photomicrographic reconstruction. B. Two months old infant. Pericellular basket enmeshing the body of one large pyramidal cell of layer V in the motor cortex. The cell body is not stained. Fine terminal ramifications form a complex of synaptic endings silhouetting the pyramidal cell body and the initial portion of the apical dendrite. (B was kindly provided by Dr. M. Marin-Padilla.) Golgi method

(Fig. 3B) but more detailed identifications require the integration of different focal planes on mosaic reconstructions made of several pictures (Figs. 2A, 3A, 4A, 5A). With camera lucida drawings made on transparent sheets it has been possible to follow tortuous axonal pathways and small bundles of fibers by superimposing individual drawings (Fig. 7).

Chemistry of the Golgi Impregnation

The chemistry of the rapid Golgi method has not been completely explained. The procedure was described, as easy as indeed it is, in a short paragraph by Golgi (1903) in 1873. It has been impossible to retrace the individual steps that led to the development of the technique, but it seems conceivable that Golgi arrived at his celebrated

Fig. 6. A. Mouse 6 months old. Low power photomicrographic picture of a portion of the hippocampus. Dense impregnation of pyramidal cells and dendrites in the strata radiatum (above) and oriens (below). Note deposition of abundant Golgi precipitate between the bodies of pyramidal cells, probably staining glial formations. B. Cat 7 days old. Low power photomicrographic picture of the visual cortex. Pyramidal cells of layer V with their basal and apical dendrites appear completely stained. Golgi method

method in the course of testing the effects of silver nitrate applied to pieces of nervous tissue hardened and fixed in Müller's fluid, a fixative widely used at that time (see also Ramón-Moliner's chapter in this Volume).

Blackstad (1965) and Stell (1965b, 1967) have shown in electron microscopic studies that the Golgi impregnation takes place *inside the cells*, filling the cytoplasmic ground substance but sparing some cellular organelles. The reaction probably consists in the formation of a lipoprotein-chrome-silver compound under the catalytic action of osmium tetroxide (Kallius, 1910; Liesegang, 1910; Cajal and de Castro, 1933); random physical and chemical (and possibly functional) factors might be responsible for the staining of some individual cells while others remained unstained.

The reactions of osmium tetroxide with tissue components have been studied (Bahr, 1954; Wolman, 1957) because of the great value of this chemical as a cytological fixative for electron microscopy. The osmium tetroxide acts as an additive fixative, probably reacting in some side-groups of proteins at the double bonds between adjacent carbon atoms (Baker, 1958; Lison, 1960). It reacts also with the double bonds in unsaturated fats, a reaction resulting in the formation of a black compound containing the osmium in various possible linkages (Wigglesworth, 1957; Pearse, 1961).

The admixture of other oxidants, as for instance potassium dichromate in the rapid Golgi method, modifies the action of osmium tetroxide. In the presence of such additional oxidants the reduced tetroxide is immediately re-oxidized, and thus takes the effect of an oxidative catalyst (Bahr, 1954; Wolman, 1957; Lison, 1960). It has been shown, moreover, that mixtures of osmium tetroxide and other oxidants prevent those compounds which are known to react with the tetroxide alone to reduce it, except for some specific organic compounds liberated in the process of myelin sheath degeneration. This has been of particular interest for an understanding of part at least of the mechanism of the Marchi reaction (Weisschedel and Jung, 1939; Wolman, 1957; Pearse, 1961).

As far as the Golgi method is concerned, the effects of both osmium tetroxide and potassium dichromate on tissue components should be considered jointly. Chromium salts in general react with protein chains, effecting a binding linkage. The effects may be different depending upon the chromium salt and the pH of the solution (Gustavson, 1956; Baker, 1958). The effects of potassium dichromate on lipids have been studied in particular by Lison (1960). A large number of unsaturated lipids can be rendered insoluble in lipid solvents by potassium dichromate. This effect is elective for phospholipids and sphingolipids; Lison (1960) attributes it to the presence of a high number of unsaturated conjugated double bonds in these compounds. In the process of oxidation the double bond most distant from the carboxyl group is easily oxidized to a peroxide which subsequently forms a polymeric substance. In the double bond nearest the carboxyl group the process passes from the peroxide stage through hydroxyketone to diketone. Isomerization of this compound may yield certain basic substances taking up the chromium as an oxide.

In the rapid Golgi method the deposition of the black substance inside nerve cells might occur as follows: Osmium tetroxide is able to form at the double bonds an osmium-cyclic compound which through hydrolysis would pass to the diol form. In the presence of potassium dichromate, the free osmium dioxide would immediately

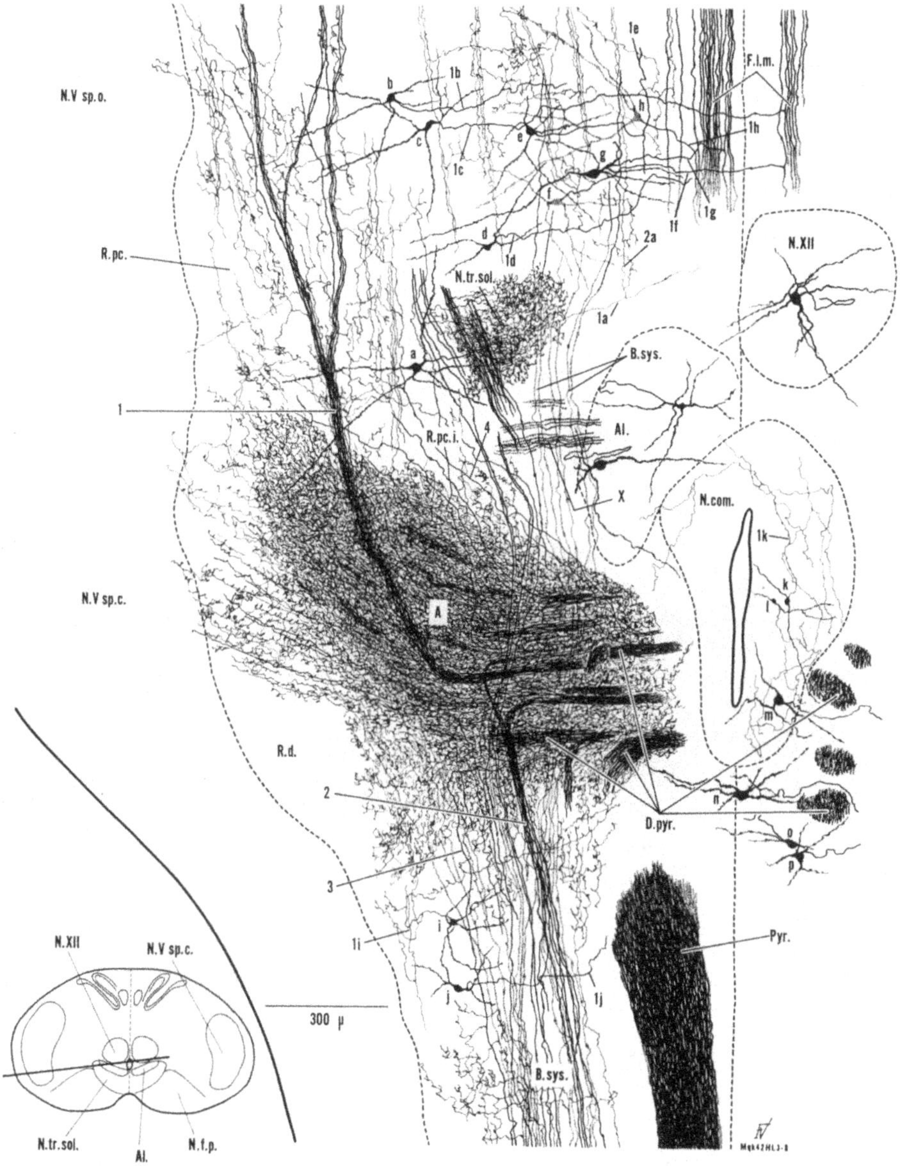

Fig. 7. Mouse 12 days old. Camera lucida drawing made from six adjacent horizontal sections passing immediately ventral to the nuclei gracilis and cuneatus of the medulla oblongata. A recurrent ascending pyramidal bundle (*1*) could be followed and reproduced by integrating observations in the individual sections. Various types of reticular formation cells and details of the decussating pyramidal tract (*D.pyr.*) entering the dorsal funiculus of the spinal cord (*Pyr.*). The plane of the sections is indicated in the diagram. Golgi method. (From F. Valverde, 1966; reproduced with permission of Springer-Verlag)

be reconstituted to the tetroxide. The diol might be subsequently oxidized to a diketone which is able to take up the chromium and thus forms a chromium compound which will react with the silver nitrate. The result will be the formation of

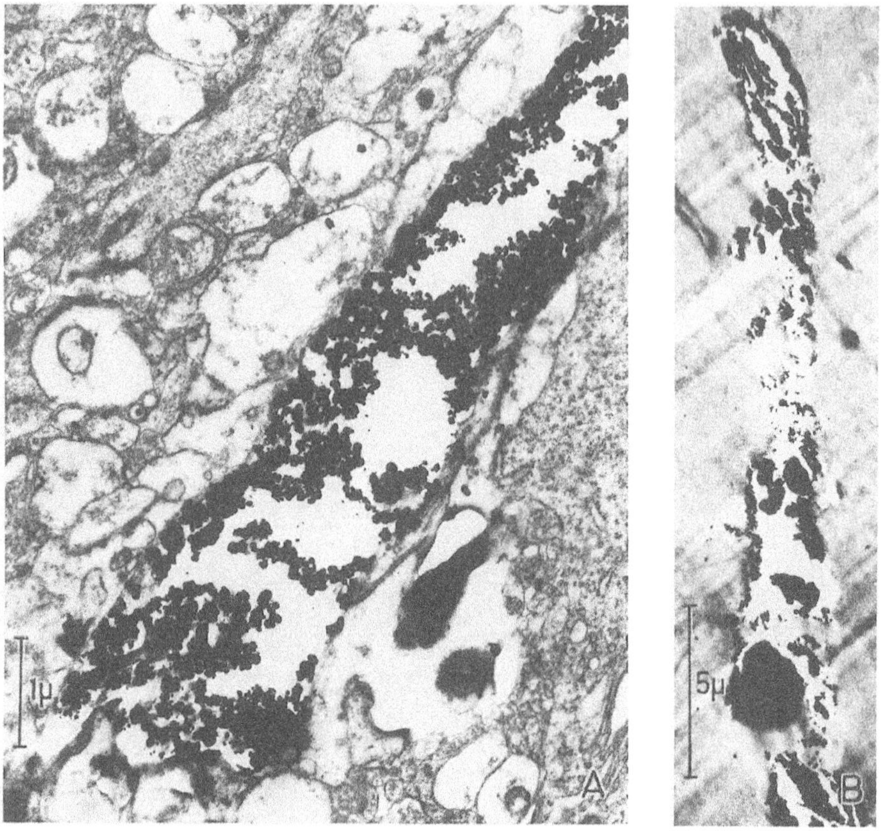

Fig. 8 A and B. Electron micrographs of Golgi-impregnated structures. A. Mouse 21 days old. Osmium-dichromate perfusion. A portion of a thick dendrite appears, filled with electron-dense granules, the Golgi precipitate. Layer V of visual cortex. Note preservation of non-impregnated structures. B. Mouse 12 days old. Portion of a Golgi-impregnated cerebellar climbing fiber. Electron-dense material fills the fiber. No adjacent non-impregnated structures can be recognized. This specimen was prepared by demounting an old Golgi section

a lipoprotein-chromium-silver compound of unstated composition inside the nerve cells and their processes.

Some of our Golgi sections were used for electron microscopic analysis. Selected regions from thick (100—300 μ) sections were trimmed to smaller pieces, dehydrated with acetone and embedded in Araldite. Ultrathin sections cut on a Reichert ultramicrotome and stained with lead citrate were examined in a Zeiss EM-9 electron microscope.

Fig. 8 A is an electron micrograph from the mouse visual cortex (layer V) showing a portion of one Golgi-impregnated large dendrite. The specimen was taken from the brain of a mouse 21 days old, perfused with the osmium-dichromate solution and prepared for electron microscopic examination immediately after the third silver bath. Fig. 8 B, an electron micrograph from the molecular layer of the cerebellum of a mouse 12 days old, shows a portion of one impregnated climbing fiber. This specimen was prepared for electron microscopic examination by de-mounting a Golgi section made four years ago. Previous light microscopic examination of both sections of Fig. 8 had shown the quality of impregnation to be equally excellent in both specimens. The electron microscopy, however, revealed an excellent preservation of non-impregnated structures in one specimen (Fig. 8A), whereas a complete fading of non-impregnated structures is evident in Fig. 8B.

In summary, the Golgi precipitate appears to be a solid, electron-dense irregular mass of granules of varying sizes and shapes filling the impregnated structures. In several cases we have found a finely-arranged precipitate between two opposing cell membranes, an observation suggesting that the formation of the precipitate may start either intra- or intercellularly, spreading out later until a membrane intervenes (Blackstad, 1965).

Comparative Structural Analyses

The Golgi method revealed its power through the well-known studies performed by our classics to investigate the fundamental structure of nerve cells. In the period between the two World Wars the method was used progressively more rarely, and it seemed to be an almost forgotten tool when, shortly after the Second World War, a resurgence of interest in this incomparable method was triggered by the introduction of various novel anatomical and physiological research techniques. Although numerous valuable data and insights have already resulted from this renewed application of the Golgi method, it can be said that some classical studies still remain valid and unsurpassed in almost their entire content. Examples of such unexcelled studies are the pioneering investigations of Cajal on the cerebellum, hippocampus and olfactory bulb. The spectacular progress achieved by modern Neurophysiology is firmly based upon these studies. The Golgi method has still many possibilities for correlation with other neurohistological procedures, especially with the Nauta and electron-microscopic techniques. The latter of these methods is needed in particular to maintain a dialogue with Golgi observations concerning the structural details of neuronal inter-relations.

The analysis of brain organization by means of pattern classification and statistical quantification of different neuronal parameters has achieved encouraging results. As an aid in this approach the Golgi method is certainly unsurpassed. By statistical analysis of a series of measurable parameters in the cerebral cortex, the cortical neurons have been shown to obey exact quantitative rules. Bok (1936, 1959) described a number of approaches to the study of the complex cortical organization, and formulated certain laws from data obtained by measuring numerous morphological details such as the size-relations between perikaryon and nucleus, the branching pattern of the dendritic trees, and the fibre network in general. Sholl (1953, 1956, 1959) in an attempt to assess the principles of neuronal interconnections performed several original studies of the dendritic branching pattern, and established, among others, a functional relationship between the number of intersections per unit area and the distance from the centre of the perikaryon: the connective field of the neuron. Sholl's (1953) representation of quantity gradients in the number of dendritic intersections per unit area by the use of concentric circles centered in the cell body has been of considerable theoretical and practical interest. This and other neuronal

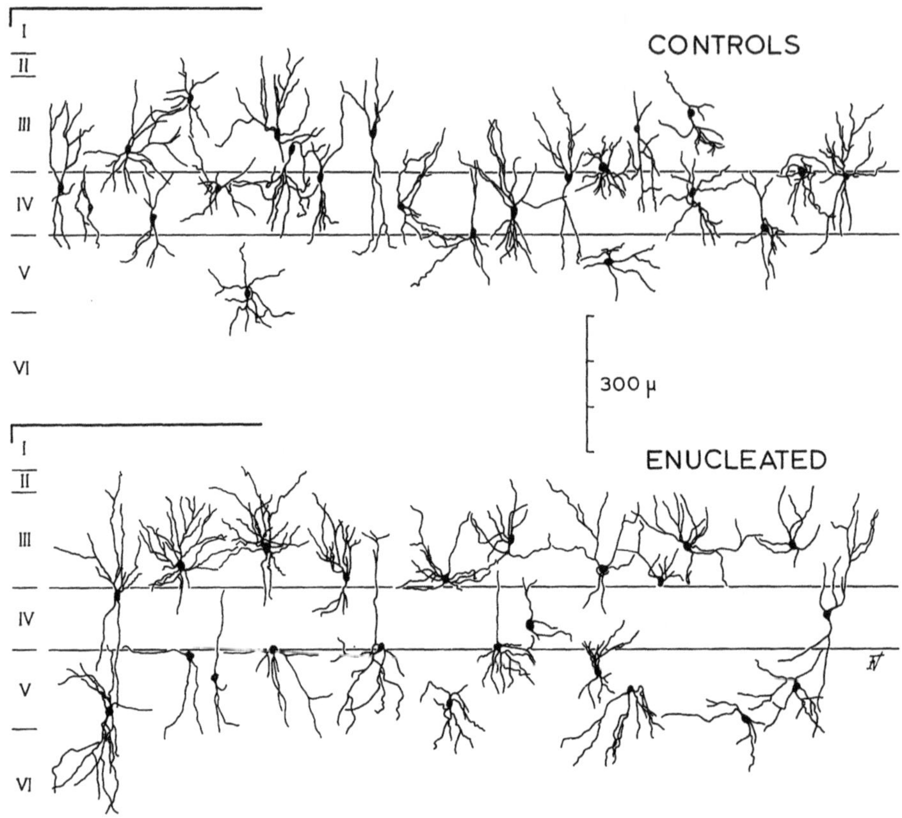

Fig. 9. Composite drawing of stellate cells in the area striata from mice 48 days old. Stellate cells in normal mice appear with bodies and dendrites distributed randomly through layers V, IV and III (Controls). In mice enucleated at birth an orientation of dendrites avoiding layer IV is evident (Enucleated). The axons were not traced. Golgi method, camera lucida drawing. (From F. Valverde, 1968; reproduced with permission of Springer-Verlag)

parameters measured in Golgi preparations have been found valuable in the analysis of unexpected relations (Mitra, 1955; Eayrs and Goodhead, 1959; Ramón-Moliner, 1961a; Schadé *et al.*, 1964) not only in studies of the evolution and ontogenetic development of behavioral complexity but also in the analysis and quantification of the structural effects of sensory deprivation.

In the following paragraphs some applications of the Golgi method to the study of morphological changes in neural structures elicited by sensory deprivation will be discussed.

In mice blinded by eye enucleation at birth it has been possible to demonstrate alterations in the normal dendritic orientation of stellate cells of the visual cortex (Valverde, 1968). The dendrites of stellate cells in normal subjects radiate in all directions through layers III, IV and V (Fig. 9, Controls). In mice 48 days old that had undergone a unilateral eye enucleation at birth, the dendrites of stellate cells in the contralateral striate cortex appeared oriented in a peculiar manner which was most evident in composite drawings of large numbers of these cells (Fig. 9, Enucleated). The stellate-cell dendrites in this case showed a clear tendency to "escape"

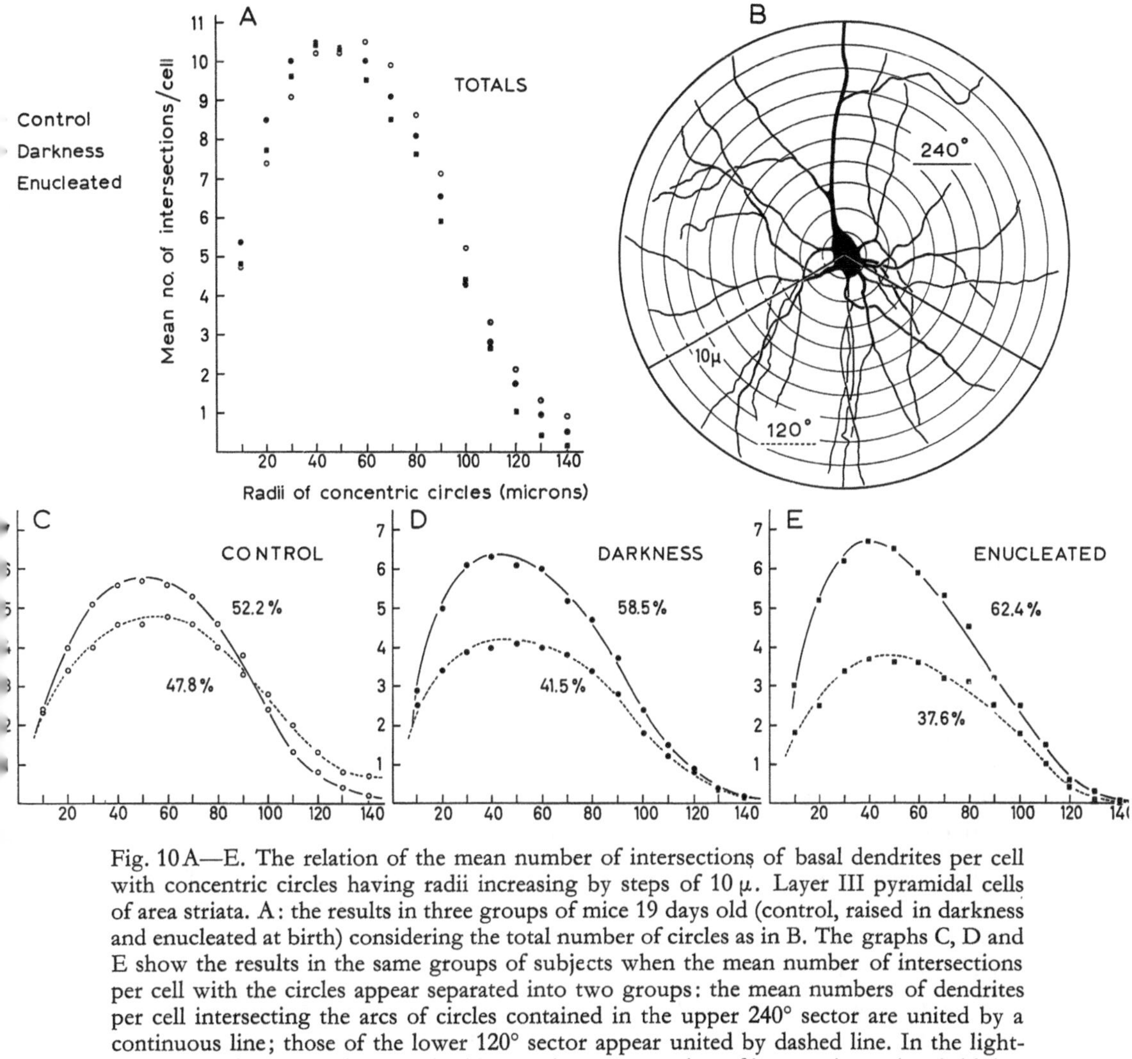

Fig. 10A—E. The relation of the mean number of intersections of basal dendrites per cell with concentric circles having radii increasing by steps of 10 μ. Layer III pyramidal cells of area striata. A: the results in three groups of mice 19 days old (control, raised in darkness and enucleated at birth) considering the total number of circles as in B. The graphs C, D and E show the results in the same groups of subjects when the mean number of intersections per cell with the circles appear separated into two groups: the mean numbers of dendrites per cell intersecting the arcs of circles contained in the upper 240° sector are united by a continuous line; those of the lower 120° sector appear united by dashed line. In the light-deprived and enucleated groups in this case the mean number of intersecting points is higher in the upper 240° sector, while the means corresponding to the lower 120° sector had diminished correspondingly

from layer IV, i.e. to avoid the zone in which the major termination of specific (in this case, geniculo-cortical) afferents takes place, as if they were "searching" for alternative afferent relations outside layer IV. This change in dendritic orientation revealed by the Golgi method strongly suggests that dendrites, instead of being passive structures, are actively growing and changing neuronal formations which must accommodate to changing functional demands, or die.

Comparable alterations in the orientation of dendrites were later found in the basal dendrites of pyramidal cells of layer III of the visual cortex in 19-day old mice that had been either raised in darkness since birth or blinded by eye enucleation at birth. The results of a pilot study are given in Fig. 10.

The number of intersections of dendrites with a series of concentric circles centered in the cell body and increasing in radius by steps of 10 μ were plotted (Sholl, 1953) for individual cells. The mean values of the number of intersections per cell as a function of distance-from-cell body are represented in Fig. 10A, for control, dark-reared and blinded mice 19 days old. The results show no significant difference in the distribution of the mean values between the three groups of subjects. Nevertheless, a significant variation in the orientation of dendrites in dark-raised and blinded mice with respect to the control group is revealed if the series of concentric circles is divided into an upper sector of 240° and a lower sector of 120° (Fig. 10B). In the graphs of Fig. 10C, D and E, the mean number of basal dendrites intersecting the concentric arcs of circles contained in the upper 240° sector have been united by continuous lines and those intersecting the concentric arcs of circles contained in the lower 120° sector appear united by dashed lines. The control groups (Fig. 10C) shows a distribution of dendrites of an approximately half-and-half ratio between the upper and lower intersecting group of dendrites. In both the dark-reared and blinded groups a higher mean number of intersections with the concentric arcs of circles in the upper 240° sector is found. The difference is especially pronounced in the blinded group (Fig. 10E).

The significance of the variation in the orientation of the basal dendrites of the layer III pyramidal cells of the visual cortex toward the upper sector is not clear, but it may have the same explanation that was suggested above for the dendrites of stellate cells in 48-day old mice subjected to unilateral eye enucleation at birth. The basal dendrites of pyramidal cells of layer III are receptor sites of specific afferent fibers. Functional as well as anatomical abnormalities of the specific afferent connection could be thought to originate a series of compensatory mechanisms reflected morphologically by variations in dendritic orientation.

The anomalous orientation of the basal dendrites observed in the foregoing experiments appears to be restricted to the visual cortex. In any case, no comparable deviations from the normal pattern were found in the temporal cortex of animals visually deprived since birth.

The existence of a statistically significant decrease in the mean number of dendritic spines per 50 μ length segments of the apical shafts of layer V pyramidal cells in the visual cortex in dark-raised mice was first reported by Valverde (1967). Changes in the morphology of the dendritic spines have been described by Globus and Scheibel (1967b) in young rabbits visually deprived for the first 30 days of life. Coleman and Riesen (1968) showed that stellate cells in layer IV of the visual cortex of cats reared in the dark have small dendritic length and few dendrites in comparison with normal control animals.

The effects of unilateral neonatal eye enucleation on the histology of the visual centers has also been studied further with the Golgi method. In rabbits subjected at birth to either eye enucleation or lesion of the lateral geniculate body, Globus and Scheibel (1967a) found a consistent decrease in the number of spines over the central portion of the apical dendrites of pyramidal cells. Our own studies in the mouse (Valverde, 1968) have shown that unilateral eye enucleation at birth produces a significant decrease in the number of spines in those apical dendritic segments that are located in layer IV of the contralateral visual cortex. It is of interest to note that the observed difference in number of dendritic spines from normal was greater in young animals than in those permitted to survive neonatal surgery for a long time.

It was found for the apical dendrites of pyramidal cells that the mean number of dendritic spines per unit segment of 50 μ length increases exponentially with the distance from the cell body (Valverde, 1967). A mathematical model which defines

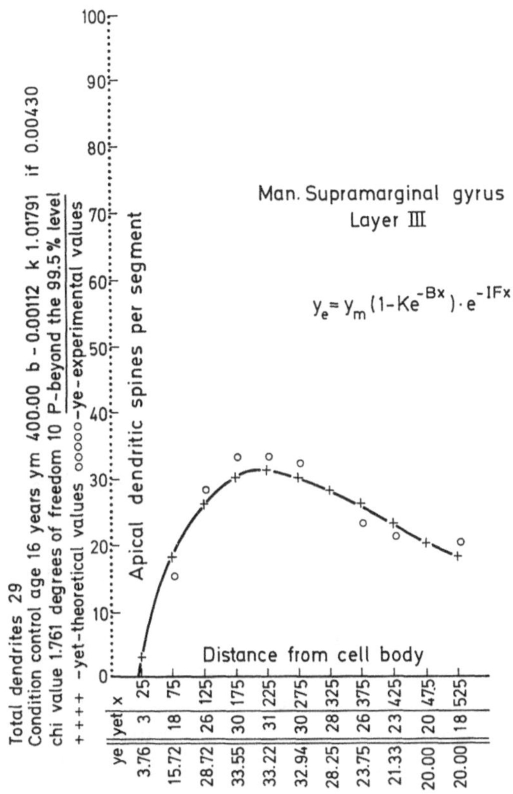

Fig. 11. Experimental and theoretical distribution of the mean number of dendritic spines per segment of 50 μ along the apical shafts of the layer III pyramidal cells of the supramarginal gyrus in man. *YE*, value of the experimental ordinate according to microscope countings of spines. *YET*, value of the theoretical ordinate according to the equation given. When the difference between *YE* and *YET* is less than one spine per segment, only the asterisk is printed. Data sheet from IBM 7070

the distribution of spines along the apical dendrites has been developed recently (Valverde and Ruiz-Marcos, 1969; Ruiz-Marcos and Valverde, 1969). The model consists of a principal equation which is given in Fig. 11, and describes the distribution of the spines in terms of three coefficients IF, B and K whose values have a functional relation with the age of the animal. The principal equation of the model has been derived by data-processing of microscopic counts of spines in Golgi preparations through a series of programs written for an IBM 7070. The model satisfies the distribution of spines, and its evolution with age, for the apical dendrites of pyramidal cells of layer V in the visual cortex of mice 10 to 180 days old raised in normal conditions and in complete darkness. The principal equation of the model

(Fig. 11) is valid also in defining the distribution of dendritic spines in the visual cortex of mice subjected to unilateral eye enucleation at birth, and the distribution of spines along the apical dendrites of various cortical areas of hamster, cat and man. The number of dendritic spines increases with the age of the subject and the distribution varies according to the values of the parameters of the model.

Fig. 11 reproduces a data sheet from one of the programs of this study to which reference is made for further details. The data sheet represents the distribution of spines along the apical dendrites of layer III pyramidal cells of the supramarginal gyrus of a 16 year-old boy. The mean number of dendritic spines per dendrite segment of 50 μ length is plotted against the distance from cell body. The experimental mean values obtained by light-microscopic counting of spines in Golgi preparations (YE) are represented by dots. The theoretical distribution according to the equation of the model (YET) is represented by the sequence of asterisks. In this case the computer approximated to within 99.5% the theoretical distribution formulated by the equation of the model.

Light-microscopic visualization of dendritic spines so far has been possible only with the Golgi method. Our studies on the number and distribution of spines along the apical dendrites of cortical cells have depended on the use and interpretation of optimal Golgi material. The characteristics of the spine distribution along apical dendrites still await evaluation. Recent electron-microscopic observations by Colonnier (1968b) have indicated that only a small number of synapses involve the inter-spine surface of cortical dendrites, and hence, that the majority of synaptic contacts on cortical cells are established on the dendritic spines.

All these facts emphasize the importance of the study of the distribution, arrangement and temporal evolution of dendritic spines as specific places in synaptic transmission. The spines are postsynaptic structures whose physical integrity depends not only upon the intactness of the axons with which they stand in synaptic contact (Colonnier, 1964; Globus and Scheibel, 1967a; Valverde, 1968; Valverde and Esteban, 1968), but apparently also upon a normal pattern of impulse traffic arriving over such afferent fibers (Valverde, 1967; Ruiz-Marcos and Valverde, 1969). It is only by virtue of the Golgi method that these — as well as most other — details of cortical organization have come to light.

Acknowledgments

This study has been supported by a grant (Ayuda de Investigación, 1968) from the Juan March Foundation, Spain, and by a grant from the Sociedad Española de Industrias Químicas y Farmacéuticas S.A., División Farmacéutica Lepetit.

I am grateful to Mrs. Eva V. Valero for her excellent histological work, and to Miss M. Estrella Esteban for her assistance in the statistical analysis of the data.

Special thanks are due to Prof. M. Marin-Padilla, Department of Pathology, Dartmouth Medical School, Hanover, N.H. who kindly read the manuscript and provided two of the illustrations of this account from his own Golgi material.

Discussion

Dr. Cowan: I have two questions, Dr. Valverde. The first one concerns your experimental finding of a loss of dendritic spines after eye enucleation and also after rearing animals in the dark: Do you think that the spines are in fact not present (perhaps because they failed to develop or because they developed and subsequently degenerated and were removed); or, do you think that they are simply not impregnated for some reason which, at present, we do not understand?

Dr. Valverde: I really cannot answer the question, but I would think that both possibilities are tenable.

Dr. Cowan: The reason I ask whether you think they may not be impregnated is related to Powell's findings in the olfactory bulb. Are you familiar with them?

Dr. Valverde: Yes, I read his paper (Powell, 1967) and recall that he attributed the apparent absence of dendritic spines in animals subjected to sensory deprivation to a resistance of these structures to the Golgi impregnation. However, I believe that the phenomenon is related to a shift in position of the dendrites rather than to a staining problem. In other words, I think he may have been dealing with a situation much like the one I have just described, namely, the re-orientation shift of basal dendrites in response to eye enucleation. One first gets the impression that there are fewer basal dendrites in these animals, but careful quantitative work reveals that they have merely shifted their position.

Dr. Cowan: My second question relates to your method of computing the number of dendrites and their changes with age. Do the apical dendrites themselves not increase in length? If they do, are you not, by taking uniform 50 μ segments, introducing a bias? Would it not be more appropriate to take the total length of each dendrite and divide it into, say, nine or ten segments, and then calculate the number of spines per segment?

Dr. Valverde: You mean that an error may have been introduced by comparing the same size segments in 10-day old animals and in 180-day old ones in which the cortex is much thicker and the dendrites correspondingly longer. In our studies, we have not considered the superficial ramifications of the apical dendrites in the blocking of the distribution. It is the region of superficial ramifications which increase with age and we have concerned ourselves only with the portion of the apical dendrite between the soma and the point where the superficial ramifications begin. This portion is of about the same length in the two age groups.

Dr. Cowan: Yes, but is it not conceivable that the terminal portion of the dendrites in the older animals is derived from the same portion that formed the last segment in the younger ones? What I'm concerned about is, are you in fact counting the number of dendritic spines on identical segments?

Dr. Valverde: I am not sure about this.

Dr. Van Buren: We have been wrestling with the application of some of the Golgi techniques in larger animals, primates and man; and it would seem reasonable to us that if you are dealing with subcortical structures, you must employ a perfusion technique for the primary fixation and then go on to silver. We have tried with some success the Hortega-Graves method in which a chloralhydrate-dichromate-formalin fixation is used. I'm wondering whether you have had experience with fixation of subcortical structures, the thalamus in particular.

Dr. Valverde: We have some experience with perfusion fixation with osmium-dichromate, having used this in a number of mice in an attempt to obtain better preservation of the tissue. The usual osmium-dichromate mixture is used for the perfusion. The circulatory system is first washed out with a saline solution buffered at pH 7.2. About 50—70 cm^3 of osmium-dichromate is then passed through the heart. The resulting fixation is very good, with complete blackening of the brain; but when we compare these Golgi sections with the usual immersion fixed material, we don't find enough difference to justify the procedure.

Dr. Van Buren: Do you get axonal staining also?

Dr. Valverde: With the perfusion fixation, yes, and we also get a light deposit of osmium on the myelin, for instance, in the pedunculus cerebri.

Dr. Karten: I believe that Dr. Kent Morest has been making the point that perfusion fixation is particularly beneficial if you want to see the relationship of the neuroblast processes to the external limiting membrane. Have you noticed this in your embryonic material?

Dr. Valverde: We have not studied this in embryonic material.

Dr. Ramón-Moliner: First of all, I would like to congratulate Dr. Valverde on this presentation. I think that there is a very pertinent point to the Golgi technique that becomes apparent from this experimental study of the dendritic spine distribution. We sometimes refer to this technique as if it were useful only for observation purposes and not for experimentation. One could go even further and regard it more as a source of aesthetic pleasure than as a scientific tool. But if Science is to be regarded as the ability to predict more or less accurately the outcome of an experiment, we must conclude that Dr. Valverde has proven the Golgi method to be a very scientific technique, suitable for experimental purposes.

There is another point that I'm not so happy about, and that is the use of the Golgi technique for studying synaptic relationships. I do not think that the Golgi is the technique one should use to study contacts between dendrites and axons. In fact, on the basis of random, one should expect contacts to be established between dendrites, axons, glial processes, and even vessels, with a frequency much greater than the one observed in Golgi preparations (Ramón-Moliner, 1967). As I have said before, one of the peculiarities of this technique is that it appears to avoid the staining of adjacent structures. One very clear example is what happens in the cerebellum, where you can stain either the baskets or the bodies of the Purkinje cells but never both simultaneously. You may sometimes find both in different parts of the same section, but never in contact with each other. In other words, it is as if one stained structure entailed the non-stainability of the adjacent one. It is for this very reason that I am a little bit reluctant to use the Golgi technique for the study of synaptic contacts. It is true that sometimes you see such contacts, but the other point may be made: would it not be very much stranger if the contacts were never seen at all? After all, when we look at the E.M. picture, we see everything in close contact with everything else. I think that on the basis of mere random, such contacts should be expected to appear in Golgi material more often than they actually do (Ramón-Moliner, 1967).

Dr. Valverde: This is one of our main problems. In our studies of the distribution of dendritic spines, we do not see the terminal endings on the spine, but we ought to see at least the postsynaptic structures. There is a recent paper by Colonnier (1968a) on synapses in the cerebral cortex of the cat in which he mentions that the vast majority of the contacts on the dendrites involve the spines. Therefore, when we study the patterns of dendritic spine distribution, I think that our findings could be applied to the terminals because what we saw in the dendritic spine arrangement is much the same as a photographic negative of the constellation of their afferent terminals.

Dr. Nauta: Dr. Ramón-Moliner, is perhaps one of the reasons for your question the fact that electron microscopy has made us painfully aware of the fact that membrane-to-membrane contact does not necessarily mean synaptic contact? We may find, for instance, a synaptic end bulb which lies in apposition to a cell soma or a dendrite, and yet has its functional contact with another structure which adjoins its opposite side.

Dr. Ramón-Moliner: Yes. If we see the dendritic spine as well as the axon terminal that impinges upon it, the question is whether or not we are dealing with an active synaptic contact. It could be either and, in my opinion, only electron microscopy can demonstrate if the contact has the specific features of a synapsis.

Dr. Valverde: I am afraid that no direct answer is possible at the moment.

Dr. Peters: Perhaps what we are talking about is what Van der Loos in recent papers (1964) has described as dendro-dendritic contacts and axo-axonic contacts on the basis of profile configurations in both Golgi and electron microscopic preparations. He feels very strongly that these contacts are functional contacts.

Dr. Valverde: I don't agree with the interpretation about the contact of descending processes of the granule cells in the olfactory bulb, because I think the descending processes are no dendrites, but are actually axons.

Dr. Peters: We in fact have been looking at that very point and the granule cell processes have the structure of dendrites all the way down to the cell body. There is no doubt about that.

Dr. Valverde: They may look like dendrites and yet function as axons.

Dr. Cowan: Dr. Valverde, concerning the termination of specific afferents from the geniculate, did I understand you to say that you think most of these end in the visual cortex upon short axon neurons and not directly upon the apical dendrites of pyramidal cells?

Dr. Valverde: We mentioned both possibilities, but there is a good deal of experimental evidence now for direct contacts with apical dendrites. As Dr. Ramón-Moliner mentioned, it is very difficult to determine exactly where these specific afferents terminate.

Dr. Nauta: Dr. Cowan, would you say there is good evidence for direct contacts of geniculo-cortical axons with apical dendrites of pyramidal cells?

Dr. Cowan: I don't know of any wholly convincing evidence for the termination of geniculo-cortical axons but Jones and Powell, at Oxford, have electron microscopical evidence for the termination of thalamo-cortical fibers from the ventro-basal complex upon apical dendrites in the somatic sensory cortex.

Dr. Voneida: This short question relates to the animals which have been visually deprived. Have you tried the situation in which the visually deprived animals are later exposed to normal light, and then checked for dendritic spines? It would seem to me that your method might enable you to determine to what extent the spine deficiency is reversible.

Dr. Valverde: We have not done that yet, but this is one of the first problems that we would like to solve.

Dr. Nauta: Certainly, from the findings of Hubel and Wiesel, you would at least expect some part of the deficiency to be irreversible.

Dr. Voneida: Yes, the technique lends itself extremely well to questions of this sort which have strong functional and physiological overtones.

The Golgi-Cox Technique

E. Ramón-Moliner[1]

Introduction

At a time when an avalanche of new techniques are being applied to the morphological analysis of the nervous system, it may appear out of place to reconsider the application of one of the oldest metallic stains used in Neurohistology. Nonetheless, the past decade has witnessed a remarkable resurgence of interest in the Golgi methods, a development that may reflect the more or less explicit suspicion that the image provided by these techniques may be, after all, the one that most distinctively characterizes the nervous system, particularly when compared with the histological picture of non-nervous tissues. As is true of all of Histology, the greatest single source of new fundamental information concerning the structure of brain tissue has been the electron microscope. But the great value of combining electron microscopic examination with the analysis of Golgi-stained material has also been proven (Blackstad, 1963; Fox *et al.*, 1967; Pappas *et al.*, 1966). Electron microscopy, for example, has permitted observations concerning details of synaptic relationships that lie well beyond the reach of the Golgi method or any other light-microscopic technique. On the other hand, but for the Golgi method, it is doubtful that a panoramic conceptualization of the nervous system could ever have been attained. It is important to realize that *the degree of information provided by any form of microscopy is by no means uniformly proportional to the degree of magnification or resolution provided by the technique*. For example, a magnification in the order of several millions, if technically possible, might permit the visualization of the elementary particles of matter: electrons, protons, etc. But it would be ludicrous if, confronted with such pictures, one would draw the conclusion that there is no essential difference between the material structure of the brain and, for example, that of a piece of wood. On the other hand, in view of the fact that the brain can be regarded as a signal-processing structure, one may be tempted to point out its similarities to an electronic apparatus. In that case, one may regard the picture provided by Golgi-stained material as the one which better accounts for that similarity, at least from the morphological point of view. The image of tangled and overlapping neural processes produced by the Golgi method is indeed more reminiscent of electronic wiring than are histological brain pictures provided by any other technique.

History

The present author is indebted to Dr. S. O. E. Ebbesson for providing the reference to a book by Retzius (1933) which contains the following account of the circumstances that led to the discovery of the Golgi method:

[1] This work was supported by the Medical Research Council of Canada and by grant no. DA-49-007-MD-1005 from the U.S.Army Research and Development Command.

"Through one of Golgi's assistants, I have learned that his staining method, the so-called Golgi method, was discovered unexpectedly when he was trying to impregnate with silver the innermost membrane of the pia mater that Key and I had previously described as pia intima. At the time, he was using material that had been hardened in a solution of potassium dichromate. By chance, he observed that there were occasional nerve cells stained in dark brown, in the adjacent brain substance. He proceeded then to experiment with the silvering of previously chromated material and, eventually, found that mercury solutions could also be used."

The use of mercury was proposed by Golgi (1879) only a few years after he had described his original method based on the precipitation of silver chromate (Golgi, 1873). Golgi's new method consisted in keeping pieces of nervous tissue in potassium dichromate or Mueller's fluid for several months, then in a solution of mercuric chloride; thick sections cut from tissues so treated were mounted without cover slips. Later, Golgi found that the stain could be intensified by a treatment of the sections with dilute sodium hydroxide or sodium carbonate (Golgi, 1891). Tal (1886) instead recommended 1% sodium sulphite, and Pal (1887), dilute ammonia, whereas Cajal (1933) used a saturated solution of sodium sulphite. Mondino (1885) and Flatau (1895) published techniques almost identical with the original.

A significant improvement of Golgi's mercury method was made by Cox (1890; 1891) who introduced the use of a mixture of potassium dichromate and mercuric chloride, with potassium chromate (K_2CrO_4) added to moderate the exceedingly acid reaction of the solution. So good were the results of the Cox modification that both Cajal (1909) and Lorente de Nó (1934) in their studies of the hippocampus occasionally obtained better neuronal impregnation with this method than with the original rapid Golgi technique.

Further modifications of Golgi's mercury method were not developed until many years later. In 1959, Bertram and Ihrig published a remarkable variant in which the initial chromation is followed by an immersion of the tissue in a mercuric nitrate solution. This process is comparable to Golgi's (1873) original silver chromate technique in the sense that the impregnation is carried out in two stages, the final one of which involves the nitrate ion. The present author (Ramón-Moliner, 1958) reported that the addition of potassium tungstate to the Cox mixture appeared to improve the consistency of the reaction; he also recommended alkalinization of whole blocks rather than sections. Even though this block-alkalinization procedure yields a relatively coarse impregnation (Figs. 2 and 4), it will be described below because of its simplicity of execution. However, when more delicate impregnations are required (Figs. 1 and 2), it is preferable to use a later revision of the process (Ramón-Moliner *et al.*, 1964) in which the alkalinization of whole blocks is replaced by the Van der Loos (1956) procedure of alkalinizing celloidin sections with alcoholic ammonia (see also Turcotte and Ramón-Moliner, 1965).

In optimal Golgi preparations only a small number of cells are impregnated, while the majority of neurons remain unstained and are usually completely invisible. This sparseness of impregnation may cause serious problems of orientation, first of all because it eliminates the criterion of comparative cell density, one of the characteristics most commonly relied upon in cytoarchitectonic delineation. Attempts to counterstain Golgi sections by Nissl-type methods have therefore often been made. Golgi-Cox sections, however, consistently have been found nearly refractory to routine Nissl methods, and only recently has a technique been developed (Van der

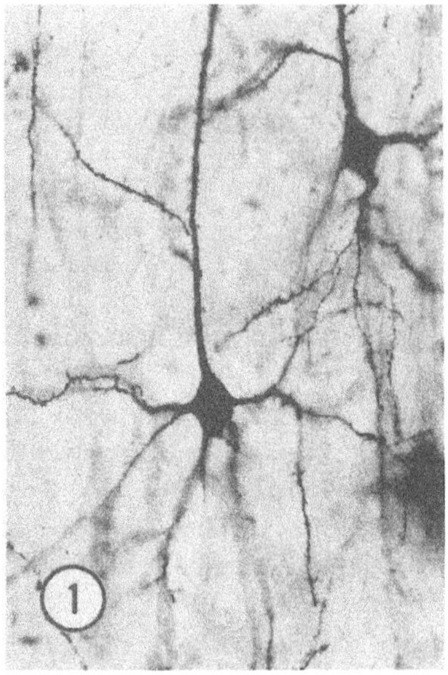

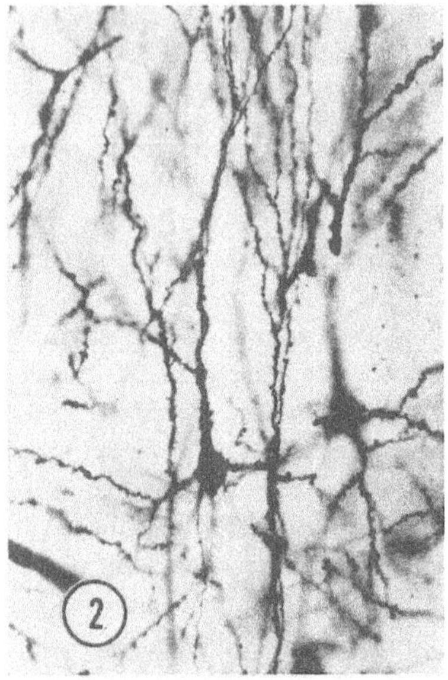

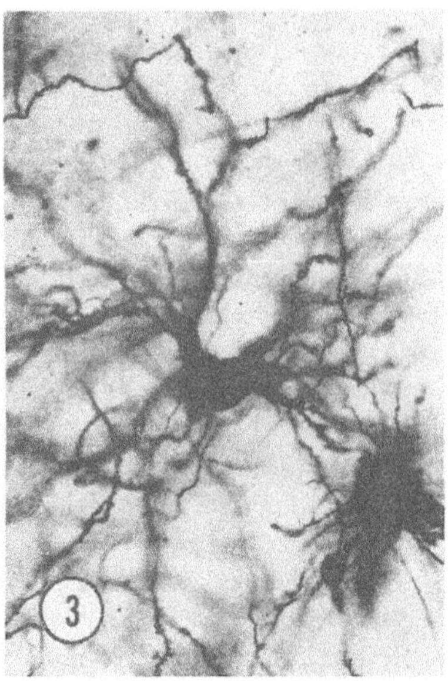

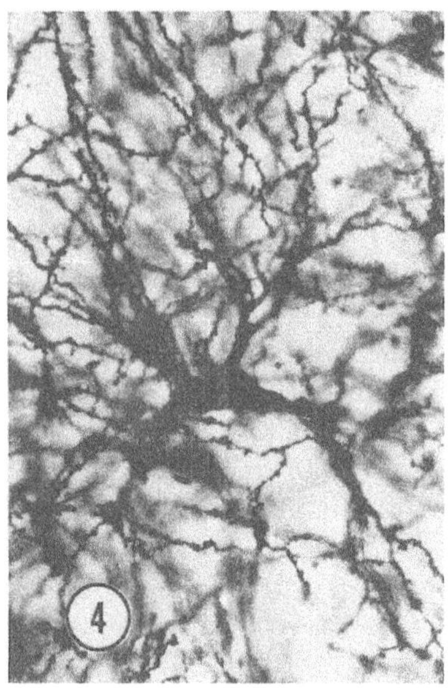

Fig. 1. Pyramidal neurons of cat cerebral cortex. Alkalinization of section with alcohol-ammonia. Note that the impregnation is more delicate in this case than in Fig. 2. (×250)

Fig. 2. Pyramidal neurons of cat cerebral cortex. Alkalinization of the block of tissue in lithium hydroxide-potassium nitrate solution prior to celloidin embedding. The dendritic processes are coarsely impregnated, but the over-all impregnation appears to be more complete than it is in Fig. 1. (×250)

Fig. 3. Thalamic neuron of cat. Alkalinization of section with alcohol-ammonia. (×400)

Fig. 4. Thalamic neuron of cat. Alkalinization of the tissue block with lithium hydroxide-potassium nitrate, prior to celloidin embedding. (×400)

Loos, 1956) which permits a satisfactory counterstaining of unimpregnated cell bodies by a pretreatment of the sections with alcoholic ammonia. The author and his associates have come to regard the introduction of alcoholic ammonia as an important contribution to the Golgi method, not only because it permits subsequent counterstaining with Nissl-type techniques (Figs. 5—8) but also because it has proved to be superior to other current methods for the purpose of alkalinizing celloidin sections.

General Principles of the Golgi Method

The original techniques submitted by Golgi (1873, 1879, 1891), together with all the modifications that have been subsequently proposed can be grouped into two main categories (a) those leading to the deposit of silver chromate (Golgi, 1873), usually referred to as "Golgi methods" (the one involving chromation in a mixture containing osmium tetroxide is frequently called the "rapid Golgi"), and (b) those producing a deposit of metallic mercury and, probably, complex oxides of mercury (Golgi, 1879, 1891; Cox, 1890, 1891), commonly named "Golgi-Cox methods". Both variants share four essential features that differentiate them from other metallic stains: (1) Only a few cells are stained out of the large number of cells present in the tissue. (2) In the few cells that take the stain, the metallic deposit is massive and includes the soma or perikaryon as well as the ramifying processes that originate from it. (3) The remaining unstained structures compose a transparent background against which the stained cells stand out very clearly. (4) There are no staining gradations: the structures are either completely opaque to light or perfectly transparent.

The practical execution of the Golgi-Cox method, especially when combined with techniques for counterstaining, consists of a fairly lengthy series of manipulations that will be described sequentially in a later section of this paper. Many of the details of that description may strike the unacquainted reader as quite arbitrary. On the contrary, however, each procedural detail has been arrived at only after untold numbers of trials by workers in many laboratories. All of those who contributed such technical details were confronted not only with the problem of how to obtain optimal impregnation of neurons in blocks of brain tissue, but also with challenging difficulties in obtaining microscopic sections from such blocks without impregnation-fading, and without disintegration of tissue. The following pages have been written in order to acquaint the reader with the nature of these problems and thus, to facilitate his understanding of the subsequent description of the Golgi-Cox procedures routinely followed in the author's laboratory.

The essential steps involved in the Golgi-Cox method are as follows: (1) Pieces of freshly removed nervous tissue are placed for about one month in a solution containing potassium dichromate and mercury bichloride as the basic ingredients. (2) The tissue is then sectioned on a microtome. (3) The sections are treated with an alkaline solution and, finally, (4) properly mounted for microscopic examination.

Following their exposure to the alkaline solution, many nervous structures appear intensely stained in black against a practically colorless background. Without the use of alkalinization such structures can also be seen, although with less clarity, as "ghosts". The chemical composition of the material that makes the cells look like ghosts is not clear but it could be a complex compound of bivalent mercury. That of the black substance which is formed under the action of alkalinization could be a mixture of metallic mercury and some complex oxides of the metal. Bertram and Sheppard (1964) contributed an analysis of the possible chemical reactions involved in this process.

Choice of Animals and Tissue Preparation

The best results are obtained in young animals, but the newborn should be avoided (see Results). Animals can be sacrificed with ether or sodium pentobarbital. Perfusion with the impregnating mixture is not recommended since it results in a marked tendency of blood vessels to stain. The pieces of tissue should be no thinner than 2 mm, as the superficial layers often fail to impregnate. On the other hand, they should not be too thick because in such case the deeper regions are poorly stained. The author has adopted 5—6 mm as a satisfactory standard thickness for slices of the comparatively large brains of cat, and rabbit. The mouse brain, by contrast, can be placed into the impregnating fluid undivided. The amount of the fluid should be at least twenty times the volume of the pices of tissue. In our experience, poor results are obtained in material previously fixed in formalin.

The freshness of the tissue is a matter of great importance. If the staining is not begun until a few hours after the removal of the brain, incomplete or negative results are the rule. Probably for this reason, excessively large pieces of tissue always show a central region of poor impregnation, caused in all likelihood by autolytic changes developing before the fixative has had time to penetrate to the depth of the block.

The Impregnating Solution

As will be explained in more detail in a subsequent section, the impregnating mixture is made by mixing *separately prepared* solutions of mercuric chloride and potassium dichromate (Ramón-Moliner, 1958). When these two solutions are mixed a slight amount of precipitate is formed. This precipitate, probably mercuric chromate, is practically insoluble in water. Generally, the higher the proportion of chromate in the mixture, the greater the amount of precipitate. Sometimes, however, in spite of having added much chromate, no precipitate appears. The amount of precipitate appears to depend on the method of preparing the solution of mercuric chloride: if the salt was dissolved in cold water, the amount of precipitate is greater than if it had been dissolved in boiling water. Apparently, mercuric chloride, a homopolar compound, undergoes molecular rearrangement when heated. It is not certain that the formation of precipitate has any significant effect upon the quality of the final impregnation.

Golgi (1879) and Cajal (1933) emphasized the importance of the temperature during the impregnation period. According to the latter, the optimal range is between 20° C and 26° C; below 15° C it would be very difficult to obtain good results. It seems unnecessary to keep the tissue in the incubator at 36° C as advised by Mann (1902); besides, this procedure probably increases the amount of precipitate.

A study of the possible interfering or promoting effects of adding various substances to the original Cox solution (Ramón-Moliner, 1958) has shown that certain chemicals known to improve the results of Golgi's silver chromate method severely impair the outcome of the Golgi-Cox process. The chemicals in question are osmium tetroxide, chloral hydrate, and potassium chlorate. Potassium tungstate, by contrast, has been found to be a useful additional ingredient of the Cox fluid.

The action of potassium (or sodium) tungstate cannot be compared to that of osmic acid in Golgi's silver chromate technique. The salt has no pronounced hardening properties, although at the end of the impregnation the pieces are slightly harder than when no tungstate is used. Tungstate seems to promote the reaction that has been started by the effect of dichromate, and prevents the phenomenon of partial cell impregnation that sometimes occurs when the Cox mixture alone is used. By adding osmic acid to the chromating fluid it is possible to stain with the silver chromate (i.e. the rapid Golgi) technique structures that are never stained without it. By contrast, the addition of sodium or potassium tungstate to the Cox mixture appears to facilitate the impregnation of structures that are stained only rarely when the original formula is used. Tungstate, in other words, tends to widen the range of reliable application of the method, but does not increase the variety of stainable cellular elements. Rather than to contribute a qualitative improvement to the staining process, it reduces the chance of failures. This effect is particularly evident in brain tissues of young animals. The results of the standard Golgi-Cox mixture in such material are notoriously irregular but, when good, they are far superior to those obtained in adults as far as clearness of background and delicacy of axonal plexuses are concerned. The addition of tungstate to the mixture greatly increases the incidence of such superior results. The use of tungstate in old animals, however, leads to overstaining.

Dehydration, Embedding and Sectioning

Once the impregnation process is completed, two alternative procedures can be followed: (1) alkalinization of the block, followed by celloidin embedding, cutting and mounting; or (2) celloidin embedding followed by sectioning, alkalinization of the sections, counterstaining (optional), and mounting.

Celloidin embedding of tissues impregnated by Golgi-Cox method was recommended by Flatau (1895) and Bremer (1910). Both authors followed routine procedures of celloidin embedding, apparently without encountering the fading of impregnation reported by Cox and experienced by many after him. In large pieces of tissue requiring many hours of dehydration in alcohol and several months of celloidin infiltration, the impregnation may be effaced completely. As most of the fading appears to take place during the stage of alcohol dehydration, Sholl (1948) recommended dehydration in acetone instead of alcohol. In the author's experience, this modifications is useful for small tissue blocks, but larger pieces requiring longer exposure to acetone become too brittle for satisfactory sectioning.

The hardness of Golgi-Cox material prohibits the use of any embedding medium save celloidin, and it is therefore understandable that intensive efforts have been made to overcome the difficulties presented by fading during the process of celloidin embedding. Two ways of minimizing fading have resulted: one consists in thorough alkalinization of the impregnated tissue before celloidin embedding, the other in the use of an ether-alcohol mixture for dehydration, followed by celloidin embedding at low temperatures.

Alkalinization of tissues impregnated with the Golgi-Cox mercury bichloride-potassium dichromate mixture, as mentioned above, is an essential step without which only partially opaque (i.e. black) impregnation is obtained. Equally important is the fact that, in contrast to this reduced, black form of mercury, *the prodromal, more transparent mercury compounds formed in neurons during the impregnation process are progressively destroyed or leeched out during prolonged exposure to alcohol at room temperature.* To prevent such loss of impregnation, either one of two procedures may be followed: the alkalinization of whole blocks of tissue, prior to embedding, and the alkalinization of sections obtained through modified celloidin-embedding technics that will not cause fading.

The steps that follow alkalinization of whole blocks of tissue (Ramón-Moliner, 1958, see below), present no special problem. They can be carried out according to routine methods of celloidin embedding. In the author's experience, the addition of 4% ethylene glycol to the celloidin solutions improves the results. Provided the blocks, following alkalinization, are properly washed in slightly acidified water and then in tap water, as described below, no failure should occur.

When greater delicacy of the impregnation or/and counterstaining are desired, alkalinization must be performed on sections rather than on the block. This necessitates celloidin-embedding methods that will not cause fading of the impregnation (Ramón-Moliner, 1961, 1964). As mentioned above, the dehydration in alcohol appears to be the most harmful part of the embedding procedure. In the course of experiments designed to find a suitable substitute of alcohol, the author has found that *the use of a 2:1 mixture of ether and alcohol greatly reduces the loss of impregnation.* Unlike acetone, recommended by Sholl (1948), this mixture does not cause excessive tissue hardening. The celloidin solutions are, likewise, made in a mixture of ether and alcohol in the same proportions. The infiltration with celloidin at temperatures of approximately —10° C is of additional help in preventing fading; it also reduces ether evaporation.

The final embedding stage leading to the hardening of the celloidin consists in sudden immersion in chloroform. An even effect of chloroform can be achieved by the use of *light cardboard containers with flexible sides* (Ramón-Moliner, 1961). If, instead, rigid containers are used, the film of hardened celloidin which is formed on the surface will crack, with the result that celloidin and chloroform will mix irregularly.

It should be emphasized that prolonged infiltration in celloidin appears to interfere with the affinity of the tissue for basic dyes. In general, the shorter the tissue has been soaked in celloidin, the better the results of counterstaining. The long celloidin-infiltration times recommended in the Procedures section are necessary for large (4—6 mm thick) slices of the brains of cat, dog, or monkey. If only a small volume of brain tissue is to be investigated, the duration of the celloidin infiltration can be shortened in order to obtain better counterstaining. Other embedding techniques suggested for the Golgi-Cox method (Sholl, 1948) will probably also give good results, provided the pieces of tissue are not too large.

Attaching Celloidin Sections to the Slides

The processing of loose sections through many solutions is time-consuming and often results in damage and loss of sections as well as in uneven staining. These difficulties can be obviated by first attaching the sections to slides. Such mounting, however, requires precautions because the familiar albumen-glycerol adhesive fails to provide a bond that will resist the alkalinization process, even when combined with celloidin coating (Ramón-Moliner *et al.*, 1964).

To obtain an adequate bond it is necessary to use slides previously coated with egg albumen only (no glycerol) and allowed to dry thoroughly. Slides are coated by immersion in dilute albumen, followed by draining and drying, and hardening of the albumen layer overnight at approximately 60° C. The dilute albumen is prepared by dissolving the white of one egg in 500 ml of distilled water to which 1 ml of strong ammonia has been added. Commercial powdered albumen also gives good results when 2.5 g of the dry substance is used instead of a fresh egg white. After the solution is completed, the froth should be removed from the surface with a cloth.

Before being attached to the slides, the sections are immersed in a mixture of methyl benzoate, benzyl alcohol and chloroform (Ramón-Moliner, 1961). The first two constituents are non-volatile and, in the absence of chloroform, dissolve the celloidin. However, the presence of an adequate concentration of chloroform prevents softening of the celloidin and thus permits several sections to be mounted on a single slide, provided this is done before chloroform evaporates. This evaporation naturally takes place during the subsequent exposure of the slides to the air, with the result that the remaining liquids dissolve the celloidin. When the celloidin has become soft, the slides are dipped and vigorously moved in three changes of toluene, a treatment that hardens the celloidin again and keeps the sections adherent to the slide. *Fading will eventually occur if methyl benzoate and benzyl alcohol are not completely washed out in toluene.*

It is important that the sections, while in the methyl benzoate-benzyl alcohol-chloroform mixture, be kept at the bottom of the dish. This prevents the softening which would result from contact with the surface of the fluid where the chloroform concentration is lower. The sections should be soaked in the mixture for at least three minutes, and the same minimum time is required for the development of a proper bond during the air exposure. *Shriveling and cracking of the sections during exposure to the air indicate that the sections have not been soaked long enough in the methyl-benzoate mixture.*

Alkalinization and Stabilization

After a few weeks of impregnation in the Golgi-Cox fluid, sections taken from the block with a safety razor blade can be observed directly under the microscope. They show some scattered, apparently incompletely stained cells against a yellow background. If such sections, without any further treatment, are properly dehydrated and mounted, many cell processes and sometimes cell bodies appear filled with a substance which despite its transparency is clearly visible because its refractive index is much higher than that of the background. When exposed to an alkaline solution, this substance becomes opaque so that under transmitted light it appears black.

Apparently two kinds of opaque compounds are formed during alkalinization (Ramón-Moliner, 1958). One of these is stable and is not affected by alcohol or mounting media. The other compound fades easily under the same circumstances. Strong and prolonged alkalinization promotes the formation of the stable compound, whereas weak or short alkalinization favors the unstable material. However, alkalinization is not the only factor involved, for it appears that a high ratio of stable-versus-unstable compound can also result from long exposure of the tissue to the Golgi-Cox impregnation mixture. Flatau (1895) and Bremer (1910) used exceptionally long impregnation times, up to several months, and it may be significant that they experienced little or no fading despite the use of routine celloidin embedding procedures. Neither did they observe the fading of cover-slipped preparations which has exasperated so many other users of the method and has led to a widely held belief that Golgi-Cox sections should never be mounted under cover glass (Golgi, 1879; Cox, 1890, 1891; Cajal, 1933), unless the impregnation is stabilized by aftertoning with gold (Robertson, 1900). In all likelihood, the cover slip has the effect of retarding the evaporation of the solvent from the mounting medium, thus allowing it progressively to dissolve the less stable mercury compounds.

It appears preferable to stabilize the Golgi-Cox image by intensive alkalinization, rather than by the use of excessively long impregnation times which may lead to overstaining. The alkalinization procedures to be described below have enabled the author to preserve cover-slipped Golgi-Cox preparations for years with little or no fading.

Block-alkalinization cannot be done properly by merely soaking the block in one of the alkaline solutions used by Tal (1886), Pal (1887) and Cajal (1933). The unavoidably long exposure to variously concentrated solutions of sodium hydroxide, sodium sulfite or ammonia leads to intense tissue swelling and, although the cells become black, the "background" is also blackened and filled with a "dusty" precipitate. This difficulty has been overcome recently by the finding that *potassium nitrate added to the alkalinizing agents in a high concentration prevents both swelling and background blackening*. By its mild oxidizing properties, this salt prevents the reduction of mercury in the "background" without interfering with the blackening of impregnated cells. Besides, its extreme solubility permits the preparation of solutions of high osmotic pressure which counteract the swelling effect of the alkali. Even so, sodium hydroxide solutions may impair the delicacy of the impregnation. The latter is better preserved when lithium hydroxide is used instead. This chemical also causes less tissue swelling than does sodium hydroxide.

After alkalinization, the blocks must be washed in distilled water to remove the potassium nitrate which otherwise would precipitate during dehydration. During this washing process the osmotic pressure of the mixture decreases and the tissue swells. This can easily be prevented by adding a small amount of acetic acid to the distilled water.

Alkalinization of whole blocks of tissue prior to embedding should not be done if counterstaining is desired. In this case, it is better to apply the alkalinizing process to individual sections. Ammonia, first introduced by Pal (1887), is the agent of choice in this procedure. Alcoholic ammonia has less tendency to detach the celloidin sections from the slide and, unlike other alkaline agents, does not eliminate the affinity of the non-impregnated neurons for basic dyestuffs (Van der Loos, 1956).

Counterstaining

One of the advantages of the Golgi-Cox method lies in the possibility of counterstaining the sections. This permits a combined picture in which the perikarya of non-blackened nerve cells appear stained by a basic dye.

The advantages of combining a dye stain with the metallic impregnation for purposes of orientation was recognized long ago. Cajal (1933) hinted that such combined staining is possible in Golgi-Cox material, but did not describe his procedure. Anderson (1954) recommended alcoholic thionin as a practical counterstain. This method yields useful results in material impregnated according to the Golgi modifications described by Fox *et al.* (1951), Porter and Davenport (1949), Davenport and Combs (1954), Ramón-Moliner (1957) and, probably, in all procedures leading to the deposit of silver chromate in formaldehyde-fixed material. However, Anderson's method fails when applied to Golgi-Cox material, in which the impregnating substance is either metallic mercury or some insoluble complex of the metal. The Van der Loos (1956) technique for counterstaining Golgi-Cox sections was therefore a significant innovation. This method, in which methylene blue is the dyestuff, produces good results in the cerebral cortex; the author, however, has found it less reliable in the brain stem or in other regions rich in tracts of myelinated fibers.

In addition to methylene blue recommended by Van der Loos, thionin, toluidine blue, gallocyanine, methylene blue chloride, neutral red, cresylechtviolet, Darrow red, and safranin 0 have been tested, using a wide range of variations in the time of exposure to the stain, the concentration of the dye, and the pH and molarity of the buffer (Ramón-Moliner *et al.*, 1964). In general, the optimal pH was found to range between 3.2 and 3.8. In later experiments (Turcotte and Ramón-Moliner, 1965) it was found that a considerable improvement in the overall quality of the stain could be obtained by adding alcohol to the staining fluid. In our present method, cresylechtviolet is used exclusively. Neutral red and Darrow red can also be used, but these two dyes give less contrast and have a greater tendency to stain the celloidin.

Mounting and Preservation

The tendency of Golgi-Cox preparations to fade when mounted under cover slip has been noted and discussed above. Such fading invariably occurs, irrespective of the mounting medium, when the sections have been inadequately stabilized. The author has observed little fading in cover-slipped Golgi-Cox sections that had been alkalinized as described below, and mounted in synthetic resins. After several years, the only change observed was a slight loss of definition in the appearance of dendritic spines.

Procedures

Depending on the moment when alkalinization is introduced, two procedures can be followed, viz. a relatively rapid one which permits a complete series of cat brain to be obtained in about one month, and a slower procedure which may require up to three months to complete. If no counterstaining is desired, and if a certain loss in the delicacy of the impregnation is not a serious objection, the rapid procedure is recommended. Otherwise, the slow procedure should be followed. Both techniques are described below. The user will find it advantageous to acquaint himself with the underlying technical principles discussed in preceding sections of this paper.

Impregnation

The impregnation stage is the same for both the rapid and the slow procedure.

Pieces of fresh nervous tissue no thicker than 6 mm, and if possible no thinner than 3 mm, are stored for 20—30 days, in the dark and at room temperature, in a solution made up by mixing A and B, as follows:

Solution A

Potassium dichromate	1 g
Mercuric chloride	1 g
Distilled water	85 ml

The solution is boiled for 15 min, and water is added to compensate for evaporation. Once cooled to room temperature, the fluid is mixed with Solution B:

Solution B

Potassium chromate	0.8 g
Potassium (or sodium) tungstate	0.5 g
Distilled water	20.0 ml

The slight turbidity which sometimes accompanies the mixing of A and B can be disregarded. The solutions should be freshly prepared and be kept in the dark. Special attention must be paid to the purity of the mercuric chloride. Brain tissue of young animals should be kept in the impregnation fluid for 20 to 30 days; this time must be shortened to two weeks for material from older animals.

Rapid Procedure

1. After rinsing in water, the pieces of tissue are transferred from the impregnating fluid to a solution made up of:

Lithium hydroxide	0.5 g
Potassium nitrate	15 g
Distilled water	100 ml

The amount of liquid must be about one hundred times the volume of the tissue. Small brains such as the frong's require immersion in this fluid for approximately 4 hr., the brain of a mouse 10 or 12 hr., 6 mm thick slices of cat brain 24—30 hr.

2. The blocks are washed in distilled water containing 1 ml of acetic acid per 500 ml of volume. This solution should be changed several times; large pieces of cat brain require 3 or 4 such changes every 6 hr. At this stage, the tissue is very brittle and should be handled with care.

3. The blocks are washed in abundant tap water (but not in running water) for 2 hr.

4. Dehydration in alcohol. The large 5—6 mm thick slices of a cat brain should be kept at least 6 hr. in each of the following concentrations: alcohol 40%; alcohol 60%; alcohol 80%; alcohol 90%; absolute alcohol (three changes); alcohol-ether (equal parts).

5. Celloidin embedding, following standard techniques.

6. Sections are cut at 50—150 micra thickness. As a rule, no sectioning difficulties are encountered. However, should the tissue have become too brittle as a result of excessive alkalinization, the following maneuver can be tried: before cutting each section, the face of the block is covered with a thin layer of 1% celloidin by means of a brush. When this layer has dried, the section is taken and, if necessary, covered on the back side with another celloidin film.

7. As they are obtained, the sections are processed one at a time through the following solutions: (1) 70% alcohol, 1 min; (2) 95% alcohol, 2 min; (3) a 3:1 mixture of absolute alcohol and chloroform, three changes; (4) toluene, two changes. One at this stage, the sections can be stored in serial order for several days between pieces of filter paper soaked in toluene. This is possible because the material has already been alkalinized. *In the absence of alkalinization, toluene causes fading.*

8. The loose sections are immersed in a freshly prepared mixture of methyl benzoate, 50 ml; benzyl alcohol, 200 ml; and chloroform, 150 ml. The sections should be kept on the bottom of the dish: near the surface they tend to soften and may even disintegrate.

9. After a minimum of 3 min soaking, each section is transferred quickly by the aid of a small brush to a *pretreated slide* (see preceding section "Attaching celloidin sections to the slides") which is held partially submerged in the fluid. Orientation on the slide must be completed before softening occurs.

10. The slide is exposed to the air for 3—5 min, or until the chloroform evaporates and the celloidin softens.

11. The slide is now dipped and washed thoroughly in 3 changes of toluene, which hardens the celloidin and thus anchors the section to the slide.

12. Mounting, under cover glass, using a synthetic mounting medium having a refraction index of approximately 1.61.

Slow Procedure with Optional Counterstaining

1. Following impregnation in the Golgi-Cox fluid, the tissue blocks are rinsed briefly in water and transferred to a mixture of

Absolute alcohol	one part
Ether	two parts

The dehydration usually requires the use of three volumes, in each of which the blocks are kept for 4—8 hr. The familiar dehydrating sequence of graded alcohols causes fading and should therefore be avoided.

2. The blocks of tissue are next transferred to the first celloidin solution which is prepared as follows:

Celloidin	30 g
Ether	250 ml
Absolute alcohol	250 ml

Dissolve in shaker, and then add, while stirring:

Absolute alcohol	100 ml
Ethylene glycol	40 ml
Ether	450 ml

The pieces of tissue should be kept in this solution for ten days *in the freezing compartment of a standard refrigerator* (—10° C).

3. Second celloidin infiltration, also in the freezing compartment. This second solution is prepared as the first one, but, instead of 30 g, contains 50 g of celloidin per 1,000 ml of solvent mixture. The blocks should be kept in this celloidin solution for five days.

4. Third celloidin solution, containing 80 g of celloidin per 1,000 ml of solvent mixture. The blocks should remain in this solution for five days, also in the freezer. Considerable shorter permanence in the celloidin solutions is possible in small pieces of tissue if good counterstaining is desired.

All the celloidin solutions should be kept in the dark. Pellets of calcium carbonate on the bottom of the bottles will minimize the risk of free nitric acid formation.

5. In order to obtain the blocks of hardened celloidin, it is advisable to use *cardboard containers with flexible sides.* Such containers can be made by cutting a strip of light cardboard, the end of which are stapled together to form a ring or bracelet (see Ramón-Moliner, 1961, Fig. 1). By means of celloidin dissolved in equal parts of alcohol and ether, this cardboard ring is sealed to a relatively heavy plate of metal or glass, which thus forms the floor of the container. When the carboard ring has become adherent, a thin layer (2—3 mm) of 10% celloidin is poured to cover the bottom of the container. After pouring this bottom layer of celloidin, its surface is hardened by blowing on it for a few seconds. The container in then placed in pure chloroform for one hour, after which time the celloidin should be hard enough to receive and support the tissue blocks (Ramón-Moliner, 1961, p. 230). Without this bottom layer of hard celloidin the tissue would be in direct contact with the floor plate, and tissue fragments would adhere to the latter when the celloidin blocks are detached. The container is next filled with 10% celloidin *in equal parts of ether and alcohol.* The pieces of tissue can then be placed in the container and appropriately oriented. Then, after a surface film has formed, the container is returned to cold chloroform.

6. Twelve to 24 hours before sectioning, the hardened block is removed from its container, trimmed, quickly washed in a mixture of equal parts of alcohol and ether (about one or two minutes), stuck to a wooden block the recipient surface of which is amply covered with 10% celloidin, and placed in 70% alcohol at $-10°$ C. After six hours, the 70% alcohol should be renewed.

7. The sections are cut at 100—150 μ but, when counterstaining is desired, it is of advantage to cut sections of alternating thickness, for example, 60 μ and 120 μ, keeping the series of thinner sections for counterstaining.

8. As they are obtained, the sections are processed, one at a time, through the following solutions: (1) 70% alcohol, 1 min; (2) 95% alcohol, 2 min; (3) a 3:1 mixture of absolute alcohol and chloroform, three changes, and (4) toluene, two changes.

Non-alkalinized sections should not be stored for long periods of time in toluene as this causes fading. If for any reason the continuation of the procedure must be postponed, the sections can be stored between pieces of filter paper soaked in chloroform.

9.—12. The sections are attached to slides following steps 8—11 outlined in the description of the rapid method. To prevent the detachment of the sections during the subsequent alkalinization, these steps should be followed punctually.

13. In order to darken and stabilize the mercury impregnation, the slides with the sections properly attached are placed in racks. If a standard rack for ten slides is used, no more than five slides should be processed at a time.

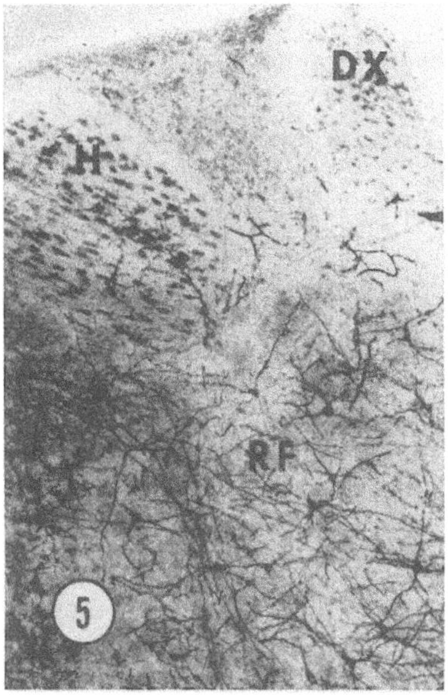

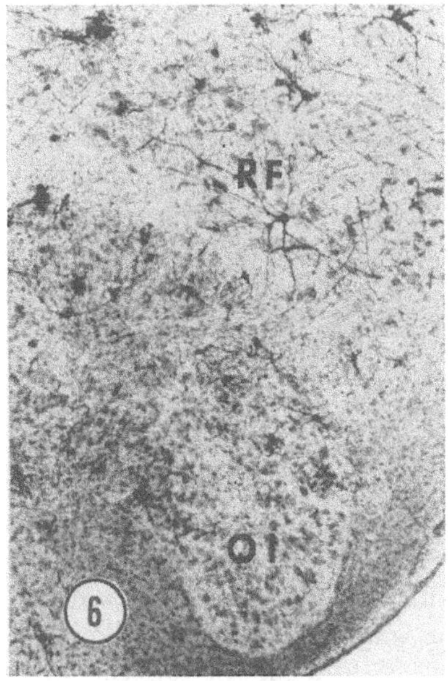

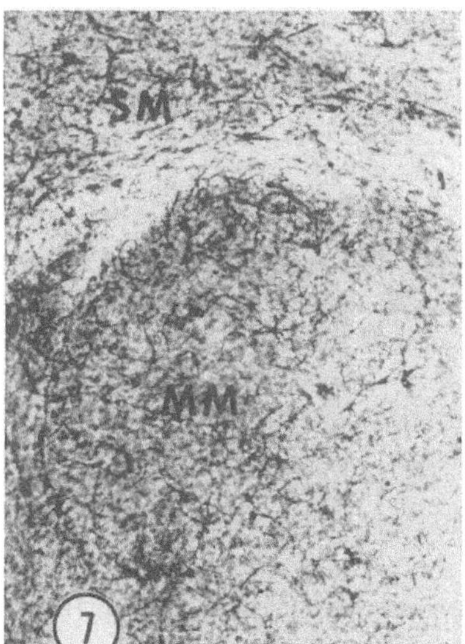

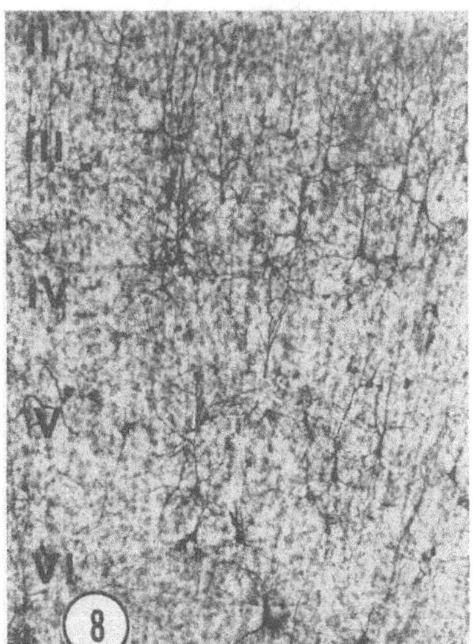

Fig. 5. Medulla oblongata of cat. Counterstained Golgi-Cox section. In this section, as well as in the ones shown in Figs. 6, 7 and 8, the nerve cells that have not taken the Golgi stain can be seen faintly tinted with cresylechtviolet. The dye also tends to stain the background differentially, and thus permits the visualization of landmarks which would not be evident in a non-counterstained section. *H* hypoglossal nucleus; *DX* dorsal motor nucleus of vagus; *RF* reticular formation. (×25)

Fig. 6. Medulla oblongata of cat. Counterstained Golgi-Cox section. *OI* inferior olivary nucleus; *RF* reticular formation. (×25)

Fig. 7. Mammillary region of cat. Counterstained Golgi-Cox section. *MM* medial mammillary nucleus; *SM* supramammillary nucleus. (×40)

Fig. 8. Posterior ectosylvian cortex of cat. Counterstained Golgi-Cox section. (×40)

14. The racks are passed through descending grades of alcohol: 95%, 2 min; 70%, 2 changes, 2 min each; 40%, 2 min.

15. The sections are transferred for 2 min to a mixture of: absolute alcohol, 80 ml; strong ammonia (28% NH_3), 120 ml.

16. Sections are placed for 1 hr in a mixture of: absolute alcohol, 140 ml; strong ammonia (28% NH_3), 60 ml.

17. The sections are passed successively through 70% alcohol and 40% alcohol (2 min each), and are then washed thoroughly in running tap water for 10 min.

18. Sections are soaked in 1% sodium thiosulfate for about 10 min, or until the yellowish color of the background has faded.

19. The sections are washed in running tap water for 10 min. Sections not to be counterstained can now be dehydrated, cleared in toluene, and mounted under cover glass, using a synthetic mounting medium.

20. The sections to be counterstained, preferably 60 μ thick or less, are transferred for two minutes to 25% alcohol, and next placed into the counterstaining solution which is prepared as follows:

2 gm of cresylechtviolet are dissolved in
185 ml of 1 M acetic acid

and the following solution is added:

1 M sodium acetate	15 ml
Distilled water	400 ml
Ethyl alcohol	200 ml

Sections should be kept in the dye solution for 90 min.

21. Without rinsing in water (which appears to impair the subsequent differentiation) the sections are differentiated in 50, 70 and 95% alcohol (about 45 sec each), followed by two volumes of a 3:1 mixture of absolute alcohol and chloroform, and two volumes of toluene.

22. Mounting under cover glass, using a synthetic resin.

Results

Throughout the history of the Golgi methods investigators have shared the experience that not all parts of the brain are impregnated with equal ease. The hippocampus is by far the most consistently well stained structure; excellent impregnations are obtained in this region even in very young animals in which often no other structure attracts the stain. The cerebral cortex and brain stem nuclei begin to be stainable by the Golgi-Cox method, almost at the same time, in the two-week old kitten. The method, as a rule, yields poor results in animals less than 15 days old. In the author's dendroarchitectonic studies of the cat's brain stem, the ideal time for optimal impregnation was found to coincide with the time just before myelination appears to be completed, i.e. around twenty days of age. Beyond that time, the cortex remains fairly stainable, and good impregnation of cortical neurons can be obtained even in animals older than 2 months.

The Golgi-Cox method is unexcelled in the clarity and sharpness with which it impregnates neuronal perikarya and dendrites. In optimal material, nerve cells and their dendrites appear in jet black against an almost colorless background devoid of

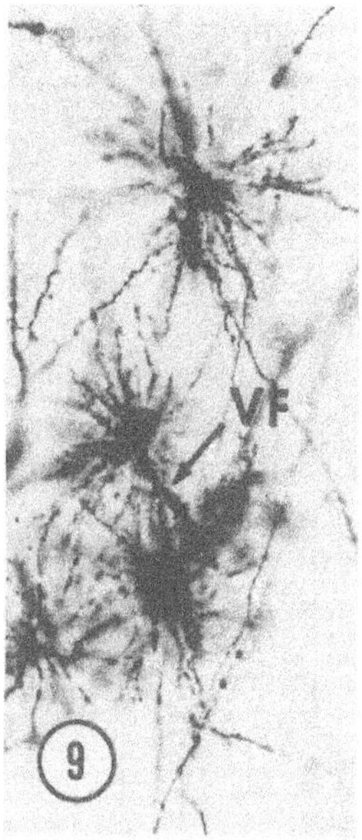

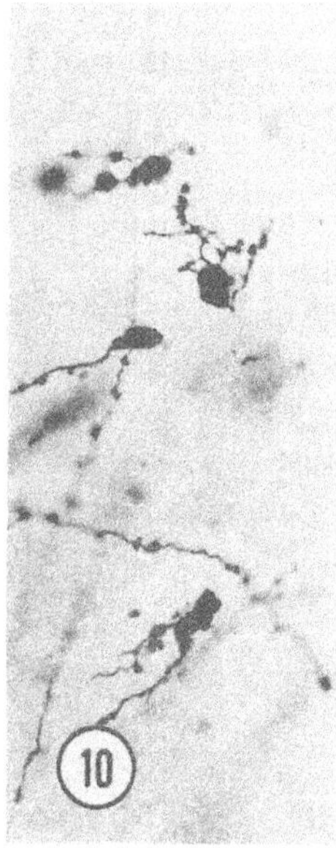

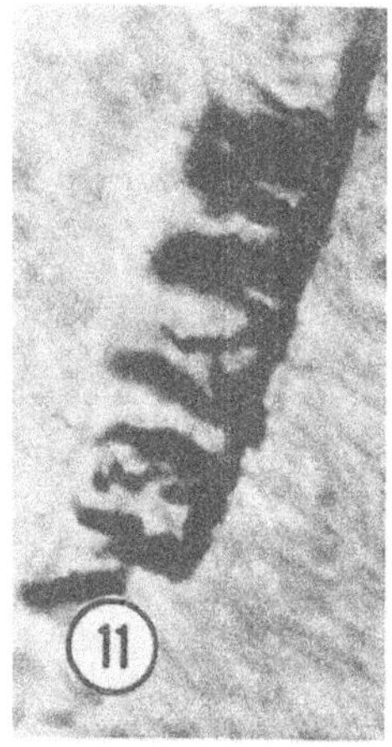

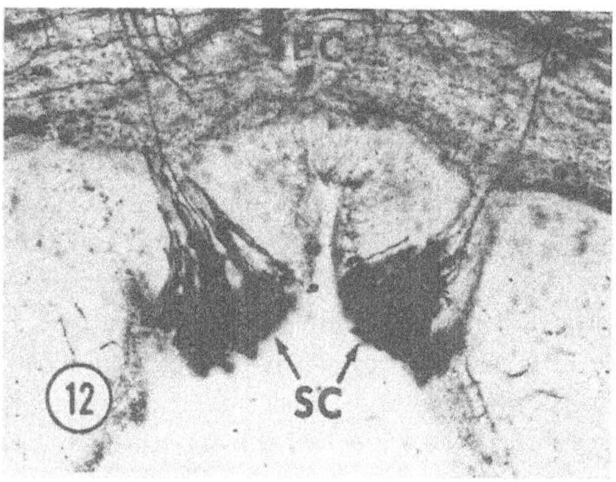

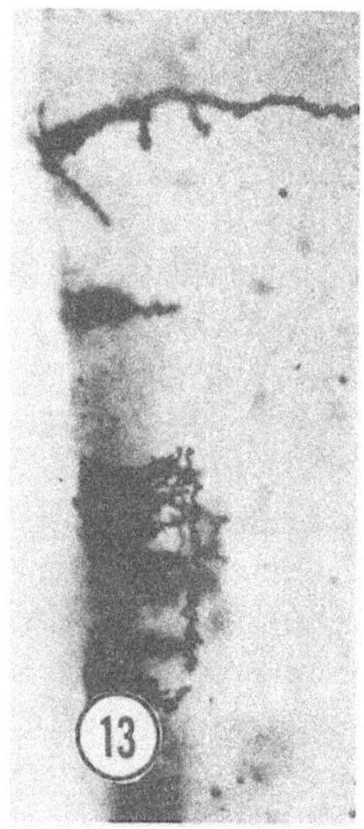

Fig. 9. Astroglia. These elements are impregnated more rarely by the Golgi-Cox method than by other Golgi variants. *VF* vascular foot. (×250)

Fig. 10. Oligodendroglia. Impregnation of this form of glia by the Golgi-Cox method is rare. It is promoted by the addition of tungstate to the impregnating solution. (×250)

Fig. 11. Arteriole. The perivascular structures could be either smooth muscle fibers ot pericytes (perivascular microglia?). (×400)

Fig. 12. Region of the posterior commissure and cerebral aqueduct of cat. *PC* posterior commissure; *SC* subcommissural organ. (×250)

Fig. 13. Wall of third ventricle of cat, showing impregnated ependymal cells. (×400)

any precipitate (Figs. 1 through 4). In most cases, the initial part of the axon is also stained, and some axons at least are almost always impregnated in compact fiber bundles such as the pyramidal tract, the mammillothalamic tract and the fasciculus retroflexus. Terminal axon arborizations, however, are only occasionally impregnated.

The Golgi-Cox method is therefore less suitable than the original "rapid Golgi" ("Golgi osmic") method for the visualization of terminal axon arbors and synaptic end structures. Non-neuronal structures can also be stained (Figs. 9 through 13). This, however, occurs only rarely.

The counterstained sections show all unimpregnated nerve cells stained as in the usual Nissl picture, although with lesser intensity (Figs. 5 through 8). As a rule, the cytoplasm is evenly stained and the Nissl substance seldom appears in the form of separate granules. Scattered about the field are nerve cells which have taken the metallic impregnation and display black cell bodies and dendrites, quite often with part of the axon. The Nissl picture in counterstained Golgi-Cox preparations is not sharp enough for the study of cytological detail, but entirely adequate for routine cytoarchitectonic delineation and thus, for its original purpose of identification of the cell group to which a particular Golgi-stained neuron belongs.

In some cases, the method fails completely and yields no impregnation of nerve cells, glia, or blood vessels, nor even any form of precipitate. Such failures occur almost exclusively in newborn animals, in embryos, and in old animals when too thin slices have been cut. Naturally, no impregnation can be expected if the hardening period is shortened excessively, to 2 or 3 days for example. In order to establish the optimal impregnation time, one may occasionally examine thin hand-cut fragments from the block that have been treated with an alkaline solution. One should bear in mind, however, that the superficial regions of any block tend to be under-stained, and that even when the examination of surface fragments suggests a failure, the deeper regions may be well-stained. In other cases, much precipitate is found. If only precipitate is present, one of the following causes should be suspected: a) chemicals of poor quality, particularly mercuric chloride; b) stale solutions, especially likely when the fluids have been exposed to light for long times; c) the fragment was taken from a region which is always difficult to stain. In the last-mentioned case, mercury, instead of being attracted by the nerve cells, is irregularly precipitated. The presence of any well-stained neuron negates the possibility that poorly prepared solutions were used. Among the structures difficult to stain are: the superficial region of the molecular layer of the cerebellum, the retina, the olfactory bulb and the spinal cord. The thalamic nuclei and lower brain stem of cats older than two months are usually poorly impregnated. Failures attributable to causes inherent in the tissue itself are manifested by either excess or lack of affinity. The latter condition is frequently encountered in the thalamus of the mouse which is often poorly stained even in cases in which neighboring structures, such as the striatum, show excellent impregnation. Excess of affinity may appear in the form of simultaneous staining of dendrites and the associated axonal endings. When this occurs, the nerve cells appear covered by a thick, mossy layer of filaments. This is a common imperfection in cerebral cortex tissue of old animals that has been kept too long in the impregnating mixture; it is promoted by the use of tungstate. The brains of cats and mice older than three months should not be exposed to the impregnation fluid longer than two weeks. Finally, and quite unpredictably, the reaction may be completely shifted to capillary vessels, leaving the nerve cells unstained. The use of animals too old, and tissue blocks too large, appears to promote this form of failure, as also does perfusion with the impregnation fluid.

Theories on the Mechanism of the Golgi Stain

The peculiar sparseness of impregnated neurons and the "all or nothing" characteristics of the staining have raised much speculation with respect to the mechanism of the Golgi stain. According to one theory, the functional activity of the cell during the last moments of its life would either predispose it or render it immune to impregnation. Another theory holds that microscopic mechanical injuries, caused by handling of the tissue before staining, would be responsible for the unequal affinity of the various cellular components.

The first hypothesis would seem to be endorsed by the results of Bertram and Sheppard's (1964) experiments. These authors found that, following long-lasting unilateral antidromal stimulation of the hypoglossal nerve, motor neurons of the ipsilateral hypoglossal nucleus were impregnated in larger numbers than those of the control side. One should not forget, however, that there are many other non-neural structures which are also stained in a scattered manner with the Golgi method: capillary vessels, secretory canaliculi of the liver and of the salivary gland, collagenous fibers, endothelial cells, dead cells of the exfoliative epithelia, etc. If the stainability of a cell depends on its electrical state during the last moments of life, an unequal electrical history would have to be postulated for all of these non-neural tissues as well as for nerve cells.

According to another hypothesis, mechanical injury to the cell is a prerequisite for its staining. Nissl-stained sections quite commonly reveal a smaller or larger number of slightly shriveled, dark-stained neurons even in tissues fixed under apparently optimal conditions (e.g. by perfusion with formalin promptly after death). Formerly often interpreted as evidence of brain pathology, the "dark neuron" phenomenon has more recently been shown to result from artefactual conditions during fixation and tissue removal (Cammermeyer, 1960). It now appears likely that dark neurons are nerve cells that have been "squashed" prior to staining. It is, however, doubtful that they are the "Nissl equivalent" of Golgi-stained cells, for in Golgi sections counterstained by the Nissl method many dark neurons may appear that have been left unstained by the Golgi procedure. Naturally, from the study of such sections one cannot ascertain whether the cells that have taken the Golgi stain were originally dark ("squashed") or intact nerve cells. Therefore, it is still possible that the category of the dark cells might be larger and include the class of the Golgi-stained cells. However, experiments by Morest and Morest (1966) involving perfusion of the brain with fixatives prior to its removal have shown that the improved preservation of the tissue thus achieved is accompanied by an improvement of the Golgi stain, a fact that militates against the hypothesis that links stainability to mechanical damage of the nerve cells. Moreover, even if "squashing" were the primary conditioning factor, it would still remain to be explained how it is that certain cells are "squashed" and others not. In other words, by invoking the notion of mechanical damage to the cell, the explanation of the peculiar sparseness of the Golgi stain is only procrastinated.

On the basis of certain personal observations, a probabilistic interpretation could be suggested. It should be emphasized that in the Golgi method *the staining of any particular cell seems to condition a non-stainability of adjacent cells.* For example, synaptic complexes are rarely if ever revealed by the Golgi method. Although there are exceptions to the rule, one can state that in most cases of successful impregnation, if the presynaptic component of a synaptic junction (bouton terminal) is stained, the postsynaptic component (dendrite or soma) has not taken the stain. In the cerebellum, for example, both somata of Purkinje cells and axonal "baskets" that are known to surround these somata may be stained. However, rarely if ever is the soma of a Purkinje cell stained together with the particular axonal "basket" that surrounds it, a finding suggesting that the staining of any of these two structures conditions a non-stainability of the other. By the same token, the distribution of precipitate or stained capillaries in Golgi-Cox sections is generally confined to those regions in which nerve cells have failed to take the stain. By contrast, where nerve cells are stained in large numbers, impregnated capillaries and precipitates are scarce.

There thus appears to be something in the Golgi mechanism that could be compared to the exclusive "OR" gates of logical machines: "This cell or its neighbor, but not both"; "either well-stained nerve cells or capillaries and precipitate, but not both" (this property is particularly evident in the Golgi-Cox modification); "either neuronal perikarya and dendrites, or axons, but not both" (this is common in rapid-Golgi material); "either glia or nerve cells, but not both" (which tends to occur in the formaldehyde modifications).

These observations indicate that the Golgi method tends to stain the whole of any cytoplasmic space enclosed by a membrane, and that this membrane is the site of a "flip-flop" mechanism which may allow the impregnation of one, and only very seldom both, of the two phases separated by it.

Discussion

Dr. Frontera: I have experimented fairly extensively with various modifications of the Golgi technique. For example, I have used Dr. Ramón-Moliner's tungstate modification in adult cat brains and, after standard celloidin embedding, obtained preparations with almost crystal-clear background. I have also modified the Fox *et al.* and the Ramón-Moliner chlorate-formaldehyde techniques with some improvement in the staining as well as the clarity of the background. Some of this material is over 5 years old, and is still in perfect condition.

Dr. Nauta: Dr. Frontera, could you describe this modification?

Dr. Frontera: Essentially, the modification consists of using a slightly alkaline formalin for the perfusion-fixation and slightly alkaline silver nitrate for the staining in the Fox *et al.* method. For the Ramón-Moliner's chlorate-formaldehyde technique, the modification consists of perfusion-fixation with slightly alkaline formalin followed by the latter's own fixative without formaldehyde.

Dr. Valverde: Does this modification stain glia?

Dr. Frontera: Yes, it does too. It is sometimes possible to follow the fiber tracts by the arrangement of the glia in them.

Dr. Nauta: Dr. Frontera, we would be glad to include a description of your procedure in this discussion.

Dr. Frontera: There are really two procedures. These are referred to as variant "A" and variant "B". They were summarized in an abstract in 1964 and are fully described below:

A. Frontera's (1964) variant of the Fox, Ubeda-Purkiss, Ihrig and Biagioli (1951) procedure for obtaining a Golgi-type impregnation of nervous tissue, as modified by Bertram and Ihrig (1957 and 1958):

1. Fix the brain in *situ* by pulsating-perfusion with $1^1/_2$ or 2 liters of the following fluid (prepared fresh):

40% Formaldehyde	100 ml
Distilled water	900 ml

M/1 sodium hydroxide, 10—12 ml to bring the pH to 7.2, then add monosodium acid phosphate monohydrate

($NaH_2PO_4 \cdot H_2O$)	4 g
Disodium acid phosphate (Na_2HPO_4)	6 g

2. Remove brain. Cut into 3—5 mm thick slabs in the desired plane and immerse at least three days in the same fluid used for perfusion.

3. For the following steps (chromating and silvering) it is recommended that rigid, plastic "baskets" with perforated partitions to separate the individual slices and well perforated sides and bottom be made. One of these trays, big enough to fit inside a large glass staining dish or other suitable dish of approximately one liter capacity, is used for chromating, and the other for silvering the slices. The slices are placed in the individual partitions of the tray and the tray placed inside the staining dish. The tray should not rest directly on the bottom of the staining dish (or any other glass container with cover), so that a teflon-covered cylindrical stirring bar may be placed under it and the whole set-up placed over a magnetic stirrer to provide continuous, smooth agitation of the chromating or silvering solutions without damaging the slices of tissue.

4. Prepare a fresh chromating solution containing:

Zinc chromate ($ZnCrO_4$)	60 g
Formic acid	32 ml
Distilled water to	1,000 ml

The pH of this solution should be 3.1, so that more formic acid may be added, if needed.

Remove the slices from the fixative and place them on paper toweling, drying both surfaces without pressing. Place the slices in the individual partitions of the basket. Drop the teflon-covered stirring bar on the bottom of the dish. Place the basket with the slices inside the dish over the stirring bar. Place the dish on the magnetic stirrer, preferably in the dark, and pour the chromating fluid into the dish. Cover the dish and set the stirrer for a smooth, continuous agitation. Change the slices to freshly prepared chromating fluid daily for three days. To do this, drain the old fluid from dish; carefully place the slices (in serial order if this is necessary for serial sectioning) on a paper towel and rinse the basket and dish three times with distilled water. Replace basket in dish over stirring bar; replace slices in basket; place dish with basket and slices over the magnetic stirrer; add the chromating solution and set the magnetic stirrer in motion.

5. After three days chromation, the slices are again dried on paper towels and placed in the individual compartments of the "silvering basket". This is placed, as before, inside a "silvering dish" over a teflon-covered stirring bar. Prepare a fresh silvering solution as follows:

10% silver nitrate in distilled water (stock)	75 ml
Distilled water, to	1,000 ml

Add approximately 0.5 ml of the M/1 sodium hydroxide. Shake well and check the pH, which should be 7.2, add more M/1 sodium hydroxide if necessary. The solution should be opalescent. Disregard any dark precipitate formed. Shake the solution well and add to the slices on the dish.

Cover the dish and set the stirrer in motion. Try to keep the dish protected from light. (I do the chromating and silvering on a shelf inside a locker.) Repeat the same procedure after 24 hours, removing the slices to paper toweling and brushing (not rubbing) the surface vigorously with a camel-hair brush. Rinse the dish, basket and stirring bar three times with distilled water; replace the slices in the basket; add a freshly prepared solution and keep on stirring for another 24 hours.

6. At the end of two days in the silver solution, the slices are rapidly embedded in paraffin, as indicated by Fox *et al.* (1951) and sectioned at 80 micra. I have embedded them in celloidin (Parlodion) through ethanol and ether. A convenient time schedule is: 50% ethanol, 4 hours; 70% ethanol, overnight; 85% ethanol, 2 hours; two changes of 95%

ethanol of two hours each; followed by three changes of absolute ethanol of no less than 2 hours each, and one of which should be overnight; mixture of equal parts of absolute ethanol and ether, two changes of 2 hours each. The slices are then placed in thin (4%) celloidin and gradually embedded.

7. Paraffin or celloidin sections are collected in 70% ethanol. Care should be taken to maintain the order, if serial studies are intended. The sections are then individually passed through two changes each 85% and 95% ethanol; one change absolute ethanol (rapidly, in case of celloidin sections); through a terpineol-xylene mixture, and through three changes of xylene. They are then mounted with synthetic resin and cover glass.

8. The brain slices (cat or monkey) should not be over 5 mm thick. The use of reagent grade chemicals throughout is imperative. The usual precautions of chemical cleanliness required when working with silver solutions are stressed. In my experience, sections covered with "Permount" mounting medium, but no cover glass, have the tendency to become brittle, trap air and become useless if allowed to stand for prolonged times without a cover slip.

B. Frontera's (1964) variant of the Ramón-Moliner (1957) chlorate formaldehyde modification of the Golgi method is as follows:

1. Fix the brain *in situ* as explained in variant "A", step 1.

2. Remove brain, cut into 3—5 mm thick slabs in the desired plane and place in the following solution for three days, changing fluid daily:

5% Potassium dichromate	40 ml
5% Potassium chlorate	20 ml
20% Chloral hydrate	30 ml
Distilled water, to	100 ml

The volume of liquid used should be twenty times the volume of the tissues.

3. Transfer slices to a 3% potassium dichromate solution, changing it twice daily for two days.

4. Remove slices to paper toweling to dry the surface. Place the slices in 1% silver nitrate solution at room temperature, in the dark and, preferably, agitated with a magnetic stirrer, in which case the use of plastic basket as described in variant "A", step 3, is recommended. Leave the slices for 24 hours, remove to paper toweling, brush well and place into fresh 1% silver nitrate solution for another 24 hour period, in the dark and using the magnetic stirrer.

5. Dehydrate through ethanol and ethanol-ether and embedded in celloidin, as in variant "A", step 6.

6. Sections are cut in the sliding microtome at 50—100 micra and collected, cleared and mounted as in variant "A", step 7.

7. Apparently, the most significant modifications introduced here are the perfusion-fixation with buffered formalin on the alkaline side, and the omission of the formalin in Ramón-Moliner's (1957) solution. The use of continuous agitation, as in variant "A", may help to provide better interaction between tissue surface and fluid. The dehydration through ethanol and celloidin embedding facilitates the handling of the sections. The use of terpineol-xylene mixture to clear the sections avoids the strong acidity of the carbol-xylene mixtures and provides good dehydration.

Dr. Valverde: Dr. Ramón-Moliner, do you believe that the beading of the axons that you showed represents a natural phenomenon?

Dr. Ramón-Moliner: I find it difficult to answer that question because I have never looked at a live axon. The thing that strikes me, in looking at Golgi-Cox preparations, is that some brains will be full of beaded axons and other brains will have none or very few. Such variations from one brain to the next make one wonder if beading of axons may not have something to do with the fixation, the impregnating mixture, or the dehydrating technique. If these variables are not entirely responsible for all the beading, one or more of them could possibly potentiate a tendency to beading.

Dr. Valverde: Dr. Peters, do you believe that these thickenings are actually synaptic points?

Dr. Peters: I would not be able to say.

Dr. Fox: I think that in some instances the swellings are indeed real and not artifacts. The parallel fibers of the cerebellum, for instance, show constrictions and dilations in the Golgi picture, and have the same appearance in electron micrographs. The same is true for the terminal fibers in the caudate nucleus, putamen and globus pallidus. I, therefore, have a strong feeling that in many instances the beading is not an artifact at all. It is also noteworthy that these swellings are found in areas loaded with synaptic vesicles.

Dr. Konigsmark: Coers and Woolf, in studying terminal innervation of muscle fibers with the methylene blue stain, found axonal swellings in these distal fibers. These authors also believe that the swellings are not artifacts.

Dr. Valverde: Do you know if the deposit of the substance is inside the cells?

Dr. Ramón-Moliner: I think that Van der Loos showed it was inside. The reaction is definitely a cytoplasmic one, no a membrane reaction.

Dr. Nauta: Dr. Ramón-Moliner, in a cross section of a dendrite, do you invariably find the whole profile filled, or do you sometimes find a ring-like encrustation?

Dr. Ramón-Moliner: I have never seen a ring.

Dr. Valverde: We have electron microscopic evidence that the Golgi precipitate in the "rapid" technique is deposited within the nerve cell.

Dr. Blackstad: My experience in electron microscopy of Golgi material is restricted to the rapid Golgi method; I have not studied the Cox variant. In rapid-Golgi material, I have found that brown or black precipitate can be present almost anywhere, i.e., fill any type of cell or cell process, or selectively stay in extracellular spaces. Sometimes dendrites and boutons are outlined in a manner reminiscent of the net-like structures that had been seen around hippocampal pyramidal cells. I am inclined to believe that the latter were really extracellular precipitate on the surface of the pyramidal cells and boutons. I think that everything that has been said about the localization of the Golgi precipitate is correct in the sense that the precipitate can be anywhere.

Dr. Fox: I would like to ask Dr. Valverde a trivial question: How do you cut your sections? I believe Rámon y Cajal cut his by hand.

Dr. Valverde: My technicians cut with the sliding microtome, but I prefer to cut the tissue by hand using a razor blade.

Dr. Karten: The problem of cutting rapid-Golgi blocks has bothered us a great deal and we finally mastered a method that is very similar to the one used by Dr. Valverde. We recently tried something else which may be of some interest. We took a goldfish brain that had been processed according to Dr. Valverde's modification of the rapid Golgi and embedded it directly in Epon-Araldite.

Dr. Fox: How large is that brain?

Dr. Karten: It is about 1 cm long and perhaps 4 mm wide. After polymerization and blocking we cut it with steel knives on the ordinary rotary microtome at 4—40 μ. The difference in thickness seemed to determine a curious difference in general appearance. In 40 μ sections the background looks exactly as usual, but at 10—15 μ, the background is considerably lighter and the cytology vastly superior. The axon plexus, for example, can be seen in much greater and sharper detail than is the case with the usual 40 to 120 μ sections. We could not detect any fading as a result of theis embedding procedure. It would be useful if one could also do a Richardson stain on these rapid Golgi preparations for better identification of cells. Has anyone here tried this on Epon-Araldite embedded material?

Dr. Blackstad: Dr. Stell has recently published a paper on electron microscopy of Golgi preparations (Stell, 1965, 1967), and he has done this type of counterstaining. He used the rapid Golgi method.

Dr. Valverde: One of the reasons for hand-cutting sections is that it allows one to decide quickly which piece should be continued in dichromate and which should not, because the moment you pass the dehydration procedure you have lost many hours. The impregnation may also suffer by this.

Dr. Ramón-Moliner: We have recently been impressed by the acceleration of the infiltration that can be obtained by adding approximately 4% of ethylene glycol to the celloidin solutions. We are still in the experimental stage and feel that higher concentrations, provided they are compatible with the alcohol-ether mixture, may be even more advantageous.

Dr. Sidman: I have yet another minor technical point. We, and I think others as well, have used aldehydes other than formaldehyde as an initial fixative prior to the Golgi-Cox method. For the mouse brain, fixation in acrolein followed by immersion in glutaraldehyde, followed by the Golgi-Cox is an excellent procedure. One can put the intact adult mouse brain through this procedure, embed in celloidin, and cut serial sections. The method has been particularly useful for us in the study of mutants in which one region of the nervous system may be much smaller than normal or in which the cytological features of interest lie close to the tissue surface. For example, in some mutants the cerebellum is about 1/5th of the normal size and the cortical cells are particularly affected, so that we must impregnate the outermost zone of this tissue or lose virtually the entire image of the cerebellar cortex.

Dr. Ramón-Moliner: Do you know the pH of the glutaraldehyde?

Dr. Sidman: The glutaraldehyde is buffered to a pH of 7.3 to 7.4.

Dr. Frontera: Returning to Dr. Ramón-Moliner's earlier comment about why some cells stained and others didn't, and the feeling that it may be due to the agonal state of the cell at the time of fixation, I would like you to consider the fact that when one experiments with different fixatives at different pH, one gets a certain range of pH with better results than others. This leads me to think of the possibility that it may be the metabolic state of the cell itself, the pH of the cell, that determines whether a cell is to be stained or not. I also do not completely agree with Dr. Ramón-Moliner that impregnation of two contiguous neurons never occurs. I have seen instances of such neighboring impregnation.

Dr. Ramón-Moliner: How do you explain the following experiments: If you add strong alcohol to an albumin solution, you obtain a milky coagulate, a suspension of a multitude of albumin particles. Then you mix those particles with albumin which has not been treated with alcohol and heat the mixture until it coagulates. You end up with particles of albumen coagulated by alcohol embedded in albumen coagulated by heat. When you stain this material with the Golgi method, some of the particles of alcohol-coagulated albumen are stained and others are not.

Dr. Frontera: The two things are not mutually exclusive, Dr. Ramón-Moliner. I think that you have a good point, but that does not preclude the other possibility because protoplasm is much more complex than albumin.

Dr. Ramón-Moliner: How do you account for the fact that the dead cells of exfoliative epithelia are also stained in an unequal manner by the Golgi method? Some cells take the stain and others do not, always following an all-or-nothing pattern (the same applies to non-neuronal elements, as shown in Fig. 9 through 13).

Dr. Frontera: Do they conduct?

Dr. Ramón-Moliner: They might. But I believe that we should not forget the non-specificity of the reaction. It is not like the gold chloride technique, which not only stains selectively the glia, but confines itself to astroglia, or like the silver carbonate mixtures that may, under one condition, show one type of glia and under different conditions, another type. So you can actually direct these techniques. You can guide them, to a certain extent. By contrast, one of the peculiarities of the Golgi technique is the difficulty of guiding the reaction. In fact, it is impossible. You can decide whether to use the rapid technique, or the technique proposed by Fox, or the Golgi-Cox, and that is about all you can do. Once you have placed the tissue in one of the corresponding impregnating mixtures, you cannot provide any further guidance. On the other hand, in the silver carbonate technique, for example, you can direct the silver staining process to microglia, instead of oligodendroglia, merely by blowing or not blowing on the surface of the ammoniacal silver solution.

Dr. Valverde: I recall that Dr. Blackstad (1965) has described a very curious experience a few years ago. He put a drop of silver nitrate on a chromated section on a slide and observed the reaction in the microscope. Can you tell us about this. Dr. Blackstad?

Dr. Blackstad: This was a ten-minute episode five years ago. I cut a slice of chromated tissue placed it in dichromate solution under a cover slip and then added silver nitrate to the tissue on the slide. I saw a dendrite with spines becoming stained like a worm creeping into the tissue.

Dr. Valverde: I have tried a hundred times to repeat that experiment, but have never seen it. There would be distinct advantages to an application of the Golgi technique to sections.

Dr. Ramón-Moliner: It is indeed strange that the Golgi process requires the use of blocks of tissue instead of sections. There seems to be but one exception, the retina. In terms of thickness, the retina could be considered as equivalent to a section, but in order to stain it with the Golgi method, you either have to roll it, as Cajal did, or embed it in a block of other tissues. It is nearly impossible to impregnate the more usual artificial tissue membranes that we call sections. Do you have any experience to the contrary?

Dr. Blackstad: No, I do not. However, I have on occasion stained slices almost as thin as sections. By surrounding them with a film of gelatin and then letting them float in the liquids, it is possible to process a considerable number of specimens in a short time. This technique may be useful for testing new procedures.

Dr. Ramón-Moliner: In any case, it is most curious that proper impregnation cannot be achieved in thin sections floating freely in the chromating and silvering solutions.

The Fixation of Central Nervous Tissue and the Analysis of Electron Micrographs of the Neuropil, with Special Reference to the Cerebral Cortex

Alan Peters

Introduction

One of the greatest difficulties encountered by electron microscopists studying the central nervous system is the fixation of the tissue. This problem has delayed detailed analyses of the central nervous system in mammals to such an extent that only recently have the descriptions of the individual components of this tissue been completed. By contrast, in other tissues of the body this same stage was arrived at early in the present decade. Even today, a student of the fine structure of the nervous system cannot be certain of obtaining good preservation of his tissue and there are still some parts of the brain and spinal cord that are successfully resisting attempts to study their morphology. These parts are the ones that contain a high proportion of myelinated nerve fibers.

In addition to examining some of the factors involved in obtaining successful fixation of central nervous tissue, the other purpose of this report is to provide a description of the various profiles that may be encountered in an examination of the neuropil of normal animals. It is hoped that both these topics will be useful to those embarking upon initial studies of the brain and spinal cord, since no similar accounts appear to be available elsewhere in the literature. With the exception of one electron micrograph from the olfactory bulb, all the illustrations are from the cerebral cortex of the rat, but the description is given in such a way that it should be generally applicable to any part of the neuropil from an adult animal.

Preparation of Central Nervous Tissue

The most successful and useful method for fixing the brain and spinal cord for electron microscopy is vascular perfusion. This technique was described in 1962 by Palay, McGee-Russell, Gordon and Grillo, and by Pease. Palay *et al.* (1962) used osmic acid solutions as the fixative, while Pease (1962) employed pure formaldehyde. Prior to the use of the perfusion technique, electron microscopists interested in the central nervous system exposed the brains and spinal cords of anesthetised animals, quickly excised the tissue and cut it up into small fragments in a pool of fixative. Not only does such a procedure lead to distortion of the soft nervous tissue, but it frequently produces severe artifacts because of the anoxia which obtains during the time that the tissue is being removed and penetrated by the fixative. Another serious problem is that although the superficial parts of the central nervous system may be amenable to fixation by this procedure, the removal of specific portions of the deeper structures is virtually impossible. Most of these difficulties are obviated by perfusion techniques in which the fixative is delivered rapidly via the blood vessels to all parts of the central nervous system. If respiration is continued until the last possible moment anoxia is also avoided and specific parts of the brain can be removed at a leisurely pace once the tissue has hardened. This means that any pieces of tissue that are required can be positively identified and their orientation readily determined.

Although the original osmic acid fixative described by Palay *et al.* (1962) has been largely superseded in recent years by the introduction of aldehydes as primary fixatives, particularly formaldehyde (Pease, 1962) and glutaraldehyde (Sabatini, Bensch and Barnett, 1963), the principles of procedure described by Palay *et al.* (1962) are still generally adhered to. Furthermore, the excellent fixation achieved by this group of workers is a standard that we still try to attain. At this point, it may be asked what constitutes good fixation, and in answer to this question the statement of Palay *et al.* (1962) is still valid. The important characteristics of good fixation are a "large area of intact tissue, in which clefts, shrinkage or retraction spaces, explosions, vacuoles, disrupted limiting membranes, and dehiscences all are absent. Each cellular component fits precisely against its neighbors according to the architectonics of the region, axons against dendrites, neurons against neuron, capillary endothelium against basement membranes and that against neuroglia, myelinated nerve fibers, and, throughout the whole tissue, the amazingly complex feltwork of the neuropil forming a close contexture that fills every interstice between the cells and their major processes."

Briefly, the technique used by Palay *et al.* is as follows. The animal is anesthetised and respired artificially with a gaseous mixture of 5% CO_2/95% O_2. The chest cavity is opened and vasodilation is produced by a 1% solution of sodium nitrite injected directly into the heart. Along with the sodium nitrite a small quantity of heparin is also introduced into the blood stream to prevent the blood from clotting. Respiration is continued with the chest cavity open and a clamp is placed on the descending aorta just before the tip of the heart is amputated to allow a cannula to be introduced into the common aorta. As soon as the cannula is introduced, the blood is washed out with a small volume of a balanced salt solution and this is immediately followed by the fixative which is allowed to flow for about 30 min. Sufficient pressure for the perfusion is achieved by suspending bottles containing the fixative and washing-out solutions at a height of about 5 ft. above the animal, and care should be taken to ensure that no air bubbles enter the vascular system (see Cammermeyer, 1967a). This can be largely avoided by allowing the fixative to drip from the end of the cannula before it is inserted.

As pointed out above, the type of fixation obtained by this method has set standards that must be met or improved by any procedure. However, the use of osmic acid as a primary fixative has serious disadvantages. These are the increasingly high cost of osmic acid, of which about 150 ml of a 1% solution are required to fix the brain of a rat, the slow rate of penetration of osmic acid into the tissue and the toxicity of its fumes. Formaldehyde and glutaraldehyde are far more convenient primary fixatives. They are less toxic, have faster penetration rates and are much cheaper, so that the loss of tissue by a bad perfusion is a less traumatic experience for the investigator than when osmic acid has been used.

In our laboratory we have found that when aldehydes are employed, it is still convenient to follow the principles of procedure outlined by Palay *et al.* (1962). Many investigators have tended to omit artificial respiration of the animal and to avoid anoxia by carrying out their surgical approach as rapidly as possible (e.g. Bodian and Taylor, 1963). Although the use of artificial respiration as an absolute necessity is debatable, we still prefer to use it for the simple reason that the operator can carry out the steps prior to the fixation at a leisurely pace, so that mistakes are less liable to occur. We still use heparin, but we have tended to omit the preliminary washing out of the blood with balanced salt solution for, unlike osmic acid, aldehydes do not coagulate the blood. Also, sodium nitrite is still utilized as a vasodilator, although we usually inject this solution, along with the heparin, into the femoral vein before the chest cavity is opened.

At variance with the above, Reese and Brightman (1968) inform me that they have long since abandoned artificial respiration. They feel that the inevitable tugging on the carotid sheath during the cannulation of the trachea may cause a reflex constriction of the cerebral vessels. Also, since an increase in CO_2 tension of the blood produces an effective dilation of the cerebral vessels, they feel that some small amount of anoxia may be an advantage in effecting a good perfusion.

Despite the many attempts that have been directed towards this goal, it is impossible, even today, to outline a rigid protocol for the fixation of the central nervous system that can be guaranteed to produce good preservation. Even if the fixation procedure is standardized, there are some variables that cannot be controlled. In particular, the state of health of the animal and the experience of the investigators should not be ignored, so that two animals apparently perfused under identical conditions can give quite different results. Also, the vascularity of different parts of the central nervous system differs, so that while one part may be well preserved, the fixation of an adjacent structure may be poor.

Some of the factors to be taken into account in the preservation of central nervous tissue are considered below, but it can be fairly stated that hardly any two investigators agree about the best procedure to be used. Indeed, even when two independent investigators follow the same procedure it seems possible for them to obtain quite different results. Consequently, in the end each person has to independently evolve his own formula for success.

1. Composition of the Primary Fixing Solution

In recent years, we and other groups of investigators (for example, Karlsson and Schultz, 1965; Schultz and Karlsson, 1965; Sotelo and Palay, 1968; Vaughn and Peters, 1966; Reese and Karnovsky, 1967; Reese and Brightman, 1968) have used buffered solutions containing glutaraldehyde and formaldehyde as primary fixatives for the central nervous system. It is extremely important however, that each of these aldehydes is pure. Otherwise artifacts are produced.

Glutaraldehyde as obtained from the manufacturer is in the form of 25% or 50% solutions which are impure and commonly have pH values as low as 1.5 to 2.5.

Before use as fixatives, they must be purified and this is achieved by one of two methods:

(a) Repeated treatment with an activated charcoal (see Fahimi and Drochmans, 1965a) such as Norit-A (Fisher Scientific Company) or (b) Distillation of the solution under vacuum (see Fahimi and Drochmans, 1965a, 1968).

Once either of these two treatments has been carried out, the purity of the resulting solution of glutaraldehyde is checked with a UV spectrophotometer. Commercial and impure glutaraldehyde has two absorption peaks, one at 235 mμ and one at 280 mμ. The peak at 235 mμ is due to impurities that produce fixation artifacts. These impurities may be either the acidic products of glutaraldehyde oxidation (Fahimi and Drochmans, 1968) or acrolein (Fahimi and Drochmans, 1965). When the purification is complete, the 235 mμ peak is lost, leaving only the 280 mμ peak of glutaraldehyde. The concentration of glutaraldehyde after the purification procedure is determined by the absorbance of the solution at 280 mμ, for there is a direct relationship between this absorbance and the concentration of glutaraldehyde. In our experience, failure to purify the glutaraldehyde leads to a number of defects in the fixed tissue. Amongst these are mitochondrial swelling, dilation of the cisternae of the Golgi apparatus and endoplasmic reticulum, the production of a fine granular precipitate, and the formation of myelin figures in the cytoplasm. However, these latter figures may sometimes occur even when purified glutaraldehyde is used.

The specific effects arising from the use of impure formaldehyde are not known to us, since we have always used formaldehyde prepared from paraformaldehyde according to the method recommended by Pease (1962). The paraformaldehyde is added to a volume of water heated to 60° C on a hot plate and vigorously stirred. Sodium hydroxide (1N or 5N) is then added slowly, drop by drop, until the cloudiness disappears. Formaldehyde prepared in this way is preferred to the commercially available solutions because, as pointed out by Pease (1964), such solutions contain between 11 and 16% of methyl alcohol.

Many workers have used solutions containing only glutaraldehyde or only formaldehyde for their perfusions, but in our experience the results of such a procedure have not been attractive. Formaldehyde alone, in concentrations of the order of 4% and contained in buffered solutions, seems to produce some shrinkage and it does not preserve the cytoplasmic organelles as well as glutaraldehyde. This applies particularly to the microtubules, whose relative absence makes the neuronal processes appear empty. According to Schultz and Case (1968), acrolein is also a poor fixative for microtubules, for fixation by this aldehyde alone results in an almost total loss of these organelles. On the other hand, glutaraldehyde by itself, in concentrations of between 3 and 5%, often leads to a swelling of the cisternae of the neuronal cytoplasm and does not effect good preservation of the myelin sheaths.

These defects can be minimised by using solutions containing both formaldehyde and glutaraldehyde. After many trials with solutions containing different proportions and concentrations of these two aldehydes, we have found that the most generally successful fixation is achieved using the two-stage perfusion outlined by Reese and Brightman (1968). In their procedure, two solutions are used consecutively. The first is dilute and this is followed by a more concentrated one containing four times the amount of aldehydes. The two solutions have the following compositions:

Dilute solution: (1% formaldehyde — 1.25% glutaraldehyde)

Paraformaldehyde	— 1 g
Glutaraldehyde	— equivalent of 5 ml of 25% solution
0.2 M Na cacodylate	— 40 ml
Water	— sufficient to bring the final volume to 100 ml
5% $CaCl_2$	— 6 drops

Add 0.1 N HCl to adjust pH value to 7.2—7.4

Concentrated solution: (4% formaldehyde — 5% glutaraldehyde)

Paraformaldehyde	— 2 g
Glutaraldehyde	— equivalent of 10 ml of 25% solution
0.2 M Na cacodylate	— 20 ml
Water	— sufficient to bring the final volume to 50 ml
5% $CaCl_2$	— 6 drops

Add 0.1 N HCl to adjust pH value to pH 7.2—7.4

The proportions of formaldehyde and glutaraldehyde used in these solutions was first recommended by Karnovsky (1965).

For perfusion, the solutions are put into separate drip bottles of the type used for intravenous injections and warmed to about 40° C. The fixation is initiated with the dilute solution and this is perfused through the animal for about 10 minutes; to allow its escape, the right auricle is incised as soon as the perfusion is started. The concentrated solution is perfused slowly through the animal for 2 to 5 min.

If the perfusion is successful, the neck muscles of the animal soon harden, the jaw becomes locked and the tongue is cleared of blood. We have also found that in our best preparations the animal did not undergo extreme convulsions at any time during the perfusion, but remained relatively inactive.

Use of the above solutions produces good fixation in those regions of the brains of rats and cats containing few myelinated nerve fibres. It is less successful when large numbers of myelinated fibres are present in the neuropil and seemingly unsuccessful in white matter. For the fixation of white matter, a phosphate buffered solution of 4% formaldehyde and 0.5% glutaraldehyde (Vaughn and Peters, 1966) gives a somewhat better result and incidentally, this solution also gives good fixation of the brains of foetal and early postnatal animals. The higher concentration of formaldehyde seems to stabilise the myelin, although Hirano, Zimmerman and Levine (1966) have obtained good results in white matter by using a solution containing 5% glutaraldehyde and no formaldehyde.

The variables that can affect the fixation obtained be perfusion of the central nervous system with aldehyde solutions may be briefly considered under the following headings:

a) Osmolarity

When fixation is carried out with osmium tetroxide it is important for the fixing solution to be nearly isotonic with the tissue fluids, i.e. 300—350 milliosmoles, but with aldehydes, hypertonic solutions give the best results (Sabatini, Bensch and Barnett, 1963). Fahimi and Drochmans (1965b), found that concentrations of glutaraldehyde between 1.7 and 3.0% in a 0.1 M phosphate buffer produced acceptable fixation. The osmolarities of these solutions varied between 400 and 630 milliosmoles.

Low concentrations of glutaraldehyde (0.5%) gave poor fixation and very high concentrations (37%) produced shrinkage. Changes in the tonicity of the fixative brought about by varying the molarity of the buffer had much more pronounced effects on the swelling and shrinkage of the tissue than the changes effected through varying the concentration of the glutaraldehyde. Schultz and Karlsson (1965) found somewhat similar results in their experiments involving alterations in the concentration of either the buffer or the aldehyde and so did Maunsbach (1966) who studied fixation of the kidney.

b) Buffer and pH value

Both phosphate and cacodylate buffers may be used with equal success in fixatives containing aldehydes. The only factor that makes phosphate buffers less preferable than cacodylate is that the addition of much $CaCl_2$ leads to the formation of insoluble calcium phosphate. Reese and Brightman (1968) also consider that the use of phosphate buffers leads to more broken membranes than cacodylate buffers. Veronal-acetate buffers should not be used with aldehydes, for Holt and Hicks (1961) state that formaldehyde reacts with this buffer to produce a substance with no buffering capacity within the physiological range of pH values. Bicarbonate buffers also appear to be unsatisfactory, for their use with aldehydes apparently leads to an almost complete loss of microtubules from the cytoplasm (Schultz and Case, 1968).

Generally, the pH range employed for fixation is 7.2—7.5. Karlsson and Schultz (1965) noted that at low pH values (pH 2.0), the perfused tissue had a flocculent appearance in electron micrographs and that at high pH values (pH 11) many of the details of the structure were obscured.

c) Temperature

No advantage seems to be gained by using aldehyde fixatives at the low temperatures that are used for those containing osmic acid. Indeed, the best results with aldehydes are obtained by perfusing the fixative at body temperature, for this seems to avoid constriction of the cerebral blood vessels.

2. Osmication of Aldehyde Fixed Material

Although aldehydes fix the tissue, for electron microscopy it is necessary to carry out a post-fixation with osmic acid. The main purpose of this secondary treatment is to introduce an electron dense substance into the tissue, but the osmium may also help to stabilize the structure. In our laboratory we have tended to delay the handling of tissue after aldehyde perfusion until the next day and to put the animal into a seal plastic bag overnight. On the other hand, Reese and Brightman (1968) remove the brain from the skull after two hours, and leave it in the concentrated aldehyde solution overnight. Some delay in cutting up the brain into the small pieces necessary for osmication is certainly desirable. This gives the tissue time to harden, so that the chance of distorting the components is minimised. Indeed, in the cerebral cortex for example, premature handling frequently leads to the production of "dark" neurons (Mugnaini, 1965).

Before osmication, the tissue is usually cut into small pieces that are no more than 2 mm thick. These pieces are then thoroughly washed in a solution containing the buffer used in the aldehyde fixative. In the case of cacodylate buffers, a good washing solution is 0.2 M cacodylate buffer at pH 7.4. This has an osmolarity of about 400 milliosmoles, and three or four washes with this solution are sufficient. The purpose of the wash is to remove excess aldehyde from the tissue. If the aldehyde is not removed, it will reduce the osmic acid and result in a deposit of fine granules within the tissue.

Depending upon the size of the blocks, osmication is usually carried out for 1—2 hours, and again the buffer used for the perfusion is generally employed as the carrier for the 1—2% osmic acid, although Hirano, Zimmerman and Levine (1966) have successfully used Dalton's solution for the osmication of white matter. The osmium-containing solution should be isotonic with the washing solution, since osmotic changes in the tissue can occur even at this stage. If the solution is hypotonic, then the osmolarity may be adjusted by the addition of dextrose or sucrose (for example, see Palay, Sotelo, Peters and Orkand, 1968).

Some investigators (Robertson, Bodenheimer and Stage, 1963; Sotelo and Palay, 1968) have preferred to carry out a preliminary osmication by perfusing the aldehyde fixed brain or spinal cord with an osmic acid solution before it is removed from the animal. The results of this procedure are good, but the volume of osmic acid required for the secondary perfusion is quite large, and a further osmication is still necessary after the tissue is removed.

3. Dehydration and Embedding

After osmication, the pieces of tissue should again be rinsed thoroughly with the washing solution and then dehydrated before embedding in plastic. The choice of plastic is a personal preference and does little to affect the end result. Variations in the appearance of the tissue can however, be produced by the use of different dehydrating agents. Thus, acetone, which was used by Karlsson and Schultz (1965) and Schultz and Karlsson (1965) in their experiments varying the parameters in tissue preparation, tends to bring about close approximation between the outer faces of adjacent plasma membranes (Bondareff, 1966). On the other hand, with the more commonly used methyl or ethyl alcohols, the extracellular space is better defined.

Analysis of Electron Micrographs of the Neuropil

The grey matter of the central nervous system contains neurons and neuroglia, each of which have processes that emanate from their perikarya to become intermingled and interconnected in the neuropil. At first sight, these processes appear to be disposed randomly, but a careful analysis shows this impression to be incorrect. Rather, they are related in a very specific and precise manner, as would be expected from our knowledge of the functioning of the nervous system, although the pattern of the connections varies in different locations. While the perikarya of the neurons (Fig. 7, *N*), astrocytes (Fig. 7, *As*) and oligodendrocytes (Fig. 6, *O*) are relatively easy to identify, their processes pose more of a problem, and it is with the identification of these processes that the present account is concerned. Since the part of the brain that we have been most recently studying is the cerebral cortex, this tissue will be mainly used to illustrate the following account.

In any neuropil, the most commonly encountered profiles are those of axons and dendrites. Since neurons usually have many dendrites but only one axon, it might be expected that most of the neuronal profiles in the neuropil would belong to dendrites. However, this is rarely the case, for dendrites do not project far from the cell body and with the exception of neurons such as the Purkinje cells of the cerebellum, they do not undergo prolific branching. In contrast, axons may travel for long distances so that a given neuropil contains not only the axons of the indigenous neurons, but also ones derived from other areas of the central nervous system. These axons are both myelinated and unmyelinated ones. Some of the axons may merely pass through the area, but the ones that are terminating there, whether from indigenous or distant neurons, usually undergo quite prolific branching and consequently they form many profiles.

Dendrites within the neuropil are usually of larger diameter than axons, and their cytoplasm often contains many of the organelles that are present in the perikarya from which they extend without any sharp morphological transition. When dendrites are sectioned transversely their profiles are characterised by a content of regularly arranged microtubules (Fig. 1, D_1—D_4; Figs. 2—4, D). The microtubules are each about 220—240 Å in diameter and are frequently arranged to form almost crystalline arrays. Mitochondria are also commonly encountered in dendrites (Figs. 1 and 2, *mt*), but it is unusual for many neurofilaments to be present. One notable exception is in the dendrites of the anterior horn cells of the spinal cord, in which bundles of neurofilaments do occur (see Wuerker and Palay, 1968). The relative content of granular (Fig. 1, *ER*) and agranular endoplasmic reticulum, as well as clusters of free ribosomes (Figs. 1 and 2, *r*) in a dendritic profile, largely depends upon the level at which the dendrite has been sectioned. All of these organelles are common at the bases of dendrites, which may also contain extensions of the Golgi apparatus (Fig. 1, *G*) of the perikaryon, but they become less frequent as the distance from the perikaryon increases. This is illustrated in Fig. 1, that shows four profiles of apical dendrites (D_1—D_4) of pyramidal neurons sectioned at various distances from the cell body. In some large dendrites, such as those of the anterior horn cells, small Nissl bodies may be encountered, but in the cerebral cortex of the rat, even at a short distance from the perikaryon the apical dendrites of the pyramidal cells have hardly any organised Nissl bodies within them (Fig. 1, D_1 and D_2). An exception occurs at the angle of bifurcation of a dendrite, for a Nissl body is frequently located at such a site.

In longitudinally sectioned dendrites, the regular spacing of the microtubules is a less useful criterion for identification (Figs. 3, 4 and 5, D_1), particularly when small diameter profiles are involved, for small axons also contain microtubules. However, the unmyelinated axons only rarely achieve the thickness of even the smallest dendrites and do not contain ribosomes beyond the initial segment (see Palay, Sotelo, Peters and Orkand, 1968). Other criteria that may be employed to identify dendrites are the somewhat irregular contours of their profiles (Figs. 1—5, D and D_1), which are in marked contrast to the smooth ones of axons, and the fact that dendrites often have axon terminals (Figs. 1 and 4, *At*) synapsing upon their surfaces. This latter feature is not unique to dendrites however, for as will be considered later, synapses can also occur between two axons (axo — axonal synapses).

In the neuropil of some parts of the central nervous system like the anterior horn of the spinal cord, most of the axon terminals surround and synapse with the stems

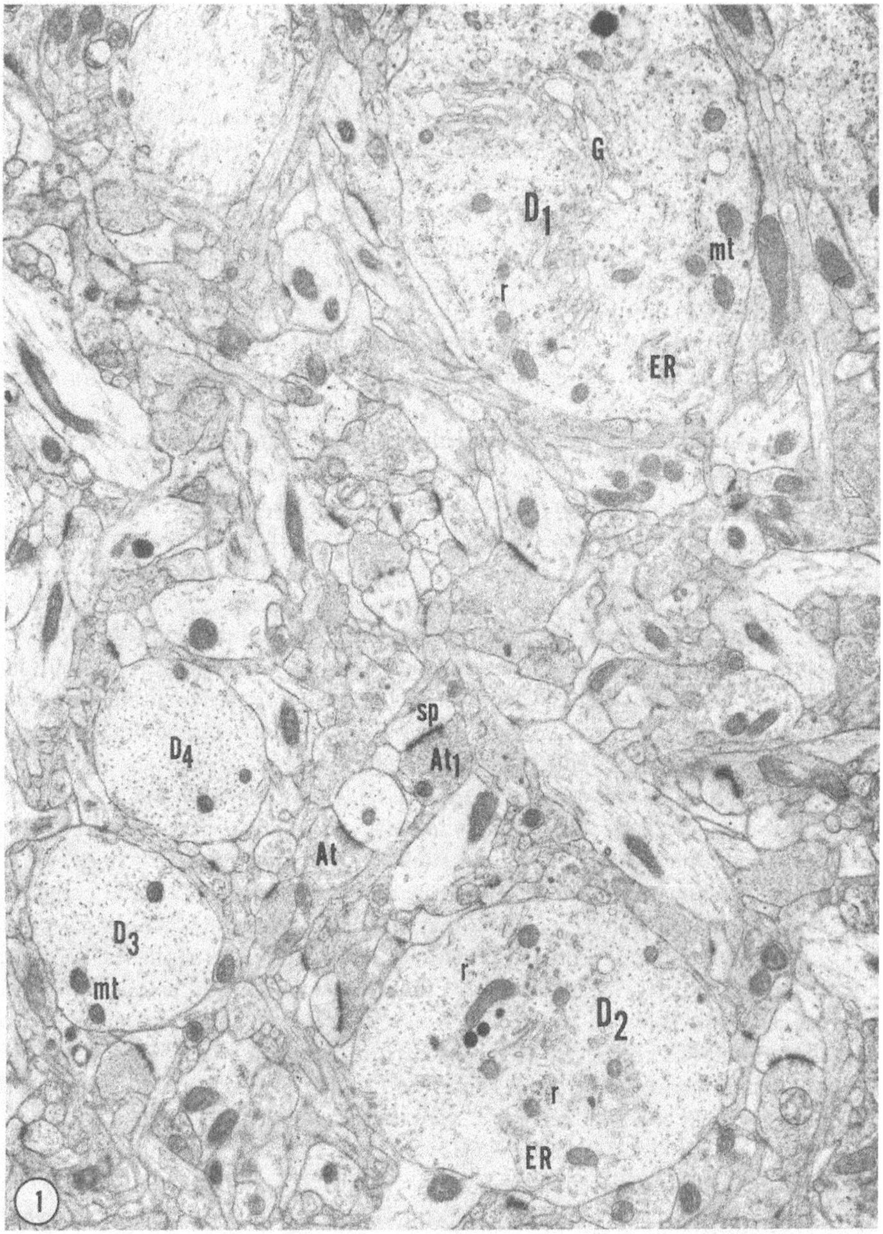

Fig. 1. Parietal Cortex of the rat. This and all subsequent electron micrographs of the cerebral cortex are from layers II and III and have been sectioned parallel to the surface of the cortex, so that the apical dendrites of the pyramidal cells are transversely sectioned. The tissue was fixed by perfusion with the aldehyde solutions recommended by Reese and Brightman (1968), post-fixed in osmic acid, embedded in Araldite and the sections stained on the grid with lead citrate and uranyl acetate. In this figure are four profiles of apical dendrites (D_1—D_4) sectioned at different levels. Sections passing through the bases of the dendrites (D_1) contain profiles of the Golgi apparatus (*G*), mitochondria (*mt*), endoplasmic reticulum (*ER*), ribosomes (*r*) and microtubules within their cytoplasm, but the only commonly occurring organelles in the more distal portions of these dendrites (D_3 and D_4) are microtubules and mitochondria. Both these apical and the basal dendrites have axon terminals (*At* and At_1) synapsing upon both their stems and the spines (*sp*) that emerge from them. ×15,000

of the dendrites, so that the neuropil has a relatively ordered appearance. Few such configurations occur in the cerebral cortex, in which most of the axon terminals synapse with the spines or thorns that project from the dendritic stems. Although some of these spines may be observed to arise directly from the dendritic stems (Figs. 3 and 4, sp_1) the majority appear as small isolated and rounded profiles (Figs. 1 to 4, *sp*). However, the identification of these small profiles as those of dendritic spines is relatively easy. No microtubules enter the spines, but their cytoplasm is filled by a very characteristic feltwork of fine fibrils. Furthermore, the spines are usually synapsed upon by axon terminals (Figs. 1—4, At_1) which contain many synaptic vesicles and whose size usually matches that of the attached spine. These features are shown by the spines of the cerebral cortex and cerebellum for example, and also by the much larger ones that emanate from the dendrites of some thalamic neurons to enter into synaptic glomeruli (Peters and Palay, 1966; Pappas, 1966; Guillery, 1967). The spines of the cerebral cortex also possess a distinctive organelle: the spine apparatus (Gray, 1959). This organelle (Fig. 4, *sa*) is composed of flattened cisternae that alternate with plaques of dense granular material. The portions of axons that give rise to axon terminals are usually unmyelinated and their isolated profiles may be easily confused with those of dendrites, for like dendrites their most common cytoplasmic organelle is the microtubule. However, as pointed out above, except for the initial segment (Figs. 2 and 3, *I*), axons do not contain ribosomes. The unmyelinated axons (Figs. 2 and 3, *A*) have relatively smooth contours and are generally much thinner (0.1—0.2 microns) than dendrites. Only the dendritic spines attain the small diameters that may be achieved by axons and as already described, these have fine fibrils in their cytoplasm. An additional fact is that unmyelinated axons frequently travel in company with others of the same kind, so that they are often contained within groups (Figs. 2 and 3, *A*). Dendrites never form such groups, but instead they traverse the neuropil individually.

With the prominent exception of the olfactory bulb (Fig. 5), the most reliable characteristic that can be used for the identification of axons in the neuropil of most regions of the brain is the presence of synaptic vesicles in the cytoplasm. These vesicles, which usually range in size from 200 to 400 Å, may be spherical or ellipsoidal (elongated). Although there may be some correlation between the function of an axon terminal and the type of vesicle it contains, to some extent the shape also seems to depend upon the fixative that has been used. Ellipsoidal vesicles are only rarely seen in material fixed primarily in osmic acid. They are more common after fixation with aldehydes, but even then their frequency of occurence seems to be variable. For example, Colonnier (1968) finds them consistently in his cerebral cortex material, but in our preparations they appear to be less common. Why this should be is presently unknown.

Particular concentrations of synaptic vesicles are present in the axon terminals (Figs. 1—4, *At* and At_1) which may occur as enlargements either along the course of an axon (boutons en passant), or at the terminations of an axon (boutons en terminaux — Fig. 4, At_2). It is these enlargements that take part in the formation of most of the chemical synapses within the neuropil, that is synapses at which the passage of an impulse is mediated through a chemical transmitter. In addition to the presence of synaptic vesicles within the presynaptic component, a synapse is also characterized by a junction at which the plasma membranes of the two neuronal

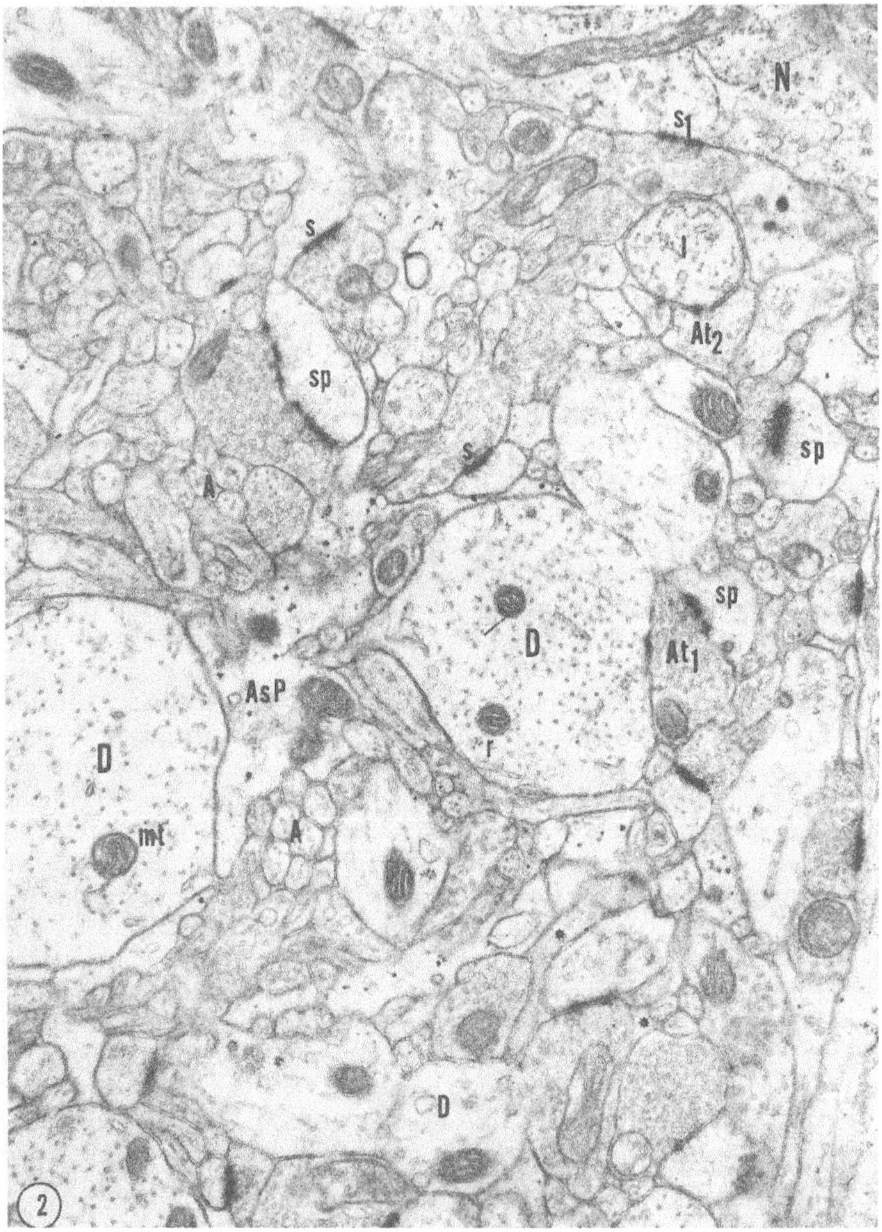

Fig. 2. Parietal Cortex of the rat. In this field three dendrites (*D*) are indicated. The spines (*sp*) arising from the dendrites form synaptic junctions (*s*) with terminals (At_1) formed by small axons (*A*). Surrounding some of these synapses are sheet-like extensions (*) of the astrocytic processes (*Asp*). At the top of the field is a synapse (s_1) upon a neuronal perikaryon (*N*). Below this synapse is the initial segment (*I*) of an axon and this is also forming a synapse with an axon terminal (At_2). ×13,500

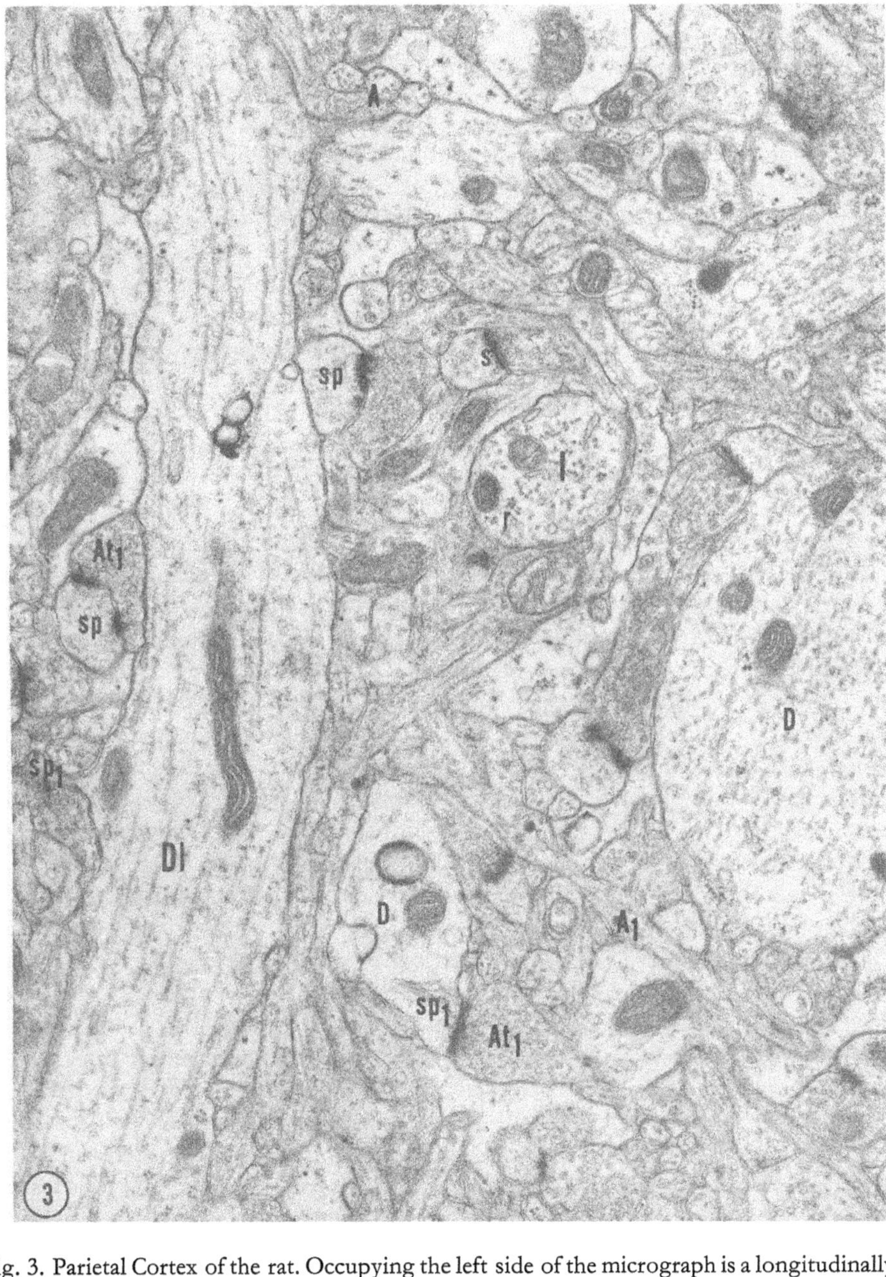

Fig. 3. Parietal Cortex of the rat. Occupying the left side of the micrograph is a longitudinally sectioned dendrite (D_1), while two other dendrites (D) in the field are either transversely or obliquely oriented. Two spines (sp_1) are seen to emerge directly from dendrites, but other spines (sp) appear as isolated profiles. These spines form synapses (s) with terminals (At_1) of small axons (A and A_1). Also included in the field is the initial segment of an axon (I) that contains ribosomes (r) and rows of microtubules. ×25,000

components partaking in the synapse lie parallel to each other and have some electron density which may be disposed either symmetrically (Figs. 2 and 4, S_1) or asymmetrically (Figs. 2 and 3, *S*) on the cytoplasmic faces of the plasma membranes. This combination of synaptic vesicles with the dense material has been referred to as a "synaptic complex" (Palay, 1958). Dense material may also be present in the extracellular space between the membranes forming the synaptic junctions.

The present evidence suggests that all chemical synapses are polarized. That is, an impulse will pass in only one direction, from the pre- to the postsynaptic element. In the neuropil, it is most common for the presynaptic component to be an axon and the postsynaptic component a dendrite. However, axo-axonal synapses do exist, but so far they have been described only between an axon terminal and the initial segment of an axon (see Fig. 2, At_2 and *I*, and Peters, Proskauer and Abramof, 1968) and between two axon terminals. In this latter case, the post-synaptic component of the axo-axonal synapse itself forms a synapse with a third, non-axonal component. This arrangement has been described in a number of locations, including the spinal cord of the rat and cat (Gray, 1962), the posterior column nuclei of the rat (Valverde, 1966) and the lateral geniculate nucleus (Colonnier, Guillery and Szentagothai, 1964; Peters and Palay, 1966) and it is thought that it forms the basis for presynaptic inhibition. In such situations, the presynaptic component is considered to be the one in which the synaptic vesicles are located in close proximity to the synaptic junction.

In chemical synapses between two neuronal processes, the possibility cannot be ignored that both processes are dendrites. At present however, in the brains of mammals dendro-dendritic synapses have only been described in the olfactory system (see Reese and Brightman, 1965; Rall, Shepherd, Reese and Brightman, 1966; Price, 1968) where the finding of synaptic vesicles in otherwise typical dendrites was unexpected. An example of such synapses is shown in Fig. 5 which shows the mitral cell dendrites (*Dm*) forming synapses (*S*) with the spines (*sp*) or gemmules derived from the dendrites (*Dg*) of granule cells. Not only are the synapses unusual because both components are dendritic, they are also unusual in that synaptic vesicles are present on each side of the synaptic junction. Indeed, in some sections it is apparent that a junction may have two synaptic complexes with opposite polarities as defined on the basis of the location of the synaptic vesicles (see arrows). Rall, Shepherd, Reese and Brightman (1966) have postulated that the mitral-to-granule cell synapses are excitatory and the granule-to-mitral cell synapses are inhibitory in function.

Intercellular junctions that bear a strong resemblance to synapses, but which have no intimately associated synaptic vesicles, also occur throughout the neuropil. They commonly occur not only between neuronal components, but also between neuroglial ones. These junctions are often short and usually have a symmetrical disposition of a small amount of dense and fibrillar material on each side. Because of the absence of associated vesicles and because such junctions may also occur between neuroglial profiles, they are considered to have a purely adhesive function. In view of their punctate form, Palay (1967) has named these junctions *puncta adhaerentia*.

It is of interest that in the synaptic glomeruli of the lateral geniculate nucleus (e.g. see Peters and Palay, 1966; Colonnier and Guillery, 1964) the amount of fibrillar material associated with the plasma membranes taking part in the *puncta adhaerentia* depends upon the nature of the neuronal processes. Thus, the cytoplasm of the dendrites has more fibrillar material than that of axons, but this situation is unusual.

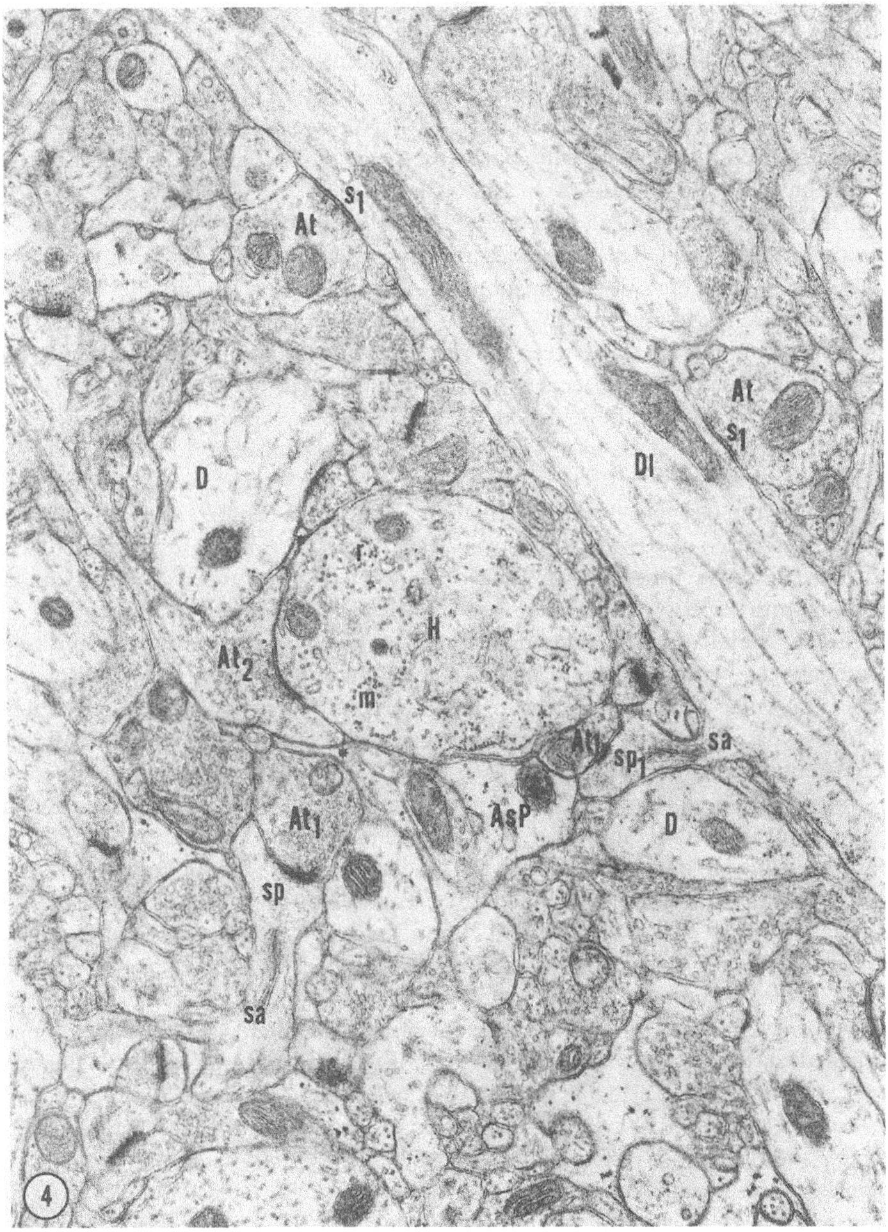

Fig. 4. Parietal Cortex of the rat. Crossing diagonally across the micrograph is a longitudinally oriented dendrite (D_1), which has axon terminals (*At*) synapsing (s_1) upon its surface. Other axon terminals (At_1) form synapses with the spines (*sp* and sp_1), some of which contain a spine apparatus (*sa*). In the centre of the field is an axon hillock (*H*) with ribosomes (*r*) and clusters of microtubules (*m*) in its cytoplasm. Synapsing with this hillock is an axon terminal (At_2) which arises as an enlargement of a thin axon. Note that this synapse is surrounded by sheet-like astrocytic processes (*). Two transversely sectioned dendrites (*D*) are also indicated. ×28,000

It must also be noted that at a number of synaptic junctions, particularly those between axon terminals and large dendrites, neuronal perikarya or axon initial segments, *puncta* are interspersed with the synaptic complexes. The two may be distinguished from each other however, by the disposition of the dense material and the presence or absence of closely associated synaptic vesicles. At the *puncta*, the dense material is disposed symmetrically, but at the synaptic complexes it is asymmetric in its distribution.

In addition to synapses and *puncta adhaerentia*, another type of cell junction is also present in the neuropil. This resembles the *zonula occludens* described by Farquhar and Palade (1963), in which plasma membranes of two adjacent elements lie parallel to each other and are closely apposed. In the mammalian central nervous system such junctions commonly occur between the processes of astrocytes, particularly between the plasma membranes of adjacent perivascular and subpial end-feet. In most electron microscope preparations, the outer leaflets of the two membranes forming the junction appear to be fused. On more careful inspection it is apparent that the faces of the plasma membranes are separated by a gap of about 20 Å. The patency of this gap is shown by the fact that it can be penetrated by horseradish peroxidase introduced into the ventricles of the brain (Brightman and Reese, 1967). In lower vertebrates, and particularly in fishes (Robertson, 1963; Pappas and Bennett, 1966; Bennett, Nakajima and Pappas, 1967), there is evidence to suggest that such junctions are electrical synapses. So far, junctions of the same form do not appear to have been found between neurons of the mammalian central nervous system, except within the lateral vestibular nucleus (Sotelo and Palay, 1968). In this nucleus, large terminal boutons synapse with the neurons of Deiters nucleus and at the interface between them chemical synapses, *puncta adhaerentia* and sites of close apposition are all present.

To turn again to the identification of axons, apart from the presynaptic portions of the unmyelinated axons, other axonal profiles may be encountered within the neuropil. Those of myelinated axons are readily identified by the multilayered and regular sheath of plasma membranes that enclose them (Fig. 7, *A*). Profiles of initial axon segments (Figs. 2 and 3, *I*) and nodes of Ranvier present more of a problem. The initial segments of axons have two characteristic features (Palay, Sotelo, Peters, and Orkand, 1968; Peters, Proskauer and Abramof, 1968). One is the layer of dense material that undercoats the plasma membrane and makes it appear thickened at low magnifications. The other is the fascicles of microtubules that occur within the cytoplasm and which are often arranged in rows (Figs. 2 and 3, *I*). An axon at a node of Ranvier also possesses an undercoating of plasma membrane (Peters, 1966), but the microtubules are more evenly distributed and never form fascicles. On the basis of these features, the initial segments and the nodes of axons can usually be identified and distinguished from the other unmyelinated portions of axons. Two other points worth mentioning are that ribosomes (Fig. 3, *r*) may be encountered in the initial segments of axons and that nodal axons may form synapses and so contain accumulations of synaptic vesicles and mitochondria.

An infrequently encountered portion of the neuron is the axon hillock, whose features are intermediate between those of the neuronal perikaryan and the initial segment of the axon (Peters, Proskauer and Abramof, 1968). When it is seen in transverse section, the axon hillock is often difficult to distinguish from the base of a dendrite, for both contain the same types of organelles (compare Fig. 1, D_1 and D_2

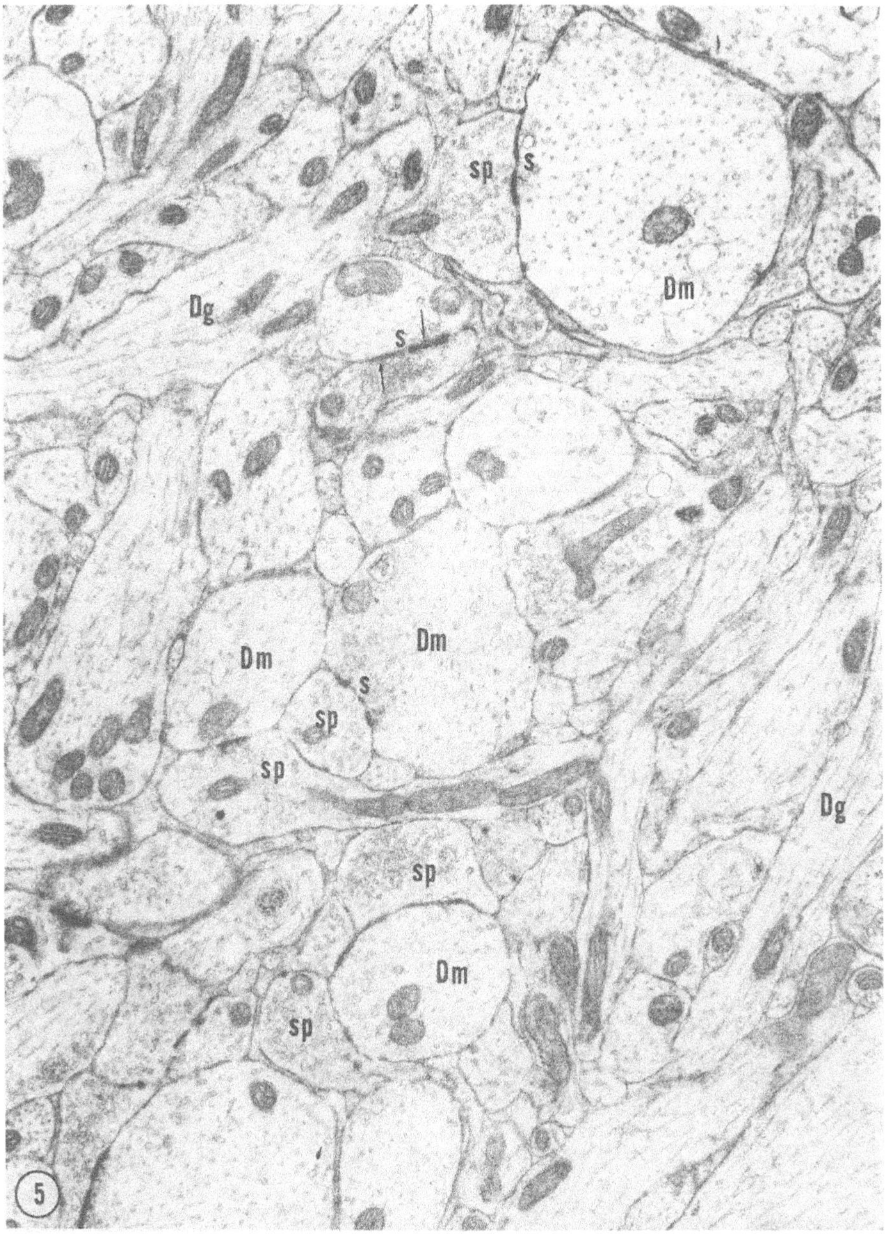

Fig. 5. External Plexiform Layer of the Olfactory Bulb of a rat. This layer contains the smooth — surfaced dendrites of mitral cells (*Dm*) and the spine-bearing dendrites of granule cells (*Dg*). The spines (*sp*) of the granule cell dendrites and the dendrites of the mitral cell synapse with each other and form junctions (*s*) that have vesicles on both sides. At some junctions, two synaptic complexes are apparent. These have opposite polarities (arrows) as indicated by the location of the synaptic vesicles. For further explanation see the text. ×22,500. (Micrograph supplied by Dr. J. W. Hinds)

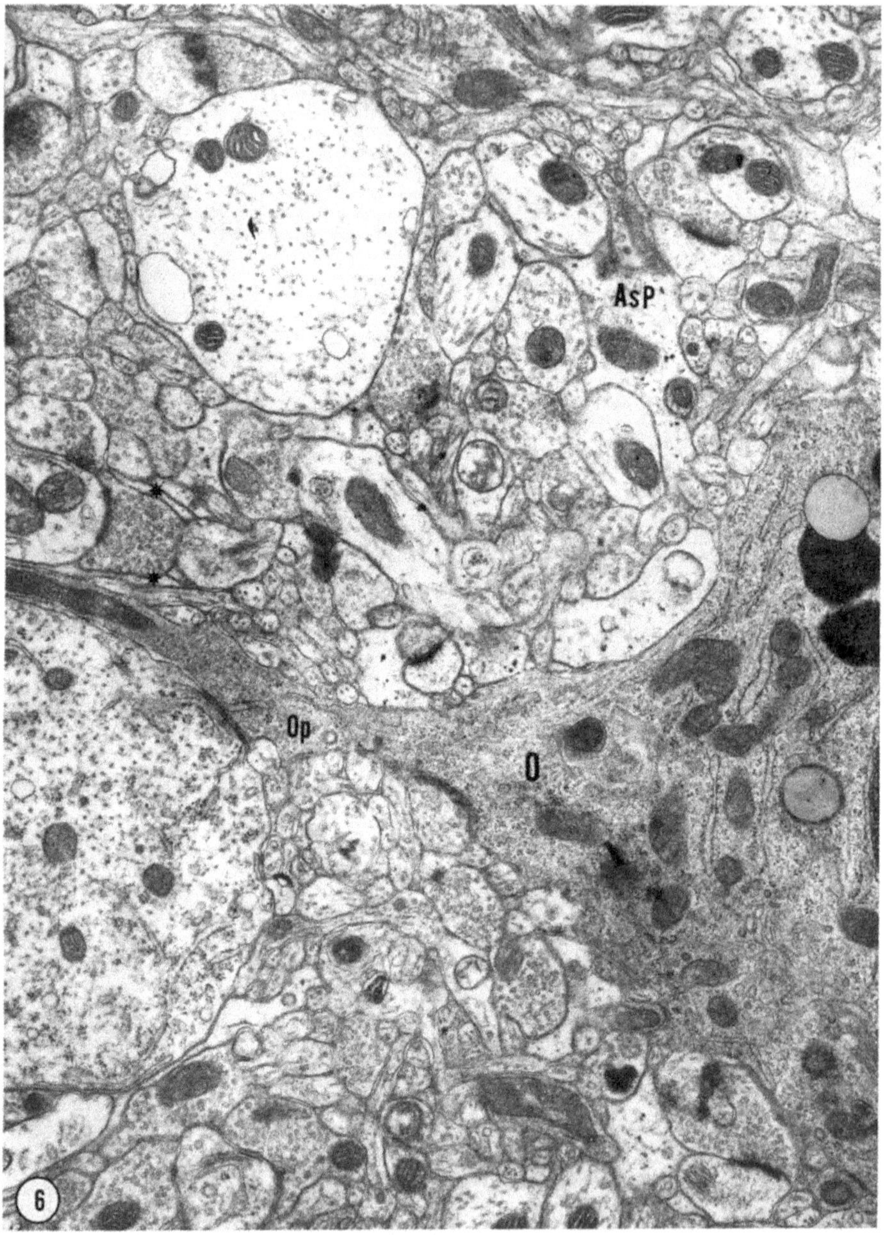

Fig. 6. Parietal Cortex of the rat. Included in this field is part of the perikaryon of an oligodendrocyte (*O*) and arising from it is a process (*Op*). The cytoplasm of the oligodendrocyte is dark and is in contrast to the lighter cytoplasm of the astrocytic processes (*AsP*) which form sheet-like extensions (*) around some synapses. ×22,500

and Fig. 4, *H*). The only feature that seems to be useful in making a positive identification is the arrangement of the microtubules. In a dendrite, the microtubules are arranged in a regular pattern as separate units, whereas in the axon hillock some of

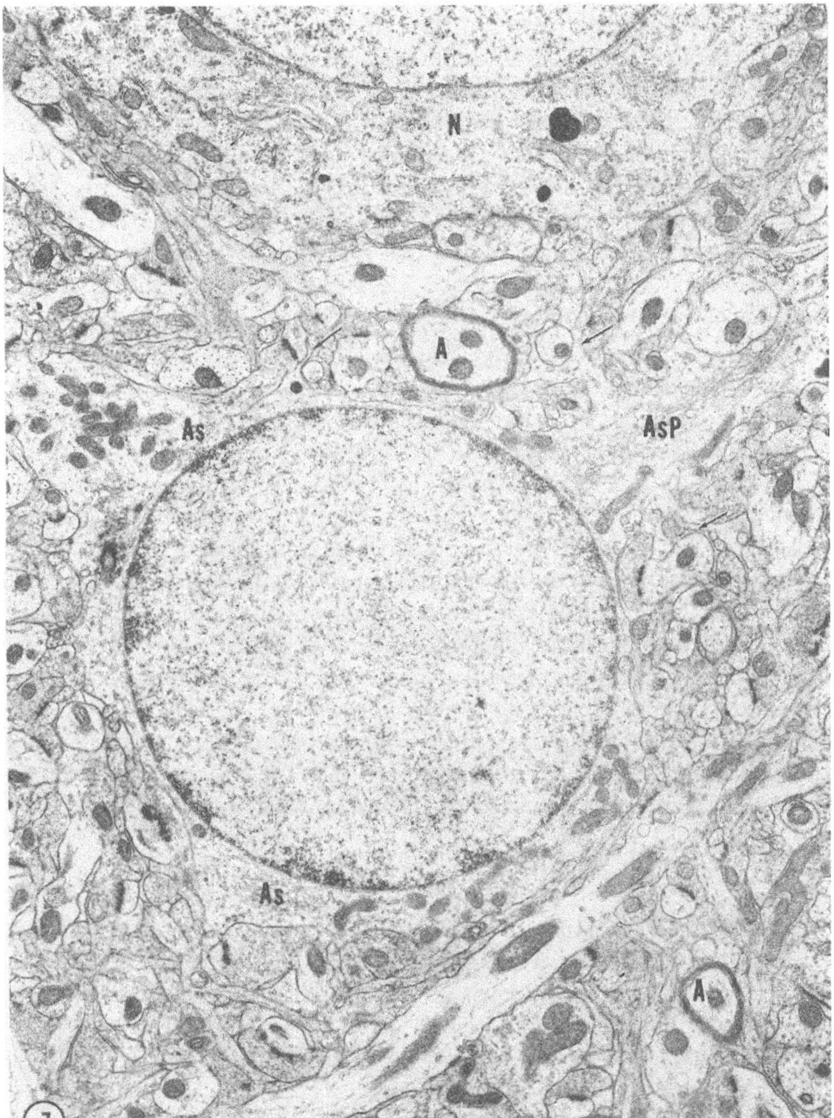

Fig. 7. Parietal Cortex of the rat. Both the perikaryon (*As*) and the processes (*AsP*) of the protoplasmic astrocyte included in this field give rise to sheet-like extensions (arrows) that invade the neuropil. Within the neuropil, profiles of myelinated axons (*A*) are indicated and at the top of the field is part of the perikaryon of a neuron (*N*). ×16,000

them (Fig. 4, *m*) are beginning to congregate into the clusters and rows that are so characteristic of the initial axon segment (Figs. 2 and 3, *I*). In the axon hillock the fasciculation of the microtubules is not complete and the undercoating of the plasma membrane, which is the other characteristic feature of the initial segment (Figs. 2 and 3, *I*), is not present.

Other profiles that are encountered in the neuropil are those of processes of astrocytes and oligodendrocytes. The cytoplasm of oligodendrocytes is dense (Fig. 6, *O*) because of the many small and tiny granules that it contains and this density is particularly apparent in extensive profiles such as oligodendrocytic processes that have been sectioned longitudinally (Fig. 6, *Op*). Since oligodendrocytes contain many microtubules that are oriented parallel to the long axes of the processes, their transversely sectioned processes may be confused with cross sections of dendrites. However, the microtubules of oligodendrocytic processes tend to be more abundant than those of dendrites and this fact, together with the granules in the cytoplasm, usually makes them distinct. Also, it would not be expected to find axon terminals synapsing with oligodendrocytic processes.

Astrocytes in grey matter have irregular shapes and mould themselves to the contours of the surrounding neuronal elements (Fig. 7, *As*). The same is true of their processes (Figs. 2, 4, 6 and 7, *AsP*) and both the perikaryon and processes give off thin, sheet-like extensions (Fig. 7, arrows) that often surround synapses (Figs. 2, 4 and 6). In some regions of the central nervous system such as the cerebellum and substantia gelatinosa, astrocytic processes may also cover the surfaces of the dendrites and neuronal perikaryon. Indeed, as indicated by Peters and Palay (1965) and by Palay (1966), in any neuropil the distribution patterns of the astrocytic processes are typical of that area. Their relations to the neuronal components of a given neuropil may perhaps even be convenient in differentiating between synapses formed by axons originating from separate sources. The cytoplasmic content is less useful than shape in identifying astrocytic processes. Bundles of fibrils are only infrequently encountered in the cytoplasm of the protoplasmic astrocytes of the rat, although they are relatively common in the cat. The processes of astrocytes frequently contain glycogen particles in their cytoplasm (Figs. 2 and 4, *AsP*).

Conclusions

From the above account it is apparent that sufficient information about the identification of the perikarya and processes of both neurons and neuroglial cells in electron microscopic preparations is now available for us to embark seriously upon the problem of how cells in the central nervous system are related to each other. Some such studies have already been initiated in a number of laboratories in different parts of the world. One fact that is becoming increasingly clear however, is that a study of normal tissue by electron microscopy is in itself generally inadequate. This phase of analysis certainly has to be passed through, but eventually resort has to be made to supporting data derived from other techniques. The reason for this is that although electron microscopic preparations have the great advantage of showing all of the components of the tissue in proper relation to one another, the sections are so thin that in effect a two-dimensional preparation is being studied. One way of entering the third dimension is through a study of a series of sections from which reconstructions can be made, and although laborious this method is perfectly adequate for small objects and localised structures (for example, see Valverde, 1966, and Karlsson, 1966a and b, 1967). But for sampling larger volumes of the neuropil, the method is not adequate. Then one can evoke the third dimension by correlating the electron microscope studies with studies of Golgi preparations. For an analysis of the origin of long projecting axons, this approach is not practical, but information can be obtained by making experimental lesions to bring about the degeneration of the afferent axon terminals whose positions can thereby be localised in electron micrographs (for example, see Walberg, 1966; Heimer and Wall, 1968). Each of these techniques has provided valuable information for the analysis of the central nervous system and examples of some of the results that can be obtained by their utilization are given elsewhere in this volume.

Acknowledgements

I wish to acknowledge the cooperation of Dr. Ita R. Kaiserman-Abramof in this study and to thank Charmian Proskauer and Larry McCarthy for their valuable assistance. The electron micrograph of the olfactory bulb was generously supplied by Dr. J. Hinds.

This work was supported by U.S. Public Health Service Research Grant NB-07016 from the National Institutes of Neurological Diseases and Blindness.

Discussion

Dr. Konigsmark: Dr. Peters, you mentioned close juxtaposition of adjoining cell membranes as a criterion for perfect fixation. To what extent do you think such a picture represents the actual width of the extracellular space during life?

Dr. Peters: This is a thorny problem at the moment and I'm afraid there is no direct answer to the question. As you know, the problem has become complicated by conflicting conclusions of physiologists with respect to the actual volume of extracellular space in the brain. Most electronmicroscopists, however, share the impression that the "best fit" of cellular profiles corresponds to a state of fixation in which the space is about 120—200 Å wide. That is to say, when spaces are wider than that, it usually looks as if the surrounding profiles would fit better if they were pushed together a bit more closely. The matter will obviously require much further study.

Dr. Sidman: In places where membranes are in direct apposition without an intervening space, how do you decide on the validity of the image?

Dr. Peters: That depends on how consistently direct contact appears. If fusion is constantly present between two particular profiles, one would be tempted to accept it as a natural phenomenon rather than an artifact. When randomly distributed, it would appear of doubtful significance.

Dr. Campbell: I was wondering how far one can generalize your criteria for identifying cellular profiles. Have you looked at other mammalian material beside rat, for example?

Dr. Peters: Oh, yes, and in many different regions of the brain. I think the criteria are applicable regardless of the species of mammal we study. The characteristics of the initial axon segment, for example, are exactly alike in the rat and cat. In both species the identical dark undercoating appears beneath the plasma membrane, and it stops at a given point regardless of whether or not the axon acquires a myelin sheath at that point.

Dr. Heimer: I am interested in your views on the comparative merits of cooling and warming during fixation. Can you specify any ill effects of cooling?

Dr. Peters: No, I can't. I think it is a purely personal thing whether you cool or warm. Our reason for warming was the thought that it might prevent shock, and we felt also that because aldehydes are such fast penetrators they would cross the membranes easily and achieve good fixation quite rapidly anyway. On the other hand, the purpose of cooling osmium solutions was to prevent the cells from becoming autolyzed before the slowing penetrating osmium could reach them. There seems to be little point in cooling an aldehyde fixative.

Dr. Heimer: Your account made me realize that we light-microscopists who are trying to improve existing silver impregnation methods may not have experimented nearly enough with different fixatives, despite the examples of a previous era. Perhaps as a result of established procedural routine, I must admit that I have come to think of formaldehyde almost as if it were a storage medium rather than a fixative determining particular chemical properties in fixed brain tissue. Incidentally, we have found glutaraldehyde-formaldehyde fixation very compatible with our silver impregnation procedures for terminal degeneration, both in frozen and Epon-Araldite material. But we might do well to pay some attention to further possibilities in the realm of fixation.

Dr. Peters: There is a curious thing about aldehyde perfusion that I had wanted to say and didn't. Namely, that we find our best fixation with aldehyde perfusion is achieved in those cases in which the animal does not develop any violent muscle spasms. With osmium fixation the opposite appears to be the case. I have no idea how to explain this.

Dr. Cowan: Dr. Peters, you indicated that you leave the brain in situ overnight following the perfusion. Do you immerse the head in the fixative at room temperature during this time?

Dr. Peters: No, after perfusion we keep the whole animal in a well sealed plastic bag in the refrigerator until the next morning. Refrigeration may not be an essential requirement; we use it mainly because we don't like to leave the animal on the bench overnight.

Dr. Cowan: You mentioned that dehydration in acetone produces large numbers of "tight-junctions" Would you like to comment about the alternative situation where the intercellular spaces appear to be rather large, as in Van Harreveld's work using the technique of freeze substitution?

Dr. Peters: I have no experience with freeze substitution, and have really no idea what might be happening under such circumstances.

Dr. Karten: Do you control the pressure of your perfusion at all? I noticed that the vessel you showed looked rather engorged, almost too nice and round and distended.

Dr. Peters: No, we do not control the pressure. We leave the perfusion bottle about 4 ft. above the animal and rely on gravity feed. We have experimented with pressure pumps, and there are certain parts of the brain that benefit from their use. The optic nerves, for example, fix much better when you use a pressure pump, but too much pressure tends to explode cells in the cerebral cortex. The difference may have to do with local variations in angioarchitecture. White matter is much more difficult to fix than gray matter, possibly as a result of the sparser blood supply and when one wants to fix white matter, the increased pressure seems to improve fixation. For gray matter we tend to use as low a pressure as possible. One can see, by the way, that in some parts of the brain where white matter and gray matter are next to each other, the gray matter is often well-fixed, while the white matter has myelin sheaths that are split and the astrocytes are very badly swollen.

Dr. Karten: Dr. Peters, you stressed the great similarity in electronmicroscopic appearance between oligodendrocytic processes and neuronal dendrites. I could not help wondering how you come to identify the one type as an oligodendroglial process instead of a dendrite.

Dr. Peters: First of all, of course, we have followed the oligodendroglial process from the cell body. Also, we have made use of the fact that in white matter there are many oligodendrocytes but no dendrites. The glial profiles in such regions are packed between the myelinated nerve fibers.

Dr. Konigsmark: For light-microscopic studies Koenig, Groat, and Windle (1945) have advocated the use of gum acacia in the perfusion fluid. Has this been tried, and is there any point in using such colloidal additions in electronmicroscopy?

Dr. Peters: Solutions containing sucrose and colloids have been used to perfuse tissue for electron microscopy. The only purpose of such additions is to achieve an optimal osmolarity of the solution. Aldehydes, for example, seem to have hardly any osmotic effect, whereas slight changes in the concentration of the buffer or of any salt can grossly affect the tissue preservation. Consequently, it seems immaterial whether proper osmolarity of the carrier fluid is achieved by salts or by organic colloids.

Light- and Electron-Microscopical Studies of Normal and Degenerating Axons

R. W. Guillery

Introduction

This review will be concerned with an evaluation of the methods that have been used for tracing axonal connections. Specifically it will deal with methods that depend upon degenerative changes in the central nervous system of vertebrates. A number of accounts have considered this subject and many have appeared during the last decade. In order to appreciate the extent to which interpretations of axon degeneration have varied during this time and for details of much of the earlier literature reference should be made to Cajal (1928), Gibson (1937), Schimert (1938), Haggar and Barr (1950), Glees and Nauta (1955), Evans and Hamlyn (1956), Bowsher *et al.* (1960), Alksne *et al.* (1966), Beresford (1966), Gray and Guillery (1966), Heimer (1967 and 1968a) and Glees and Hasan (1968).

Consideration of the earlier literature demonstrates that the limitations of the light microscopical methods have not always been fully appreciated, especially when these methods were used to determine synaptic connections rather than simply for tracing tracts to their terminal cell groups. In the following review the changes that occur in the synaptic parts of degenerating axons will be considered in particular detail, but the degeneration occurring in the non-synaptic parts of axons will also be discussed. Each of the major light microscopical techniques for staining degenerating axons will be considered under a separate heading and the results that can be seen with each type of method will be related to electron microscopical observations.

Critical evaluation of the light-microscopical methods has advanced during the past decade as electron-microscopical studies have been used to show what structures can be stained with a particular light-microscopical technique and as the details of the degenerative changes that occur at synaptic junctions have been demonstrated by electron microscopy. Perhaps the most striking result of studies of axonal degeneration has been the clear demonstration that the signs of degeneration are not the same in all axons; that the degeneration varies greatly in its time course and that all axons do not go through the same sequence of degenerative changes. Apart from the law of Wallerian degeneration, which states that the peripheral segment of a cut nerve will undergo degenerative changes, we have few rules that will allow us to generalize from observations made in one part of the central nervous system of one species to other parts, or to other species. It is to be hoped that in the next decade some such generalizations will begin to be possible.

Neurofibrillar Methods

The neurofibrillar component of nerve cells was recognized by Schultze (1871) but a clear demonstration of the neurofibrils was not possible until the methods of Apáthy (1897), Bielschowsky (1902 and 1904) and Cajal (1903) were introduced toward the turn of the century. Apáthy's method has not been used by many others, but the Cajal and Bielschowsky methods have found wide application either in their original form or in one of many modifications (see e.g. Barr, 1939; Nonidez, 1939; Bodian, 1936; Glees, 1946).

Each of these methods stains the neurofibrils clearly. However, since each also stains other components to a greater or lesser extent, one may question whether the methods should properly be called "neurofibrillar" methods. The term will be used here because all of these methods give a sharp impregnation of the neurofibrils, generally allow a ready distinction between the neurofibrillar and other components of the nervous system and give a picture of the nerve cells that depends mainly upon the distribution of the neurofibrils. Further, it is important to have a term to distinguish these methods from other reduced-silver methods, which are discussed in subsequent sections.

Cajal used his method extensively to study the changes that occur in degenerating axons (see especially Cajal, 1928). Throughout this study he distinguished between the fibrillar and nonfibrillar parts of the nerve cell. He described the neurofibrils as a distinct internal skeleton of the nerve cell (see also Bielschowsky, 1928) and showed that these fibrils undergo characteristic reactions during axonal degeneration. The distinction that Cajal made so clearly has been ignored in many more recent studies and reviews, and much of the confusion that exists in the literature has been due to this.

a) Neurofibrils in Normal Axons

In order to evaluate the fibrillar changes that occur during degeneration it is necessary to consider the distribution of neurofibrils in the normal nervous system. Neurofibrils lie in some, but not all normal axons. They are commonly seen in many of the thicker myelinated axons and there form local condensations opposite the nodes of Ranvier (see Cajal, 1911 and 1928). The fibrils stain darkly and generally stand out from the surrounding axoplasm, which stains more lightly. In some preparations, however, all components of the axon may be so heavily stained that the fibrils can no longer be distinguished from the surrounding axoplasm. In such preparations it may be difficult to tell a bundle of fine axons from a group of neurofibrils and sometimes, as has been shown for Auerbach's plexus (Richardson, 1958), the distinction can be made only with the electron microscope. Fine axons often contain no neurofibrils and regions where many fine axons are grouped together then stain only lightly or do not stain at all with neurofibrillar methods (see e.g. Ralston, 1968a).

When axons that show neurofibrillar contents under the light microscope are studied with the electron microscope it is possible to identify bundles of neurofilaments[1] within them (Gray and Guillery, 1966). Lund and Westrum (1966) and Heimer and Ekholm (1967) have shown that neurofibrillar stains deposit silver granules upon

[1] The term "neurofilament" refers to fine filaments which are 70 to 100 Å in diameter and which can therefore be identified by electron microscopy only. "Neurofibrils" refers to much coarser structures seen with the light microscope. The latter are composed of bundles of the former.

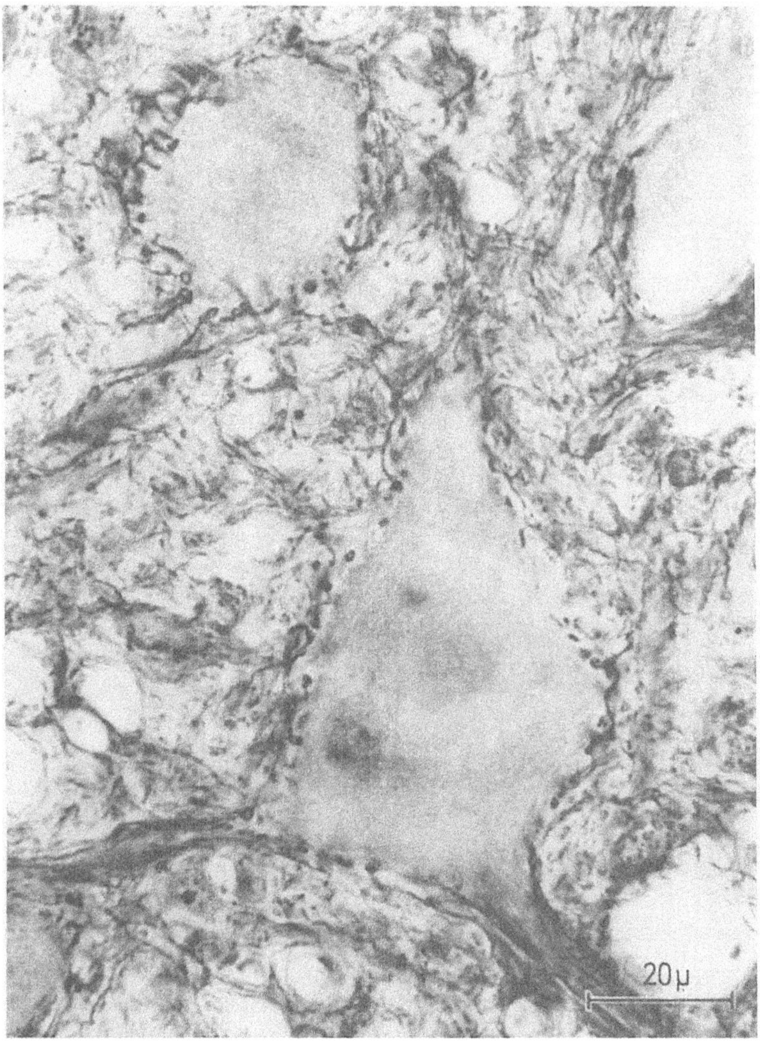

Fig. 1. Cells from the base of the dorsal horn of the lumbar cord of the cat. Note the neurofibrillar rings and clubs covering the surface of the cells in the lower central and upper left parts of the figure. The broad dendritic profile that passes out of the upper right part of the figure is virtually free of boutons. Block Cajal method, after Barr (1939). (From material kindly loaned by Dr. A. D. Loewy)

the neurofilaments and Ralston (1968a) has shown that in regions of the spinal cord where neurofibrillar stains show relatively few of the axons most axons contain microtubules and no filaments. That is, the neurofibrillar stains demonstrate the filaments but not the microtubules of the axons.

The terminal part of the axon can sometimes be stained by neurofibrillar methods. Fig. 1 shows a cell from the base of the dorsal horn, stained with a modification of one of the block Cajal methods (see Barr, 1939). The cell is richly covered with ring-shaped and club-shaped fibrillar formations which can reasonably be regarded as

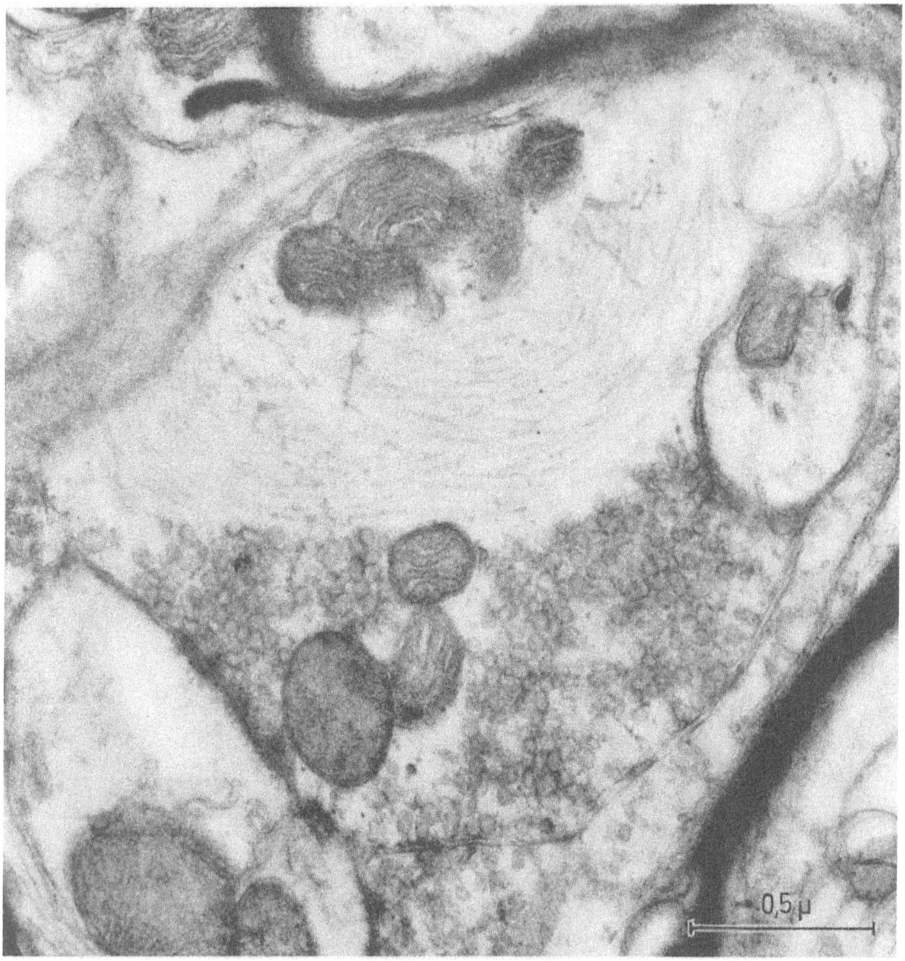

Fig. 2. A synaptic knob from lamina IV of the lumbar cord of the cat. The bundle of neurofilaments between the two groups of mitochondria can be regarded as a part of one of the ring-shaped terminal formations that are stained by the neurofibrillar methods. (Photograph kindly provided by Dr. H. J. Ralston, III)

synaptic "boutons"[2]. The identification of such synaptic boutons has played an important part in studies that have attempted to trace neuronal interconnections on the basis of degenerative changes in axons. It is therefore worth considering the arguments that lead a light microscopist to regard them as synaptic structures and inquiring to what extent material similar to that shown in Fig. 1 can lead to definite conclusions about synaptic connections.

Two important relationships are seen in these preparations. The first is that the fibrils of many of the boutons can be traced into an axon[3]. The second is that the boutons show a preferential distribution around some of the perikarya and larger dendrites. Further, in many places the surface of the boutons is closely applied to

[2] The use of this term, of "end-feet" and "synaptic knobs" is discussed by Gray and Guillery (1966), and the terminology employed there will also be used here.

and often follows the contour of the cell surface[3]. That is, the light-microscopical evidence suggests that these are indeed axonal terminations applied to the cell body and dendrites and that they resemble the axonal terminations that can be demonstrated at neuromuscular junctions (see Tello, 1906; Cajal, 1925).

Electron-microscopical studies support this interpretation. They show that there are bundles of filaments arranged as loops or reticulated formations within axonal

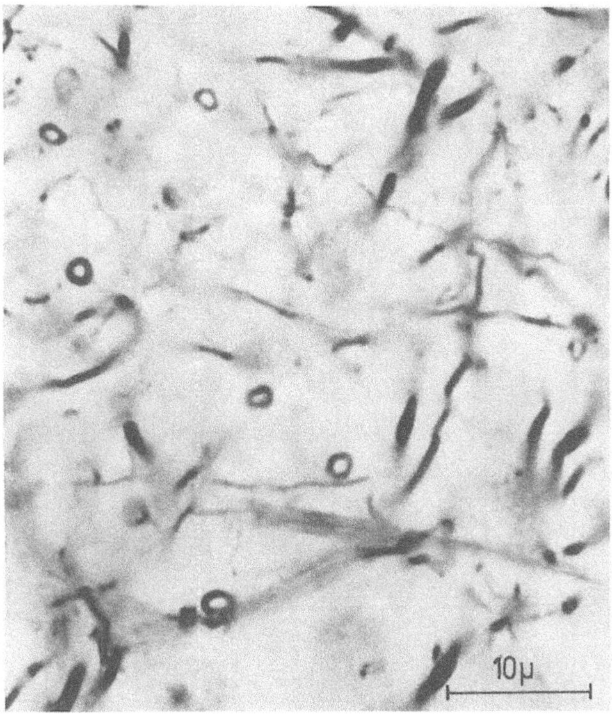

Fig. 3. Neurofibrillar rings in the hippocampus of a lizard. Notice that these rings are not related to perikarya and that the postsynaptic structures cannot be identified. Holmes method. (From material previously described by Boycott *et al.*, 1961)

swellings (Boycott *et al.*, 1961), that the filaments are argyrophilic (see Gray and Guillery, 1961; Lund and Westrum, 1966), and that the axonal swellings (the synaptic knobs) make specialized contacts with the perikarya and large dendrites (see Gray and Guillery, 1966 for details). Fig. 2, taken from the same region as Fig. 1 shows an electron micrograph of one of these synaptic knobs within which a part of one of the loops of neurofilaments is readily recognizable.

Fig. 3 shows another group of boutons, which are not so easily identified as marking the synaptic region of axons. These boutons are found in the hippocampus of lizards that have been kept at a relatively low environmental temperature for some time (see Boycott *et al.*, 1961). They are not specifically related to any postsynaptic structure that can be identified with the light microscope and a great many are

[3] The relationships described here can be distinguished more readily under the microscope by careful focussing than in the micrographs.

simply isolated rings that cannot be traced into an axon. Whereas the light micrographs leave some doubt about the interpretation of these rings, electron micrographs show that they are also formed by neurofilaments which lie in axonal swellings close to the synaptic region (Boycott *et al.*, 1961).

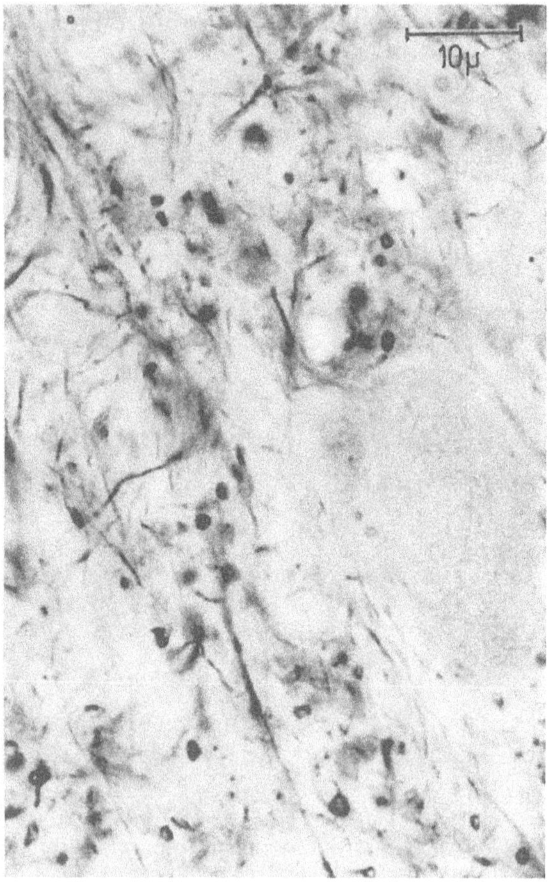

Fig. 4. Neurofibrillar rings and spheres from the dorsal lateral geniculate nucleus of the cat. A pale cell is seen at the right side of the figure. These rings do not form around the perikarya and the light microscope gives no clue as to their nature. See also Fig. 5. Block Cajal, method V. (From material previously described by Guillery, 1967a)

The neurofibrillar rings shown in Fig. 4 are smaller and less delicate than those in Fig. 3, but in terms of their relationships to other structures they can hardly be distinguished from those in Fig. 3. They lie in the dorsal lateral geniculate nucleus of the cat and have been described by Glees (1941), Hayhow (1959), Szentágothai (1963) and Smith *et al.* (1964) as axonal structures. This identification was based on their resemblance to boutons that occur in other parts of the central nervous system. However, in the lateral geniculate nucleus it has never been possible to trace one of these rings into an axon, and from electron-microscopical evidence one must conclude that they are not axonal structures at all (Guillery, 1967a). Thus, it has been shown that the axons in the lateral geniculate nucleus of the cat contain no bundles of neurofilaments arranged as rings or spheres. Nor do the axons contain any other structures that might form a basis for the neurofibrillar formations. Instead, there

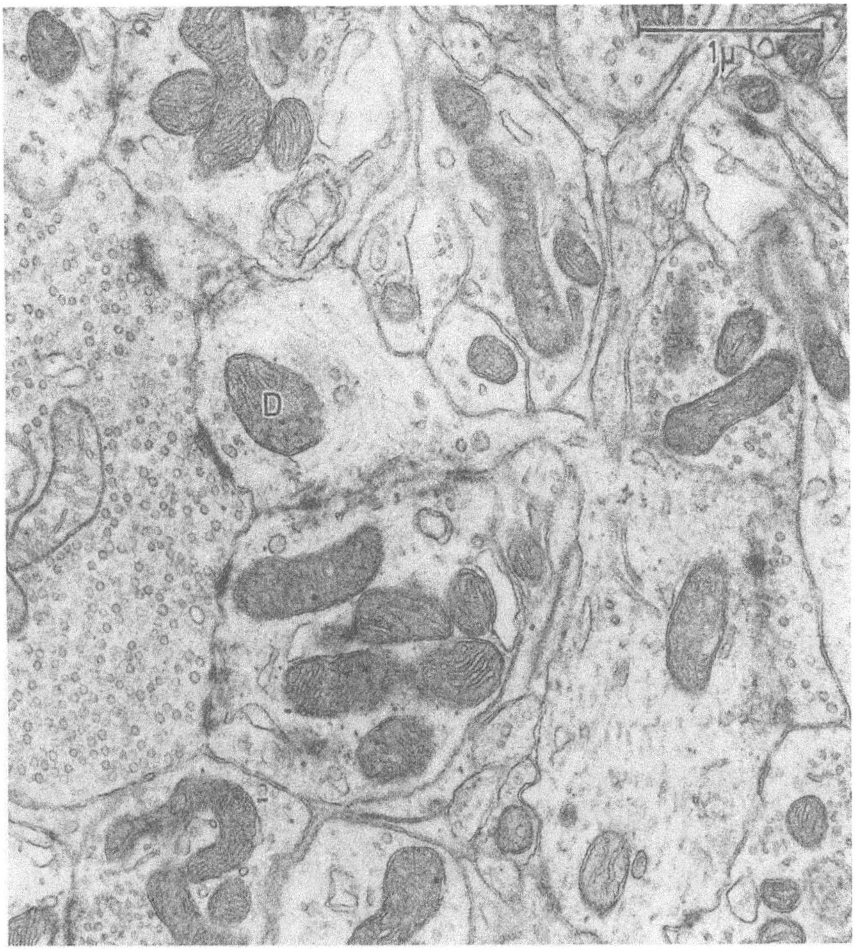

Fig. 5. A dendritic appendage (*D*) is seen in the central part of the figure. It contains a single mitochondrial profile and a small bundle of neurofilaments, which appears to form a part of a ring. To the right this appendage is linked by a slender stalk to a dendritic stem. (From Guillery, 1969)

are bundles of filaments forming ring-shaped (see Fig. 5) or spherical groups within certain dendritic profiles, and the laminar distribution of these profiles matches the laminar distribution of the neurofibrillar formations (see Guillery, 1967a).

Figs. 1—5 illustrate three distinct types of neurofibrillar structure resembling boutons. Whereas the first type (Figs. 1 and 2) can be recognized as synaptic structures on the basis of light microscopy alone, the second and third can be interpreted only on the basis of electron micrographs which show that the second (Fig. 3) but not the third (Figs. 4 and 5) form true axonal boutons.

In the central nervous system there are relatively few regions where axonal terminals can be identified on the basis of neurofibrillar methods. In many regions, as in the neocortex or striatum, the neurofibrillar methods do not stain any boutons or bouton-like structures at all because the axons contain no neurofilamentous rings or nests in their terminal parts (see Boycott *et al.*, 1960; Gray and Guillery, 1966). Even in regions where neurofibrillar boutons are stained and where, as in the spinal cord, they can be definitely identified as axonal terminals, the neurofibrillar methods generally show only a proportion of the total population of synaptic knobs. Other methods, such as those described by Armstrong *et al.* (1956) and Rasmussen (1957) show many more axonal end feet especially in relation to the finer dendrites.

In the normal nervous system the neurofibrillar methods show a specific component of the nerve cell, the neurofilaments, and these have a definite distribution in relation to certain synaptic structures. The idea that neurofibrillar stains are inconsistent or fickle, producing an unreliable "artefact" does not take account of the extent to which the structure of the synaptic region varies in different parts of the nervous system and between species. Wherever electron-microscopical studies have demonstrated a group of neurofilaments the neurofibrillar methods have shown stainable fibrils[4] and, conversely, filaments can be seen in regions where fibrils are demonstrable. This close correspondence is found not only in normal axons but also during degeneration.

b) Neurofibrils in Degenerating Axons

Cajal (1928) gave a detailed description of the changes that a degenerating axon goes through. Close to a lesion the "traumatic degeneration" produces certain well marked changes. One of the most striking of these is a neurofibrillar hypertrophy that occurs in the free ends of some of the cut fibers, relatively close to the cut. This produces clubs, rings and reticulated terminal swellings. It is of interest that these neurofibrillar changes, which in many respects resemble the changes that take place in the synaptic terminals of cut nerves (see below) occur in the segment of the nerve that remains connected to its perikaryon and also in the segment that has been separated from the perikaryal region (see Fig. 6).

Farther from the lesion an axon that has been separated from its cell body undergoes secondary, or Wallerian degeneration. According to Cajal (1928) the fineness of the neurofibrillar impregnation in these fibers is lost at an early stage, the fibrils take on a granular appearance and the axon becomes beaded. Later on the beaded fibers break up into globular or fusiform masses which are more or less deeply stained by the fibrillar methods. It is important to note that these masses, which often lie far from the synaptic terminals, do stain with the fibrillar methods. Whether they represent degeneration products of the neurofibrils or some other argyrophilic component (see e.g. Beresford, 1966) has not been determined.

[4] There is one apparent exception to this statement. In the nerve cells of the lateral vestibular nucleus of the rat Sotelo and Palay (1968) have described membrane-bound bodies which are densely packed with neurofilaments. From their micrographs one would expect to find extremely dense argyrophilic bodies in these cells. They found no such bodies, only some unstained zones, in their block Cajal material. Since dense argyrophilic bodies can be stained in these cells when paraffin sections are stained on the slide (Guillery, unpublished observations) there must be some doubt about this apparent exception.

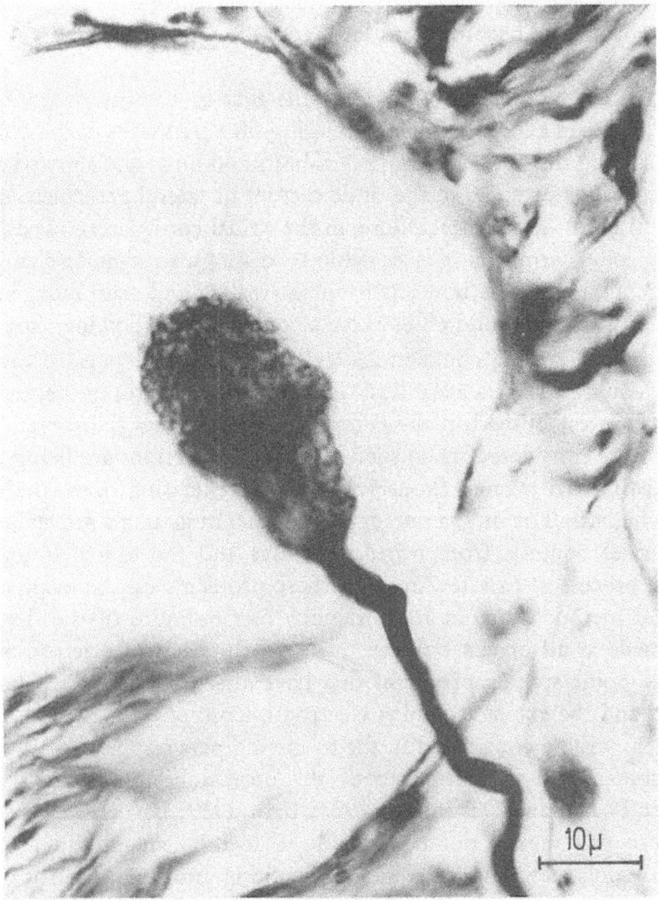

Fig. 6. Traumatic degeneration in the optic chiasm of the cat 5 days after a lesion was placed immediately rostral to the chiasm. Note the dense neurofibrillar network within the large swollen terminal bulb. Paraffin Nauta method with no suppression of the neurofibrillar component. (See Guillery *et al.*, 1961)

The structure of the axonal swellings that occur close to the lesion has been studied with the electron microscope by Lampert and Cressman (1964) and by Lampert (1967). These swellings contain a great many mitochondria, dense bodies and vesicular elements. Some, but not all of the swellings also show an increase in neurofilaments. The changes that occur close to the lesion in the central nervous system do not differ significantly from those that occur in the peripheral nervous system (see e.g. Blümcke *et al.*, 1966) except that they are slower in the central nervous system, where the reactive axonal bulbs appear to survive for several weeks. Electron-microscopical observations of the Wallerian degeneration will be considered in a later section (see p. 101).

Cajal did not describe the changes that occur in degenerating axon terminals in the central nervous system. Tello (1907) and Cajal (1928) did, however, study the changes that occur during degeneration at the neuromuscular junction and they showed that in some of these axon terminals the neurofibrils increase, stain more darkly than normal and become tortuous before the axon finally undergoes a granular disintegration.

Cattaneo (1923) appears to have been the first to use the neurofibrillar hypertrophy that occurs in axon terminals for tracing fiber pathways in the central nervous system. He studied the optic pathways of rabbits and birds and showed degenerating fibers as small rings or clubs in the optic tectum or lateral geniculate nucleus. Hoff (1932a and b) used the same technique in the spinal cord. He showed that in many degenerating axon terminals it is possible to distinguish definite changes with the neurofibrillar methods. The boutons become swollen and stain more intensely, and the rings often turn into solid clubs three to seven days following axon transection.

This method of tracing connections had a considerable appeal because it looked as though it could provide a great deal of information regarding the precise synaptic linkages that are established in the central nervous system. However, several problems have to be considered when such synaptic connections are being studied. The first is to distinguish normal axons from the degenerating ones, the second is to determine whether all or only some of the degenerating axons are stained, the third is to tell axonal boutons from other structures and the fourth is to identify the postsynaptic processes. In a few regions these problems can be evaluated by light-microscopical methods, but in most regions one or more of the decisions either cannot be made at all or can only be made on the basis of electron-microscopical observations. Some specific problems that have arisen in the interpretation of light micrographs and the extent to which electron microscopy has helped, or may help, to clarify them will be considered in the following paragraphs.

In the spinal cord the structure of the normal boutons shows considerable variation (see Phalen and Davenport, 1937; Barr, 1939; Barnard, 1940; Haggar and Barr, 1950) so that it may not be possible to identify an individual degenerating bouton in a region that contains some degenerating and some normal boutons. The light-microscopist is then limited to identifying the small proportion of boutons that represent the extreme of degenerative change and describing their specific synaptic sites. Electron micrographs of such regions show some synaptic knobs in which there is a markedly abnormal network of neurofilaments (see Ralston, 1968b) and these correspond to the grossly abnormal boutons seen under the light microscope. However, the electron micrographs also show a great many more synaptic knobs which are clearly degenerating but which contain no neurofilaments. This non-filamentous degenerative change is discussed below (p. 98); at present it is only relevant to note that there are many degenerating synaptic knobs in the spinal cord, and also in many other parts of the nervous system, which do not show an increase in neurofilaments at any one stage of degeneration and which therefore are not shown as boutons by the neurofibrillar methods.

It has been mentioned that in the normal brain there are many areas in which neurofibrillar methods show no boutons. In some of these regions, such as the tractus solitarius (Evans and Hamlyn, 1956) or the striatum of the cat (Webster, 1965) no fibrillar boutons can be demonstrated by these methods at any stage of degeneration. In other regions, such as the avian optic tectum (Evans and Hamlyn, 1956; Cowan *et al.*, 1961) or primate lateral geniculate nucleus (Glees and Le Gros Clark, 1941; Glees, 1961) a great many neurofibrillar rings and clubs appear during degeneration. In these last two regions the normal and degenerating axons can thus be distinguished from each other readily, but the light microscope does not allow a definite identification of the synaptic parts of the axons. The neurofibrillar rings and clubs look as though they might represent the synaptic portions of the degenerating axons. However, they show no specific relation to the postsynaptic elements so that they cannot be certainly identified as synaptic boutons. Only the general region of the axon termination can be determined with the neurofibrillar methods; the detailed synaptic patterns cannot be shown in these regions.

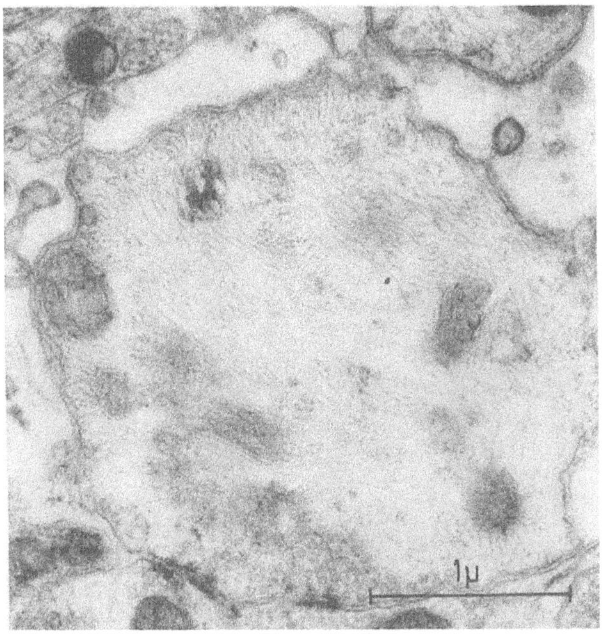

Fig. 7. Degenerating synaptic knob from the dorsal lateral geniculate nucleus of the monkey 5 days after a unilateral section of the optic nerve. Notice the rich tangle of neurofilaments which occupies most of the profile. The vesicles, which are normally distributed throughout the cytoplasm of the synaptic knobs, have been displaced towards the lower part of the figure close to the synaptic thickening. (From Colonnier and Guillery, 1964)

Both the lateral geniculate nucleus and the tectum have been studied with the electron microscope and in each it has been possible to demonstrate that the neurofibrillar rings and clubs do occur in the synaptic portions of the axons. In the avian optic tectum Gray and Hamlyn (1962) showed that there is a dramatic increase in neurofilaments in the synaptic knobs and in its time course this change corresponds to the neurofibrillar hypertrophy that occurs in the degenerating axons. In the lateral geniculate nucleus of the monkey the light microscope shows that the degenerating axons form extremely large clubs which, in many respects resemble the swollen axon segment shown in Fig. 6. Electron micrographs (Colonnier and Guillery, 1964; Guillery, 1965; Glees *et al.*, 1966) show that, during degeneration, the unusually large synaptic knobs in this nucleus become filled with filaments and that the synaptic vesicles gradually become displaced towards the synaptic thickenings (see Fig. 7). This filamentous invasion of the synaptic knobs is maximal after five to seven days but can still be seen at eleven days, when most of the vesicles have become displaced and the mitochondria are beginning to show signs of degeneration (see Fig. 8).

Degenerating axons become beaded all along their length and form globular and granular masses that can be stained with the neurofibrillar stains. With the light microscope it is not always easy to distinguish these degeneration products of non-synaptic axon segments from degenerating synaptic boutons, especially when the boutons form more or less solid black masses within which the fibrillar structure

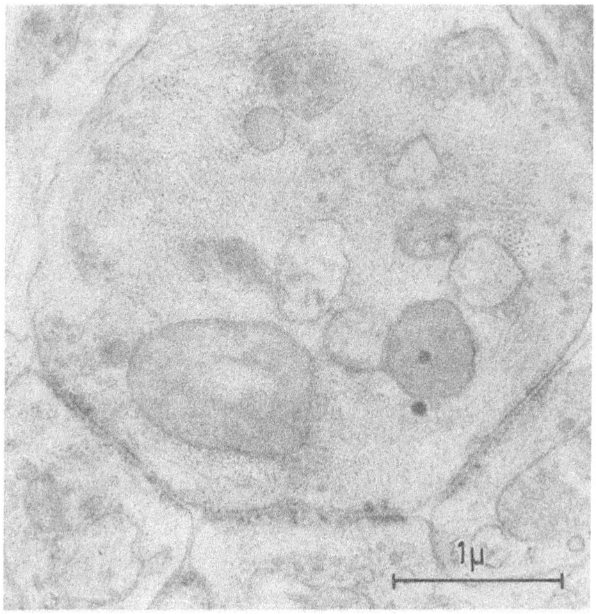

Fig. 8. Degenerating synaptic knob from the dorsal lateral geniculate nucleus of the monkey 11 days after a unilateral section of the optic nerve. Very few synaptic vesicles remain in this axon terminal. (From material previously described by Colonnier and Guillery, 1964)

cannot be seen clearly. Webster (1965), using the Glees method, showed that in the cat degenerating corticostriate fibers have this appearance and he pointed out that in this system it was impossible, with the light microscope, to distinguish the synaptic or "terminal" parts of fibers from the non-synaptic fibers of passage. In the lateral vestibular nucleus of the cat Walberg *et al.* (1958) showed the beaded and broken up degenerating axons of the vestibular nerve and interpreted these as synaptic sites of the vestibular fibers. However, more recent electron microscopical studies from the same laboratory have shown that these fragmented portions do not represent synaptic regions (Mugnaini *et al.*, 1967).

The axonal degeneration that occurs in the mamillary bodies of the rabbit after fornix lesions has been studied by Sprague and Meyer (1950) with the Glees method. These authors showed the disrupted fragments of the degenerating fornix fibers terminating within the medial mamillary nucleus. In the rat the fornix degeneration in the medial mamillary nucleus was shown as a very much denser granular deposit by Heimer and Ekholm (1967), who also used the Glees method. In such material it is not possible to identify synaptic knobs with the light microscope, and electron microscopical evidence as to what components are stained by the Glees method in this region is not available. There is some reason to think that in the mamillary bodies, and in some other regions, degenerating non-fibrillar synaptic knobs may be stained by the Glees method[5] (see below), but specific electron-microscopical evidence for this view is lacking.

In the premamillary hypothalamus it is common to see fibers that entirely resemble degenerating axons (see Cowan and Powell, 1956; Szentágothai *et al.*, 1968). These are beaded and often appear fragmented when stained with one of the modifications of the Bielschowsky method. Occasionally they have been wrongly regarded as produced by a specific lesion placed in the brain for experimental purposes. However, these fibers can be stained in many apparently normal animals, and one can doubt that they represent degeneration at all. Thus, Szentágothai *et al.* (1968) have suggested that they are neurosecretory fibers, and that the fragmented appearance of the fibers is dependent upon silver impregnation of neurosecretory droplets.

When the degenerative reactions that can be demonstrated by neurofibrillar methods are considered in terms of electron-microscopical observations two points emerge clearly. The first is that the neurofibrillar methods stain a component of the nerve cell (the neurofilaments) which shows marked changes in some degenerating axons and axon terminals and shows no changes in others. That is, these methods reveal that the degenerative reaction is not the same in all parts of the central nervous system. The differences are not obviously related to axon caliber, or to axon length, nor have they been related to any particular features of perinuclear region of the nerve cells. These different axonal reactions are discussed further below.

The second point is that the neurofibrillar methods only rarely allow a clear identification of degenerating synaptic structures. Many degenerating axons are not stained by the methods. Sometimes the degenerating axons are stained but cannot be clearly distinguished from normal axons, at other times the synaptic regions cannot be distinguished from the non-synaptic regions. Because the neurofibrillar reaction within axons is not the same in all parts of the brain and even shows differences between species for homologous parts of the brain[6] it is not possible to use the electron microscope in order to establish general criteria for identifying the degenerating synapses. At present each region must be studied individually.

The Nauta Methods

Nauta has described several staining methods (Nauta, 1950; Nauta and Gygax, 1951; Nauta and Gygax, 1954; Nauta, 1957) but only the two described in 1954 and

[5] In the medial mamillary nucleus of the rat the Nauta method does stain degenerating synaptic knobs (see Guillery and Ralston, 1964; Heimer and Ekholm, 1967; and see p. 92). The degeneration shown in this nucleus by the Nauta method resembles that shown by the Glees method (Guillery, 1956; Heimer and Ekholm, 1967) and one is tempted to conclude that in this region both methods are staining the same components. However, since Heimer and Ekholm (1968) found that in the rat, the Glees reaction occurred over a very limited time span while the Nauta reaction was positive over a more extended period there must be some doubt about this conclusion.

[6] These species differences are strikingly shown in the dorsal lateral geniculate nucleus, where the degenerative reaction seen in the rat (McMahan, 1967) differs markedly from that seen in the monkey (Colonnier and Guillery, 1964; Glees *et al.*, 1966) and the reaction seen in the cat (Szentágothai *et al.*, 1966) lies between these two extremes.

1957 will be considered specifically as "Nauta methods" here. The earlier procedures are modifications of the Bielschowsky method. They give a relatively dark impregnation to degenerating axonal elements, but also stain much of the normal neuronal plexus. The methods published in 1954 and 1957 differ from the Bielschowsky method in leaving most or all of the normal fiber plexus unstained. The staining of normal nerve fibers in a Bielschowsky preparation depends largely upon the fibrillar element, and the Nauta methods produce their characteristic result by not staining the neurofibrils. The Nauta methods, therefore, do not stain the neurofibrillar boutons (see Glees and Nauta, 1955).

The absence of neurofibrillar boutons in Nauta preparations led to the more general conclusion that the Nauta method does not stain the synaptic portions of axons at all, and a rather awkward terminology was introduced which referred to the pericellular degeneration seen in Nauta preparations as "preterminal" while "terminal" was reserved for boutons recognizable in neurofibrillar preparations (see e.g. Bowsher *et al.*, 1960).

Some descriptions did not follow this usage (Sprague, 1958; Szentágothai, 1958; Ha and Liu, 1963). These observers considered that the Nauta method does stain degenerating synaptic and terminal segments of axons and that it is possible to describe the synaptic relationships of degenerating axons on the basis of the Nauta method. However, in most regions the light-microscopical study of Nauta sections does not allow a distinction between synaptic and non-synaptic portions of an axon. Degenerating axons become beaded and disrupted and the Nauta methods stain the beaded fragments dark brown or black, with no discernible internal structure. It is generally not possible to determine whether any particular fragment represents a terminal segment, a segment making a synapse *de passage* or a non-synaptic axon merely passing through the region.

Fig. 9 shows portions of two laminae of the dorsal lateral geniculate nucleus of the cat stained with the Nauta method. Degenerating retino-geniculate fibers from the contralateral cut optic nerve are shown. The fibers form a dense pericellular plexus in the upper layer (lamina A) and form a sparser plexus in the lower layer (lamina A1). The fibers that enter lamina A all come from the optic tract which lies beyond the lower border of the figure and, therefore, they must pass through lamina A1. There are good reasons for believing that none of these fibers synapse in lamina A1 (Meikle and Sprague, 1964; Bishop, 1965) but the Nauta preparation itself does not allow one to arrive at this conclusion. The degenerating fibers that pass through lamina A1 form a loose network around the cells and often appear to make intimate contact with the cells of this lamina (see Fig. 10). The relatively dense pericellular degeneration in lamina A suggests that most contralateral fibers terminate there. However, in this system there is no way in which one can distinguish synaptic from non-synaptic parts of the degenerating axon and the same difficulty applies to the interpretation of Nauta preparations in many other parts of the brain.

In a few regions it has been possible to stain degenerating axons at an early stage (3—7 days) before the fragmentation is very far advanced, and to compare the branching pattern in the terminal parts of the degenerating axons with the terminal pattern that one can see in a normal brain prepared by one of the Golgi methods. In this way it can be shown that the terminal parts of axons in the cerebellum (Brodal and Grant, 1962; Brodal and Høivik, 1964) in the lateral cervical nucleus (Ha and

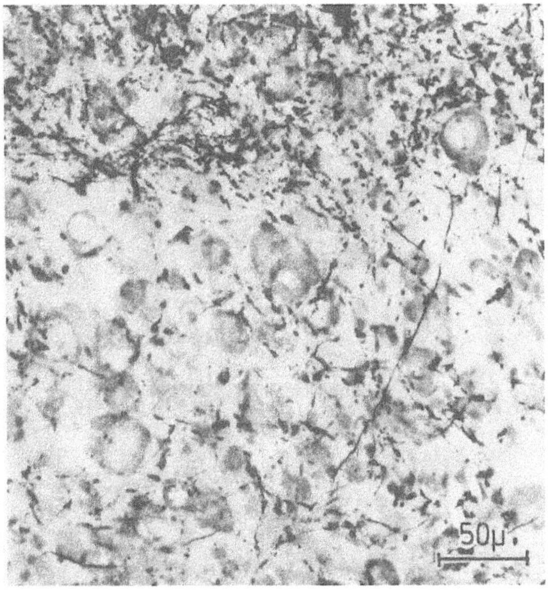

Fig. 9. Nauta preparation (Nauta and Gygax, 1954) from the dorsal lateral geniculate nucleus of the cat 7 days after contralateral section of the optic nerve. Lamina A is in the upper, lamina A1 in the lower part of the figure. (See text and see Fig. 10)

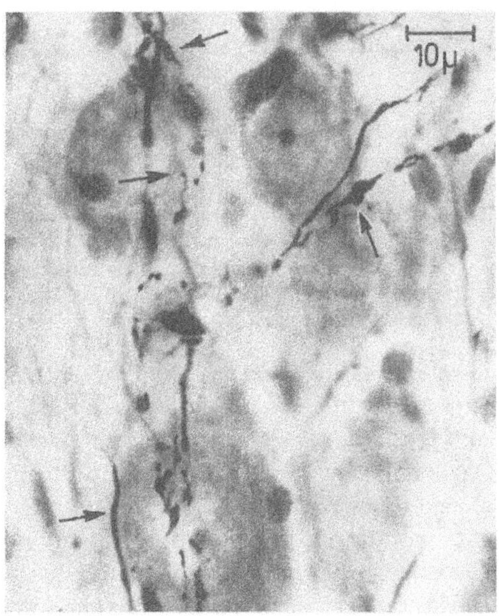

Fig. 10. Three cells of lamina A1 from the section shown in Fig. 9. At this higher magnification one can see the close relationship between the degenerating axons and the perikarya (arrows). With the light microscope it is impossible to determine whether or not these axons form synaptic contacts with the cells. (See text)

Liu, 1963) and in the lateral geniculate nucleus (Guillery, 1967b) are stained by the Nauta method.

This type of demonstration is only possible where the terminal part of the axon has a branching pattern that is easily recognized. In other parts of the brain the electron microscope must be used to determine whether or not the synaptic knobs are stained. This has been done by Guillery and Ralston (1964), Lund and Westrum (1966) and Heimer and Ekholm (1967) and it has been shown that in the spinal cord

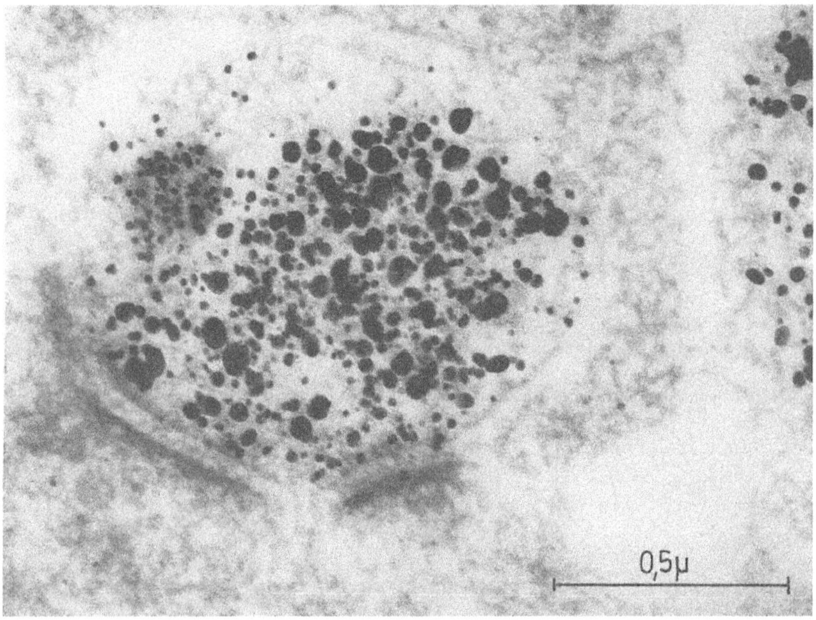

Fig. 11. Electron micrograph obtained from a Nauta preparation (Nauta and Gygax, 1954) of the spinal cord of a cat with unilateral dorsal root section and 5 days postoperative survival. The silver that is deposited by the staining procedure can be seen in the synaptic knob. The synaptic knob can be identified by the two synaptic thickenings in the lower part of the figure. The tissue was stained with phosphotungstic acid, so that the postsynaptic thickening appears more marked than the presynaptic thickening. (From material previously described by Guillery and Ralston, 1964)

of the cat and in the superior colliculus and mamillary bodies of the rat the Nauta method does deposit significant amounts of silver in the synaptic knobs, enough for these to be visible with the light microscope (see Figs. 11 and 12).

These studies were done by staining tissue according to the Nauta method and then preparing the stained tissue for electron microscopy. The Nauta procedure can either be carried out on sections cut from unfrozen tissue or on the usual frozen sections. The former technique was used by Guillery and Ralston (1964) and has the disadvantage that on the relatively thick slices that are obtained the Nauta impregnation is much more irregular than usual. Great care has to be taken to make sure that the small pieces of tissue prepared for electron microscopy come from an area in which the Nauta staining is of acceptable quality. Thus, they must be free of the dust-like artefact that Nauta sections sometimes show (see Glees and Nauta,

1955; Nauta, 1957), should not contain areas of myelin cuffing (see below) and the degenerating fibers should be clearly shown. Precautions should be taken to see that the silver is not lost from the tissue during processing for electron microscopy (see Lund and Westrum, 1966).

If frozen sections are used, the sections are likely to contain more areas of damage produced by ice crystals (see Lund and Westrum, 1966), but the chances of producing poor, irregularly stained Nauta sections are reduced. None the less, it is worth preparing tissue for electron microscopy only if the quality of the Nauta sections can

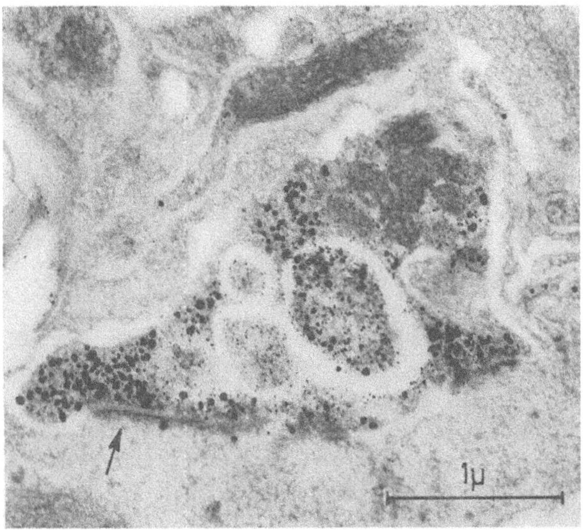

Fig. 12. Electron micrograph obtained from the material illustrated in Fig. 11. A postsynaptic thickening can be seen in the lower left part of the figure (unlabelled arrow), and the substance of the synaptic knob shows a clumping suggestive of degeneration

be judged under good optical conditions[7]. Provided that in the Nauta sections the normal fibers are not stained to a significant extent one can reasonably interpret the axonal profiles that contain quantities of silver as degenerating axons.

This technique is laborious and the poor quality of the tissue preservation (see Figs. 11—13) gives electron micrographs that are often difficult to interpret. At present the method has been used only to determine the capacity of the Nauta method for staining different portions of the nerve fiber; it has not been used for studying patterns of synaptic connections. It could be used for this purpose but, in general, direct electron microscopical observations have proved more valuable.

Figs. 11, 12 and 13 show that the Nauta methods deposit silver in degenerating axons and these figures indicate that the myelin which surrounds the degenerating axons is not stained. Nauta (1957) pointed out that with the light microscope it is possible to see areas in some sections where the myelin has stained. Occasionally the myelin is seen as a characteristic double line (see Fig. 14, *m*), but more commonly it forms cuffs or bracelets around the fibers. When such cuffs are formed around

[7] Guillery and Ralston (1964) embedded their Nauta sections in thin, flat pieces of Araldite. The parts of the section that were relatively flat and close to the surface of the plastic could then be checked with the highest powers of the light microscope.

particularly large fibers one can see that they are made up of complex whorls or spirals (Fig. 14 large arrows), whereas around thinner fibers they appear as smaller, more or less solid structures (Fig. 14 small arrows). When a section which shows such myelin cuffing is examined with the electron microscope one sees that the silver is deposited around the axon in a pattern which suggests that the silver lies in the cytoplasm of an oligodendroglial cell at a node of Ranvier (Fig. 15).

This deposition of silver in relation to myelin is not selective for degenerating fibers. It occurs around normal axons, and when it is seen in a Nauta section it

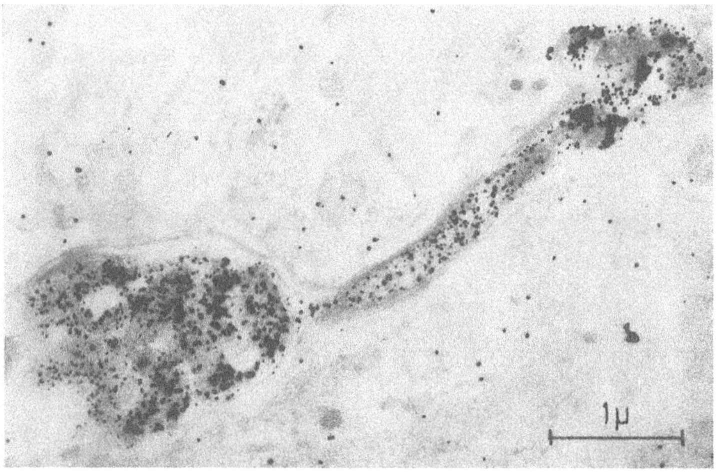

Fig. 13. Electron micrograph obtained from the material illustrated in Fig. 11. Note that the silver is deposited in the axoplasm of the beaded degenerating fiber, but that the myelin is not stained

should be treated as an artefact not related to the axonal degeneration (see Nauta, 1957). This may explain why some reports of electron microscopical observations on Nauta stained sections have described silver deposition in relation to myelin (Eager and Barrnett, 1966; Lund and Westrum, 1966) while others have not (Guillery and Ralston, 1964; Heimer and Ekholm, 1967). The Nauta methods can stain myelin and related oligodendroglial cytoplasm, and this is important for understanding the chemical affinities of the method[8], but the Nauta sections that one uses in order to trace degenerating axons should not include silver deposition in relation to myelin. Fortunately, the myelin staining can be easily recognized with the light microscope.

In most regions the degenerating neural processes that are stained by the Nauta methods can be clearly recognized as axons. Nauta (1957) has stressed that one of the important advantages of the Nauta method is that "the silver impregnated fragments typical of fully developed axon degeneration exhibit a more characteristic shape than the droplets of degenerating myelin stained by the Marchi method. Thus, the silver picture is usually less equivocal, especially where fibers of fine caliber are involved."

[8] The affinities of the Nauta method extend beyond degenerating axoplasm and oligodendroglial cytoplasm. Perivascular tissue is very commonly stained. Whether the breakdown products of myelin can be stained with the Nauta method is not known because in electron micrographs of fibers that show advanced degeneration the break-down products of the myelin cannot be distinguished from the break-down products of the axoplasm.

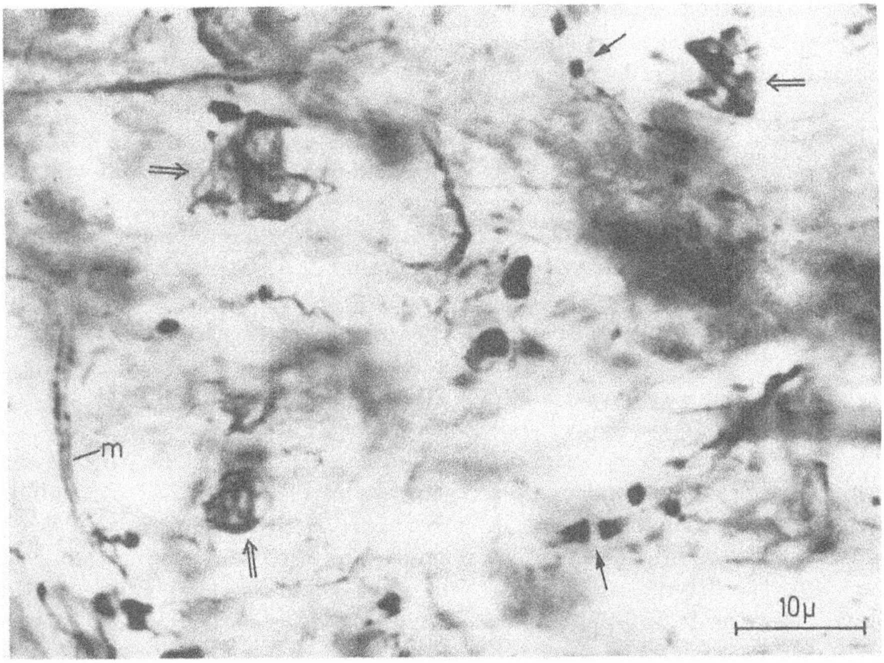

Fig. 14. Nauta-Gygax (1954) preparation from the brain stem of the cat. The region shown here contained no degenerating fibers but a great deal of myelin artefact. The coarse whorl-like silver deposits can be seen around the thicker fibers (large arrows) and smaller cuffs are present around the thinner fibers (small arrows). To the left of the figure the double line that is seen when myelin is stained has been labelled "*m*"

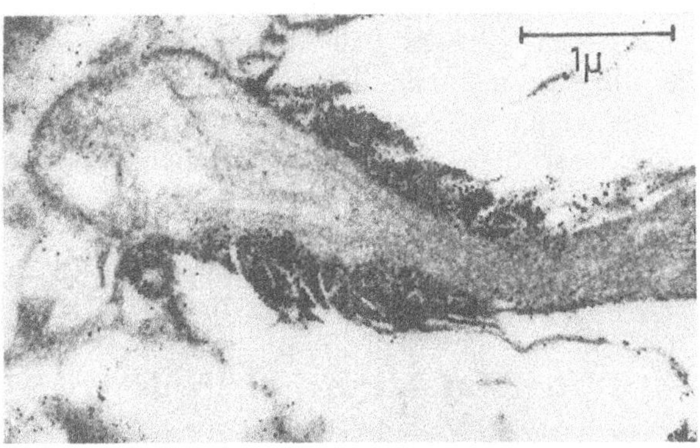

Fig. 15. Electron micrograph obtained from material similar to that illustrated in Figs. 11 to 13. The silver is deposited around the nerve fiber in a pattern that would appear as a myelin cuff when seen with the light microscope

When the Nauta method produces a fine dust-like, apparently random, precipitate in a section this precipitate should generally be treated as an artefact not related to the fiber degeneration (Nauta, 1957). However, there are several reports which show that occasionally a dense dust-like deposit of silver may be formed specifically in regions that are undergoing rapid or extensive retrograde changes (Guillery, 1959;

Cragg, 1962; Powell and Cowan, 1964; Montero and Guillery, 1968). Comparison of Nauta sections with adjacent Nissl sections shows that the dust is more or less confined to regions of cellular degeneration and that the dust appears at about the same time as the cellular changes (see Powell and Cowan, 1964; Montero and Guillery, 1968).

At present these regions of retrograde change have not been examined with the electron microscope to determine whether the dust is deposited in relation to any particular part of the degenerating cells. It is possible that the dust represents degenerating dendrites (see especially Grant and Westman, 1968; Grant and Aldskogius, 1967) or that it is due to axonal synaptic knobs which end on the degenerating cells and which are undergoing transneuronal degeneration (see Cowan, this Volume). It appears that the dust is more readily stained in the rabbit and rat than in the cat and that other methods may stain it more consistently than the Nauta methods do (see Fink and Heimer, 1967; Montero and Guillery, 1968).

A number of observations of degeneration patterns in different parts of the nervous system suggest that there are some axons or parts of axons which cannot be impregnated with the Nauta method. It has been stated already that the Nauta methods do not stain the neurofibrillar boutons. The large degenerating clubs that occur in the lateral geniculate nucleus of the monkey (see above) cannot be stained with the Nauta method (Glees, 1961) and the axonal degeneration that can be seen in the avian optic tectum in Nauta preparations does not reach its full development until after the neurofibrillar degeneration has disappeared (Evans and Hamlyn, 1956).

In some parts of the brain (e.g. the hippocampus, the pyriform cortex or the superficial parts of the superior colliculus) the Nauta methods show little or no degeneration while the Glees method, the earlier Nauta methods (see Nauta and Gygax, 1951) or the methods of Fink and Heimer (1967) show a relatively dense, dust-like deposition of silver granules (see Heimer, 1967, and see below).

In summary, the Nauta methods stain degenerating axons, including their terminals. However, they do not stain all axonal systems in the central nervous system and leave some terminals unstained. These methods thus indicate, like the neurofibrillar methods, that all axons do not go through the same sequence of degenerative changes. The Nauta methods have proved extraordinarily useful for tracing fiber pathways to their terminal cell groups and in this they are far more reliable than the neurofibrillar methods. However, like the neurofibrillar methods they only rarely give useful information about details of synaptic connections.

The Fink-Heimer Methods

These methods have been developed recently and have been discussed by Fink and Heimer (1967), Heimer (1967 and 1968a), Heimer and Peters (1968) and by Heimer in this volume. They stain the degenerating axons that can be shown by the Nauta methods, they suppress the staining of normal fibers, and they also stain the fine dust-like deposit that the Glees and early Nauta-Gygax (1951) methods show in some parts of the brain.

In the thalamus the Fink-Heimer methods stain the retrograde dust that was discussed above (see Fink and Heimer, 1967). In the cerebral cortex and superior colliculus they stain a slightly coarser dust in the superficial laminae while the Nauta methods show relatively few degenerating fibers there. The dust-like deposit characteristic of the Fink-Heimer methods generally shows no morphological features which allow a light microscopist to determine what elements have been stained. Only occasionally is it possible to recognize that the particles are deposited in fine axons or in glial elements. There is no light-microscopical criterion by which the retrograde dust can be certainly distinguished from Wallerian changes. In some regions a comparison with Golgi material suggests that the dust, at least in part, represents degeneration in fine, richly branching terminal axon segments. This is seen, for example, in the superior colliculus, where the dust that one sees after optic nerve section lies in a region that is normally occupied by rich and delicate terminal branches of the optic nerve fibers (see Cajal, 1911). However, the possibility that the methods can show the "synaptic distribution" of a fiber system (see e.g. Heimer, 1968b) must be interpreted conservatively. Individual synaptic relationships can not be demonstrated by these methods, since there are no morphological clues by which the synaptic portion of an axon can be identified and since the postsynaptic elements are not shown. The general area within which a fiber system terminates can often be shown, and only in this gross sense can the methods be regarded as giving information about synaptic relationships.

In order to interpret the Fink-Heimer sections it is necessary to demonstrate what elements are stained by the method and this demonstration must eventually depend upon electron microscopical analyses. Heimer and Peters (1968) have studied the silver-stained sections with the electron microscope. They have shown that some of the silver is deposited in degenerating synaptic knobs and that some lies in degenerating myelinated and unmyelinated non-terminal axon segments. They do not show silver deposited in glial elements although the light microscope suggests that glia can be stained (Fink and Heimer, 1967). The electron micrographs shown by Heimer and Peters (1968) contain only very little silver, and one must doubt whether the amount of silver shown in their electron micrographs could produce the dense granules that their light micrographs show. Heimer and Peters (1968) used weak silver solutions in order to reduce the background staining in their sections, but they do not describe the effect that this had on the light-microscopical appearances. Their results indicate that the Fink-Heimer method impregnates degenerating axons and axon terminals. They do not show what proportion of the dust-like particles that one sees with the light microscope represent degenerating synaptic knobs, degenerating non-synaptic axons or other elements.

The observations described by Heimer and Peters (1968) were not concerned with a comparison of the Nauta and Fink-Heimer methods. The light-microscopical observations suggest that there are some degenerating axons and axon terminals which can be stained only by the Fink-Heimer or early non-suppressive Nauta methods and that there are others which can be stained by the Fink-Heimer and also by the suppressive Nauta methods. If this is so then the two groups of axons may go through different degenerative changes and one can reasonably expect that there may be associated fine structural differences. These remain to be studied[9].

[9] In the cat the Nauta and Fink-Heimer methods both show the same pattern of optic nerve fiber degeneration in the dorsal lateral geniculate nucleus. In the superior colliculus, however, the two methods produce significantly different results. This may mean that the retinogeniculate axons differ from the retinotectal fibers in their degenerative reaction or that the glial reaction in the two nuclei differs sufficiently to affect the staining properties of the axons.

Electron-Microscopical Observations

a) Degenerative Changes in Synaptic Knobs

From the material presented in the previous sections it can be concluded that in many parts of the brain light-microscopical methods can not give clear evidence about synaptic relationships of degenerating pathways[10]. For such evidence the electron microscope must be used. The identification of synapses on electron micrographs has been discussed previously by Gray and Guillery (1966). Here, only the type of synaptic contact that shows a group of synaptic vesicles crowding close to a localized membrane thickening will be discussed, and this combination of features will be considered diagnostic of a synaptic junction[11].

In order to trace out the detailed patterns of synaptic interconnections that exist in particular pathways one can use either of two methods. The first is to identify an axon in the course of degeneration and describe its synaptic linkages; the second is to note the axonal elements that are missing when the degenerative process is complete. Only the former method, which has been more widely used and which gives results that are more easily interpreted will be considered here. Provided that one can identify the degenerative change and the synaptic junction in the same axonal process the method can give very precise information about neuronal linkages.

Two types of degenerative change have been generally recognized in axon terminals. The first is illustrated in Figs. 7 and 8, and has been described above in relation to the neurofibrillar hypertrophy that occurs in degenerating boutons. In many parts of the brain this change is relatively slow. The increase in filaments (or the new formation of filaments) that occurs in the synaptic knobs during the first three to seven days leaves many of the knobs almost or entirely filled with a dense tangle of filaments and there is no difficulty about distinguishing these degenerating axons from normal ones. Although there is a decrease in the number of synaptic vesicles during this time the surviving vesicles tend to crowd towards the synaptic thickening, so that the features characteristic of a synaptic junction can be recognized readily. This type of change has been described in the avian optic tectum (Gray and Hamlyn, 1962), the lateral geniculate nucleus of the monkey (Colonnier and Guillery, 1964; Glees *et al.*, 1966) and of the cat (Szentágothai *et al.*, 1966; Smith *et al.*, 1964), in the spinal cord of the cat (Ralston, 1968b), the cerebellar mossy fibers (Smith *et al.*, 1966) and the cerebello-vestibular fibers of the cat (Mugnaini and Walberg, 1967) and also in the efferent component of the auditory (Smith and Rasmussen, 1965) and optic nerves (Dowling and Cowan, 1966).

The second type of degeneration is seen more commonly than the first and the most noticeable change is an increased electron-opacity of the whole synaptic knob (see Figs. 16 and 17). The vesicles and the mitochondria form an ill-defined dark mass and even at an early stage one can no longer determine whether the vesicles are crowded towards the synaptic thickening. The presynaptic part of the thickening may also be difficult to see and one has to rely on the postsynaptic thickening (see

[10] Exceptions to this are found only where Golgi preparations show readily recognizable axon pattern in normal material, or where the synaptic knobs can be shown intimately related to a well defined postsynaptic structure.

[11] Electrically transmitting synapses which show a different type of junctional specialization (see Gray and Guillery, 1966) will not be considered in the present review, since there is no evidence available about the degenerative changes that occur at such junctions.

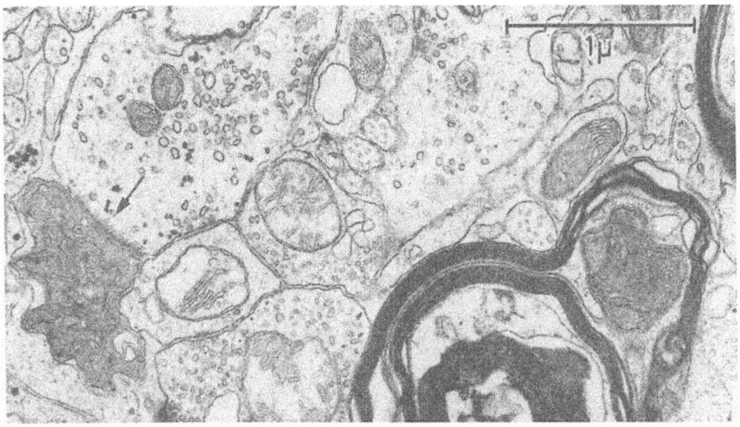

Fig. 16. Axonal degeneration in the dorsal lateral geniculate nucleus of the cat 6 days after unilateral section of the optic nerve. A degenerating, darkened synaptic knob is seen at the left of the figure. The synaptic thickening can be seen at the arrow. A degenerating myelinated fiber is seen in the lower right part of the figure. Note that its axoplasm shows the type of darkening that is also seen in the synaptic knob, while the myelin is forming a complex system of outpocketings

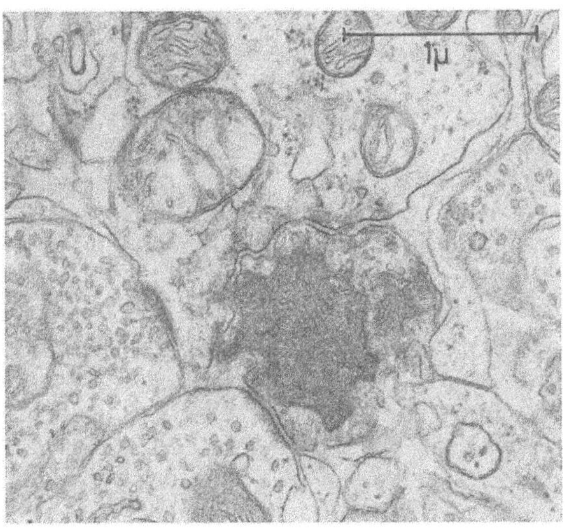

Fig. 17. A degenerating synaptic knob from the dorsal lateral geniculate nucleus of the cat obtained six days following section of an optic nerve. The darkening reaction is not far advanced and a few irregular and apparently shrunken vesicles can still be seen in the upper portion of the knob. However, the identification of this as a synaptic knob depends upon the presence of the synaptic thickening in the lower left part of the knob

Figs. 16 and 17) to identify the presumed synaptic site. It is important that the postsynaptic thickening be identifiable, for without it one is usually unable to describe any synaptic relationship[12]. Since axons undergo a similar darkening reaction along

[12] Lund (1969) has pointed out that the thickening opposite the degenerating axoplasm is not necessarily diagnostic of a synaptic junction. In regions where symmetrical desmosome-like contacts (see Gray and Guillery, 1966) are formed by axons, one of these contacts would be indistinguishable from a synapse, provided that the darkening reaction in the axon was reasonably advanced.

much of their course (see p. 102 and Figs. 16, 18) and since vesicles cannot be seen easily in the darkened knobs there is no other way of establishing the synaptic sites. This second type of degenerative reaction has been described in a great many parts of the brain (e.g. Colonnier and Gray, 1962; Walberg, 1963 and 1964; Colonnier, 1964; Ralston, 1965 and 1968b; Alksne *et al.*, 1966; Szentágothai *et al.*, 1966; Mc Mahan, 1967).

In some parts of the brain only the second type of degeneration (darkening) can be seen, and there is no sign of a filamentous reaction (Colonnier, 1964; Laatsch and Cowan, 1967; McMahan, 1967). In some regions the filamentous reaction pre-

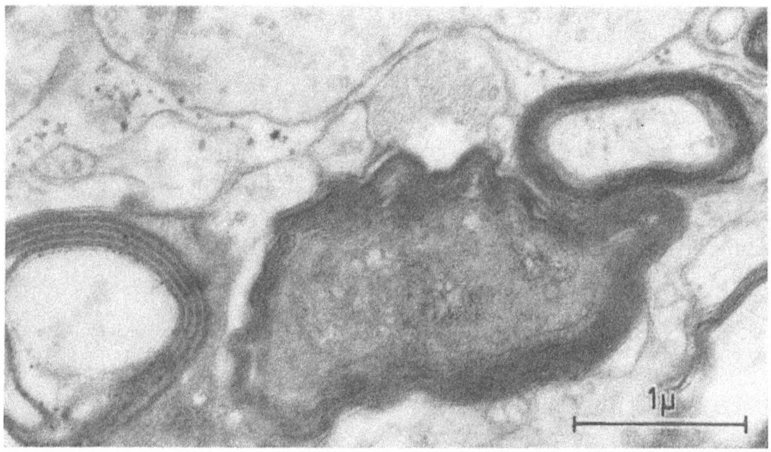

Fig. 18. Degenerating axons from the dorsal lateral geniculate nucleus of the cat obtained six days following section of an optic nerve. Notice the dark axoplasm in the center and the abnormal pattern of myelin lamellae at the left

dominates and the darkening has either not been seen at all or only to a small extent (Gray and Hamlyn, 1962; Colonnier and Guillery, 1964; Glees *et al.*, 1966; Dowling and Cowan, 1966). Finally, there are regions where both types of reaction occur concurrently (Smith and Rasmussen, 1965; Szentágothai *et al.*, 1966; Smith *et al.*, 1966; Mugnaini and Walberg, 1967; Ralston, 1969). Szentágothai *et al.* (1966) have suggested that the filamentous reaction is merely an early stage of the darkening reaction and that the latter should be regarded as a more reliable sign of axon degeneration. Ralston (1969) finds that in the ventrobasal complex the filamentous reaction occurs very early and is over when the darkening reaction is the dominant change, and Lund (1969) has described a similar relationship in the superior colliculus of the rat. It is possible that some axons which go through the filamentous reaction subsequently become dense dark masses (see e.g. Mugnaini and Walberg, 1967). However, such a sequence has not been clearly demonstrated in any part of the brain and it is necessary to recognize that the two types of degeneration described above are quite distinct and readily recognizable forms. Thus, axons which show the filamentous reaction lose most of their vesicles before they finally disintegrate, while the darkening that occurs in synaptic knobs generally appears to involve the vesicles *in situ.* Also, there are some regions where only one type of degeneration is seen. The rate at which the filamentous change appears is variable. In some systems it is seen as an early fleeting change (Szentágothai *et al.*, 1966; Ralston, 1969; Lund, 1969) but in other systems it is the dominant or only recognizable change for several days.

It has been stated above that the filamentous reaction is seen with the light microscope as a neurofibrillar hypertrophy. In contrast, the darkening of the synaptic knobs has been described in regions where the Nauta or Fink-Heimer methods stain degenerating axons and where terminal neurofibrils cannot be seen. Heimer and Peters (1968) have shown that the Fink-Heimer methods deposit some silver in the dark degenerating axons and it would appear that the two types of degeneration that can be recognized with the electron microscope are also recognizable in terms of their reaction to light microscopical staining methods.

The light-microscopical observations that have been reviewed suggest that there may be at least three types of axon degeneration. That is, some axons can be stained by the Nauta methods, some cannot be stained by these methods but can be shown by the neurofibrils that they contain, and there are some which are not demonstrable in either of these two ways but which are stained by the Fink-Heimer methods.

Electron-microscopical observations suggest that, in addition to the darkening and the filamentous reaction described above, there may be a third type of reaction. Thus, de Robertis (1958), Taxi (1962), Birks *et al.* (1960) and Hunt and Nelson (1965) described a clumping of the synaptic vesicles in the axon terminals without an increased electron-opacity, and Smith *et al.* (1966) also described a similar change in some of the cerebellar mossy fibers.

It is tempting to consider that the three types of change distinguishable by light microscopy correspond to the three types seen with the electron microscope. However, it seems more probable that the Nauta and Fink-Heimer methods both stain axons that are undergoing the darkening reaction, and that there may be two chemically distinct forms of this reaction. Thus, in the dorsal lateral geniculate nucleus of the cat the Nauta and Fink-Heimer methods show the same degenerative pattern and the predominant change seen here with the electron microscope is the darkening reaction. One would then be forced to conclude that, at present, there is no specific light-microscopical method for demonstrating the clumping of the vesicles.

b) Wallerian Degeneration in Fiber Bundles and Fibers of Passage

Wallerian degeneration in non-synaptic parts of nerve fibers involves two components: the axon and the myelin. The Marchi method stains the degenerating myelin and is a useful method for tracing myelinated pathways (see Jakob, 1912; Beresford, 1966). However, at present there are no electron-microscopical studies which show whether the Marchi reaction occurs upon any particular morphologically recognizable degeneration product or whether the artefactual deposit that so often occurs in Marchi preparations is related to any one type of structure. The reasons why the Marchi reaction is not the same in all parts of the central nervous system (see Swank and Davenport, 1934) are not fully understood.

Although the Marchi method can show degeneration within three days of axon section (Combs, 1951; Glees, 1948) and electron micrographs show clear signs of axonal degeneration at this stage (see McMahan, 1967; Mugnaini and Walberg, 1967; Bignami and Ralston, 1968) the myelin generally degenerates more slowly than the axon. During degeneration the myelin forms irregular lamellar patterns (see Fig. 18) and commonly forms irregular sheets and outpocketings (Fig. 16). In the central nervous system the myelin does not usually show the rapid disintegration that is seen peripherally. Some reports indicate that in the central nervous system myelin which shows no clear sign of degeneration may still be found surrounding axonal debris (Walberg, 1965) or even a glial cell (Gray and Hamlyn, 1962) after relatively long survival periods (112 and 30 days respectively). Whether this myelin has survived for such long periods in a more or less normal condition or whether it is newly formed in the region of axonal degeneration (see e.g. Schultz and Pease, 1959; Colonnier and Guillery, 1964; McMahan, 1967; Bignami and Ralston, 1968) remains to be determined. The results of such a determination would clearly be relevant to interpreting the light microscopical findings on rates at which the myelin degenerates (see below).

The evidence that is available about axonal degeneration between the lesion and the terminals will be considered only briefly here. Most commonly, the changes that have been described in electron micrographs include a darkening of the axoplasm similar to the darkening seen in many synaptic knobs (see e.g. Walberg, 1965; McMahan, 1967; Lampert, 1967; Ralston, 1968b; and see Fig. 18). It has been pointed out already that for this reason darkened profiles cannot be regarded as synaptic unless a well-defined postsynaptic thickening is shown. Further, one might anticipate that the darkened axoplasm of non-terminal axon segments will show the same reaction to light-microscopical staining methods as do the darkened terminals, and the observations of Heimer and Peters (1968) suggest that in some parts of the brain this is so.

Rates of Axonal Degeneration

Knowledge of the rates at which nerve fibers degenerate is of some interest for those who wish to choose the best survival period for staining degenerating fibers in a particular tract. It is also of interest for deciding how the temporal patterns of degeneration that one sees should be interpreted in terms of fiber connections. For example, there are reasons for thinking that the axonal degeneration occurring between a cut and the axon terminals precedes the axonal degeneration that may occur between the cut and the perikaryon (van Gehuchten, 1903; Powell and Cowan, 1964; Guillery, 1967b).

Most light-microscopical observations agree in showing that coarse fibers degenerate more rapidly than fine fibers, but recent electron-microscopical observations have suggested that some fine axons may degenerate early. Since it has been shown in the previous sections that there are at least two types of terminal degeneration, which occur at different rates and which appear to be independent of fiber caliber, the problems related to rates of axonal degeneration merit brief consideration.

Cajal (1928) stated that the thickest fibers degenerate first, being beaded already after thirty hours and largely disintegrated by the fourth to eighth day. He found that the medium-sized fibers were in a beaded stage after seven days and became disrupted between the fourteenth and eighteenth days. The fine fibers were preserved for as long as eight or ten days before undergoing degenerative changes. Guillery (1959 and 1961) has reported that with the Nauta method many fine fibers also show no clear sign of degeneration until eight or ten days have passed, although other fine fibers degenerate earlier (see Guillery, 1967b).

Rates of degeneration in the optic nerve and pyramidal tract were studied by van Crevel and Verhaart (1963a and b) on the basis of transverse sections stained with the Häggquist method. The absolute rates of degeneration that they give are not readily compared with those given by Cajal, since a fiber cut in longitudinal section can be recognized as degenerating at a stage when the majority of transverse sections through the same fiber would appear normal. However, the relative rates that van Crevel and Verhaart give agree with Cajal's observations in showing that the coarse fibers degenerate first. Further, van Crevel and Verhaart found that the debris of the fine fibers was cleared first and the coarser debris of the thick fibers was cleared last. A similar result was reported in the peripheral nervous system by van Gehuchten and Molhant (1910).

Several electron-microscopical observations have shown that some fine fibers degenerate earlier than coarse fibers (Mugnaini and Walberg, 1967; McMahan, 1967; Ralston, 1968b). However, this sequence has been reported in gray matter close to the fiber terminations, not in fiber tracts. It is thus possible that the fine, rapidly degenerating axon segments represent the earliest change in fine terminal ramifications of relatively coarse fibers. More detailed studies of rates of degeneration in axon bundles and in terminal regions may demonstrate that all coarse axons degenerate more rapidly than fine axons and that the terminal segments degenerate before the rest of the cut fiber (see Lubińska, 1964). However, such a simple generalization may not prove possible. The existence of at least two quite distinct types of terminal degeneration suggests that factors other than fiber caliber may determine rates of degeneration (see also Ebbesson's chapter in this Volume).

Summary and Conclusions

It can be concluded that, in general, the light microscopical methods cannot be used to show the synaptic relations of degenerating axons. This is possible in only a limited number of regions, where the axon terminals can be clearly recognized by their characteristic morphology and where their special relationship to well defined postsynaptic structures can be demonstrated. Electron micrographs allow a more reliable identification of synaptic relationships, but even these present difficulties when degenerating axons are studied, for the features that characterize a synaptic junction are often lost at an early stage of the degenerative process.

Both light and electron microscopical methods allow a ready identification of degenerating axons. However, with each method it is necessary to know the extent to which the normal axonal structure varies. Only those individual elements that are clearly beyond the normal range, or those regions in which an unusual number of axons lie at one end of the range can be regarded as degenerating.

Light and electron microscopical methods demonstrate that the degenerative reaction of axons is not the same in all fiber systems. The various components of the axoplasm undergo different alterations and the changes occur at different rates. It has been argued that two distinct types of change can be clearly recognized and that there is some evidence for other types. Although the morphological methods show only a few distinct types of degeneration the possibility that more types occur cannot be excluded. In this review the morphological observations have not been related to biochemical or histochemical studies of axon degeneration. Only limited comparisons are possible at present but one can expect that an understanding of the chemical changes that occur during degeneration will clarify the variety of morphological changes that can be seen.

Acknowledgements

This research was supported by Grant NB 06662 from the National Institute of Neurological Diseases and Blindness, National Institutes of Health. The skillful technical assistance given by Mrs. E. Langer and Mrs. B. Yelk is gratefully acknowledged.

Discussion

Dr. Peters: With reference to Dr. Heimer's paper, it would seem important to emphasize that Dr. Heimer and I used significantly more dilute concentrations of silver in order to have as little "background" impregnation as possible.

Dr. Heimer: In fact, our aim was to put as little silver in the degenerating boutons as possible and still have it contrast with the background. We hoped that this would allow us to locate the locus of the silver deposit most accurately.

Dr. Guillery: I find it amazing that the amount of silver you show in your electron micrographs could have given you a light-micrographical picture which showed a recognizable deposit of silver. If you compare the amount that you have with the amount that we had (Guillery and Ralston, 1964), there is a tremendous difference.

Dr. Heimer: I think that we, too, had a considerable amount of silver in the degenerating axoplasm. One should remember that we are dealing with ultrathin sections, and if one were to put several such sections on top of each other, one would indeed observe a much more abundant accumulation of silver in the bouton than would seem to be there when studying ultrathin sections.

Dr. Cowan: But still there seems to be considerably less than Guillery and Ralston saw in their thin sections.

Dr. Grant: Is it possible that the silver in the tissue may be redistributed to other structures as a result of the osmium treatment?

Dr. Guillery: This is a point that Lund and Westrum (1966) made. In fact they gold-toned their Nauta sections before osmium treatment. They found that without this the silver was shifted. We did not osmicate our sections at all, but treated them in PTA. Dr. Heimer, I wonder what your experience was here?

Dr. Heimer: We did not use the Lund-Westrum procedure and took no precautions in this respect. I think the relevant point is that we found considerably more silver in the degenerating axons than in normal tissue components.

Dr. Guillery: It is important to know if some silver was lost, important particularly in the context of Dr. Moliner's question. Did you say that you never saw a degenerating terminal without silver?

Dr. Peters: Yes. One further reason we used very dilute silver solutions was that we were afraid too much silver in the tissue would result in disintegration of the section through the heat generated by the electron beam. I am sure that if we had used higher silver concentrations we would have gotten much more silver granules than we showed here.

Dr. Heimer: I think that both Dr. Guillery and I are happy that we have the electron microscope when we use the silver impregnation methods. However, I am not altogether pleased with Dr. Guillery's idea that we should classify the boutons according to the light microscopic technique we use. What I think Dr. Guillery means is that when we see axonal swellings in Golgi preparations or in methylene blue preparations, we can regard them as a special kind of synaptic end-structures. I think you have called them synaptic knobs. If we use the so-called mitochondrial methods, including, for instance, Rasmussen's methods, we call them end-feet; and if you use the so-called neurofibrillar stains, we call them boutons. The consequences of this could very well be that boutons stained with the Glees method would be called neurofibrillar boutons. Conversely boutons stained with the Nauta-Gygax modifications would be non-neurofibrillar. I think that is what is implied in this classification. Degenerating end-structures can be demonstrated in the prepyriform cortex, after olfactory bulb removal, by the original Nauta procedure, the Guillery paraffin method, or the Glees method. Yet, Westrum tells me he has not found any electron-microscopic evidence of a neurofibrillar type degeneration in this cortex. Using any of the silver methods, be it the Nauta-Gygax or the Glees method, it is clearly evident that the techniques often require modification in order to produce optimal results. It is very similar to cooking: the real goodies are often difficult to produce. Since there appear to be as many modifications as there are neuroanatomical laboratories, it would seem risky to try to classify the boutons from the point of view provided by the method one is using. From an historical point of view, I would like to point out that when it was claimed that the Glees method stained degenerating synaptic structures and the Nauta-Gygax method only the preterminal fibers, many of us began to identify degeneration products in strict adherence to these pronouncements. When the Glees method was used, we identified degenerating terminals or terminal degeneration, and when studying Nauta-Gygax preparations, we talked about preterminal degeneration. And then Dr. Guillery, among others, showed that this was not correct and that the Nauta-Gygax method, too, can impregnate degenerating synaptic terminals. We should be wary of further pitfalls of this sort when trying to classify synaptic end-structures according to the silver impregnation technique used to demonstrate them.

Dr. Blackstad: In connection with this discussion about electron-microscopic studies of degenerating axons, an interesting method which deals with myelinated fibers and not with terminal structures may be mentioned. We know that we can determine a caliber spectrum for a given fascicle of fibers, which of course, is of some interest to the physiologist who relates it to conduction time and so on. Now it is feasible to obtain a fiber spectrum of a fascicle intermingled with other tracts in the white matter, by measuring degenerating axons which appear dark within their myelin sheaths. As far as I know, there is only one paper on this topic, and that was Nafstad's (1967) study on the entorhino-hippocampal pathway which he thus characterized in terms of fiber spectrum. He also made control studies to show that there was no identifiable shrinkage of these degenerating fibers in animals which had survived section of the hippocampal commissures for 3—4 days.

Dr. Guillery: Could I make one brief comment about this. Is there any information available about where the darkening appears first? When a fiber degenerates and becomes beaded this is relevant to the alterations that you describe. That is, is the darkened piece between the beads and thus, perhaps, thinner than normal or does it represent the beads so that it is thicker than normal? Have you looked at longitudinal profiles in the sections?

Dr. Blackstad: No, I think that one must describe Nafstad's paper as a pilot attempt. His observations were made at a very early stage when there was as yet little obvious beading. I do not think, however, that this would invalidate his observations.

Dr. Guillery: I would like to come back to Dr. Heimer's comment. According to the terminology that Gray and I suggested (Gray and Guillery, 1966) it is necessary that one recognize the fibrillar nature of a structure before one labels it "bouton". There are no "nonfibrillar boutons" in this terminology. The degenerating debris that can be demonstrated in the prepyriform cortex by the various silver techniques may or may not represent degenerating synaptic knobs; there clearly are no fibrillar boutons amongst this debris. Westrum's evidence would support this interpretation of the light micrographs[13].

There are many regions where light-microscopical preparations do not allow one to decide whether a particular, densely argyrophilic structure is fibrillar or not. The problem arises more commonly in degenerating material than it does in normal material. Even in the lateral geniculate nucleus of the monkey, where many of the degenerating retinogeniculate axons show clear fibrillar boutons, there are many large, dark masses that cannot, on the basis of light-microscopical examination alone, be identified as boutons. The electron microscope shows that the dark masses do represent synaptic knobs filled with filaments.

There can be little doubt that different methods stain different components of the axoplasm. If it were not so, all methods would give the same result.

---

[13] (Note added in proof): The same point can be made about Walberg and Mugnaini's (1939) criticism that "the presence of neurofilaments does not appear to be a necessary prerequisite for a Glees impregnation of ... boutons".

Selective Silver-Impregnation of Degenerating Axoplasm

LENNART HEIMER

Introduction

Rapid advances in the neurological sciences require an increasingly detailed knowledge of the neuron circuitry of the central nervous system, in general, and of synaptic functions in particular. It is therefore fortunate that in many fiber systems, at least, the ultrastructural features of axon degeneration are easy to recognize. As a consequence, the electron microscope has become an extremely useful tool for a detailed study of the synaptic relationships of central nervous pathways in experimental material (see review article by Alksne *et al.*, 1966). The fact that degenerating axons and axon-terminals can be identified in the electron microscope, however, has by no means made the light-microscope techniques for tracing fiber tracts obsolete. The synoptic dimension essential for a correct understanding of the topographic and quantitative relationships between different nervous structures can be attained only at the light microscopic level, and a correct sampling of tissue for electron-microscopic examination of synaptic details can often be achieved only after a previous mapping of the fiber system by light-microscopic techniques. The light microscope will therefore continue to be of fundamental value in the study of the neuron circuitry of the central nervous system, regardless of whether attention is focussed on the general topography of a fiber system or on its area of distribution.

The course and termination of long nerve fibers are determined most reliably by studying the anterograde degeneration induced by experimental or pathological lesions of nerve fibers or their parent cell body. This degenerative process, also known as Wallerian degeneration after Augustus Waller who in 1850 showed that section of a nerve led to fiber degeneration throughout the amputated distal segment of the nerve, formed the basis of the Marchi method (Marchi and Algeri, 1885), the first histological technique to permit the experimental tracing of fiber tracts in the central nervous system. The Marchi technique, however, stains degenerated myelin, and therefore has been of little or no use for the tracing of thinly myelinated and unmyelinated fibers, including the terminal arborizations of all fiber categories. A more accurate experimental determination of the distribution areas of fiber systems had to await the development and subsequent refinement of the reduced-silver impregnation methods which stain axoplasm rather than myelin sheaths.

The Reduced Silver Impregnation Methods of Cajal and Bielschowsky

Silver was used in histological work as early as the middle of the nineteenth century (Krause, 1844; Recklinghausen, 1860) and Fromman first used it for the impregnation of nerve fibers in 1864. However, reduced-silver methods were not widely and systematically employed until after the development of greatly improved silver impregnation techniques by Cajal and Bielshowsky early in the present century. During the past two decades, modifications of the Cajal and Bielschowsky processes have come into widespread use in the experimental tracing of degenerating central nervous fiber connections.

Around the turn of this century the nature and function of the so-called neurofibrils intrigued anatomists and physiologists alike. Cajal's intensive search for a reliable impregnation method for these filamentous structures resulted, first in a short technical note on the silver impregnation of axons (1900), and afterwards in the development, in 1904, of three impregnation formulae, differing from each other by the use of different fixation procedures. The impregnation itself was the same in all formulae and consisted in the impregnation of tissue blocks for several days in a silver nitrate solution at 30—35° C, followed by reduction in formalin-hydroquinone or formalin-pyrogallol.

After his first disclosure in 1902 of an ammoniacal silver method reminiscent of a technique described earlier by Fajersztajn (1901), Bielschowsky in 1904 published his well-known paper entitled: "Die Silberimprägnation der Neurofibrillen" in which he described in great detail his silver impregnation procedures for both frozen sections and whole tissue blocks. In the Bielschowsky method the tissue, after an initial impregnation in silver nitrate, is transferred to an ammoniacal silver nitrate solution and finally reduced in dilute formaldehyde. This "double-impregnation" procedure of Bielschowsky has been the immediate forerunner of most of the modern experimental silver methods.

For several decades the reduced silver techniques were used almost exclusively for the tracing of normal fiber tracts, and were employed only occasionally for the study of degenerative changes in nerve fibers and synaptic axon endings. Although Cattaneo as early as 1923 used silver impregnation in tracing the degenerating retinal projections in rabbits and birds, Hoff (1932a and b) is generally given credit for the first use of the Cajal block impregnation method in a systematic experimental study of interneuronal connections. Despite Cattaneo's and Hoff's promising examples, reduced silver techniques did not come into wide use for the tracing of experimentally induced axon degeneration until about fifteen years later. This delay of popularity is undoubtedly attributable in large part to the difficulties in standardizing and controlling metallic impregnation of biological material. By sustained efforts, however, more reliable and sensitive silver impregnation methods gradually became available, and at present it is possible to trace relatively fine-fibered conduction systems and delineate their synaptic fields in most areas of the central nervous system with a high degree of accuracy.

Current Experimental Silver Methods

The Glees Method

Although numerous silver techniques have been described for different embedding media, the most commonly used methods during the last two decades have been those of Glees and Nauta-Gygax, both of which were originally described for use on frozen sections. The Glees method (1946) has been used successfully especially in studies of spinal cord, brainstem and cerebellar connections (e.g., by Glees and

Zander, 1950; Glees *et al.*, 1951; and by the Oslo school: Brodal, 1957; Walberg, 1956; Grant, 1962; Nyberg-Hansen, 1965, to choose a few examples at random), but also in the study of several forebrain systems such as the central olfactory (LeGros, Clark and Meyer, 1947; Meyer and Allison, 1949; Adey, 1953; Allison, 1953) and visual connections (Campos-Ortega and Glees, 1967a, b; Glees, 1961) of various mammals.

Although the Glees method — adapted for use also on paraffin sections (Marsland *et al.*, 1954) — has long been considered the method of choice for the study of degenerating boutons, it generally has been less successful in areas in which a majority of boutons terminaux, according to later electron microscopic observations, lack neurofibrils. The Glees method has been of limited use, for example, in the study of terminal degeneration in the cerebral cortex (e.g. Glees, Meyer and Meyer, 1946), as well as in most other forebrain regions. The fact that the Glees method has a preference for impregnating neurofibrils does not necessarily warrant the conclusion that all boutons seen with the method contain neurofibrils (Heimer, 1967; Walberg, 1964; Walberg and Mugnaini, 1969). The mechanisms underlying silver impregnation appear to be quite labile, and the intracellular attraction sites are difficult to determine and may vary according to the mode of fixation, length of survival time, and other factors. The rather ambiguous term "neurofibrillar method" should therefore be used with reserve, and be given only tentative classifying value with respect to synaptic endstructures.

The use of the Glees method for the study of terminal degeneration[1] in the hypothalamus (Adey and Meyer, 1952; Wall, Glees and Fulton, 1951) was criticized by Cowan and Powell (1956), who reported evidence that the hypothalamus and some neighboring regions in apparently normal monkeys may contain Glees-positive structures simulating experimentally induced terminal degeneration. A similar form of pseudo-degeneration was described in Glees preparations of the cerebellum by Brodal and Grant (1962). On the other hand, Adey *et al.* (1958) in a subsequent study failed to confirm the pseudodegenerative phenomena described by Cowan and Powell. This controversy clearly exemplifies the difficulties inherent in studies of synaptic endstructures by the exclusive aid of the light microscope. In certain cases at least, it may be advisable to obtain electron-microscopic verification of light-microscopic evidence suggesting synaptic relationships. The specific problems encountered in the interpretation of silver-impregnated experimental material will be discussed more fully later.

Although the Glees method has produced valuable results in the hands of experienced investigators, it has not come into general use in more than a few laboratories. One reason for this is undoubtedly the circumstance that the Glees method has so far been effective only in certain regions of the central nervous system. The successful application of any silver method, however, often requires persistent experimentation in order to adjust the procedure to locally variable requirements, including the matter of optimal post-surgical survival time. The optimal survival time for degenerating synaptic endstructures, for instance, is often considerably shorter than that for the degenerating parent axons. It therefore is quite possible that the potentialities of the Glees method as yet have not been fully exploited.

[1] The term, terminal degeneration, is used to denote Wallerian degeneration in both the terminal axon arborizations and their synaptic endings (for a discussion on this subject, see Heimer, 1967).

There is, however, another circumstance that restricts the usefulness of the Glees method. The heavy concomitant impregnation of normal axons and axon terminals tends to obscure the degenerating axons and their terminals, especially in more diffusely organized fiber systems. These difficulties of interpretation, inherent in all of the fully-impregnating silver methods became obvious already in the first systematic studies of silver-impregnated normal boutons terminaux (Bodian, 1937; Phalen and Davenport, 1937; Minckler, 1940), performed following the pioneering attempts to identify degenerating boutons in experimental specimens (Hoff, 1932; Gibson, 1937). The search for a process that could selectively impregnate the degenerating axons and axon terminals, at the exclusion of normal nerve fibers, thus became a problem of high priority which found its solution in the development of the so-called "suppressive" Nauta-Gygax method.

The Nauta-Gygax Methods

In 1950, W. J. H. Nauta published a paper in the Schweizerisches Archiv für Neurologie und Psychiatrie in which he described a Bielschowsky modification capable of impregnating the terminal degeneration in such notoriously difficult regions as the neocortex and the hypothalamus of the rat. This method, often referred to as the "original Nauta procedure" and later described in more detail by Nauta and Gygax (1951), consists of a treatment of frozen sections of formalin-fixed material in ammonia-alcohol before impregnation in a silver-pyridine solution, followed by an ammoniacal silver nitrate solution and reduction in slightly alcoholic dilute formaldehyde acidified with citric acid. In spite of the great advantages of the original Nauta procedure for the demonstration of terminal degeneration, the method has been used by only a few investigators (Blackstad, 1956; Torvik, 1956; Lohman, 1963; Raisman *et al.*, 1965, 1966; White, 1965). Although the original Nauta procedure was described for use on frozen sections, it has been applied also to paraffin-embedded material (e.g., Blackstad, 1958; Guillery *et al.*, 1961; Lohman, 1963; Raisman *et al.*, 1965). The general experience, however, appears to be that silver impregnation of terminal degeneration is more inconsistent and somewhat less sensitive in paraffin material than in frozen sections. Moreover, the main advantages of paraffin-embedding can also be obtained by encasing the formalin-fixed tissue in albumin or albumin-gelatin prior to sectioning on the freezing microtome (see chapter by Ebbesson in this Volume).

As is the case with the Glees method, the study of sections stained by the original Nauta technique is often complicated by heavy impregnation of normal fibers. The development of the so-called "suppressive" Nauta-Gygax method (Nauta and Ryan, 1952; Nauta and Gygax, 1954), which affords a selective suppression of the argyrophilia of normal axons, therefore marks the beginning of a new chapter in the history of silver impregnation. The suppression of normal-fiber impregnation, achieved by a pretreatment of the sections in a sequence of phosphomolybdic acid and potassium permanganate, greatly facilitates the identification and tracing of degenerating fibers. Later modifications of the suppressive Nauta-Gygax method include the introduction of Laidlaw's ammoniacal silver carbonate solution instead of the ammoniacal silver nitrate solution (Chambers, Liu and Liu, 1956), and the replacement of phosphomolybdic acid by uranyl nitrate (Nauta, unpublished). The advantages offered by the

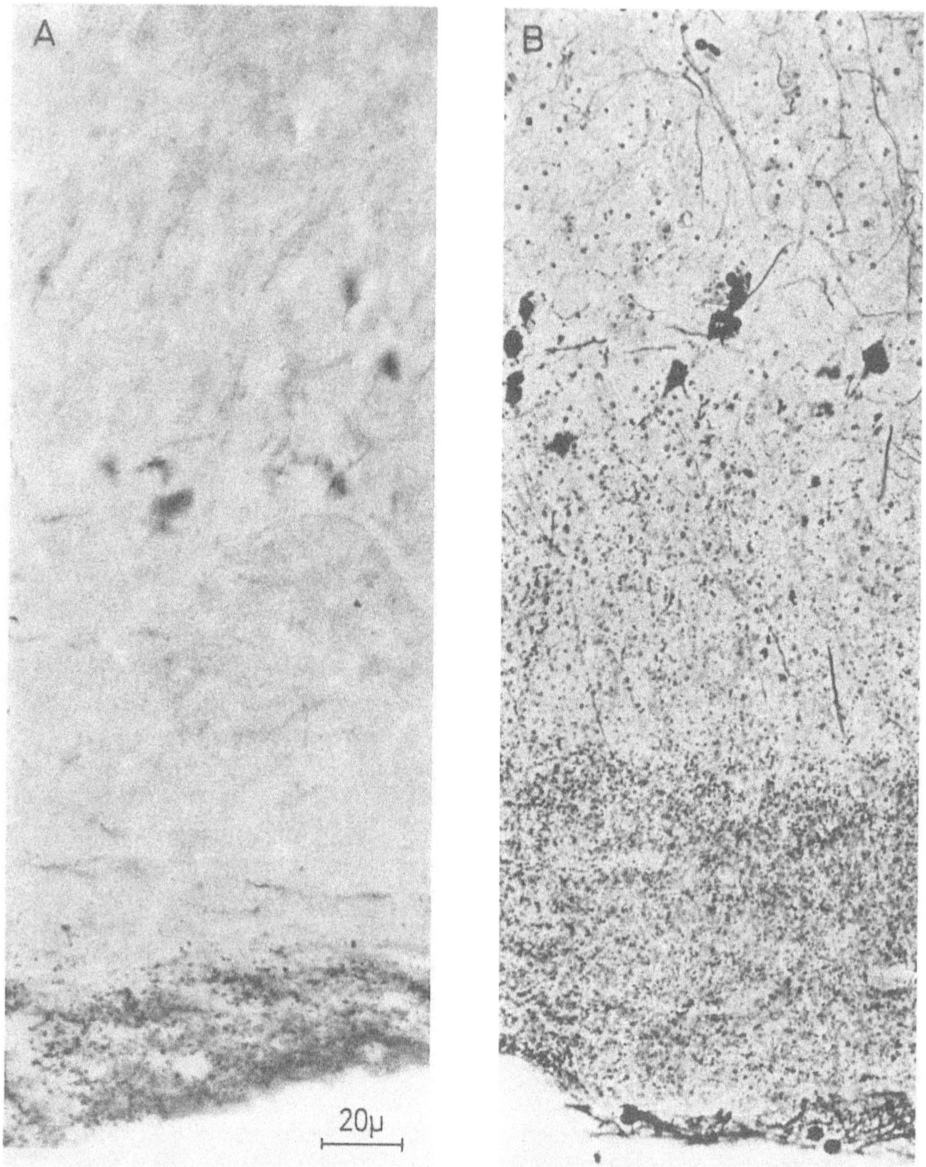

Fig. 1A and B. Degenerating olfactory tract fibers in the olfactory cortex of a rat whose ipsilateral olfactory bulb had been removed 5 days before sacrifice. Whereas preparations stained by the "suppressive" Nauta-Gygax method (A) show axon degeneration only in the most superficial part of the plexiform layer, a slightly modified original Nauta-method (no pretreatment; soaking in 100 parts 0.3% $AGNO_3$ + 4 parts pyridine, 18 hr.; ammoniacal silver nitrate, 2 min; reducer) demonstrates a dense terminal degeneration in the superficial half of the plexiform layer (B). Some additional degenerating olfactory bulb fibers are distributed to the deeper half of the plexiform layer as well as to the cell layers.

suppressive Nauta-Gygax method, especially for the tracing of axon trajectories, in the past decade have led to a veritable boom in the experimental tracing of central nervous pathways. The method, as well as the histological criteria for axon degeneration, were discussed by Nauta in 1957 (see also chapters 5 and 7 in this volume). Other critical evaluations of the use of the Glees and the Nauta-Gygax methods include those of Glees and Nauta (1955), Evans and Hamlyn (1956), Adey *et al.* (1958), Bowsher *et al.* (1960), and Walberg (1964).

It is important, however, to emphasize the limitations of the suppressive Nauta-Gygax method. Recent studies have shown that the advantage of selective staining of degenerating axons was obtained in this staining process at the expense of an incomplete visualization of the corresponding synaptic fields. Although the suppressive method can provide clear pictures of degenerative changes in the terminal axon arborizations of many fiber systems, it nevertheless may fail to show the quantity of termination, or even the full extent of the distribution area. This limitation is exemplified by Fig. 1 in which A and B represent results obtained by, respectively, the suppressive Nauta-Gygax method and the original non-suppressive Nauta procedure. Both figures show degeneration in the anterior part of the prepiriform cortex in the rat following removal of the ipsilateral olfactory bulb. Whereas the suppressive Nauta-Gygax method (A) only shows degeneration of relatively rectilinear fibers of passage in the superficial part of the plexiform layer, the original Nauta technique (White, 1965) — or a slight modification of the method (B) — clearly demonstrates a massive and well-defined laminar termination of olfactory bulb fibers in the olfactory cortex. An evaluation of the use of different Nauta-Gygax modifications in the study of some other fiber systems in the forebrain showed similar limitations of the suppressive Nauta-Gygax technique in the impregnation of terminal degeneration (Heimer, 1967).

Recent Modifications of the Nauta-Gygax Method

Difficulties encountered in the application of the Nauta-Gygax method have resulted in the publication of several modifications. A method described by Albrecht and Fernstrom (1959) as especially suitable for human nervous tissue has been used with success also in experimental animals, especially in long-fixed material. The Albrecht-Fernstrom procedure, which differs from the classic Nauta-Gygax method by a lesser selectivity for degenerating axons, was further modified by Sterling and Kuypers (1966) who succeeded in obtaining a simultaneous but differential impregnation of normal and degenerating boutons terminaux in the spinal cord of different mammals. Slight modifications of the Nauta-Gygax method have also been published by Nakamura (1963), Jacobson (1963), and Ebbesson and Rubinson (1969). Johnston and Bowsher (1969) have made an effort to combine the respective advantages of the Glees and the "suppressive" Nauta-Gygax methods in order to obtain a selective impregnation of degenerating "neurofibrillar" axon endings at the exclusion of normal boutons terminaux.

The mounting evidence that the suppressive Nauta-Gygax method in many fibers systems fails to demonstrate the full extent of synaptic distribution naturally prompted an intensive search for a staining method that would combine a selective impregnation of degenerating axons with a more nearly quantitative visualization of the corresponding axon terminals. In this laboratory such efforts have resulted in the development of two slightly different procedures (Fink and Heimer, 1967) which in studies of various fiber systems have permitted the identification of synaptic distributions which the suppressive Nauta-Gygax method had failed to disclose (Heimer, Ebner and Nauta, 1967; Heimer and Wall, 1968; Karten, 1968; Karten and Nauta, 1968; Arees and Mayer, 1968; Petras, 1968; Schneider, 1968; Lohman and Mentink, 1969; Heimer and Nauta, 1969; Hubel and Wiesel, 1969).

The main procedural variants introduced by these techniques are 1. the application of potassium permanganate without pretreatment of the sections with phosphomolybdic acid or uranyl nitrate, and 2. subsequent impregnation in a mixture of silver nitrate with uranyl nitrate (Procedure I) or pyridine (Procedure II). Of essential importance in both procedures is the use of an ammoniacal silver nitrate solution instead of Laidlaw's ammoniacal silver carbonate. Fig. 2 shows the terminal degeneration, impregnated according to Procedure I, in the superior colliculus of a hamster whose contralateral eye was removed 7 days before sacrifice. Especially noteworthy is the dense accumulation of impregnated, largely spheroidal particles in the superficial grey layer.

Procedure II prescribes a pretreatment of the sections in potassium permanganate and uranyl nitrate before impregnation in a weak silver or silver-pyridine solution. Although the application of the full sequence — potassium permanganate followed by uranyl nitrate — preceding the impregnation may make the staining procedure more hazardous, it affords more freedom to experiment with different degrees of pretreatment without a direct "competition" between the uranium and silver ions. The low-power photograph in Fig. 3 exemplifies the results that can be obtained by the use of Procedure II. It illustrates the distribution of the stria terminals in the medial hypothalamus of a rat which survived a nearly total destruction of the ipsilateral amygdaloid complex for 5 days. In the paramedian section (B), the fine degenerating fibers of the precommissural component appear grouped in several fascicles that can be followed caudalward through the medial preoptic-hypothalamic continuum to the infundibular region where the fibers disappear in a zone of dense terminal degeneration which occupies the relatively cell-poor zone surrounding the ventro-medial hypothalamic nucleus and exhibits the same morphological features that characterize terminal degeneration in sections stained following Procedure I.

Silver impregnation of degenerating axons and axon terminals is rightly considered a time-consuming and demanding task, and in spite of several attempts to elucidate the mechanism of silver impregnation, very little is known of the physicochemical principles involved in the selective deposition of reduced silver in partciular tissue components. It is therefore difficult to predict for each silver impregnation method how variations in, for example, impregnation time, or concentration and pH of the impregnating solutions, will affect the result. The outcome of changes in any one of these closely interdependent variables will depend not only on the impregnation procedure followed, but also on the physical and chemical properties of the tissue at the time of the impregnation. However, largely as a result of pragmatic experimentation, some guidelines at least have become available to the investigator attempting to cope with the more common vagaries of the methods.

The pretreatment in potassium-permanganate alone (Procedure I) or potassium permanganate followed by uranyl nitrate (Procedure II) is of essential importance

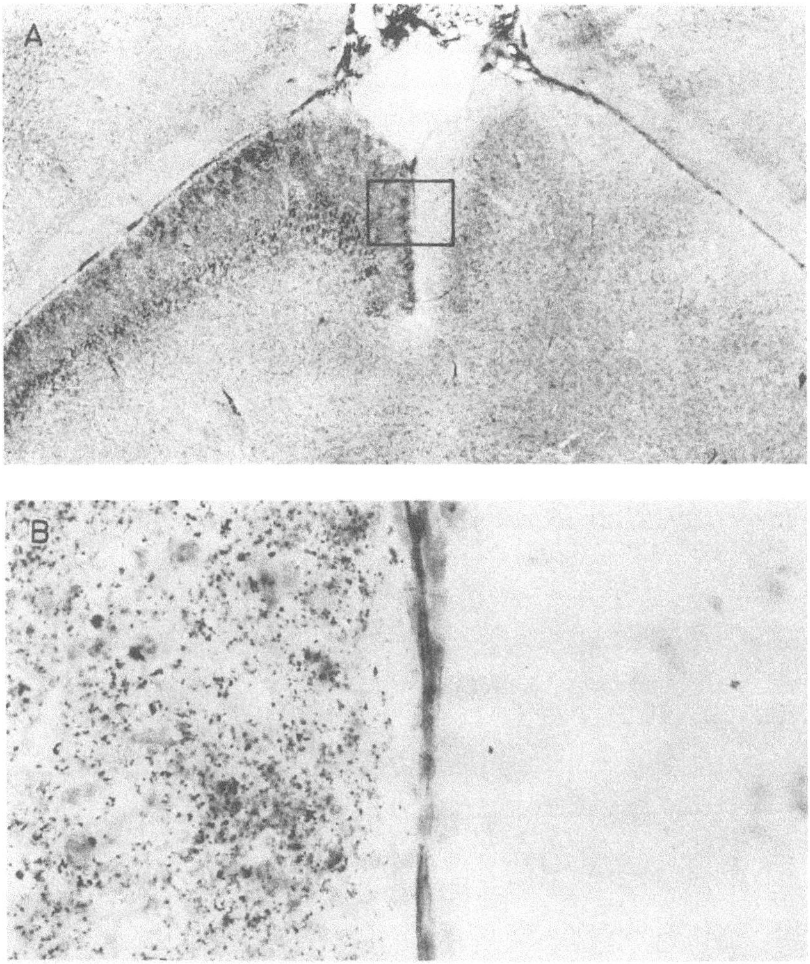

Fig. 2A and B. Massive terminal degeneration (A, left side) in the superficial grey layer of the superior colliculus in a hamster whose contralateral eye had been enucleated 7 days before sacrifice. The higher-power view shown in B and corresponding to the insert rectangle in A, shows the terminal degeneration in the form of an extremely dense accumulation of argyrophilic particles. The right side of the photograph demonstrates the absence of retinal-axon degeneration in the corresponding part of the colliculus ipsilateral to the enucleation. Although some fine degenerating fibers may be identified in the area of terminal degeneration, it is usually impossible to determine if a single argyrophilic particle represents a degenerating bouton or a terminal axon branch. Fink-Heimer method, Procedure I. (Courtesy of Dr. G. Schneider)

not only for the suppression of the argyrophilia of normal axons but also for the impregnation of degenerating axoplasm. Sometimes a prolonged pretreatment is necessary to avoid a disturbing impregnation of glia elements and other irrelevant tissue components; quite often such extended pretreatment is compatible with adequate impregnation of the terminal degeneration. In a study of the olfactory bulb projection in the rhesus monkey (Fig. 4), the best impregnations were obtained in sections which had been treated in 0.5% potassium permanganate for $^1/_2$ hour,

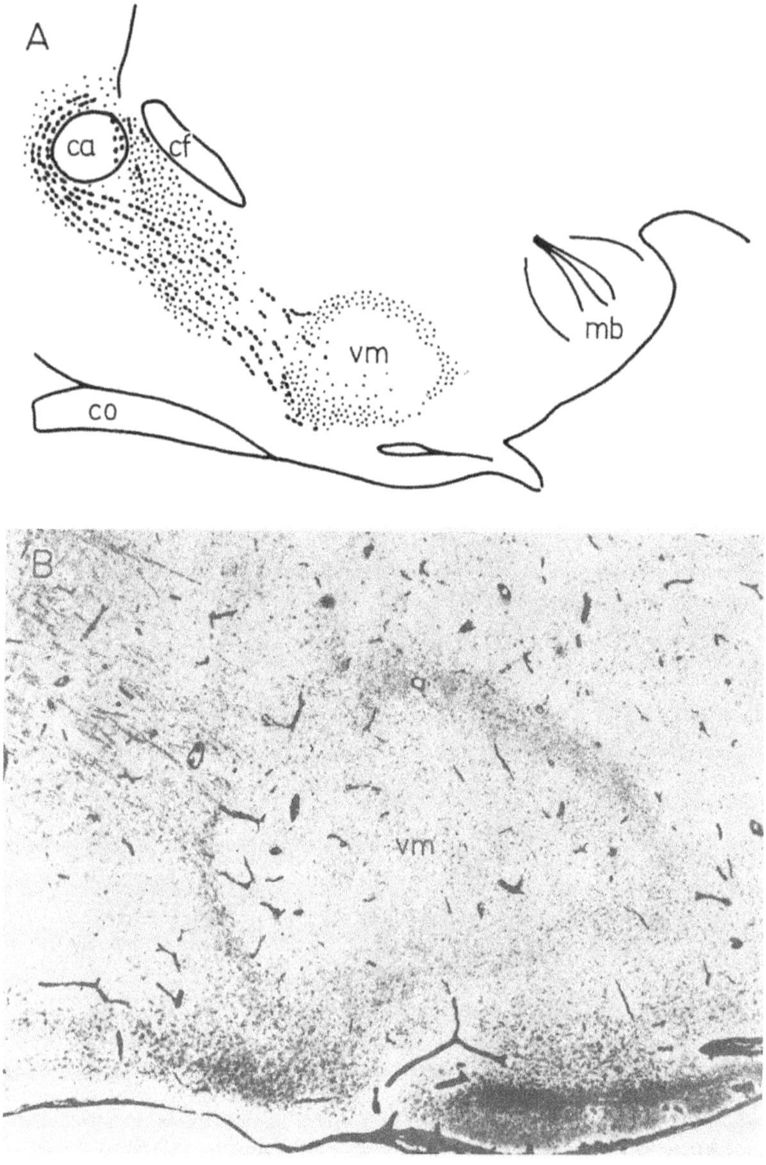

Fig. 3A and B. The upper half (A) of this figure shows the course and terminal distribution of stria terminalis fibers in the hypothalamus, charted from a single paramedian section. B: Low-power view of the tuberal region in a nearby sagittal section, showing zone of dense terminal degeneration surrounding the ventromedial nucleus. Compare Fig. 5. Fink-Heimer stain, Procedure II. Abbreviations: *ca* anterior commissure; *cf* fornix column; *co* optic chiasm; *mb* mammillary body; *vm* ventromedial nucleus. [Reproduced from Brain Res. **13**, 284—297 (1969), by courtesy of Elsevier, Amsterdam]

followed by decolorization and subsequent treatment in 5% uranyl nitrate for another $^1/_2$ hour before impregnation in silver. The argyrophilia of some fiber systems, however, shows a marked decrease even after a moderate prolongation of the

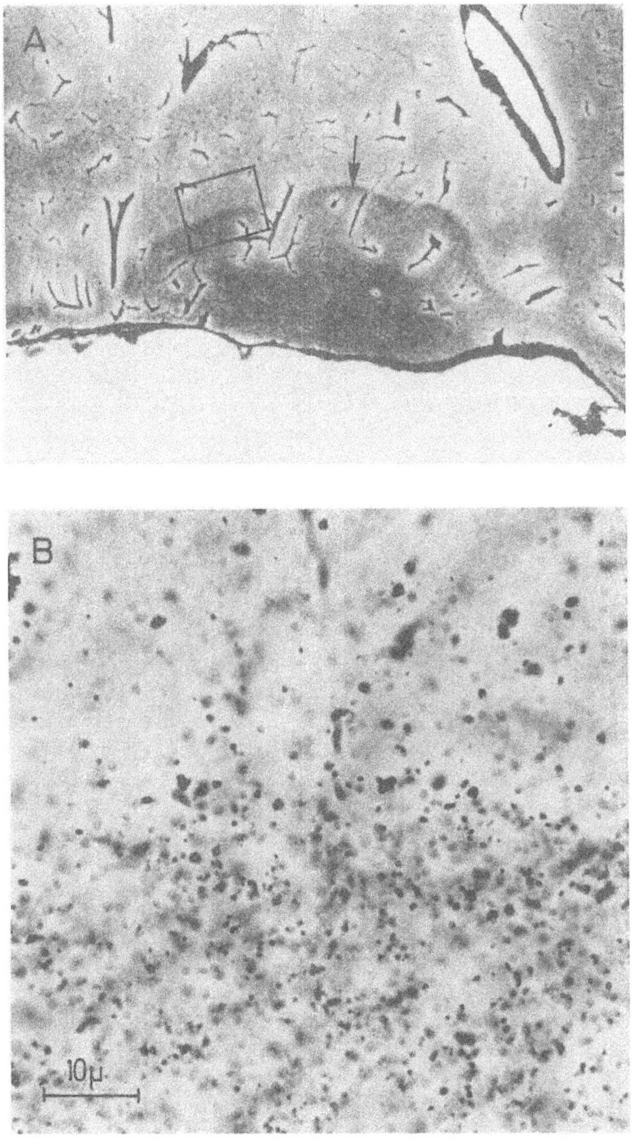

Fig. 4. A: Terminal degeneration in the olfactory cortex at the level of the olfactory tubercle in a rhesus monkey whose ipsilateral olfactory bulb had been removed 5 days before sacrifice. Extremely dense terminal degeneration appears in the superficial half of the plexiform layer; its deep border is indicated by the arrow. B: high-power microphotograph corresponding to the rectangle in A. Note sparser terminal degeneration in deeper half of plexiform layer. (0.5% $KMNO_4$, $^1/_2$ hr.; decolorization in oxalic acid-hydroquinone; 5% uranyl nitrate, $^1/_2$ hr.; soaking in: 0.3% silver nitrate, 100 parts, pyridine 4 parts, 24 hr.; ammoniacal silver nitrate, 3 min; reducer)

pretreatment. In some instances it was found necessary to delete the hydroquinone-oxalic acid treatment which follows the permanganate step. This was the case in an experimental study of the shark olfactory system (Ebbesson and Heimer, 1970), in

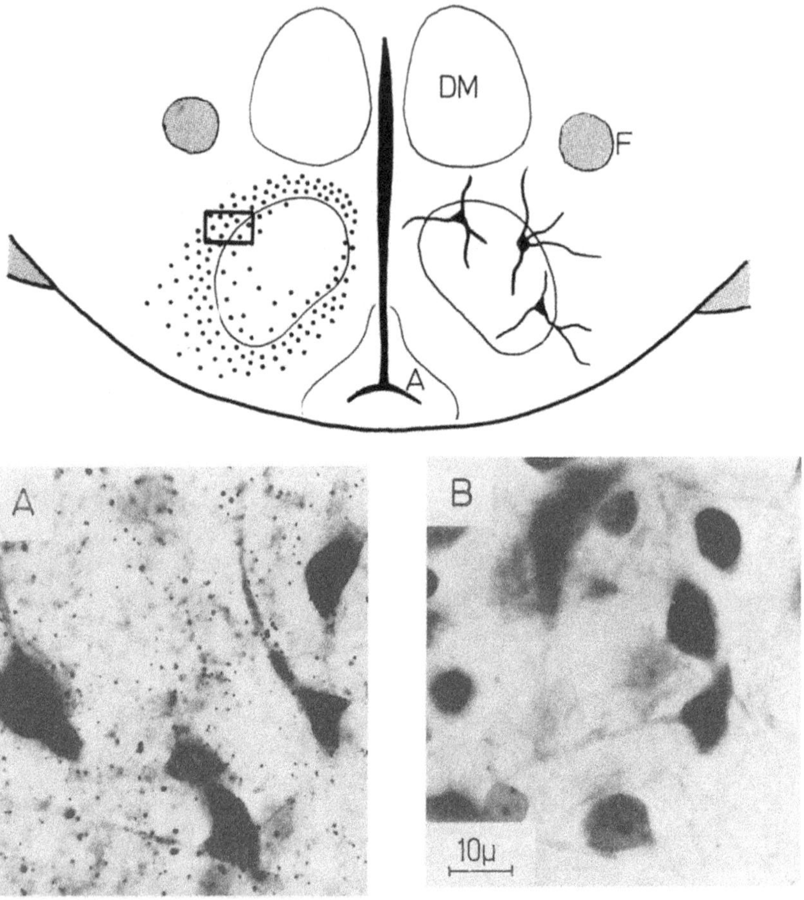

Fig. 5 A and B. Photograph A shows terminal degeneration near the lateral border of the ventromedial nucleus in a rat sacrified 4 days after ipsilateral amygdalectomy. Approximate position of the field shown is indicated by rectangular frame in upper drawing. B shows the corresponding field on the opposite side. The dense zone of stria terminalis endings surrounding the ventromedial nucleus is indicated by dots in the upper drawing. Since numerous dendrites are known to radiate from the nucleus into this zone of terminal degeneration (right side of upper drawing), the findings indicate that the connections of the stria terminalis with the ventromedial nucleus are largely of an axo-dendritic nature. Fink-Heimer stain, Procedure I, with impregnation in Solution A overnight, Solution B omitted. [Reproduced from Brain Res. **13**, 284—297 (1969), by permission of the Publishers]

which it was necessary to delete not only the bleaching procedure, but sometimes also the permanganate treatment in order to obtain a satisfactory impregnation of the degenerating afferents from the olfactory bulb.

Another factor of some importance is the silver-impregnation time. Although one to two hours impregnation seems to be adequate in most cases, 24—48 hours was found necessary in a study involving the rat's hypothalamus (Arees and Mayer, 1968) as well as in several of our own studies (Figs. 4 and 5). On the other hand, an extremely short impregnation time — 10 min in 0.3% silver nitrate (Procedure II) with frequent shaking of the sections — was used successfully in a recent study of the geniculo-striate connection in the monkey by Hubel and Wiesel (1969). Impregnation of the section at incubation temperatures of about 37° C is often recommended in the literature. Although this precaution seems to have little if any advantage in most instances where the different Nauta-Gygax procedures are applied to frozen sections, it occasionally has been used to good effect in our laboratory.

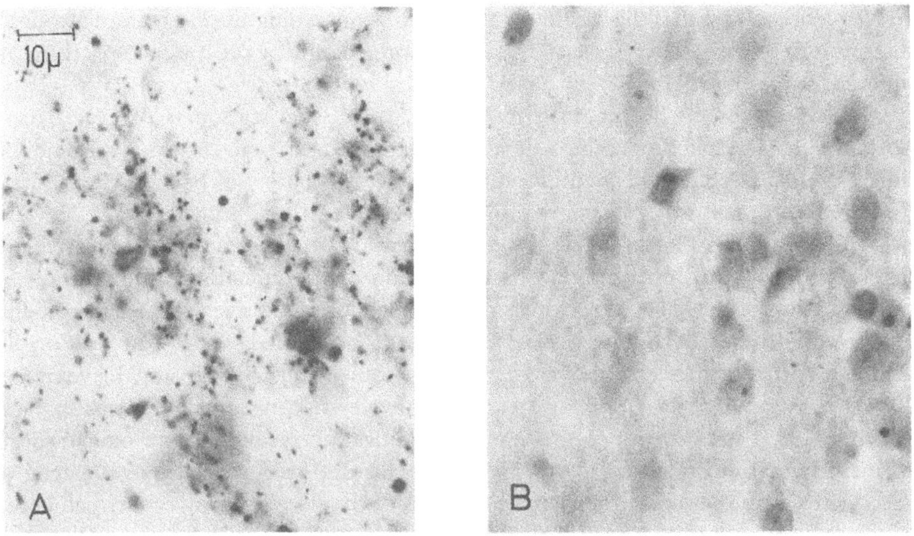

Fig. 6. A: Microphotograph showing dense terminal degeneration in Rexed's lamina II of the spinal substantia gelatinosa in a rat whose ipsilateral dorsal root S_1 had been sectioned 36 hours before sacrifice on the animal. B: corresponding zone of the spinal grey matter contralateral to the root section. [Reproduced from Brain Res. **6**, 89—99 (1968), by permission from the Publishers]

Recent advances in the light-microscopic identification of terminal degeneration have resulted not only from more sensitive silver impregnation procedures, but also from an increased awareness of the importance of the postoperative survival time. The optimal survival time varies, not only from one animal species to the other (see present chapter by Ebbesson), but also between different fiber systems in the same species. Furthermore, the study of synaptic fields usually requires a shorter survival time than is optimal for the impregnation of degenerating axons of passage. This was noted by Blackstad (1956), who found that the degenerating terminals of the hippocampal commissure in the rat were best impregnated in cases of 3—4 days survival time, whereas degenerating fibers of passage could be impregnated optimally only after survival times of about 10 days. The conditions seem to be similar in many other fiber systems. In cases of dorsal root section in the rat, for instance, the argyrophilia of the degenerating synaptic terminals in the substantia gelatinosa was found to reach its optimum after 2—4 days survival time (Fig. 6). As early as 7 days after the operation most of these structures could no longer be impregnated, although the degenerating parent fibers traversing more ventral laminae were well-stained. Some mammalian fiber systems, on the other hand, may require several weeks degeneration to reach their optimal argyrophilia. For example, a system of fine-calibre axons originating in the bulbar area of the rabbit and crossing to the opposite side through the anterior commissure was found optimally impregnated only after postoperative periods of 20—25 days (Lohman and Mentink, 1969). The importance of the survival time was recently illustrated by Schneider's (1968) finding that different subgroups of degenerating axon terminals of the retinal projection in the hamster became impregnable at various times after eye-removal. The variability of the optimal survival time for the silver-impregnation of degenerating axoplasm has important practical consequences. It implies that optimal information concerning the course and terminal distribution of any given fiber system can be obtained only by the use of a spectrum of different survival times.

As is true for all histological procedures, the tissue fixation is of crucial importance in silver impregnation. Cajal already noted that silver was deposited in different tissue elements depending on which one of a variety of fixation procedures had been used, and many later investigators — among them Bodian (1937), Davenport (1934), and Holmes (1943) — drew attention to the role of the fixative in determining the final distribution of silver within the nervous tissue. For the silver impregnation of degenerating axoplasm formaldehyde has proved to be an excellent and reliable fixative. Although there has been an occasional report that fixation in acidified formalin increases the argyrophilia of fine axons (Beech and Davenport, 1933), most investigators prefer neutral or even slightly alkaline formalin (see Holmes, 1943) and attempt to prevent the acidification resulting from oxidation of formaldehyde to formic acid by storing the stock solution of formalin over calcium-carbonate or magnesium carbonate. The results of the silver methods described here, however, were found to be but little affected by the range of variation in acidity that can be expected in making up fresh 10% formalin by diluting commercial (approximately 40%) formaldehyde solutions ten times with distilled water. In our laboratory, therefore, no attempts are made to approximate neutrality of the formaldehyde fixative.

The specimens should be stored for at least 1—2 weeks in 10% formalin before sectioning. Beyond this statement, it is difficult to specify the optimal fixation time. Eager and Barrnett (1966), using the suppressive Nauta-Gygax method, demonstrated a close relationship between the duration of formalin fixation and the required length of potassium permanganate treatment: prolonged storage in formalin was found to reduce or even eliminate the need for permanganate. Eager and Barrnett also described loss of stainability of the fine fibers in the hypothalamus and the cerebellar cortex in some mammals after long storage in formalin. However, prolonged fixation does not necessarily impair argyrophilia. It is not uncommon to find the staining quality of frozen sections markedly improved after prolonged storage of the sections in formalin; some of our most successful impregnations of fine terminal axon ramifications by the Laidlaw modification as well as with the Fink-Heimer procedures were obtained in sections which had been stored for a year or even longer in formalin. Some practical notes regarding fixation and freeze-sectioning are discussed in the Appendix.

Interpretation of the Terminal Degeneration

Although Nauta, as well as Blackstad who was one of the first to use the original Nauta method, seem to have been in no doubt about the capacity of the method to impregnate degenerating axon terminals, there has been a widespread reluctance to accept the argyrophilic particles seen in successful original Nauta sections as true "terminal degeneration". That the multitude of argyrophilic particles seen in original Nauta and Fink-Heimer sections can indeed indicate the area of distribution of the fiber system in question has been documented in several experimental studies comparing the light-microscopic results with the localization of degenerating axon terminals revealed by electron microscopy. In one such sequence of studies, conclusions regarding the distribution of the hippocampal commissure in the rat (Blackstad, 1956) as demonstrated by the original Nauta method were confirmed in detail by electron-microscopic studies by Blackstad and his collaborators (Alksne *et al.*, 1966; Nafstad, 1967) as well as by other investigators. Similar corroborating electron-microscopic studies have shown the validity of the Fink-Heimer method in the study of terminal degeneration in the substantia gelatinosa (Heimer and Wall, 1968) and in the hypothalamus (Heimer and Nauta, 1969).

Naturally, such combined light- and electron-microscopic studies are unable to identify accurately the site of deposition of the black metallic silver seen in silver-impregnated material. However, the results of a recent electron-microscopic study of previously silver-stained sections from the hippocampus and the prepiriform cortex of the rat strongly suggest that many of the black bodies seen in the previous figures indeed represent degenerating axon terminals (Heimer and Peters, 1968). Fig. 7A shows a light-microscopic picture of the terminal degeneration appearing in the olfactory cortex of a rat whose ipsilateral olfactory bulb was removed 3 days before sacrifice. The silver-stained section shows an extremely dense accumulation of small argyrophilic particles in the superficial half of the plexiform layer. The electron micrographs (Fig. 7B and C) were taken from a neighboring section impregnated by the same procedure, then post-fixed in osmic acid and embedded in Epon-Araldite for ultra-thin sectioning. Degenerating axon terminals appear in these electron micrographs as membrane-bound electron-dense profiles intimately related to postsynaptic thickenings. These profiles exhibit varying degrees of deformation and dissolution of internal structure. The small silver particles scattered throughout the tissue presumably correspond to the brown "background" coloration characteristic of silver sections in the light microscope. Much larger metallic globules are found in the axoplasm of degenerating fibers (Fig. 8) and boutons (Fig. 7B—C). Because of the loss of fine detail in the degenerating structures it is impossible to relate this argyrophilia to any particular morphological component of the axons or terminals. The relatively massive accumulation of silver in the degenerating axoplasm suggests the likelihood that degenerating fibers and boutons would appear in the light microscope as the most darkly stained tissue components.

It is necessary to deal in some detail with the question of how apparently random distributions of argyrophilic particles appearing in the light microscope can be identified with terminal degeneration. This question is particularly appropriate because almost no single one of the black-stained corpuscles shown in the photographs, when found isolated, could be readily recognized as part of a degenerating axon. Sporadic black spherules, encountered in nearly every silver-impregnated section of normal brain tissue, usually remain unexplained. The identification of terminal degeneration, therefore, must be based on pattern characteristics at least as much as on the morphological features of single components. Naturally, for the emergence of a discernible pattern a certain minimal distribution density is required.

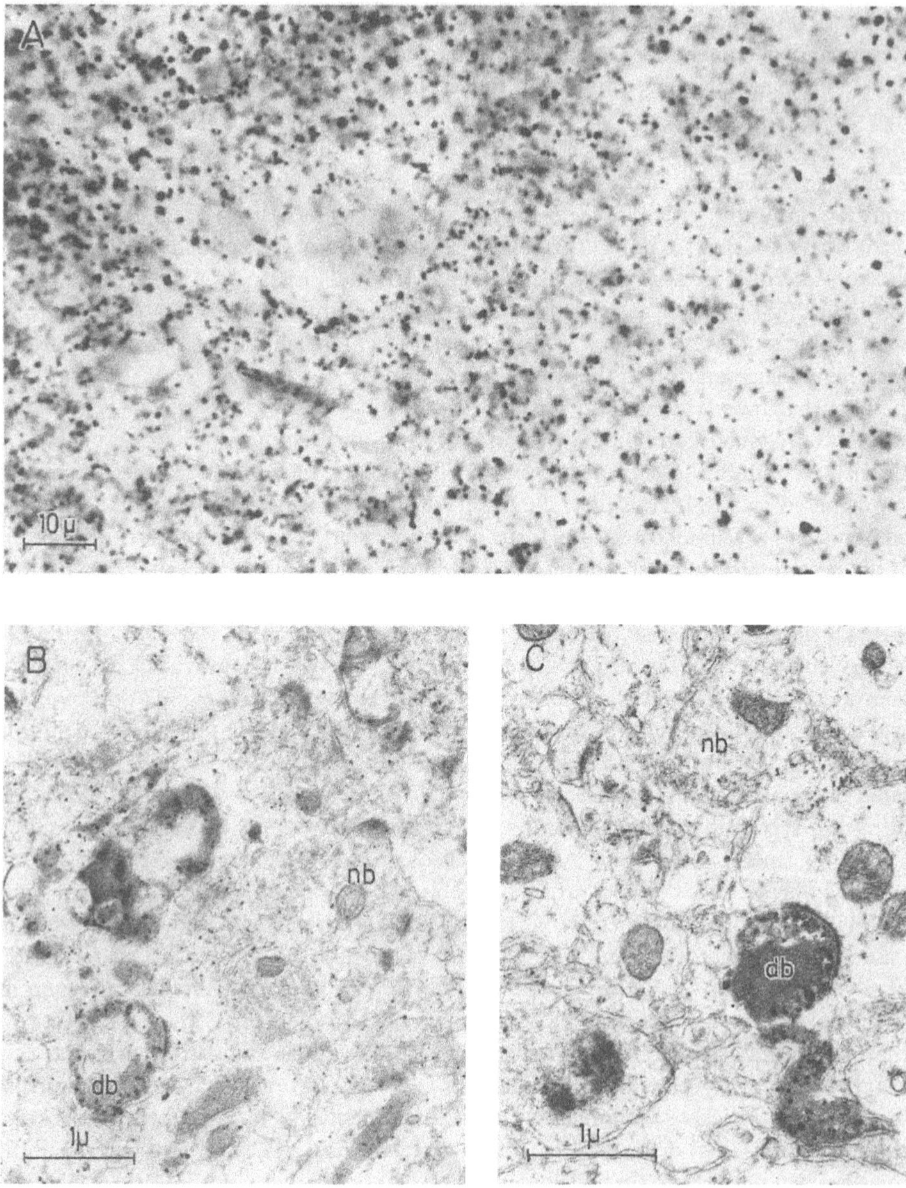

Fig. 7A—C. Terminal degeneration of the plexiform layer of the olfactory cortex in a rat whose ipsilateral olfactory bulb had been removed 4 days before sacrifice of the animal. Material impregnated by the Fink-Heimer Procedure II. A is a light-micrograph; B and C electron micrographs from neighboring hand-cut and silver-stained sections. Although the preservation of the tissue at the ultrastructural level has been adversely affected by the silver impregnation procedure, it still allows the identification of degenerating boutons as somewhat deformed, electron-dense profiles associated with postsynaptic membrane densities. The silver deposits appear in the form of electron-opaque spherical particles within the tissue. The particles, although scattered throughout the tissue, are considerably larger and more numerous within the degenerated axoplasm. Abbreviations: *nb* normal bouton; *db* degenerating bouton. [Reproduced from Brain Res. **8**, 337—346 (1968), by permission from the Publishers]

In regions where this requirement is met, the following characteristics must be noted before even a tentative identification of terminal degeneration is possible:

1. The Argyrophilic Particles are Present in the Substance of the Section. Black particles lying on one or both surfaces of the section almost certainly represent a silver precipitate formed during the final stages of silver impregnation. In most cases, the impregnated particles signifying true terminal degeneration are distributed in approximately equal density throughout the thickness of the section. Occasionally, however, they are sparser or even absent about the middle reaches of the section's

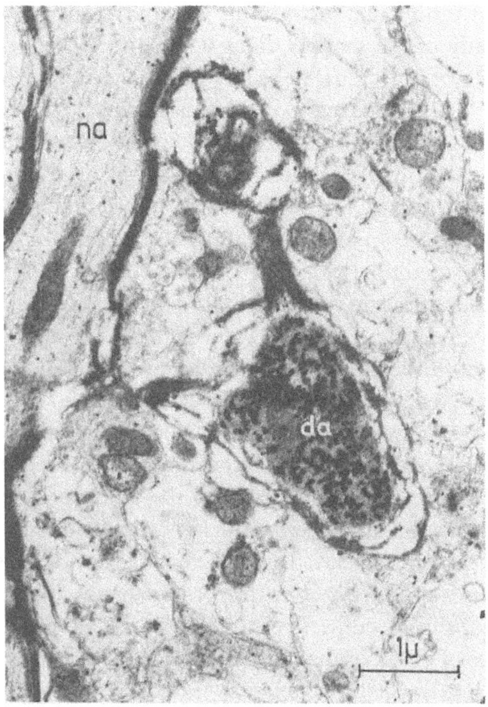

Fig. 8. Electron micrograph of a previously silver-impregnated section (Fink-Heimer, Procedure II) from the olfactory cortex of a rat whose ipsilateral olfactory bulb had been removed 3 days before sacrifice. The cytoplasm of the degenerating axon (*da*), which presents an electron-dense amorphous appearance comparable to that of the degenerating boutons in Fig. 7, is heavily loaded with large silver particles. *na* normal axon. [Reproduced from Brain Res. **8**, 337—347 (1968), by permission from the Publishers]

thickness. This is often seen in extra-thick frozen sections, and sometimes in frozen sections that have been impregnated for only short times.

2. The Distribution of the Argyrophilic Particles is Confined to One or More Reasonably Well-Defined Areas. Mode and width of fiber distribution vary widely among individual fiber systems, and no further generalizations can therefore be made. However, the region in question should have distinctive characteristics when compared with adjacent areas or with the same area on the opposite side. High quality of the sections is obviously of great importance in this matter: no conclusions should be based singly on findings in sections showing grossly uneven impregnation, poor tissue preservation, or diffuse argyrophilic grain.

3. The Particles are Somewhat Variable in Size and Shape. Although most are approximately spheroidal, more irregular shapes are common (see, for example, Fig. 4) and tend to become more so in later stages of degeneration. Variations in *size* of the particles are often quite marked and can be used to distinguish terminal degeneration from certain even-grained silver deposits such as the "fine grain" to be discussed below, an argyrophilic reaction apparently associated with retrograde cell changes. The variations in particle size characterizing terminal degeneration may be attributable to (a) natural variation in the size of synaptic end-bulbs within one and the same fiber system; (b) the circumstance that by no means all of the particles represent degenerating boutons terminaux, for metallic silver is deposited also in terminal axon branches which usually are of smaller diameter than are their end-bulbs, and probably, (c) inequal filling of individual degenerating structures with metallic silver. The same considerations may apply to the variations in the *shape* of individual particles, but here an additional factor must be taken into consideration, namely, a progressive deformation of degenerating structures by invading or encircling astrocytic processes. Apparently depending on the fiber system studied, such deformations may already be quite pronounced in electron micrographs of terminal degeneration of no more than a few days standing.

Naturally, the identification of terminal degeneration is easiest when well-impregnated fine degenerating terminal axon branches can be recognized in the area. Such axonal end-ramifications are identifiable by a distinct and often tortuous alignment of impregnated particles (see, for example, Fig. 2B). In early stages of the degenerative process, however, it is not unusual to find an abundance of boutons terminaux impregnated in a region that shows no clear evidence of fiber disintegration. This is a common finding particularly in the study of fine-fibered conduction systems. Degenerating axon branches may become identifiable in such regions at later stages of degeneration, but it is possible that some fiber systems have terminal arborizations too fine to produce a light-microscope image (i.e., to permit adequately massive agglomerations of metallic silver to form in the axoplasm) at any stage.

The identification of the terminal distribution areas of degenerating nerve fibers traditionally has been governed by Monakow's requirement that the fiber system be traceable in continuity into the suspected area of termination. This requirement applies to the contemporary silver methods no less than it did to the Marchi technique for which it was formulated. Even though it may necessitate repetitive experiments, it is of utmost importance to ascertain that the region in question is indeed invaded by degenerating axons. However, should the demonstration of such continuity consistently fail despite the use of a wide variety of survival times, the possibility shonld be considered that the fiber system in question is composed of axons of such small caliber that only their terminal enlargements can be made visible in the light microscope by silver impregnation.

The study of central nervous fiber connections by means of the reduced-silver impregnation techniques here described has, in a sense, a rather limited objective, namely, to trace the path of the fiber tract and to determine its area of termination. Although the localization of the terminal degeneration in relation to the cytoarchitectonic picture of the region may occasionally permit some tentative conclusions as to the nature of the synaptic contact (e.g. axosomatic or axodendritic), light-microscopic techniques as a rule can do little more than identify the synaptic fields of a fiber system. Detailed study of the synaptic relationship is possible only by the aid of the electron microscope, for only at the ultrastructural level can synaptic endstructures and postsynaptic profiles be identified with certainty.

Sources of Error

Although there is hardly any doubt as to the capacity of the above-mentioned techniques to impregnate degenerating axon terminals, an attitude of suspicion vis-à-vis accumulations of relatively uncharacteristic black corpuscles is understandable and even necessary. It is certain that the silver methods here described are capable of impregnating degenerating axons and their synaptic terminals but it does not necessarily follow that all of the black-impregnated structures found in the sections represent degenerating axons. Nearly all components of brain tissue are potentially argyrophilic, and the possibility of fortuitous staining of non-axonal tissue constituents must always be kept in mind. Several sources of error, such as silver precipitates on the surface of the section, myelin impregnation, and incomplete staining of normal fibers, have been mentioned earlier in connection with the suppressive Nauta-Gygax method (Nauta, 1957). To be added to this list are the following common forms of silver deposition encountered particularly often in sections impregnated by the Fink-Heimer techniques.

1. Perivascular Connective Tissue. Although the impregnation of connective tissue fibers in the vascular adventitia is often so massive as to make blood vessels stand out in stark contrast to their surroundings, a less complete impregnation of reticular fibers around blood vessels may result in pictures confusingly similar to the impregnation of degenerating axons (Fig. 9A).

2. Glia Elements. Irregular impregnation of glia elements (Fig. 9B) is sometimes prominent enough to preclude any statements as to presence or absence of degenerating axons. Such fortuitous glia-impregnation may occur anywhere, but is seen most commonly at or near the brain surface, especially in the superficial strata of the cerebellar and cerebral cortex. It usually can be prevented, or at least reduced, by either (or both) more extended pretreatment in potassium permanganate or the use of a higher concentration of silver nitrate in the initial impregnation.

3. Pericapillary Particles. Possibly also related to glia is a common silver deposit appearing in the form of an accumulation of round black corpuscles in a small and sharply defined field surrounding a capillary blood vessel (Fig. 9C). In shape, size and distribution pattern these particles are often virtually indistinguishable from those signifying early stages of true terminal degeneration. The only distinctive feature that can help to prevent misinterpretation is the circumscript, limited and isolated nature of these pericapillary accumulations.

4. Cellular Deposits. Another common source of misinterpretation is presented by a peculiar form of corpuscular silver deposit associated with neuronal cell bodies. It appears in the form of more or less round black shapes dotting the cell body (the exact site of this silver deposit has not yet been determined in the electron microscope). This form of silver impregnation (Fig. 9D), particularly common in the intralaminar, anterior and mediodorsal thalamic nuclei, the habenular nuclei, the hypothalamus and the substantia nigra, can easily be mistaken for degenerating axosomatic axon terminals, the more so since the sizes of the corpuscles fall within the size range of boutons terminaux. Its bilateral occurrence, as well as the nearly

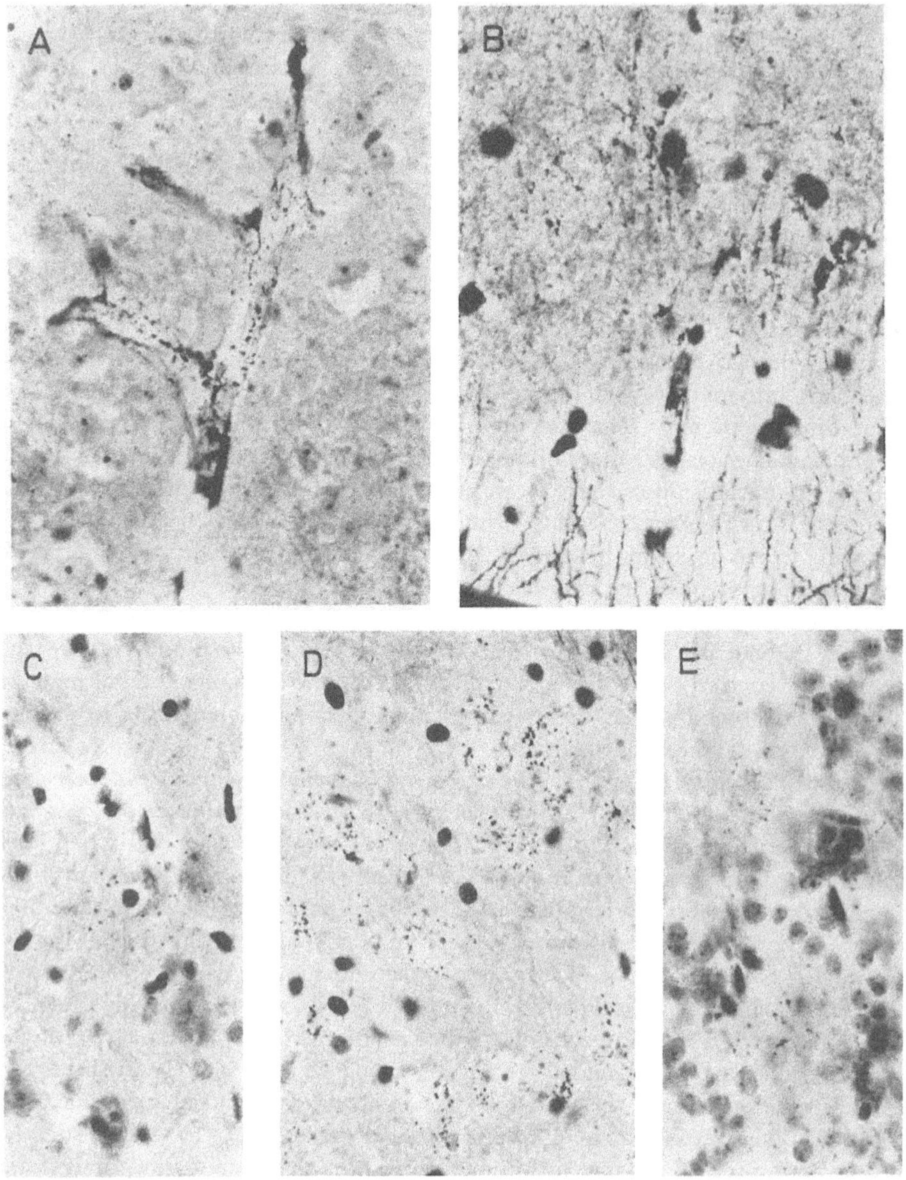

Fig. 9A—E. Microphotographs of some common sources of error. A: Impregnation of reticular fibers around a blood vessel in the cerebral cortex of a rhesus monkey. B: Irregular impregnation of glia tissue in the olfactory cortex of a rhesus monkey. C: Pericapillary particles in the lateral thalamic nucleus of a rat. D: Silver deposits in relation to neurons in the parafascicular nucleus of the rat thalamus. E: Irregular silver deposits in glomeruli of the rat olfactory bulb. Magnification 500 ×

complete absence of similar black corpuscles in the intervals between cell bodies usually permits a distinction of this characteristic deposit from true terminal degeneration.

5. *Artefacts in the Olfactory Bulb.* Irregular, mainly extracellular silver deposits in normal olfactory bulb tissue are quite common, especially in the glomerular layer (Fig. 9E). As these artefacts may appear also in other regions of the bulb, however, they are a dangerous source of misinterpretation in experimental studies of olfactory connections.

6. An important and difficult problem, finally, is posed by the often very dense and dust-like silver grain observed in circumscript regions of the thalamus following ipsilateral decortication. The distribution of this "fine-grain" suggest that it is related to a retrograde cell reaction (see also Guillery's article in this Volume). It can usually be distinguished from true terminal degeneration by its evenly distributed, uniformly small particles.

The preceding survey of some common sources of erroneous interpretation cannot do full justice to this important problem. Silver techniques in general, and methods for terminal degeneration in particular require cautious interpretation. Success in studying terminal degeneration is not seldom dependent on the ability to avoid interpreting pseudo-degeneration and other irrelevant silver deposits as true terminal degeneration. However, thorough familiarity with these problems, gained by daily practice at the microscope, usually makes it possible to sift the wheat from the chaff, much in the same way as investigators using the Golgi method must learn to distinguish nerve fibers from glial processes.

Concluding Remarks

When Augustus Waller in 1850 discovered that section of a nerve led to degeneration of the distal segment of the nerve, the stage was set for one of the most formidable research tools in neuroanatomy. It was not until 1885, however, when Marchi and Algeri invented the stain for degenerated myelin, that the neuroanatomists could proceed with a more systematic mapping of the great nerve tracts in the central nervous system with the aid of experimentally induced Wallerian degeneration. A more detailed study of interneuronal connections became possible with the development of the reduced silver impregnation techniques by Bielschowsky and Ramón y Cajal at the turn of the century. Although these silver methods were for a long time used only for the study of fiber connections on normal material, the last decades have witnessed an intensive search for more reliable and powerful modifications of the reduced silver technique for the impregnation of degenerating nerve fibers. These efforts have been well-rewarded and it is now possible to impregnate most, if not all, degenerating axons and axon terminals with the exclusion of the normal axons. The selective impregnation of degenerating axoplasm, introduced by Nauta and Ryan in 1952, makes it easy to trace the course of even a few degenerating nerve fibers in the midst of innumerable normal fibers, and to determine, with a high degree of precision, their area of distribution.

The recent development of the electron microscope has greatly amplified the usefulness and the reliability of the silver impregnation techniques for the study of terminal degeneration. The mapping of degenerating axon terminals with the electron microscope, however, is also changing our attitude in regard to the silver methods. It is no longer tenable to draw definite conclusions about synaptic interrelations from studies of silver impregnated sections alone. Although a well-degenerated nerve fiber is often easy to recognize, a synaptic endstructure cannot be identified beyond doubt at the light-microscopic level. Most of the degenerating terminals show up as uncharacteristic black corpuscles, and the single argyrophilic particle is significant only in the context of the general pattern of argyrophilia. Furthermore, two structures which in the light microscope appear to be in contact with each other may be separated by tissue elements visible only in the electron microscope. The reduced silver methods, therefore, can at best trace the course of degenerating nerve fibers and identify their synaptic fields. A detailed study of the synaptic pattern, and the type of synaptic contact present, can only be done with the electron microscope, which, however, in turn requires adequate guidance by light-microscopic techniques.

Any one of the silver methods for the study of axonal degeneration has its special merit as well as its shortcomings. The classical "suppressive" Nauta-Gygax method is an excellent method for the study of fiber trajectories, but has proven to be of limited value for the identification of the terminal distribution areas in many regions of the central nervous system. The same is largely true for the Glees method, although this technique appears to be of special value for the impregnation of neurofilament-containing axons and synaptic endstructures. The original Nauta method, as well as the more recent Fink-Heimer modifications, have their great strength in the study of terminal degeneration. A proper evaluation of terminal degeneration, however, requires experience and is sometimes beyond the reach of the light microscope. If a fiber system distributes massively within a well-defined area, there is usually little doubt about the interpretation of the silver impregnated section. The more scanty the terminal degeneration, however, the more uncertain its identification in the light microscope. Unfortunately, very sparse degenerating axon terminals easily escape identification also in electron-microscopic studies. With this restriction, it can be said that the possibility of verifying light-microscopic observations by electron microscopy promises a considerable increase in the reliability of experimental studies by the method of anterograde degeneration. Among other advantages, this combination of techniques greatly reduces the risk of interpreting irrelevant silver deposits as true terminal degeneration.

Metallic impregnation of biological material is by its very nature a dynamic process, the physico-chemical principles of which are poorly understood. The reduced silver techniques — although widely used and highly successful methods — are therefore technically demanding, and the results obtained are not seldom dificult to evaluate. There is usually no way of knowing exactly how successful the impregnation of the degenerating axoplasm is, and negative results must therefore be interpreted with the greatest caution. It is well to remember that good suppression of normal fibers does not guarantee an adequate impregnation of degenerating fibers; nor does the fact that there is a good impregnation of terminal degeneration in one region automatically signify optimal impregnation in another region of the same brain. Whenever there is doubt as to the optimal impregnation of the degenerating axons and axon terminals, there is need to modify the original procedures, and it is indeed wise to search constantly for more optimal silver modifications. The procedures described in the appendix therefore should be regarded more as guidelines than as fool-proof recipes.

Appendix

This supplemental information is provided for the convenience of laboratory workers using the technique described by Fink and Heimer (1967).

Fixation

Fixation should be initiated, if possible, by perfusion with saline followed by 10% formalin or formalin-saline. Material to be studied is carefully removed and stored in 10% formalin for at least 1—2 weeks. Blocks to be sectioned are soaked for 2—3 days (or longer if needed) in a 30% sucrose solution.

Cutting

Blocks are frozen without previous rinsing, and sections of 15—30 μ thickness are collected and stored at room temperature or in the refrigerator in dilute (1—2%) formalin. The sugar-soaked tissue requires more thorough freezing than do blocks frozen directly from formalin, water, or 20% alcohol. Therefore, blocks are best kept surrounded by pulverized dry ice on the freezing stage.

Staining Procedure I

1. Rinse frozen sections in distilled water (3 changes — 5 min each).
2. 0.05—0.5% potassium permanganate — 5—15 min.
3. Rinse briefly in distilled water.
4. Bleach in a mixture of equal volumes of 1% oxalic acid and 1% hydroquinone.
5. Rinse in distilled water (3 changes — 5 min each).
6. Solution A: 60—90 min. A considerable lengthening of the impregnation time will occasionally improve the result.
7. Solution B: 30 min. (This step can often be deleted without noticeable impairment of the result.)
8. Rinse in distilled water (3 changes — 5 min each).
9. 3—5 min in a freshly prepared ammoniacal silver nitrate solution composed of:

2.5% silver nitrate	30 ml
strong ammonia-water	1.0 ml
2.5% sodium hydroxide	1.6—1.8 ml

10. Nauta-Gygax reducer. Agitate quickly until tissue turns brown, then transfer to a second volume of reducing solution for 1—2 min.
11. Rinse in distilled water.
12. 0.5% sodium thiosulfate — 1 min.
13. Rinse in distilled water (3 changes — 5 min each).
14. Mount from a solution composed of equal parts of 0.5% gelatin and 80% alcohol (see Appendix to the present chapter by Ebbesson).

Solutions

Solution A:

uranyl nitrate	1.0 g
silver nitrate	5.0 g
distilled water	1,000.0 ml

Solution B:

uranyl nitrate	1.75 g
silver nitrate	16.25 g
distilled water	1,000.0 ml

Nauta-Gygax reducer:

distilled water	900.0 ml
95% alcohol	75.0 ml
10% formalin	17.5 ml
1% citric acid	20.0 ml

Staining Procedure II

1. After rinsing in distilled water, soak sections for 5—10 min in 0.025—0.1% potassium permanganate.

2. Rinse and decolorize sections in an oxalic acid-hydroquinone solution (see above).

3. Rinse thoroughly and transfer sections for 5—10 min to a 2.5—5% uranyl nitrate solution.

4. Rinse thoroughly in 3 changes of distilled water and transfer sections to a 0.3—0.5% silver nitrate solution for 1—2 hours. The addition of 0.2 ml of pyridine to 10 ml of the silver nitrate solution may improve the results.

5. *Without washing*, transfer sections for 2—5 min to a freshly prepared ammoniacal silver nitrate solution composed of:

1.5% silver nitrate	20 ml
95% ethyl alcohol	12 ml
strong ammonia water	2 ml
2.5% sodium hydroxide	1.6—1.8 ml

6. Transfer sections, without rinsing, to the Nauta-Gygax reducer.

7. Rinse, and transfer sections for 1 min to 0.5% sodium thiosulfate solution.

8. If desired, counterstain by immersing the mounted sections briefly in a cresylechtviolet solution (see below: note 7).

9. Rinse, mount, dehydrate and enclose.

Notes

1. Sections should be agitated every few minutes when rinsing in water or when in chemical solutions.

2. If sections become too dark in Procedure I, add a few drops of 1% citric acid to each 50 ml of reducing solution.

3. The length of exposure to the above solutions may be varied if times stated do not give the desired results. It may, for instance, be necessary to reduce or even delete the pretreatment in potassium permanganate and uranyl nitrate in order to preserve the argyrophilia of the degenerating axons, whereas in other cases a prolonged pretreatment may be needed to suppress the argyrophilia of irrelevant tissue components. A prolonged impregnation in solutions A or B (e.g. overnight) may be of advantage in both procedures.

4. That the pH of the ammoniacal silver solution is important for the argyrophilic response was well illustrated by White (1959), and we suggest, in agreement with White, that a few preliminary trials be carried out in order to establish the optimal ammonia-sodium hydroxide ratio. Too high a concentration of sodium hydroxide results in dark sections with increased risk of precipitates, whereas too much ammonia tends to suppress the argyrophilia of degenerating axoplasm. Only freshly prepared ammoniacal silver solutions should be used.

5. The use of chrome-alum-gelatinized slides may be useful in mounting the sections, especially when counterstaining silver-stained sections. Wash plain slides and immerse in chrome-alum-gelatin solution, drain and dry. The solution is prepared by dissolving 3.0 g of gelatin in 600.0 ml water and then adding 0.3 g chromium-potassium sulfate (chrome alum).

6. The quality of silver-impregnated frozen sections often deteriorates severely when sections are allowed to dry out during the mounting procedure. A method to avoid such dessication (Albrecht, 1954) is described in the Appendix to the present chapter by Ebbesson.

7. Silver impregnated sections can be counterstained by immersing the sections in a 0.5% cresylechtviolet solution for about 3 min, followed by a brief rinse in distilled water, and dehydration in 70 and 95% alcohol. Differentiate the sections individually in a mixture of 100 ml 95% alcohol and 2 ml glacial acetic acid until the desired color is obtained, and continue the dehydration in absolute alcohol. Mount and enclose. An alternative method is, to stain sections already mounted on glass slides. To prevent the sections from becoming detached during the staining procedure it is advisable to mount them on chrome-alum coated slides (see above: 5).

Discussion

Dr. Ramón-Moliner: I want to discuss two points of your presentation that bother me, and I hope that what I am going to say will be taken as words from the devil's advocate, that is, as something that one defends without actually believing in it. First of all, what solid reasons do you have for interpreting the argyrophilic particles that you showed us as degenerating axon material instead of random precipitate? I find it difficult to find any reliable morphological criteria that could help us to make this decision. The shape of the granules is very unspecific, when compared with the appearance of degenerating axons in Nauta-Gygax preparations. In the latter, the axons are disintegrated in drop-like fragments but each fragment looks unmistakably like a fragment of a disintegrated axon: it may be elongated or globular but displays, nevertheless, a neat outline. Moreover, such fragments are aligned so that it is almost impossible to conclude otherwise. The droplets that show up in your technique are not nearly so distinctive, either in shape or arrangement, and I am led to believe that you interpret them as axon terminals only because they show up in regions where you expect the fiber systems to terminate.

Dr. Heimer: Naturally, we too have worried about how to distinguish degenerating axon terminals from random precipitate, and we too wish that there were a more objective characteristic that could help us tell the difference in the light microscope. Our confidence in interpreting certain granular spherules as evidence of terminal degeneration is based first and foremost on electron microscopic verification. As I pointed out, Alan Peters and I found distinctive silver deposits in all axon terminals that showed the ultrastructural changes typical of Wallerian degeneration, and only in such boutons, at least in the olfactory cortex. We have since sampled many more brain regions in which the light microscope showed the presence of such argyrophilic spherules, invariably (so far) with the same result. Of course you are right in questioning the light microscopic evidence that I showed you. But I could raise another question: Supposing the only form in which true terminal degeneration will ever show up in the light microscope is the one I showed you, are we to disregard the evidence for the reason that its appearance is dangerously close to that of a random precipitate? We need not, for we now have the electron microscope to verify or disprove our interpretation.

Dr. Ramón-Moliner: Thank you. My second question is related to the sharp borderlines which may appear between regions with and without degeneration. Some of your slides made me think of what would happen when you scatter a spray of oil droplets over the lower half of a sheet of filter paper. At first, the drops would be quite distinct one from the other, but after awhile they would patch, and a bit later the whole half of the sheet would be a solid oil-soaked area with a neat borderline slowly progressing up the sheet. Now, that sharp borderline is very similar to the one you showed us in some of your sections of the prepyriform cortex. I was wondering if your granules could perhaps be, not true fragments of degenerating fibers, but something like what Marc Colonnier has called "significant artefacts", in this case to be re-labeled "significant precipitate". What I mean is this: Suppose that some substance diffuses from degenerating axon terminals into the surrounding tissue, sooner or later achieving a coherent boundary. Suppose that, furthermore, this substance would so modify the tissue chemistry that silver would precipitate in the area. If this were indeed the case, the borderline would not be in any way meaningful, even though the precipitate itself would be significant in the sense that it would signal the presence of terminal degeneration. As you see, I am not questioning the research value of your technique, I am only suggesting that it should be re-interpreted.

Dr. Heimer: Actually, such sharp borders as I showed in the prepiriform cortex are rather exceptional. Also, the one I showed is more apparent than real, for a higher magnification shows a lesser density of terminal degeneration deep to the sharp zone of maximal quantity. As a further answer to Dr. Ramón-Moliner's second question, let me refer once more to our EM studies which showed the largest silver aggregations only in degenerating axon terminals, and no more than a scattering of isolated and much smaller silver particles in any other structure.

Mr. Chairman, with your permission, I should like to state my position with respect to our silver pictures a bit more generally. First of all, I would never attempt to convince myself or anybody else that any particular individual argyrophilic particle is a degenerating bouton terminal or something else. The single particle is significant only insofar as it is part of a general pattern or argyrophilia. Second, it is extremely important, and we always try, to stain the degenerating fibers into the area where we suspect we are staining synaptic structures. This is not always easy for they may be too fine to show up in the light microscope, even though their bulbous synaptic end-portions do. This leads me to the main point I want to make. It is, that the electron microscope has permitted us to change attitude with respect to light-microscopic experimental silver methods. Whereas up to ten years ago the silver methods were considered to supply conclusive evidence as to both the existence and the details of interneuronal connections, it is now quite obvious that the light microscope and our silver methods are by no means infallible in identifying certain structures as axon terminals and, moreover, usually completely incapable of identifying the postsynaptic side of the synaptic complex. As silver methods for the impregnation of degenerating axons were sharpened more and more, they also became more risky, pushed as they were closer and closer to the point of non-differential argyrophilia. With the result that it has become progressively more necessary to check our light microscopic impressions electron-microscopically. The time may come, if it has not already come, that the only — but nonetheless invaluable — purpose of the silver methods is, to show us where to look with the electron microscope for both verification and the more detailed information now expected from us by neurophysiologists.

Dr. Blackstad: In our study of the hippocampal region, we had occasion to see the Fink-Heimer procedures work very well. In one particular instance, we did not need any electron microscopic confirmation to know that we were staining boutons terminaux, namely in the case of the mossy fiber system (Blackstad *et al.*, 1969). As you know, these fibers, although very thin, have relatively enormous, bag-shaped boutons measuring up to 8 μ, but more commonly around 5 μ in diameter. In cases of lesion in the fascia dentata we could follow degenerating mossy fibers as thin trails of small fragments; their boutons appeared in almost solid black and in characteristic, often elongated shapes exactly like those seen in Golgi pictures. These boutons have been shown electron-microscopically to contain no neurofilaments; they are invisible in normal silver preparations of the Cajal and Bielschowsky type, and showed up degenerated in "original" Nauta (1950) preparations in only half our cases. In the Fink-Heimer sections some other structures were impregnated, such as connective tissue reticular fibers, but there is no doubt in my mind that we obtained a remarkably selective picture of degenerating mossy fibers and their terminals.

Dr. Guillery: First of all, I want to say how pleased I am that we have come to agree in our interpretation of these methods. My long-standing skepticism has been about the interpretation of the methods, not about their usefulness. I should like to return to Dr. Ramón-Moliner's very relevant question as to how seriously we can take the borders of the synaptic fields suggested in your preparations. Your reference to your electron micrographs raised the question that in those electron micrographs we see remarkably little silver agglomeration in comparison with what many of us see in electron micrographs taken from silver-stained preparations. Since you see so very little, I wonder whether you feel sure that there are no unstained degenerating "boutons" beyond the borders of the areas shown in your preparations?

Dr. Heimer: No, I can't be sure. To repeat, I think that quite generally we must keep in mind possibility that not all degenerated axon structures are impregnated. Also, we should not the forget that the yellow or brown background of silver sections is due likewise to the presence of silver. In most cases, in other words, silver is present everywhere in the tissue. In our electron-microscopic study of silver-impregnated sections, we have attempted to minimize the background-silvering by using more dilute solutions and shorter impregnation times than we would ordinarily use. The price to pay for this was, as could be expected, a less massive silver deposition in degenerating boutons as well, and this explains the difference between our pictures and those of yours which indeed show more silver in the impregnated degenerating boutons. However, Dr. Peters and I were interested in the question whether we could find a considerable difference quantitatively between the silvering of degenerating material and that of normal structures. The differences we found were enough to convince us that the black granules which we had seen in the silver sections were indeed degenerating axon terminals and the corresponding fine axon arborizations. As to the question whether or not terminal degeneration may be present, but unstained, outside the boundaries of our "fields", only the electron microscope can help us answer that.

The Selective Silver-Impregnation of Degenerating Axons and their Synaptic Endings in Nonmammalian Species

SVEN O. E. EBBESSON

Introduction

The recently developed methods for the selective silver impregnation of degenerating axons (Nauta and Ryan, 1952; Nauta and Gygax, 1954; Nauta, 1957) and axon terminals (Fink and Heimer, 1967), by adding precision to the determination of long axon trajectories, have contributed to the revival of comparative neurology as perhaps no other technique has. Although these methods have been used for only a short time, their application has already provided significant new information with respect to the organization of a variety of vertebrate nervous systems.

Until the introduction of these experimental methods, our knowledge of the anatomy of the nonmammalian vertebrate nervous system was based primarily on normal-descriptive studies conducted by the methods of Nissl, Weigert, Marchi, and Golgi. The Golgi technique has been, and remains, the anatomical tool *par excellence* for the determination of trajectories of short axons. Longer connections were earlier determined experimentally by the Marchi method which selectively stains degenerating myelin, and therefore is inadequate for the tracing of connections established by thin and poorly myelinated axons, as well as generally incapable of demonstrating degeneration of terminal axon branches and synaptic end structures. Because the nonmammalian nervous system is characterized by a great wealth of such thin fibers, knowledge of the circuitry of such nervous systems has been greatly delayed. This is particularly unfortunate since information about the connections of a given cell group provides important clues to the identification of interspecific homologies, a matter of fundamental importance to comparative neurology.

The Selective Impregnation of Degenerating Axons

Before the introduction of silver methods selective for degenerating axoplasm, several fiber systems of the reptilian brain had been traced experimentally by the use of Nonidez' modification of the Cajal stain (Armstrong, 1950, 1951; Gamble, 1957; Goldby and Robinson, 1963; and others). Since the method is not selective for degenerating fibers, and impregnates an abundance of normal fibers, considerable difficulties were encountered in these attempts to trace degenerating axons to the target cell aggregates. The use of the Glees (1946) modification of the Bielschowsky technique for degenerating terminals has been attended by the same difficulty, as also has a recent "on the slide" modification of the Cajal process (Scalia *et al.*, 1968).

The now popular "Nauta methods" have evolved through a number of phases that began with the Nauta (1950) and Nauta and Gygax (1951) techniques (Method 1 in Appendix) for the silver impregnation of degenerating boutons. The lack of selectivity and the consequent abundant and dark impregnation of normal fibers made it difficult to locate areas of termination with these methods, and it was not until Nauta and Ryan (1952) and Nauta and Gygax (1954, Method 2 in Appendix) introduced steps inhibiting the staining of normal fibers that the tracing of degenerating axons by silver impregnation came into wide use. In the Nauta-Gygax procedure, the selectivity of impregnation was attained by a pretreatment of frozen sections with phosphomolybdic acid followed by potassium permanganate. In both procedures, the subsequent silver impregnation step is followed by an ammoniated silver nitrate solution, but in 1956 Chambers, Liu and Liu introduced Laidlaw's ammoniated silver carbonate solution as a substitute for ammoniacal silver nitrate, and thereby enhanced the clarity of the background and the contrast value of the degenerating-fiber impregnation. Nauta included this modification in his description of the procedure that has been most widely used so far (Method 3 in Appendix).

It was only several years later that the selective silver methods were first applied to nonmammalian nervous systems. Chambers *et al.* (1956) used the Nauta technique on frog material, and a thorough study of the retinal projections in Cryptobranchus by a modification of the 1957 procedure was published by Riss *et al.* (1963). Work in nonmammalian species has required a considerable variety of procedures which will be reviewed in the following account.

In 1959, Albrecht and Fernstrom introduced a modification particularly suitable for human material. In this method, phosphotungstic acid is the principal mordant (Method 4 in Appendix). About 1962, Nauta (unpublished) recommended uranyl nitrate as a substitute for phosphomolybdic acid to a number of investigators (Method 5 in Appendix). This mordant was found to enhance the selectivity of impregnation, and has been used successfully in numerous studies of mammals and in several nonmammalian species including the Tegu lizard (Ebbesson, 1966, 1967), pigeon (Karten, 1965), shark (Ebbesson, 1967; Ebbesson and Ramsey, 1968), teleosts (Ebbesson, 1968), and frog (Ebbesson, unpublished). In our laboratory, a slightly modified uranyl nitrate procedure (Ebbesson and Rubinson, 1969: Method 6 in Appendix) was applied successfully to a wide range of animal species including the Rhesus monkey, rat, cat, European hedgehog (Ebbesson, unpublished), shark (Ebbesson and Heimer, 1968, 1970), teleosts (Campbell and Ebbesson, 1969), amphibians (Ebbesson, 1969; Rubinson, 1969), reptiles (Ebbesson, 1969; Voneida and Ebbesson, 1969; Karten and Ebbesson, 1969; Ebbesson and Karten, 1969). Because of its remarkable stability and wide range of application, this procedure has become our method of choice for the selective impregnation of degenerating fibers.

Selective Impregnation of Degenerating Axon Terminals

Useful as the methods mentioned in the foregoing account were found to be for the tracing of axon trajectories and — in most cases — the identification of areas of termination, they rarely afford an optimal impregnation of the degenerating axon terminals. It is, therefore, fortunate that recent further experimentation with the Nauta-Gygax method has led to two techniques (Fink and Heimer, 1967) that extend the selective silver impregnation of degenerating axons distally far enough to include, in many cases at least, synaptic endstructures (see chapters by Heimer and by Guillery, this Volume). Although the Fink-Heimer procedures have been applied successfully in several studies on nonmammalian forms (Ebbesson, 1967, 1968; Ebbesson and Ramsey, 1968; Scalia *et al.*, 1968), observations in a study of the olfactory system of the shark have suggested that the methods do not in all species provide optimal impregnation of degenerating nerve fibers and terminals. The most important variable appears to lie in the pretreatment sequence of uranyl nitrate and potassium permanganate. In a study of the selachian olfactory system (Ebbesson and Heimer, 1968, 1970), it was found necessary to modify the Fink-Heimer procedures, and the resulting procedure (Method 7 in Appendix) has proven useful also in other vertebrate species including teleosts (Campbell and Ebbesson, 1969; Ebbesson, 1969), amphibians (Ebbesson, 1969; Ebbesson and Camuñas, 1969), reptiles (Ebbesson and Karten, 1969), and mammals, including the Rhesus monkey.

Not surprisingly, the argyrophilic characteristics of degenerating axons vary somewhat with the species and fiber systems studied. Although the sequence of pretreatment described in Method 7 (Appendix) gave the most consistent good results, other combinations such as reversal of the potassium permanganate-uranyl nitrate sequence, or mixing of the two solutions, may be found preferable in individual cases. Among the best preparations of terminal degeneration in the shark forebrain, for instance, are some that were obtained with a ten-minute treatment in 2.5% uranyl nitrate as the only step before the silver-pyridine impregnation.

A significant departure from the original procedure consists in the deletion of the decolorizing step following the potassium permanganate treatment. Whereas the bleaching procedure in most cases has but a slight effect on the argyrophilia, it will in other cases significantly suppress the impregnation of degenerating axoplasm.

The repetition of the pretreatment sequence (Step 3) in Method 7 usually has a "suppressive" effect sufficient to make the black-impregnated degenerating axons and their terminals stand out among the light-brown stained normal nerve fibers. This light brown tinge of normal fiber tracts is not disturbing, and may even be of value for orientation. Should the selectivity for degenerating fibers prove insufficient, Step 3 can be repeated once or twice more; alternatively, bleaching of the sections after the last permanganate treatment may enhance the preferred argyrophilic response.

In some cases, however, even the pretreatment described here may be too "suppressive" in its effect. If so, a further shortening or even complete deletion of the potassium permanganate step may be necessary. This need is encountered most often in the processing of brain tissue that has been fixed in formalin for a considerable time period (Eager and Barrnett, 1967).

Almost certainly, further successful variants of the silver techniques for degenerating axons are currently used in the various laboratories engaged in comparative neuroanatomical studies. The experiences of Roth *et al.* (1969) with the silver impregnation of degenerating axoplasm in fresh-water teleosts and invertebrates appear particularly noteworthy; a summary of their as yet unpublished techniques and observations is included in the Appendix (Method 8).

It is remarkable that nearly all of the above-mentioned methods for the light-microscopic demonstration of degenerating axons are based on the Bielschowsky principle of silver impregnation which is characterized by the use of labile silver-doublesalts that are easily reduced to yield nascent metallic silver by such weak reducing agents as formaldehyde. In the Cajal principle, by contrast, nascent silver is liberated from the much stabler salt silver nitrate, a process requiring the use of much stronger reducers such as hydroquinone or pyrogallic acid. Several variants of the Cajal principle have been found, in earlier experimental studies, to permit the demonstration of degenerative changes in boutons terminaux; since these changes invariably were characterized by the appearance of swollen or otherwise distorted ring-shaped structures in the region of axon termination, it now appears likely that these techniques impregnate the neurofibrillar component of degenerating synaptic endstructures (see Guillery's chapter in this volume). None of the Cajal techniques, however, were found in these studies to be particularly useful in the tracing of degenerating axons. It is therefore noteworthy that Scalia *et al.* (1968) recently have described an "on the slide" procedure permitting the tracing of degenerating axons of the amphibian olfactory tract. It remains to be seen if this procedure, like the Bielschowsky method, can be further modified so as to afford a selective impregnation of degenerating axons.

Procedures

It is now widely recognized that every fiber system studied requires individual attention in order to obtain optimal demonstration of degenerating fibers. In addition to variation of post-operative survival times, the silver impregnation techniques themselves must often be modified if optimal results are to be achieved. A number of minor but important procedural variations have evolved in various laboratories.

In the following account, the methods currently used in our laboratory will be described in some detail. Although all of the staining methods described in the Appendix are used at least occasionally, Methods 6, 7 and 12 are employed almost routinely. The end product in most lesion experiments consist of: 1. a set of Nissl preparations, 2. a series of sections impregnated with silver to show debris of degenerating axons (Method 6 in Appendix), and 3. a series of sections impregnated more especially for degenerating axon terminals (Fink-Heimer techniques, or Method 7 in Appendix). Although both silver methods to a certain extent stain both degenerating axons and their terminals, we have found it advantageous to impregnate for each separately.

Fixation

Fixation is initiated, whenever possible, by perfusion. Terrestial vertebrates are first perfused transcardially with physiological saline at room temperature (about 250 cm^3 for a 1,500 g Tegu lizard, 50 cm^3 for a bullfrog), and then with 10% formol-saline cooled to 4—10° C (about twice the volume of saline previously given). All teleosts or elasmobranchs are perfused directly with 10% formaldehyde at approximately 25° C (about 50 cm^3 for a 20 cm Holocentrus, 4 liters for a 120 cm shark, 20 liters for a 400 cm shark).

Whenever possible, the brain is exposed immediately after perfusion but left *in situ* for several days while stored in 10% formalin, preferably in the refrigerator. This procedure allows the brain tissue to harden and to be removed with minimal risk of damage. We now store all whole brains or blocks of tissue in 10% formalin in the refrigerator until embedding.

Embedding

Most small brains require embedding in order to prevent the separation of hemisphere and brain stem components of frozen sections. At about the time of initiation of comparative Nauta studies, Snodgress and Dorsey (1963) made an important contribution by developing a method in which egg albumin is used as an embedding medium for central nervous system tissue. The medium, conveniently prepared by dissolving commercial powdered egg albumin in water, does not penetrate into the tissue, and hence, does not affect the staining properties of the material. Suitably hardened by formalin, it forms a resilient mold around the specimen.

The Snodgress-Dorsey technique has been slightly modified in at least two laboratories. Mr. Fink and Mrs. Kess (personal communication) use a combination of egg albumin and gelatin (Method 10 in Appendix), whereas in our laboratory fresh egg yolk is used instead of albumin for the reason that we have found that medium to adhere more firmly to the surface of the tissue (Method 9 in Appendix). An additional advantage of the Snodgress-Dorsey method is that it permits the embedding of several brains in one block, a procedure that may save a considerable amount of the technician's time. The egg yolk block is always marked by cutting off the corner closest to the brain with the lowest identification number.

Microtomy

Before sectioning, the blocks of tissue are placed in a 30% solution of sucrose until they sink to the bottom (this may require several days). This treatment facilitates sectioning and, more important, greatly reduces the risk of damage to the tissue caused by the formation of large ice crystals.

Our method of frozen-sectioning is essentially that described by Marshall (1940). The block to be sectioned is frozen on a rectangular brass platform which is immovably attached to a plastic or wooden block that, in turn, is tightly held in the clamp of a sliding microtome (we have found the American Optical Company microtome 860 particularly suitable for our purpose). Cooling is achieved by placing a mixture of dry ice (solid carbon dioxide) and absolute alcohol (either ethanol or methanol) in a brass trough soldered to one or two sides of the platform. In order to accelerate freezing — and thus prevent the formation of large ice crystals — the block is

Fig. 1. Stratum cellulare et medullare externum of the optic tectum of a tiger shark that had survived unilateral eye enucleation for 13 days. The fusiform swellings of the optic fibers in the tectum reflect an early degenerative change. Method 5 in Appendix. ×1,040

Fig. 2. Stratum cellulare et medullare externum of the optic tectum of a tiger shark; 19 days postoperative survival. Note condensation of argyrophilic material within a varicose swelling. Method 5 in Appendix. ×1,040

Fig. 3. Stratum cellulare et medullare externum of the optic tectum of a tiger shark; 37 days postoperative survival. Fusiform swellings are no longer apparent at this stage and have been replaced by smaller irregularly shaped argyrophilic masses. Method 5 in Appendix. ×1,040

Fig. 4. Stratum cellulare et medullare externum of the optic tectum of a tiger shark; 58 days postoperative survival. Irregularly beaded appearance of degenerating fibers of passage in the tectum corresponds to the usual "Nauta picture". Method 5 in Appendix. ×1,040

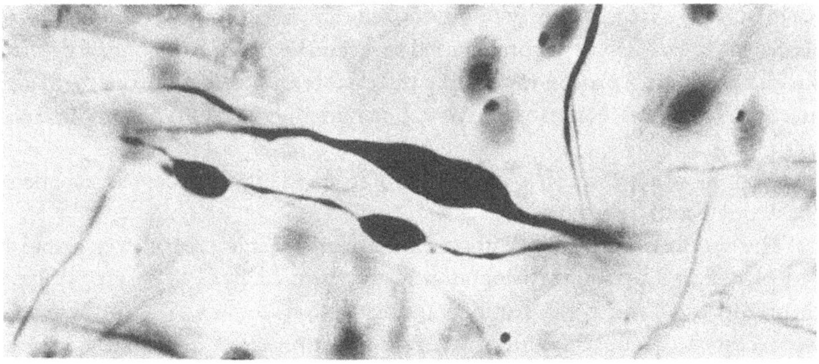

Fig. 1

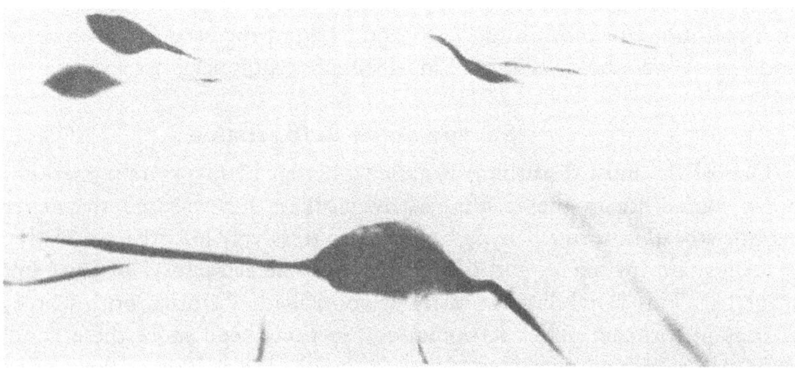

Fig. 2

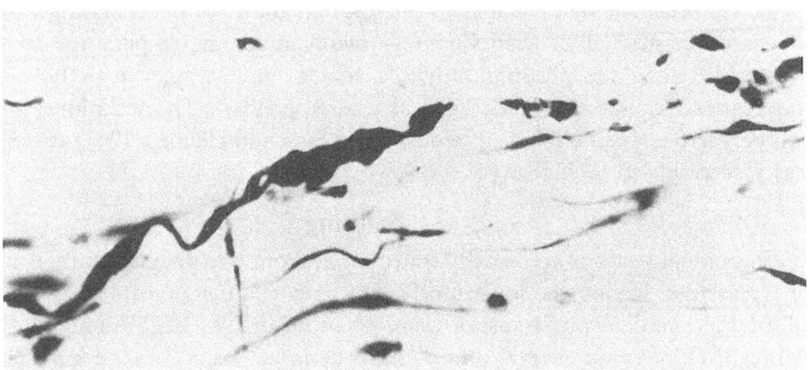

Fig. 3

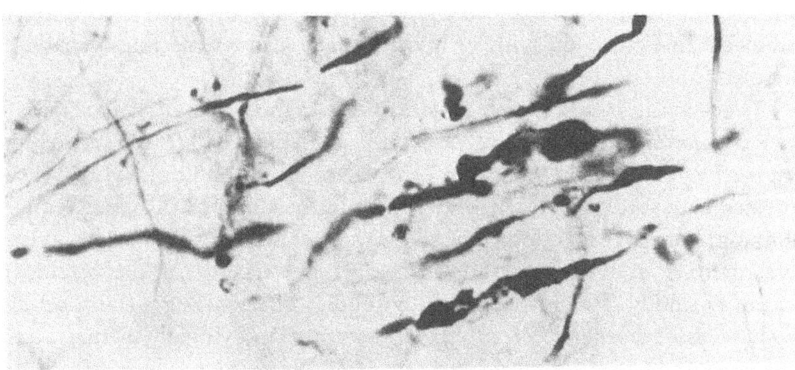

Fig. 4

covered with a large quantity of pulverized dry ice promptly after being placed on the platform. It can also be recommended to keep the block surrounded dry ice throughout the sectioning procedure: unlike material frozen directly from formalin or water, sucrose-soaked tissues can be frozen maximally without becoming too hard for sectioning.

Sections are cut at 33 μ in our laboratory. This thickness is excellent for both Nissl and Nauta preparations.

The sections are lifted from the microtome blade with a wet camel hair brush and placed in a plastic tray containing 1% formaldehyde. These trays have 24 compartments over which the sections are distributed sequentially so that each compartment comes to contain one of every 24 sections, i.e. one section per 800 μ. The sections collected in compartments 1, 7, 13 and 19 are usually processed for degenerating fibers (Method 6 in Appendix), those in 2, 8, 14 and 20 are stained by a Nissl method, and those in 3, 9, 15 and 21 are processed for axon terminals (Method 7 in Appendix, or original Fink-Heimer techniques).

Storage under Refrigeration

One of the most disturbing vagaries inherent in silver impregnation is the day-to-day variability in the results. Many workers have shared the experience that sections stored in formaldehyde become progressively less tolerant of the potassium permanganate treatment, and finally may become refractory to silver impregnation altogether. This variability can easily be controlled by storing both whole brains and sections in formalin under refrigeration. Sections kept under these conditions may retain stable staining characteristics for time periods as long as three years.

It is remarkable that changes in the impregnability of degenerating axons related to extended storage of sections in formalin at room temperature are especially noticeable when the phosphomolybdic acid or uranyl nitrate methods for axons trajectories are used (Method 3, 5 and 6 in Appendix). The techniques of Albrecht and Fernstrom (Method 4 in Appendix), and Fink and Heimer (1967) are considerably less susceptible to such changes.

Staining

Individual sections are usually transported from one dish to another with the aid of a glass rod having the lowermost 1 cm bent at an angle of 45—90°. In order to afford the simultaneous transportation of more or less large numbers of sections, Albrecht (1956), and several others, have designed section carriers made of plastic rings sawed from a tube and covered on one side by a piece of nylon hose firmly stretched and glued on with cellulose cement. Coors crucibles can be used to the same end.

Time-saving — and therefore valuable — as these procedures for mass-transportation of sections are, they are not without risks. Too great a crowding makes it impossible to prevent "stacking" of sections, a condition which easily leads to unequal exposure particularly undesirable in the ammoniacal silver-reducer sequence (ammoniacal silver solution trapped between sections may cause heavy and irregular over-staining and thereby seriously affect the quality of the sections). For this reason — and unless only a moderate number of sections are transported simultaneously — most workers prefer to process sections individually by means of a glass rod

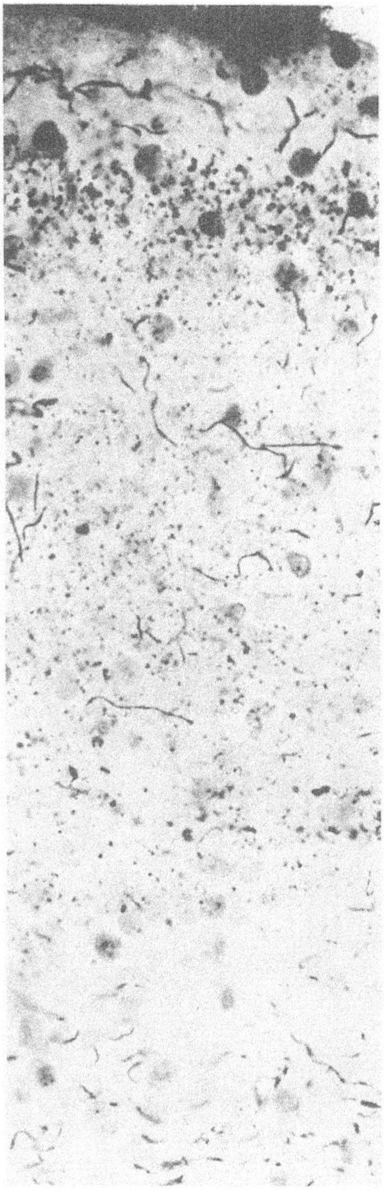

Fig. 5

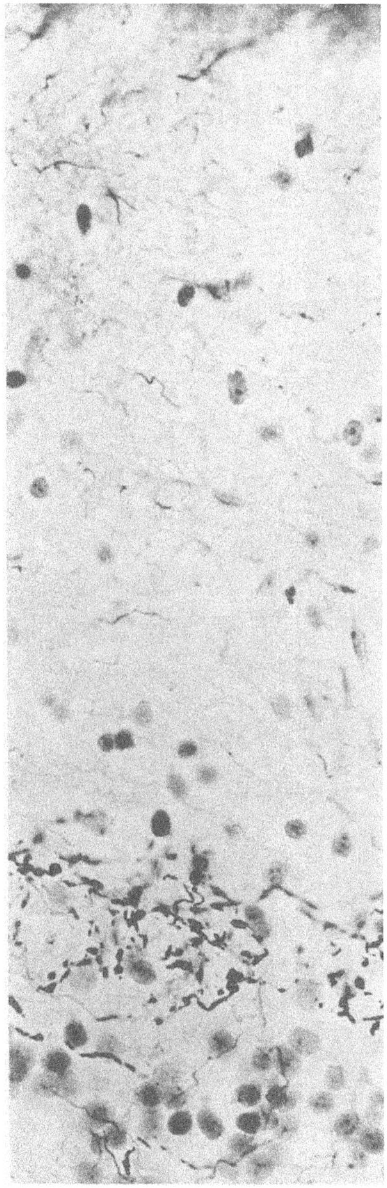

Fig. 6

Fig. 5. Stratum griseum et album superficiale of the optic tectum of a bullfrog that had lived at 25° C for 5 days after eye enucleation. The field of terminal degeneration extends throughout the layer. Modified Fink-Heimer method (Method 7 in Appendix). ×640

Fig. 6. Identical area in a section adjacent to the one illustrated in Fig. 5. Note absence of evidence of axon degeneration in the superficial (upper) regions of the layer. The degenerating fibers seen in the lower part of this photograph correspond to the third layer of incoming retinal fibers. Modified Nauta method (Method 6 in Appendix). ×640

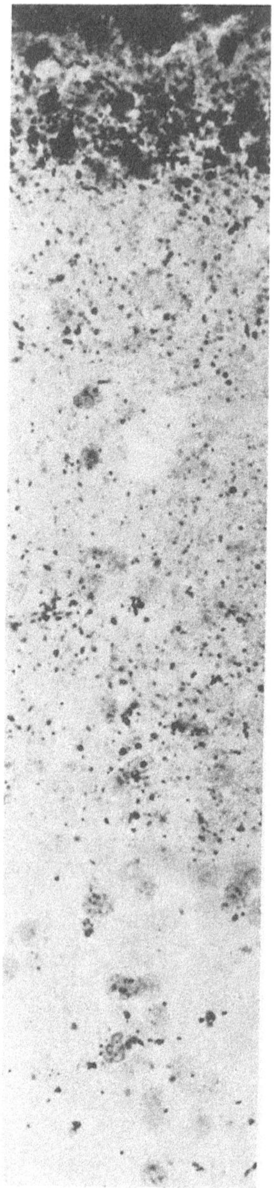

Fig. 7

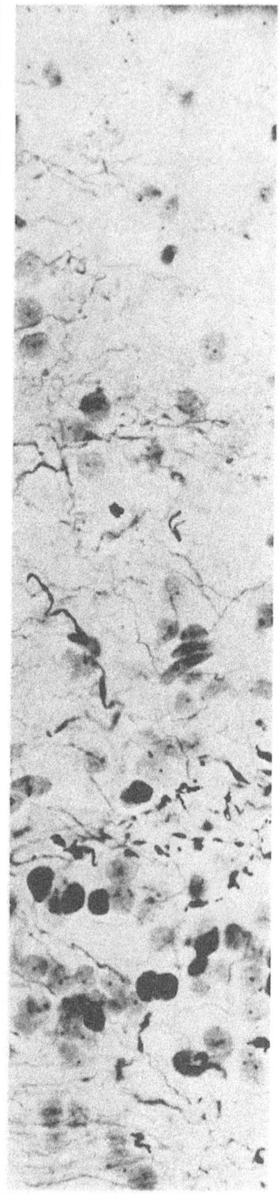

Fig. 8

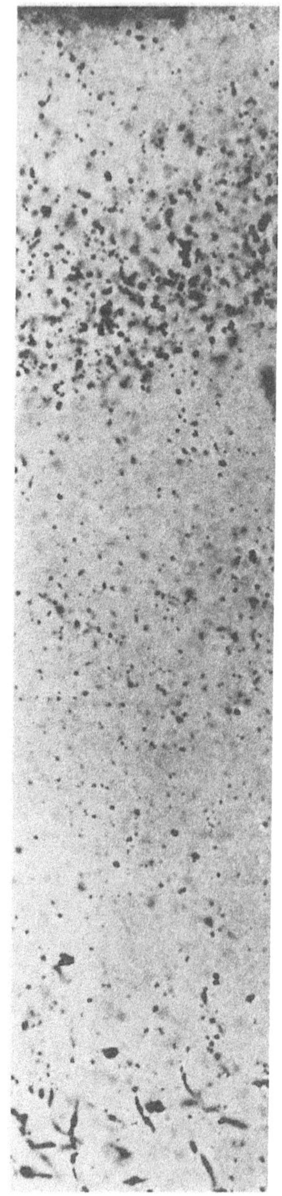

Fig. 9

Fig. 7. Left: The same region of the optic tectum in a frog that had survived the operation for 10 days at 25° C. Modified Fink-Heimer stain (Method 7 in Appendix). ×640

Fig. 8. Center: Identical area from section adjacent to the one illustrated in Fig. 7. Modified Nauta method (Method 6 in Appendix). ×640

Fig. 9. Right: Identical area in a section adjacent to the one represented by Fig. 8 and processed following the same technique except for the use of ammoniacal silver nitrate instead of Laidlaw's solution (Method 6 in Appendix). ×640 through ammoniacal silver and the subsequent reduction fluid. The use of agitating platforms eliminates some of the risks of collective transportation, but is inadequate to meet the need for completely free exposure of each individual section to the reducing fluid.

All of the sections shown in the illustrations had been processed strictly according to the various techniques described in the Appendix. However, it is quite often necessary to modify these procedures. Each individual brain requires individual attention and a trial period during which the pretreatment and other steps can be tested and changed according to the specific needs imposed by such variables as the fiber system under study, or the fixation time of the material.

It is important that all solutions be made up of reagent-grade chemicals and distilled water. All glassware should be rinsed at least 3 times in distilled water before use.

A too dark staining of the background of the silver sections can be prevented by either adding ammonium hydroxide to the ammoniacal silver solution, or increasing the amount of citric acid in the reducer.

Problems of Interpretation

Various sources of misinterpretation in the identification of degenerating axons and their terminals have been discussed by Glees and Nauta (1955) and by Heimer in the present volume. A thorough familiarity with the pseudo-degenerative phenomena described by these authors is indispensable if serious errors are to be avoided.

In the context of the present chapter, it is appropriate to emphasize the misleading pictures that can result from excessive drying of silver-impregnated sections during the mounting process. Thoroughly dessicated sections often show bizarre distortions of normal fibers, as well as accumulations of round black structures of various sizes that may represent a local agglomeration of the fine, largely sub-microscopic silver particles that normally form the brown "background" of silver-impregnated sections. The distortion of normal fibers often appears in the form of corkscrew-shaped axon fragments; combined with an accentuated spindling this phenomenon can easily suggest a degenerative axon fragmentation. Albrecht's method (Method 11 in Appendix) of mounting sections from an alcoholic gelatin solution was developed in a successful attempt to prevent such artifactual distortions.

Limitations of the Silver Methods

Useful as the modern silver methods have proved to be for the tracing of axonal connections, it must be emphasized that they are by nature subject to the limitations of light-microscopy. The formidable obstacles imposed by the limited resolution of the optical microscope are discussed in the chapters by Heimer and Guillery in this Volume. By way of summary, it can be said that it would be unreasonable to expect the silver methods: a) to reveal the postsynaptic component of a synaptic complex, even when demonstrating degeneration of the presynaptic axon endings, and b) to demonstrate degeneration of nerve fibers whose diameters lie below the resolution limits of the light microscope. The latter of these points can perhaps be illustrated and amplified by the following example. Maturana (1959) in an electron-microscopic study found 470,000 unmyelinated axons in the optic nerve of the frog *(Rana pipiens)*

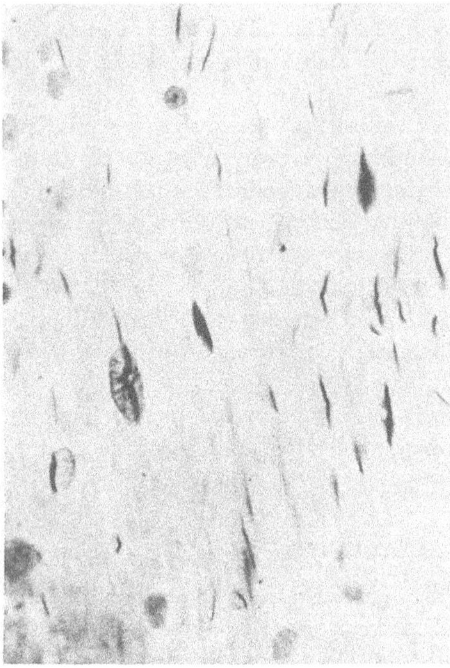

Fig. 10

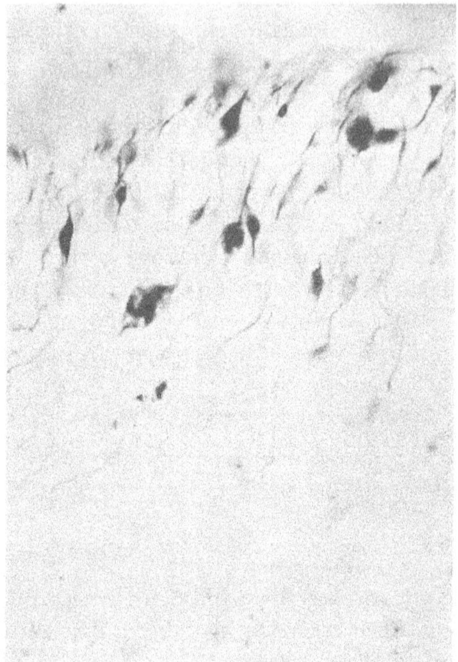

Fig. 11

Fig. 10. Early stage of axon degeneration in the posterior accessory optic tract of a bullfrog that had lived at 5° C for 35 days after eye enucleation. The fusiform swellings shown here were the only signs suggestive of axon degeneration in this animal. Modified Fink-Heimer technique (Method 7 in Appendix). ×1,040

Fig. 11. Degenerating fibers in the medial marginal optic tract in the same frog described in Fig. 10. Modified Fink-Heimer technique (Method 7 in Appendix). ×1,040

and about 440,000 retinal ganglion cells. Comparing these results with those of Bruesch and Arey (1948) whose studies by the optical microscope had indicated that the frog's optic nerve contains 15,300 myelinated and 13,700 unmyelinated fibers, it appears that about 95% of the frog's retinal efferents escape identification with the light microscope. Maturana found that although most of the unmedullated axons are between 0.15 and 0.5 μ in diameter, dimensions within the limits of resolution of the light microscope, the distances which separate the fibers are much smaller and are therefore not resolved.

Silver methods naturally depend on the formation of metallic aggregates of sufficient size and density to interfere with the visible light. Whether or not this condition is met obviously depends first and foremost upon the dimensions of the structure to be demonstrated, but it is by no means certain that all axons whose caliber lies within the resolution of the light microscope are *ergo* demonstrable by silver impregnation, in either the normal or degenerated state. Various other factors, ultrastructural as well as chemical, may determine whether the dispersion of metallic silver in degenerating axons of a particular category is compatible with light-microscopic visualization. The success of current silver methods in demonstrating any particular fiber systems may depend on such idiosyncratic factors as much as on proper execution of the techniques.

Time Course Variables

It has become evident lately that a broad range of postoperative survival times must be explored in order to ascertain the phase of axon degeneration during which argyrophilia is maximal. The relevant variables are related to: 1. interspecific differences, 2. system differences, 3. animal temperature differences, and 4. differences of degeneration rate between axons and their synaptic endstructures.

Appearances

The earliest sign of degeneration of an axon stained with the Nauta technique appears in the form of fusiform swellings (Fig. 1) within which, with further progress of degeneration, irregular condensations of impregnable material begin to appear (Fig. 2). As time progresses, the swellings disappear and the more familiar picture of irregularly shaped argyrophilic masses (droplike disintegration) emerges (Figs. 3—4).

The argyrophilic phase[1] of the degenerating axon terminals may appear before or during the sequence described above for the parent axons. In most axon systems it is quite short. Characteristically, the terminal degeneration revealed by the Fink-Heimer methods appears as a scattering of black globular particles without any individual characteristics discernable at the light-microscopic level (Figs. 5, 18). Many areas of terminal degeneration are also identifiable with Nauta techniques and are typically represented by impregnated "droplets" oriented in a manner (Fig. 12) which contrasts by its more random characteristics with the nearly rectilinear alignment of black-impregnated fragments that characterize degenerating fibers of passage.

The Effect of Temperature on the Rate of Degeneration

Gamble and co-workers (1957) observed a correlation between body temperature and the rate of Wallerian degeneration in the distal segment of the transsected popliteal nerve of the frog. In our own experiments, a similar temperature-dependence was noted in the degeneration rate of both central axons and their terminals. In one such experiment (Ebbesson, 1969) ten of twenty frogs subjected to unilateral eye enucleation were placed in an evironment of 5° C, while the other ten were maintained at 25° C. Every fifth day one animal from each group was sacrificed and the two brains were processed together, embedded in the same block (see above) for ease of comparison. In the brains of frogs kept at 5° C, very little evidence of degeneration was found. Even in cases of 35-day survival, the only degenerative signs consisted of fusiform swellings in a few fibers of the optic tract (Figs. 10, 11). No degenerative changes could be identified in the optic tectum. In the group maintained postoperatively at 25° C, by contrast, abundant signs of axon degeneration were found in the optic tectum already in the cases of 5-day survival. It is charac-

[1] The term "argyrophilic phase" is used in this account in a pragmatic sense to indicate the time period during which silver impregnation is most readily achieved *by the technique employed.*

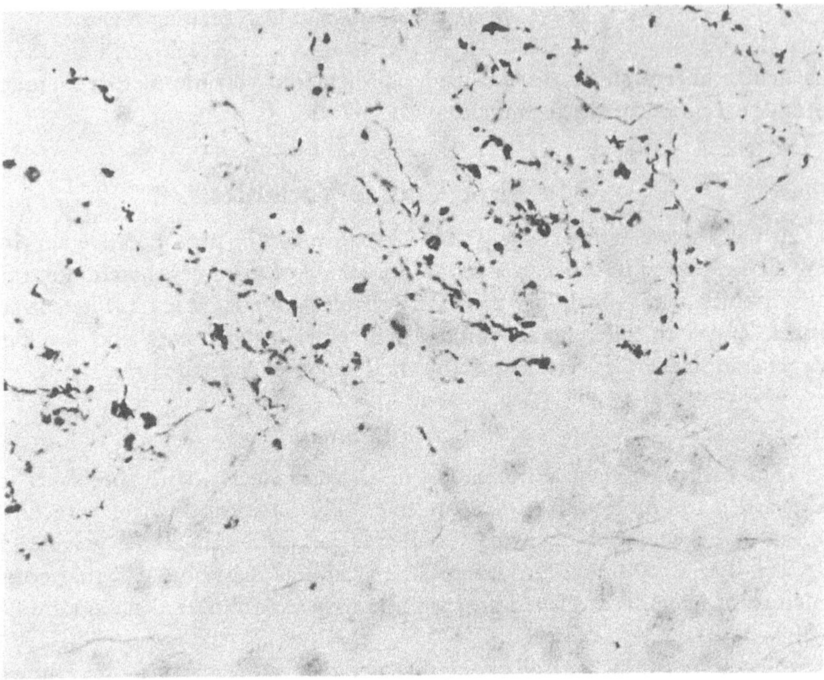

Fig. 12

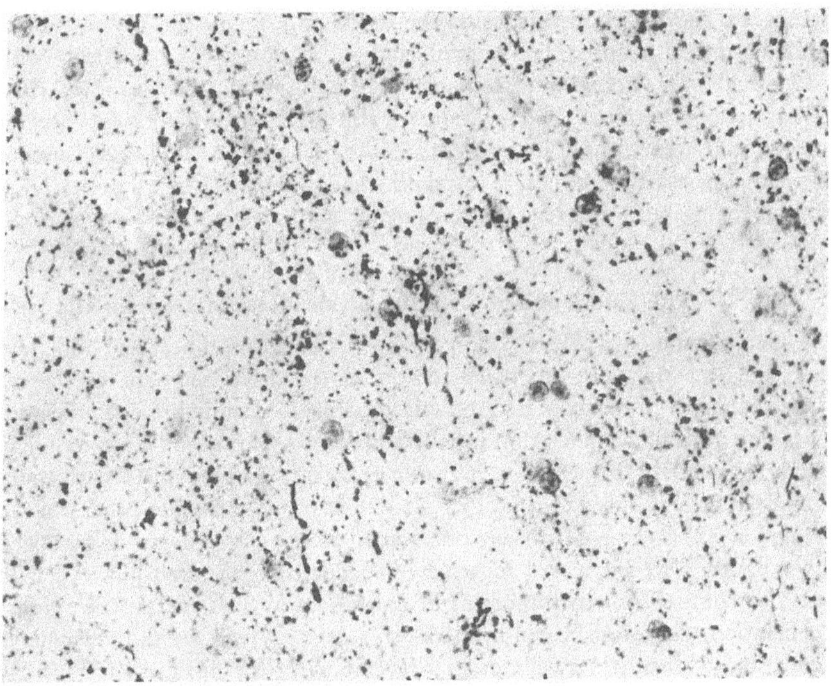

Fig. 13

Fig. 12. Degenerating fibers in the dorsal thalamus of a frog that had survived eye enucleation at 25° C for 15 days. Modified Nauta-Laidlaw technique (Method 6 in Appendix). ×640

Fig. 13. Profuse terminal degeneration in the dorsal thalamus of a frog that lived for 10 days at 25° C after eye enucleation. Modified Fink-Heimer technique (Method 7 in Appendix). ×640 teristic of the techniques used that sections prepared by the Nauta-Laidlaw procedure (Method 6 in Appendix) revealed degenerating axons only in the two deeper of the four distinct retinal-fiber layers entering the stratum album et griseum superficiale of the frog's tectum, in the same cases of short survival (Fig. 6), whereas a dense terminal degeneration appeared throughout the stratum in sections impregnated by the modified Fink-Heimer procedure described as Method 7 in the Appendix (Fig. 5). In frogs having survived surgery for 10 days at 25° C, the picture of terminal degeneration is already less than optimal (Fig. 7) and in cases of survival of 15 days (Fig. 10) and longer, terminal degeneration is no longer identifiable with certainty in Fink-Heimer sections. The Nauta-Laidlaw preparations, on the other hand, show the degenerating preterminal fibers in the deep laminae more convincingly in cases of 10-day survival (Fig. 8), and at 15 days degenerating fibers can be seen in the three deepest of the four layers of retinal fibers. Degeneration in the most superficial layer does not appear in Nauta-Laidlaw sections until 20 days and is prominent at 30 days. In cases of 30-day survival at 25° C, the deeper two laminae are beginning to show evidence of decreased argyrophilia, and by 35 days this decrease is more accentuated.

Whether the four fiber layers are composed of fibers of different size is not known, but the Nauta preparations suggest that the fibers composing the two superficial layers are of considerably finer caliber than those forming the deeper two layers. If this impression is correct, the observations reported above would suggest that, in the case of the frog's retinal efferents at least, degenerating axon terminals become demonstrable first, and are followed by the larger-caliber axons, whereas the thinner axons are the last to become impregnable by the techniques used in this study.

Further notes on the effect of environmental temperature on the rate of axon degeneration in poikilothermic animals, kindly supplied by Dr. Richard Roth, appear in the description of Method 8 in the Appendix.

Interspecific Differences

The argyrophilic phase of both degenerating axons and their terminals may appear after completely different survival times in two poikilothermic species even though the fiber systems studied are homologous and the environmental temperatures are identical.

Tegu lizards *(Tupinambis nigropunctatus)* kept at 25° C after eye enucleation were found to require a survival time of more than 20 days for the demonstration of terminal degeneration (Fig. 15) with the modified Fink-Heimer procedure. The development of demonstrable changes in parent axons required about the same amount of time (Fig. 14). On the other hand, clear pictures of terminal degeneration in the optic tectum were found in a teleost *(Holocentrus)* that had survived eye enucleation for only 8 days in water of about 25° C (Fig. 17); degeneration of stem fibers (Fig. 16) was more pronounced in this species in cases of 14 day survival (Campbell and Ebbesson, 1969). In tiger sharks (Ebbesson, 1967; Ebbesson and Ramsey, 1968) maintained at temperatures ranging between 25 and 29° C and killed 14 days after unilateral eye enucleation, only a few fusiform swellings (Fig. 1) could be demonstrated. In cases of 21-day survival an increase in the number of such configurations was found in the tectum, many of which showed a tendency to

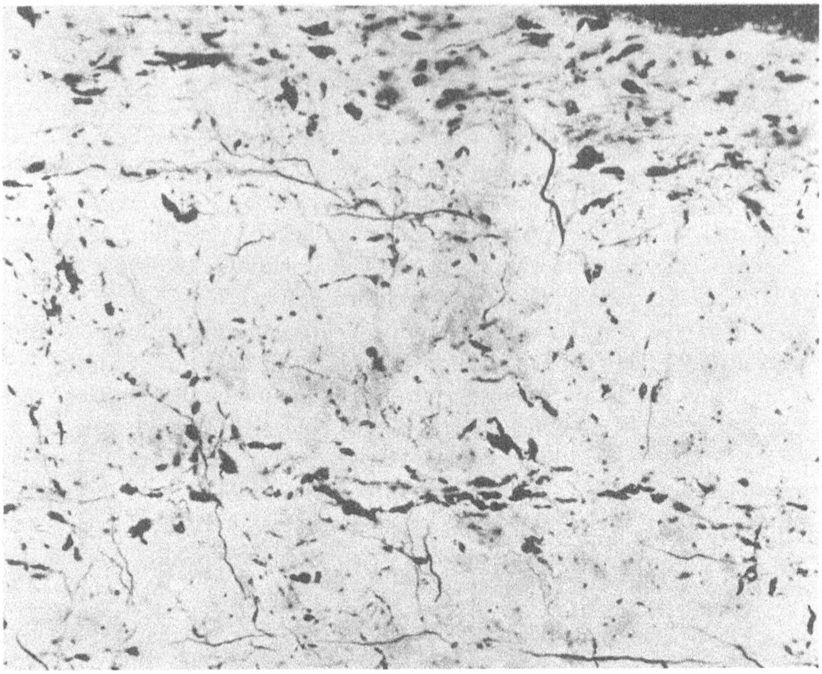

Fig. 14

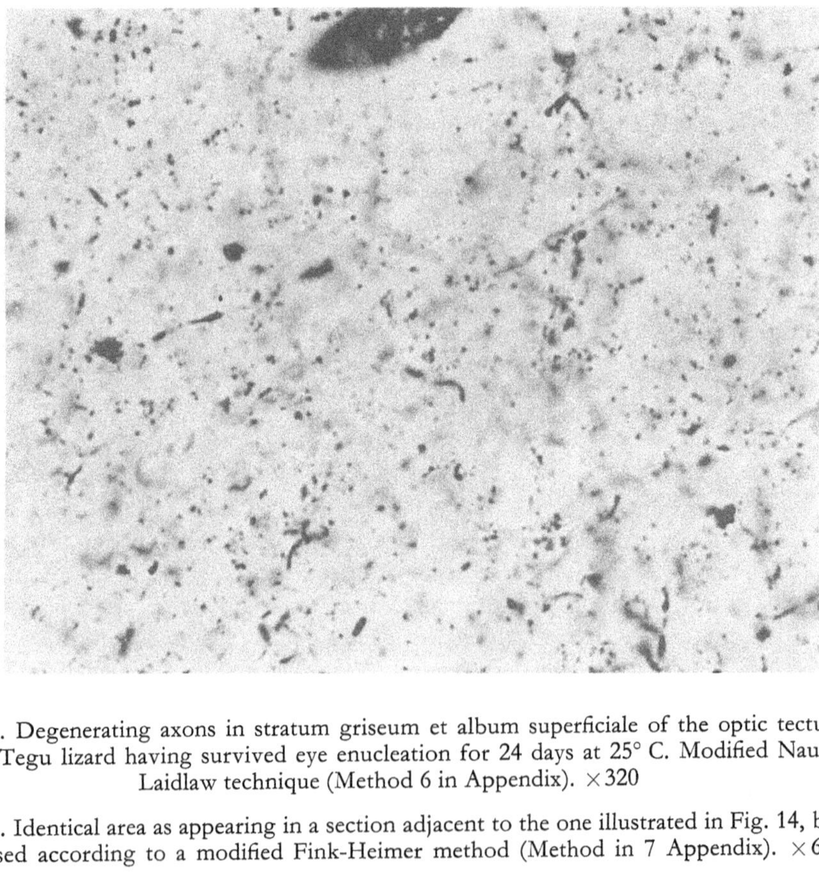

Fig. 15

Fig. 14. Degenerating axons in stratum griseum et album superficiale of the optic tectum in the Tegu lizard having survived eye enucleation for 24 days at 25° C. Modified Nauta-Laidlaw technique (Method 6 in Appendix). ×320

Fig. 15. Identical area as appearing in a section adjacent to the one illustrated in Fig. 14, but processed according to a modified Fink-Heimer method (Method in 7 Appendix). ×640

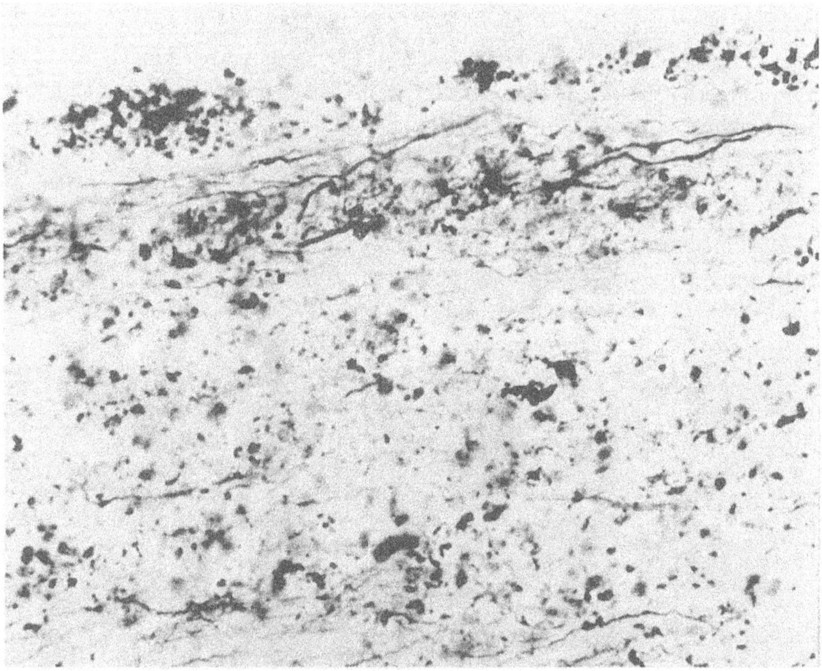

Fig. 16

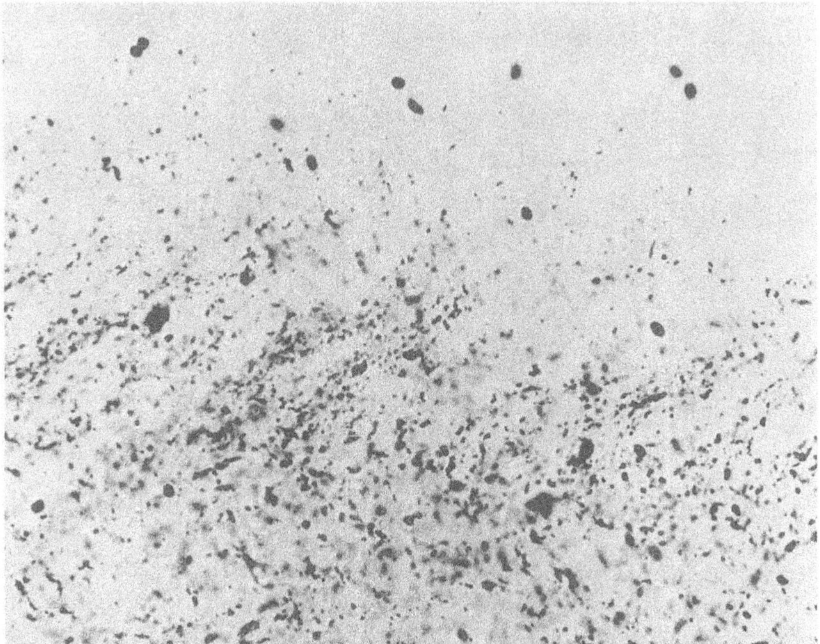

Fig. 17

Fig. 16. Degenerating axons in stratum griseum et album superficiale of the optic tectum in a Holocentrus fish that had lived at 25° C for 8 days after eye enucleation. Modified Nauta-Laidlaw technique (Method 6 in Appendix). ×640

Fig. 17. Terminal degeneration in a section adjacent to the one illustrated in Fig. 16 as revealed with the modified Fink-Heimer technique (Method 7 in Appendix). ×640 develop irregular condensations of argyrophilic material (Fig. 2). At 35 days, only irregularly shaped debris remained (Fig. 3) and after 56 days an even more accentuated picture of disintegration was observed.

System Differences

In animals of the same species maintained for identical postoperative times at the same environmental temperature, marked differences may be noted in the rate at which different fiber systems degenerate. As mentioned above, about 21 to 28 days

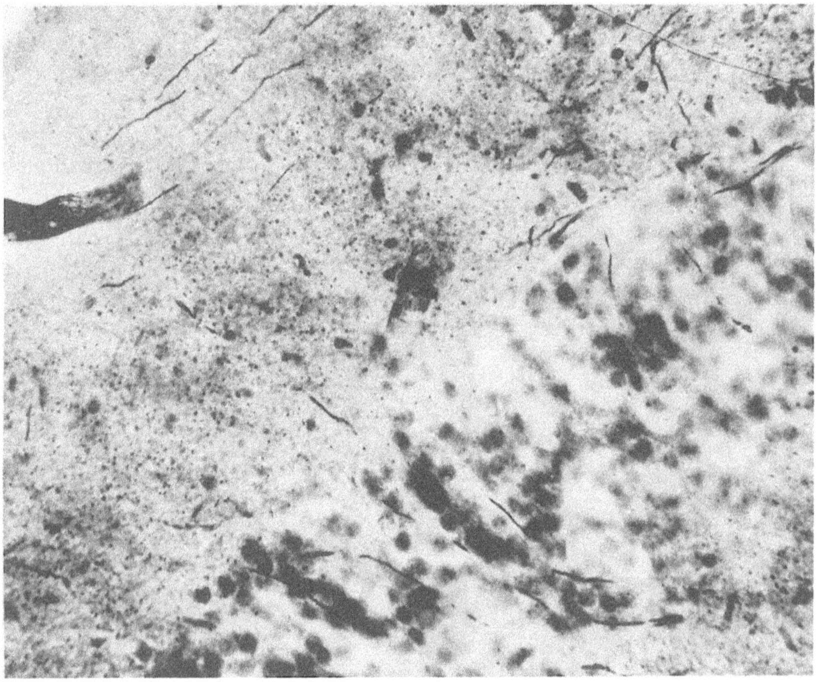

Fig. 18. Terminal degeneration superficial and deep to the granule cells of the medial cortex of the lizard after removal of the contralateral hemisphere. 28 days survival at 25° C (Method 7 in Appendix). ×640

(Fig. 15) are required in the Tegu lizard for the demonstration of terminal degeneration in the optic tectum after eye enucleation (Ebbesson and Karten, 1969), but terminal degeneration in the olfactory cortex caused by olfactory tract section can be demonstrated in this species at only 7 days survival (Karten and Heimer, unpublished observations). Terminal degeneration caused in the medial cortex of the same species by contralateral hemispherectomy requires about 28 days to become demonstrable (Fig. 18) (Voneida and Ebbesson, 1968, 1969). With respect to this experiment, it is interesting to note that without the modified Fink-Heimer technique we could not have identified the commissural afferents to the basal half of the dendrites in the molecular layer. The Nauta-Laidlaw technique stained only the degenerating fibers in the alveus and no more than a sporadic degenerating fiber in the molecular layer.

Summary

The Nauta methods for studying distant connections of neurons has revitalized comparative neurology as no other modern technique has. These methods have recently become widely used in experimental studies in nonmammalian vertebrates and even in invertebrate forms. Recent modifications appear to produce relatively consistent results in the demonstration of fibers as well as their boutons terminaux. Moreover, improved methods for handling frozen sections of small nonmammalian brains have become available.

Recent studies have revealed that the argyrophilic phase of the degenerating boutons in a given axonal system may or may not coincide with that of the degenerating fibers of passage. In most systems, the former appears earlier but is considerably shorter-lived than the latter. The rate of degeneration of different neuronal systems varies in any given species, whether homeothermic or poikilothermic.

Homologous neuronal systems degenerate at different rates in different species even under the same environmental conditions.

In poikilothermic vertebrates, the argyrophilic phase of both degenerating boutons and fibers appears earlier in animals kept in a warm environment than in such maintained at lower temperatures. This finding suggests a correlation between the rate of degeneration and the metabolic rate of the tissue.

It must be understood that the Nauta methods can only provide information about general relationships of neuronal aggregates. Dendrites of neurons, especially in poikilothermic vertebrates, often extend far from their parent perikaryon, and information from Golgi preparations are, therefore, often needed to complement the findings made in Nauta studies. Furthermore, the increasing demand for synaptologic detail virtually insures an ever-increasing use of the electron microscope in studies combining various techniques currently available to the comparative neuroanatomist. Several examples of such combined light- and electron-microscopic methods will be found elsewhere in this Volume.

Appendix

Method 1: Silver Impregnation of Degenerating Axon Terminals (Nauta and Gygax, 1951)

a) Pretreatment

1. Demyelinate sections in a mixture of 1 part NH_4OH (conc.) and 99 parts 50% alcohol for 12 hr. at 30° C.
2. Wash in distilled water.

b) Silver Impregnation

Soak sections in a mixture of: $AgNO_3$, 1.5 gm; water, 100 ml; pure pyridine, 5 ml, 12—24 hr.

c) Ammoniacal Silver Step

Transfer sections to a solution freshly prepared as follows: to 30 ml of a 1.5% $AgNO_3$ solution in 32% alcohol, add 2 ml of NH_4OH (conc.) and 2 ml of a 2.5% aqueous solution of NaOH. Allow sections to remain in this solution 2—5 min.

d) Reduction

Reduce sections until golden brown in a mixture of: 10% formalin, 13.5 ml, 10% alcohol 450 ml, and 1% citric acid 13.5 ml.

e) Fixation

1. Transfer sections to a 2.5% solution of sodium thiosulfate, $Na_2S_2O_3$, for 10 seconds.
2. Wash well in distilled water, dehydrate through alcohols, and collect sections in creosote.

f) Mounting

Mount sections on slides, blot away excess of creosote, and cover.

Method 2: Selective Silver Impregnation of Degenerating Axons (Nauta and Gygax, 1954)

a) Pretreatment

1. Wash sections briefly in distilled water, and soak in 0.5% phosphomolybdic acid for $^1/_4$—1 hr.
2. Without washing, transfer sections to a 0.05% solution of potassium permanganate for 4—10 min. Timing is highly critical: too short will cause impregnation of too many normal fibers; too long will suppress degenerating-fiber impregnation. Optimum duration varies from specimen to specimen and is best established in a trial run, treating 3 sections in which degenerating axons can be expected for 5, 7 and 9 min respectively. During this treatment, sections must be turned over from time to time in order to insure an even effect. Occasionally, results are improved by reducing the potassium permanganate concentration to 0.025%.

b) Decoloration

Assemble sections in distilled water and decolorize in a mixture of equal volumes of 1% hydroquinone and 1% oxalic acid. Duration 1—2 min.

c) Silver Impregnation

Wash thoroughly in at least 3 volumes of distilled water, and soak sections for 20—30 min in 1.5% aqueous silver nitrate. From here on, transfer sections individually.

d) Ammoniacal Silver Step

Wash briefly in distilled water, and transfer section for 1 min to an ammoniacal silver nitrate solution, freshly prepared as follows: silver nitrate, 0.9 g; distilled water, 20.0 ml. When solution is complete, add: pure ethanol, 10.0 ml; strong ammonia, 1.8 ml; 2.5% sodium hydroxide, 1.5 ml.

e) Reduction

Transfer section to a reducing fluid composed of: distilled water, 400.0 ml; pure ethanol, 45.0 ml; 10% formalin (non-neutralized) 13.5 ml; 1% citric acid, 13.5 ml. Allow section to float out evenly on the surface of the fluid. A brown color will appear within 1 min. If the stain is too light, add some sodium hydroxide to the ammoniacal silver solution for subsequent sections; if too dark, add some ammonia.

f) Fixation

Wash briefly; pass sections through 1% sodium thiosulfate for 10 sec, and wash again in 3 volumes of distilled water.

g) Mounting

Mounting is best done prior to dehydration, using a gelatin-alcohol method (Albrecht, 1954: Method 11 in this Appendix). Mounted sections are passed through graded alcohols and 3 vols. of xylene. Sections are enclosed in a neutral synthetic resin and covered with a cover slip.

Method 3: Silver Impregnation of Degenerating Axons (Nauta, 1957)

This procedure follows in detail Steps a—c of the Nauta-Gygax (1954) technique (Method 2 of this Appendix), and differs from the latter only by the use of Laidlaw's ammoniacal silver carbonate instead of the original ammoniacal silver nitrate solution.

After completing Steps a—c of Method 2, proceed as follows:

d) Laidlaw's Ammoniacal Silver Carbonate [2]

Soak section for 45—60 sec in Laidlaw's fluid, moving it about continuously with a glass rod.

e) Reduction

1. Without washing, the section is transferred to a reducing fluid, composed of: 10% formalin, 13.5 ml; 1% citric acid, 13.5 ml; 95% alcohol, 45 ml; and distilled water, 400 ml.

2. While in the reducer, the section should be waved about gently in order to keep it from sinking to the bottom of the container. After about 10 sec, transfer section to a second volume of reducer in which it is allowed to sink to the bottom.

[2] Laidlaw's solution is prepared as follows: a) Dissolve 12 gm silver nitrate in 20 ml distilled water in a 250 ml graduate. b) Add 230 ml saturated (1.33% at 9° C) lithium carbonate (C.P.) solution. Shake vigorously and let the silver carbonate precipitate settle to the 70 ml mark. c) Carefully decant supernatant fluid, refill to 250 ml with distilled water, shake and let precipitate settle again to 70 ml. Decant as before, and repeat this washing twice more. d) Again let precipitate settle to the 70 ml mark, decant, and slowly add ammonium hydroxide under constant shaking until solution has become translucent but is still markedly turbid; avoid excess of ammonia. The solution, when complete, should smell only slightly ammoniacal. e) Dilute with distilled water to a total volume of 120 ml. Filter and store in a chemically clean bottle, exposed to daylight, for at least 2 wks. before use. Unless an excess of ammonia was added, a heavy silver mirror will form on the glass of the bottle after a few days; this can be disregarded. Filter before use. The solution can be used immediately with almost equally good results if 2 drops of 20% formalin are added to 50 ml of the fluid (shake and filter). Excess ammonia, if present, can be eliminated by adding enough silver carbonate (separately prepared and washed) to turn the fluid turbid, at which stage it is filtered and stored.

The section will slowly develop a brown color which is usually complete in 1 min. It should be left in the reducer for at least 2 min. At the end of the reduction, the section should show a rather light brown color. If too dark, a few drops of ammonium hydroxide added to Laidlaw's solution (e.g. 4 drops to 20 ml of the fluid) will improve the quality of the subsequent sections. A too light stain can be counteracted by the addition of a few drops of 2.5% sodium hydroxide to Laidlaw's fluid (unless the fluid is "exhausted").

Steps f—g correspond to those followed in Method 2 of this Appendix.

Method 4: Selective Silver Impregnation of Degenerating Axons (Albrecht and Fernstrom, 1959)

This method was developed for use on human material, but is often successful in animal tissues as well, especially if material has been fixed long in formalin.

a) Pretreatment

1. If necessary (see below), the fixed tissue is placed for 2 wks. or longer in a mixture of 10% calcium chloride, 2 ml; 10% hydroquinone, 5 ml; 10% formalin, 93 ml. This step was repeatedly found to cause a general increase in tissue argyrophilia, especially desirable in previously overfixed formalin material. Frozen sections from such material likewise can often be improved by being stored in this mixture for 24 hrs. or longer.
2. Frozen sections are cut at 25—35 μ, and stored in 10% formalin, preferably in a refrigerator.
3. A convenient number of sections, not more than can be processed in one session, are treated for 0.5—1.0 hrs. in a mixture composed of equal parts of 1% oxalic acid and 1% hydroquinone. This step was found to reduce "myelin cuff" artifact most effectively.
4. The sections are washed and transferred to 0.05% chromic acid for 10 min.
5. The sections are washed and transferred to 4% hydrobromic acid for 5—6 min.
6. After washing, the sections are soaked in 1% phosphotungstic acid for 15 min.
7. The sections are washed and transferred to a 0.05% solution of potassium permanganate for 3—10 min. This step and the subsequent decoloration are not essential for subsequent staining of degenerating axons, but serve to suppress the impregnation of normal fibers. The optimal time of the permanganate treatment is subject to variation; sections of tissue fixed for a long time require a shorter treatment than sections of relatively fresh material.

b) Decoloration

Sections are decolorized in a mixture of equal parts of 1% oxalic acid and 1% hydroquinone.

c) Silver Impregnation

1. After thorough washing, sections are soaked in a 1.5% silver nitrate solution for 15—30 min.
2. Sections are washed and continued separately from here.

d) Laidlaw's Fluid

Each section is soaked for 2—4 min in Laidlaw's ammoniacal silver carbonate solution (see Method 3 of this Appendix). The section should be agitated gently while in this solution.

The remaining stages of this procedure correspond in detail to Steps e—g in Methods 2 and 3 of this Appendix.

Method 5: Selective Silver Impregnation of Degenerating Axons by the Use of Uranyl Nitrate (Nauta, Unpublished)

a) Pretreatment

1. Soak sections in a 5% aqueous solution of uranyl nitrate (uranium nitrate) for 1 hr. Treatment can be extended to 16 hr., but should not be continued much beyond this time[3].

2. Wash sections thoroughly in several volumes of distilled water for a total of not less than 15 min.

3. Transfer sections to a 0.05% solution of potassium permanganate ($KMnO_4$) for 20—60 min.

The remaining phases of this procedure correspond in detail to Steps b—g of Methods 2 and 3 of this Appendix.

Method 6: Silver Impregnation of Degenerating Fibers (Ebbesson and Rubinson, 1969)

a) Pretreatment

1. Wash sections in distilled water (2 vols.) and place in 5% uranyl nitrate for 7 hr.

2. Wash sections in three vols. distilled water.

3. Transfer sections to 0.01% potassium permanganate for 16 hrs. (overnight). The sections should be placed side by side in a staining dish, and fluid should be at least 2 cm deep. We have obtained best results by placing the staining dish on a mechanized platform that slowly and gently agitates the solution. If such a platform is not available, agitate gently every 15 min during the first hour, and then place sections in a fresh solution overnight. Rinse in distilled water once.

b) Decoloration

1. Decolorize section in a solution made up of equal parts of 1% hydroquinone and 1% oxalic acid (mix immediately before use). Rinse in distilled water 3 times as soon as sections are decolorized.

c) Silver Impregnation

Soak sections for 15—30 min in 1.5% silver nitrate.

[3] The radioactive properties of uranyl nitrate constitute no hazard to laboratory workers. However, the substance is extremely toxic, and care should be taken to prevent ingestion of even minute quantities. Wash hands after handling either the dry salt or solutions.

d) Laidlaw Solution[4]

Pass sections individually through 2 vols. (1 min in each) of a Laidlaw fluid modified by the addition of 3 ml of absolute ethanol and 3 drops of concentrated ammonia to each 15 ml of Laidlaw solution.

The remaining steps of this procedure correspond in detail to Steps e—g in Methods 2 and 3 of this Appendix.

Method 7: Silver Impregnation of Degenerating Axon Terminals (Ebbesson and Heimer, 1969)

a) Pretreatment

1. Wash sections in two vols. of distilled water, and soak in 0.01% or 0.025% potassium permanganate 3 to 5 min. Rinse once in distilled water.
2. Place sections in 2.5% uranyl nitrate, 3 to 5 min; rinse once in distilled water.
3. Repeat sequence once.

b) Decoloration

In most cases, this step can be omitted. Should decoloration be necessary, use equal parts of 0.5% hydroquinone and 0.5% oxalic acid.

c) Silver Impregnation

1. Rinse sections thoroughly in 3 vols. distilled water.
2. Transfer sections to 0.5% silver nitrate for 1—2 hr. The addition of 2.0 ml of pyridine to 100 ml silver nitrate solution may improve the results.

d) Ammoniacal Silver Solution

Transfer sections for 5 min to an ammoniacal silver nitrate solution made up of 20 ml 2.5% silver nitrate, 12 ml 100% ethanol, 2 ml 2.5% sodium hydroxide, and 2 ml concentrated ammonia.

Steps e—g of this procedure correspond in detail to Steps e—g in Method 2 and 3 of this Appendix.

Method 8: A Modification of the Nauta Technique for Freshwater Teleosts and Insects (Roth et al., 1969)

a) Fixation

1. *Fish:* Perfuse through the heart with 5—20 ml buffered fish Ringer's solution (Forster and Taggart, 1950) followed by 10—30 ml of 10% formalin made up in Ringer's solution. Expose the brain (and spinal cord if necessary) and immerse the entire animal in 10—20 times its volume of the formol-saline solution.

After being hardened *in situ* for 1—2 days, the desired part is dissected out and placed in fresh formol-saline or 10% aqueous formalin for another 1—2 days.

2. *Insects:* Flood the opened hemocoel with 25% formalin to insect Ringer's (Hoyle, 1953). After several hours of fixation *in situ*, dissect out the desired part of the CNS and store it in 10% formol-saline for 1—2 days.

---

[4] An ammoniacal silver nitrate solution can also be used advantageously. This solution consists of 20 ml 2.5% silver nitrate; 12 ml 100% ethanol; 2 ml 2.5% sodium hydroxide and 2 ml of concentrated ammonia.

b) Embedding and Sectioning

Wash tissues thoroughly in running tap water or in several vols. of distilled water and then embed in gelatin as follows:

1. Incubate tissues at 37° C in 20—25% (W/V) aqueous gelatin for 12—24 hr. at atmospheric pressure.

2. Transfer each piece of tissue to a small petri dish filled with fresh molten gelatin. Place the petri dish over a block of ice and hold the tissue in the desired orientation until the gelatin covering the bottom of the dish has hardened sufficiently to hold the tissue in place.

3. Transfer the petri dish to a refrigerator for 1 hr. or more in order to insure thorough hardening of the gelatin. Trim the gelatin close to the tissue, mark for orientation if necessary and transfer the block to chilled 25% aqueous formalin.

4. 12—24 hrs. prior to sectioning, transfer the gelatin block to 10% formalin made up with 10% ethanol. Cut frozen sections at about 25 μ and collect them in dilute (1—2%) formalin.

c) Pretreatment

Collect sections in distilled water, then transfer them to 0.05% potassium permanganate for 5—20 min.

d) Decoloration

Rinse sections in distilled water and bleach in a mixture of equal parts of 1% oxalic acid and 1% hydroquinone for about 2 min.

e) Silver Impregnation

1. Wash sections for 1—2 min in each of 3 changes of distilled water and transfer to a mixture of equal parts of 0.5% uranyl nitrate and 1.5% silver nitrate for 5 to 10 min.

2. Wash for 1—2 min in each of 3 vols. of distilled water and transfer to a mixture of equal parts of 10% pyridine and 1.5% silver nitrate for 5—15 min.

f) Ammoniacal Silver Solution

Wash for 1—2 min in each of 3 vols. of distilled water; transfer sections, individually or several at a time, to a freshly prepared ammoniacal silver nitrate solution (concentrated ammonium hydroxide, 2.6—3.6 ml; 95% ethanol, 10 ml; 0.75% silver nitrate, 20 ml) for 2—5 min, with periodic agitation.

g) Reduction

Without rinsing, transfer sections to a reducing fluid (10% ethanol, 1,000 ml; 10% formalin, 25 ml; 1% citric acid, 50 ml) for 1 min with agitation in each of two volumes.

h) Fixation

Wash in 3 vols. of distilled water, then soak in 1% sodium thiosulfate for 2 to 5 min.

i) Mounting and Dehydration

1. Wash sections thoroughly in several vols. of distilled water, and mount on albuminized slides from Albrecht's (1954) alcohol-gelatin mixture. It is necessary to allow the surface of the sections to dry before attempting to blot them with filter paper. After the sections have been mounted on slides, it is advisable to break the gelatin matrix in several places in order to avoid wrinkling of the sections by differential shrinkage of tissue and gelatin.

2. Dehydrate through 95% and absolute ethanol, clear in 2—3 vols. of xylene and cover.

Notes

1. The ammoniacal silver nitrate and reducer solutions should be changed after each batch of 8—16 sections have been processed; all other solutions may be used 2 or 3 times, provided the distilled water baths are changed after each use. After reduction, the tissues should be dark yellow to brown and somewhat lighter than the gelatin.

2. Excessive glial impregnation or randomly distributed silver deposits may be prevented in subsequent sections by increasing the ammonium hydroxide content of the ammoniacal silver solution or by the dropwise addition of 1% citric acid to the reducer.

3. For tissues which have been stored in formalin for several months, it may be necessary to use 1.0—1.5% instead of 0.75% silver nitrate in the ammoniacal silver solution. It should be noted that storage of blocks in 10% formalin for 4—6 months at room temperature or for 12—18 months at 5° C has no deleterious effect on the stainability of axonal degeneration products. In fact, it appears that such long-term formalin storage of blocks enhances the contrast between degeneration products and background. Long term storage of sections, on the other hand, yields a rather different result: even when sections are stored in 1% formalin, the demonstration of degenerating axons becoms increasingly difficult. At the moment, no explanation of this paradox is known.

Results and Discussions

1. The method outlined above selectively impregnates degenerating axoplasm. Depending on postoperative survival time, the black silver deposits may be restricted to terminals and preterminal segments of axons (short survival times) or to axons of passage (long survival times), or they may be found in both axons of passage and preterminal segments.

2. In locusts maintained at 25—30° C and in crickets maintained at 20—24° C, degenerating axons and terminals become demonstrable after postoperative survival periods of 6—12 hours; they are maximally argyrophilic at about 18—24 hours. Thereafter the amount of argyrophilic axonal debris decreases until, by 4—8 days, it is very difficult to demonstrate a single degenerating axon.

3. In goldfish of 2.5—5.0 cm, orfes of 4—8 cm, and cichlids of 4—7 cm, kept at 22—24° C, degenerating boutons and preterminal axon segments are demonstrable in great abundance after survival times of 2—10 days; the most proximal portions of degenerating axons are most easily demonstrated in cases of 10—30 days postoperative survival.

4. If consistent results are to be obtained with poikilothermic species, it is imperative that both environmental temperature and specimen size be kept as nearly constant as possible. Data from *Bufo* indicate that the Q^{10} of axonal degeneration is approximately 3.0 for the temperature range of 20—30° C (Joseph and Whitlock, 1968). Our own data from goldfish suggest a temperature coefficient of about the same magnitude for the range of 18—24° C.

5. The dependence of degeneration rate on the specimen size is particularly striking in the goldfish. Under identical environmental conditions, the degeneration of optic tract fibers in 2.5 cm specimens proceeds about 4—5 times faster than in 7.5 cm specimens.

Method 9: Egg Yolk Embedding: A Modified Snodgress-Dorsey (1963) Procedure (Ebbesson, Unpublished)

a) Prepare a paper mold of appropriate size (slightly larger than the specimen to be embedded) using an inner sheet of vegetable parchment paper and an outer sheet of aluminum foil. The latter should have 20—30 pinprick perforations in its bottom.

b) Prepare the specimen by: 1. rinsing in distilled water overnight, 2. drying the surface of the tissue by gently blotting with soft tissue paper or filter paper and 3. exposure to room air until it appears dry.

c) Using a soft brush, thoroughly moisten brain surface with fresh egg yolk, including all crevices, and immerse the specimen into the box containing egg yolk.

d) Put the box with the specimen in a refrigerator overnight.

e) Place the box in a staining dish containing sufficient 25% formalin to wet only the bottom of the box. Orient the brain properly in the box and place a cover over the staining dish.

f) The egg yolk will harden in 12—36 hrs. at room temperature and can then be removed from the container, notched in one corner for orientation purposes, and then placed in 10% formalin for storage in a refrigerator.

g) Before cutting on the freezing microtome, immerse block in 30% sucrose until it sinks. This may take 1—3 days depending on the size of the block.

Method 10: Albumin-Gelatin Embedding: A Modified Snodgress-Dorsey (1963) Procedure (R. Fink and Barbara S. Kess, unpublished)

a) Animals are perfused with formalin or formol-saline and the brains are stored in 10% formalin for at least 1 wk.

b) Brains are soaked in sucrose-formalin[5] for 3—7 days. The brain floats on the surface when first immersed in the sucrose, and the procedure is completed when it sinks to the bottom of the jar.

c) To prepare for embedding, the brain is washed very briefly in 2 vols. distilled water (not more than 2 min each), to remove any sucrose adhering to the surface.

[5] *Sucrose Formalin:* Dissolve 300.0 g sugar in 725.0 ml of distilled water. Shake vigorously until the sugar has completely dissolved. Add 100.0 ml of concentrated formalin. Store the stock solution in the refrigerator.

Surface moisture is carefully blotted off, especially in crevices, with filter paper. Dura and archnoidea are carefully removed together with the larger blood vessels.

d) Carefully position the brain in a small cardboard box[6], with the help of dissecting pins, being careful not to place any pins through brain tissue. Liquid albumin-gelatin[7], is then poured over the brain with an eye dropper, care being taken to avoid any bubbles. The albumin is left exposed to room air until the surface begins to harden slightly. At this stage, the box is placed in a large staining dish containing a small amount of concentrated formalin. By means of a small elevated platform (e.g. the lid of a small glass box, the mold is kept from touching the formalin (liquid albumin will not harden when in direct contact with formaldehyde solutions). The dish is covered and left overnight at room temperature. The next day the mold can be immersed in 20—30% formalin for several days until albumin has fully hardened.

e) Before cutting, the block is trimmed, allowing an albumin border of at least 3 mm thickness around the tissue, and marked by cutting off one corner for proper orientation of the sections. The block is then sectioned on the freezing microtome, care being taken to keep the entire block surrounded with pulverized dry ice on the freezing stage.

Method 11: Mounting Frozen Sections with Gelatin (Albrecht, 1954)

a) Wash sections in distilled water, and soak for 5 min or longer in a mixture of equal parts of 1.5% aqueous gelatin and 80% alcohol. The fluid should be at least one inch deep. If sections are swollen as a result of the preceding staining procedure, the alcohol mixture will cause shrinkage to approximately normal dimensions.

b) By the aid of a fine brush, tease a section onto a clean slide. Withdraw the slide from the fluid in a strongly tilted position in order to encourage rapid drainage of excess fluid. Loss of additional surface fluid will be promoted by carefully wiping around the section with a brush. In a short time, surface fluid will have evaporated evenly from the section which will now show a slightly matted, velvety appearance. Drying is now about to set in, and must, of course, be prevented. The following steps should therefore be performed as rapidly as possible.

c) Blot the section gently but completely with a strip of clean, smooth filter paper, using a fresh strip for each subsequent section. Blotting is best done by bending the strip down upon the section until the entire surface of the section has been in contact with the paper. Although this is usually adequate, some sections require a rapid second blotting in which gentle pressure is applied with the fingertips.

[6] Cardboard Histo-Molds of three different sizes can be obtained from the Lipshaw Manufacturing Company, Detroit, Michigan, U.S.A.

[7] *Albumin-Gelatin:* Dissolve 3.0 g of powdered gelatin in 100.0 ml distilled water in a beaker. When gelatin is completely dissolved, add 30.0 g purified albumin powder (if preparing more than this volume, add the albumin in small amounts). Place the beaker in a pan of water of 50—60° C, and stir frequently until the albumin has dissolved. Remove any froth which has formed on the surface, and store in a tightly capped jar in the refrigerator. For embedding, an appropriate amount of this stock volume is melted in a beaker placed in warm water.

d) The now almost entirely dry section is rapidly immersed in 95% alcohol which will congeal the remaining gelatin, anchoring the section to the slide.

e) If sections are to be stained after mounting, it is advisable further to insure proper adherence by coating the section with celloidin. To this end, the section is passed through absolute alcohol and subsequently immersed for 5 min in a 0.5% solution of celloidin in equal parts of absolute alcohol and ether. The slide is then removed, drained, exposed to air for several seconds, and immersed in 80% alcohol for 5 min or longer. With careful performance of the procedure, no sections will (as a rule) be lost by adherence to the filter paper. An exception is sometimes encountered in mounting thin, unstained frozen sections. In such cases, dampening the filter paper with 95% alcohol may reduce further loss of sections.

Discussion

Dr. Voneida: The important question of degeneration time deserves some further discussion. From what Dr. Guillery and Dr. Ebbesson have mentioned, there is an evident need for careful study of the relationship between degeneration time and the amount of degeneration demonstrable by silver impregnation. In mammalian systems there seems to be, as you, van Gehuchten, Verhaart, and others, have pointed out, an inverse relationship between the size of the fiber and its rate of degeneration. In a study which I began at Dr. Nauta's laboratory a long time ago, very fine efferent fibers of the cat's caudate nucleus, less than one μ in diameter, could be impregnated only after survival times of about 21 days. The remarkable thing is that the opposite appears true for some fiber systems in lower vertebrates. Richard Roth, one of my graduate students who is studying the retinal projections in freshwater teleosts, found massive degeneration of the superficial neuropil 4—5 days following eye enucleation in 2.5 cm specimens. Similar pictures of degeneration could be obtained in 7.5 cm goldfish only after 30 days survival.

Dr. Cowan: Could this finding possibly be related to the age of the fish? It is well known that retrograde and transneuronal cell reactions develop more rapidly in infant animals than in adults, and the same may well be true of Wallerian degeneration.

Dr. Voneida: Yes, I agree that age factors do play a role in the rate of fiber degeneration, but with fish, of course, it is possible to alter the sizes while keeping the ages equal.

Dr. Nauta: It is very likely that there are fibers so fine that we will never be able to show the transit axon in degeneration with silver impregnation techniques. An example of that is the nigrostriatal system which Dr. Fuxe and his colleagues demonstrated with the histofluorescence method. Many of us were, and some of us perhaps still are, very skeptical about this connection because no convincing silver evidence had become available up to that time. Since that time, however, the Fink-Heimer techniques were developed, and these new procedures have enabled Drs. Robert Y. Moore of the University of Chicago, and John Hedreen of the Walter Reed Army Institute of Research, to demonstrate a dense terminal degeneration in the rat's caudoputamen following lesions involving the substantia nigra. Quite recently I have found similar evidence of terminal degeneration in both the caudate nucleus and putamen 3—5 days after stereotaxic lesions confined to the substantia nigra in the cat. So far we have no electron-microscopic verification of this finding, but the familiar pattern of argyrophilic particles appeared only on the ipsilateral side and looked exactly like confirmed terminal degeneration in other axon systems. Despite the considerable density of the presumed terminal degeneration in the caudate nucleus and putamen, only sporadic degenerating parent axons appeared in these cases. The point I would like to raise is that possibly some fibers are so thin that their degenerative fragments cannot accumulate sufficient metallic silver to interfere with the light spectrum enough for us to see them. Unless we manage to catch the corresponding synaptic end bulbs at an optimally argyrophilic stage of degeneration, we may find no evidence of such connections in silver material.

Dr. Fuxe: In this connection I hear that Hökfelt has done a combined fluorescence-electron microscopic study using his own modification of the Richardson technique for the demonstration of dopamine storage granules at the ultra structure level. They were able to demonstrate that there was a very rapid degeneration of the nigro-striatal system. The fluorescence disappeared in 2 days, making it impossible to visualize the dopamine storage granules. It was also remarkable that the degeneration could only be demonstrated occasionally with the electron microscope. They saw evidence of some types of dark degeneration which might be due to the degeneration of the dopamine systems. I think this exemplifies the difficulties we sometimes meet. Some processes are very difficult to catch, especially when rather unspecific techniques are used. In this case, the fluorescence techniques were valuable because of their ability to demonstrate, with a high degree of specificity, the transmittor stores themselves.

Dr. Fox: Dr. Robert Schwyn, in our laboratory, has looked at the comb bundle with the EM. In addition to some myelinated fibers, it is loaded with extremely fine unmyelinated fibers and these fields have the same consistency as the EM fields in the pars reticulata of the substantia nigra. These unmyelinated fibers are of the caliber postulated for the nigro-striatal system. But what bothers me is that in our Golgi preparations the neurons of the substantia nigra are large. They resemble the cells of the globus pallidus. Cole, Nauta and Mehler (1964), and Carpenter and Strominger (1967) have demonstrated nigrothalamic connections in Nauta preparations. Do the large neurons of the substantia nigra have thick axons going to the thalamus, which give off exceedingly thin collaterals that run to the striatum? They would be unusual looking neurons.

Dr. Fuxe: Perhaps we can differentiate between different types of nerve cells in substantia nigra?

Dr. Fox: How many types are there?

Dr. Fuxe: I think there are at least two types. You have the large dopamine nerve cell type in the zona compacta, but there are also a few in zona reticulata. There must be several quite different types of neuronal systems in the substantia nigra region. There are also functional data which support this view.

Dr. Nauta: Dr. Fuxe's comment about the rapid disappearance of fluorescence reminds me of what we see in the nigro-striatal system, that is, the postoperative survival must be only a few days or we cannot demonstrate it any longer with our silver methods. I wonder if the temporal characteristics of Wallerian degeneration in any particular axon may not depend on chemical specifications as much as on the size of the axon? It might, of course, turn out that all dopamine systems have very thin axons, and that this factor is in some way related to the rate of degeneration.

Dr. Fuxe: Well, it could be related. The dopamine systems have somewhat thinner axons than the noradrenalin and 5-hydroxytryptamine systems, and also degenerate somewhat faster than the latter two systems.

Dr. Guillery: Is it possible to agree on a tighter terminology as regards "terminal", "preterminal", and "bouton" degeneration? Instead of referring to preterminal degeneration and bouton degeneration for degeneration products that lie around cells, would it not be more appropriate to call that pericellular degeneration, and then describe whether it is fine or coarse?

Dr. Heimer: I think maybe we could use the term "terminal degeneration" in the sense that Dr. Nauta introduced it almost 20 years ago, and include in that term the degeneration of the synaptic terminals. Of course, when we call these black dots "degenerating boutons", we should realize that not all of them are necessarily degenerating synaptic endbulbs. Depending on such factors as survival time or the stain modification, and others, varying proportions of these black particles could also be degenerating fibers, and axon debris engulfed by glia cells. Another point I would like to make, and that was also emphasized in Dr. Ebbesson's talk, is that we might not see everything. We tend to think that each new method is "fool-proof" and once it works for us, we sit down and use it. At the moment we tend to say that we have an optimal impregnation when we see black dots in the area with a nice suppression of normal fibers. This may not be the optimum, and we must, therefore, continue to strive for better methods.

Dr. Sidman: With reference to the problem of ascertaining the optimal time for demonstration of degenerating axons, I would like to make a brief case for the use of animals with lesions other than surgical ones. Some of the mouse mutations cause neurons of particular classes to degenerate over a span of days or weeks, and other types of lesions, e.g., certain of the ones induced by pharmacological or nutritional means, may do the same. The chances of catching some fibers at the optimal stage of degeneration at any given moment during that broader time span are rather good in comparison with the chances one has after making the surgical lesions.

Bridging the Gap between Light and Electron Microscopy in the Experimental Tracing of Fiber Connections

LENNART HEIMER

Ever since the introduction of the Marchi method, the elicitation of Wallerian degeneration has held a major place among the strategies employed in the experimental tracing of central nervous fiber connections. Contrary to the Marchi technique which they have largely replaced, the modern silver methods permit the impregnation of degenerating axoplasm, and thus the visualization, in many fiber systems at least, not only of the trajectories of degenerating nerve fibers, but also of the latter's terminal axon arborizations and synaptic endstructures.

Advantageous as these recent methods have proved to be for the study of central nervous connections, it has become quite apparent that light-microscopic techniques are inherently unable to supply pictures of synaptic relationships detailed enough to satisfy, for example, the requirements of contemporary neurophysiologists. Although the modern silver methods may enable the investigator to identify accurately the regions of synaptic distribution of most fiber systems, they fail to disclose the nature of the postsynaptic elements involved in the connections, except in those rare cases in which a conclusion can be reached by exclusion. That degenerating boutons terminaux can be identified in the electron microscope together with the associated postsynaptic structures (for a review see Alksne *et al.*, 1966) is therefore a finding of prime importance that already has led to productive combinations of experimental light- and electron-microscopic studies. Moreover, there is the promise of an effective approach to the question, of which cell a particular postsynaptic profile observed in the lightmicroscope forms part (see chapter by Blackstad, this Volume).

To be emphasized in this account is the experience that a proper selection of tissue fragments for the electron-microscopic identification of degenerating boutons terminaux is impossible without the assistance of adequate light-microscopic techniques. The sampling of brain tissue obviously would be fortuitous if it were done in the absence of an accurate foreknowledge of the synaptic distribution of the fiber system in question. Paradoxical as it may sound, the introduction of the electron-microscope in experimental-neuroanatomical studies has increased rather than alleviated the need for accuracy in the light-microscopic identification of terminal degeneration. Without an adequate preliminary mapping of terminal degeneration in the light-microscope, the investigator runs a high risk of losing information by trimming away significant areas in preparing his blocks for sectioning.

This risk of mis-sampling is especially high in those cases in which the fiber system being studied has its synaptic distribution in a region of complex geometrical shape. In such cases the investigator, lest he share the fate of Honoré Daumier's

Monsieur Babinet (Fig. 1), may have no choice but to resort to the strategy of combined light- and electron-microscopic examination of one and the same plastic-embedded specimen. This approach involves the demonstration and mapping of terminal degeneration in semi-thin sections (Heimer, 1969) taken from a large plastic-

Fig. 1. Honoré Daumier, 1808—1879. Lithograph reproduced by the courtesy of the Museum of Fine Arts, Boston, Mass. Daumier's legend to this cartoon reads: "M. Babinet prévenu par sa portière de la visite de la comète"

embedded tissue block, followed by a trimming-away of irrelevant parts of the block prior to ultrathin sectioning. In the following account this procedure will be outlined in some detail.

Silver Impregnation of Degenerating Axoplasm in Plastic-Embedded Material

Fixation

The tissue is fixed with buffered formalin (Holt and Hicks, 1961) or buffered glutaraldehyde-formalin (Karnovsky, 1965) by intravital perfusion of the animal. After careful removal of the brain or the spinal cord the appropriate part of the nervous tissue is cut in slices of about 1 mm thickness. The size of the blocks to be embedded depends upon the nature of the problem as well as the embedding technique: if a beam-capsule is used, with the closed cover providing a flat bottom, the side of the block can be as large as 5 mm, in which case the semi-thin sections may be cut with a regular glass knife on the ultramicrotome. After a suitable fixation time — usually in the range of a few hours — the tissue is prepared for ultrastructural examination, following the customary procedure of rinsing in phosphate buffer, post-fixation in 1—2% buffered osmium solution, dehydration and embedding in plastic[1].

[1] Soft Epon (mixture A, Luft, 1961) or Epon-Araldite (Epon 812 = 12.4 ml; DDSA = 27.6 ml; Araldite Resin 502 = 7.6 ml; DBP = 1.5 ml; DMP = 1 ml) has been used preferentially in our laboratory, but good results have been achieved also with Araldite (Walberg and Mugnaini, 1969), Maraglas (Rubinson, personal communication) and Durcupan.

Osmium fixation has been found to be of essential importance for successful silver impregnation of plastic-embedded tissue.

Staining Procedure

1. Sections are cut at 2—8 μ. When using a glass knife it is usually possible to flatten the sections against the surface of the knife with the help of a stiff hairbrush or jeweler's forceps.

2. The sections are transferred for 1—2 hr to a 0.5% aqueous silver nitrate solution warmed to 55—60° C and kept at this temperature throughout the impregnation. The addition of 3—5 ml of pyridine to each 100 ml of silver nitrate solution may improve the results. It is usually best to keep the sections floated on the surface of the solution: submerged, they tend to curl.

3. Without washing, the sections are transferred by the aid of a brush for about 5—10 min to a freshly prepared ammoniacal silver solution composed of:

strong ammonia water	2 ml
2.5% sodium hydroxide	2 ml
1.5% silver nitrate	20 ml
95% ethyl alcohol	12 ml

4. The sections are transferred, without rinsing, for about 5 min to a Nauta-Gygax reducing solution made up of:

distilled water	910 ml
95% alcohol	90 ml
10% formalin	27 ml
1% citric acid	27 ml

The sections are rinsed in distilled water, mounted on slides, and dried on a hot plate at about 60°. After a final rinsing in xylene, they are enclosed in a neutral mounting medium.

Results

The result of the procedure is illustrated in Fig. 2A, representing the olfactory cortex from an animal whose ipsilateral olfactory bulb had been removed 3 days before sacrifice. The picture, taken from the superficial part of the plexiform layers — with the lateral olfactory tract at the bottom of the figure — shows a heavy accumulation of argyrophilic structures, undoubtedly representing both fragments of degenerating axons as well as degenerating boutons. The section had been taken from a block embedded in soft Epon, and impregnated in a silver-pyridine solution for 2 hours at 60°. The distribution and general appearance of the terminal degeneration in Epon sections are entirely comparable to the results obtained with reduced silver techniques in frozen sections (see, for example, Figs. 1B and 4 of the preceding chapter). Fig. 2B, taken from the same section as Fig. 2A, demonstrates absence of degeneration within the multiform cell-layer of the olfactory cortex. Note the intracellular silver deposits, and also the slight argyrophilia of myelin sheaths which may make normal axons in longitudinal view appear as "railroad tracks".

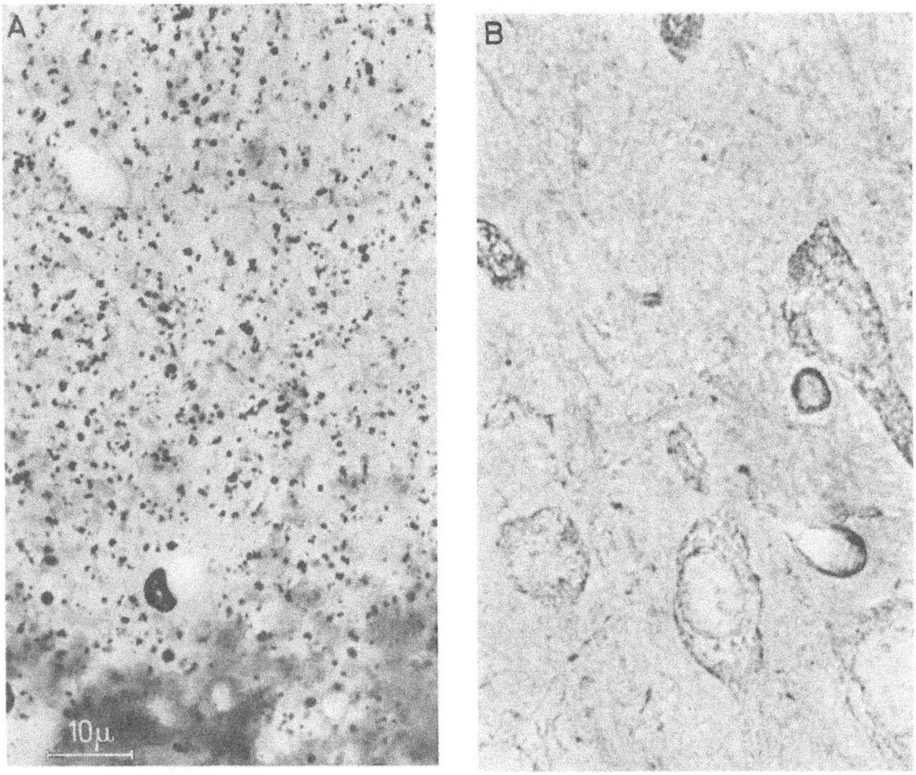

Fig. 2. A. Terminal degeneration in the plexiform layer of the olfactory cortex in a rat whose ipsilateral olfactory bulb was removed 3 days before sacrifice. The lateral olfactory tract is seen at the bottom. The 7 μ thick Epon section was impregnated for 2 hours in a silver-pyridine solution before treatment in ammoniacal silver and reducer. B. Absence of degeneration within the multiform cell layer of the olfactory cortex. Same section as in A

Although it is unreasonable to expect an unequivocal identification of any single argyrophilic particle at the light microscopic level, it is nevertheless important to know if patterns of argyrophilia such as illustrated in Fig. 2A are indeed attributable to axon degeneration. In order to test the specificity of the method for axon degeneration, the semi-thin, previously silver-impregnated Epon section photographed in Fig. 2 was therefore re-embedded on an empty plastic base following a technique described by Grimley (1965), and ultrathin sections were prepared from an area of the section containing heavy terminal degeneration. The most conspicuous ultrastructural features of axon degeneration (Fig. 3) are a general increase in electron-density of the degenerating axoplasm and a loss of ultrastructural detail within the cytoplasm. In the electron micrographs the silver deposits appear in the form of black spherical particles of varying size. Fig. 3 shows small metallic particles scattered throughout the section, but accumulations of larger black spherules are confined to the electron-dense degenerating profiles, some of which can still be identified as degenerating boutons (db) with synaptic vesicles partly preserved or a postsynaptic thickening still present. It appears reasonable to assume that these heavier agglomerations correspond to a black impregnation in the light-microscopic

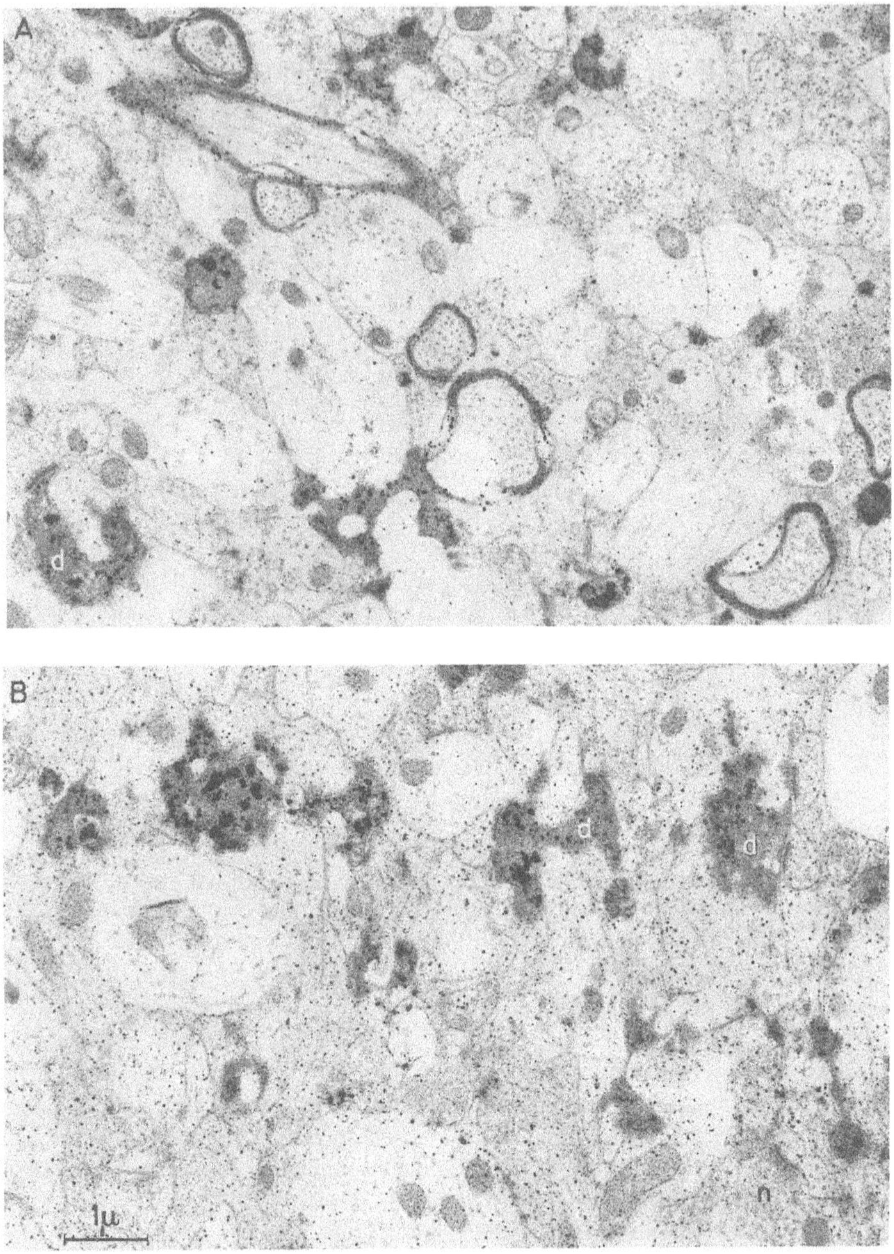

Fig. 3. Electron micrographs of silver-impregnated terminal degeneration. The ultrathin sections were prepared from the section shown in Fig. 2A following re-embedding of the section on an empty plastic base. Although silver particles are present throughout the sections, the heaviest accumulation is found within the electron-dense degenerating profiles. Abbreviations: *d* degenerating bouton; *n* normal bouton

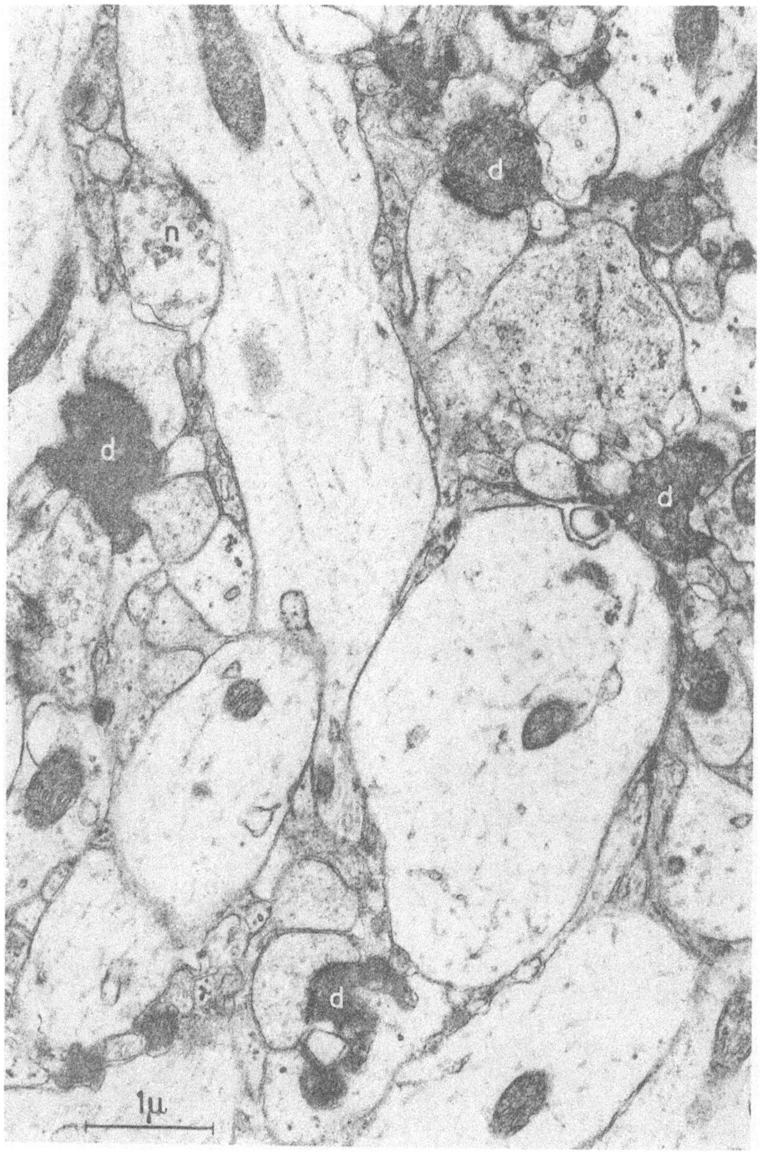

Fig. 4. Electron micrograph showing degenerating boutons (*d*) in the plexiform layer of the olfactory cortex. The ultrathin section was prepared from the same block as the sections shown in Figs. 2 and 3. *n* normal bouton

picture, whereas the more diffusely scattered finer silver particles might account for the brown and often faintly granular "background" characteristic of most silver-impregnated sections (Fig. 2A).

Fig. 4 shows a regular electron micrograph from the superficial part of the olfactory cortex. It was prepared from an ultrathin section, not previously silver-impregnated, cut from the same block from which the sections shown in Figs. 2

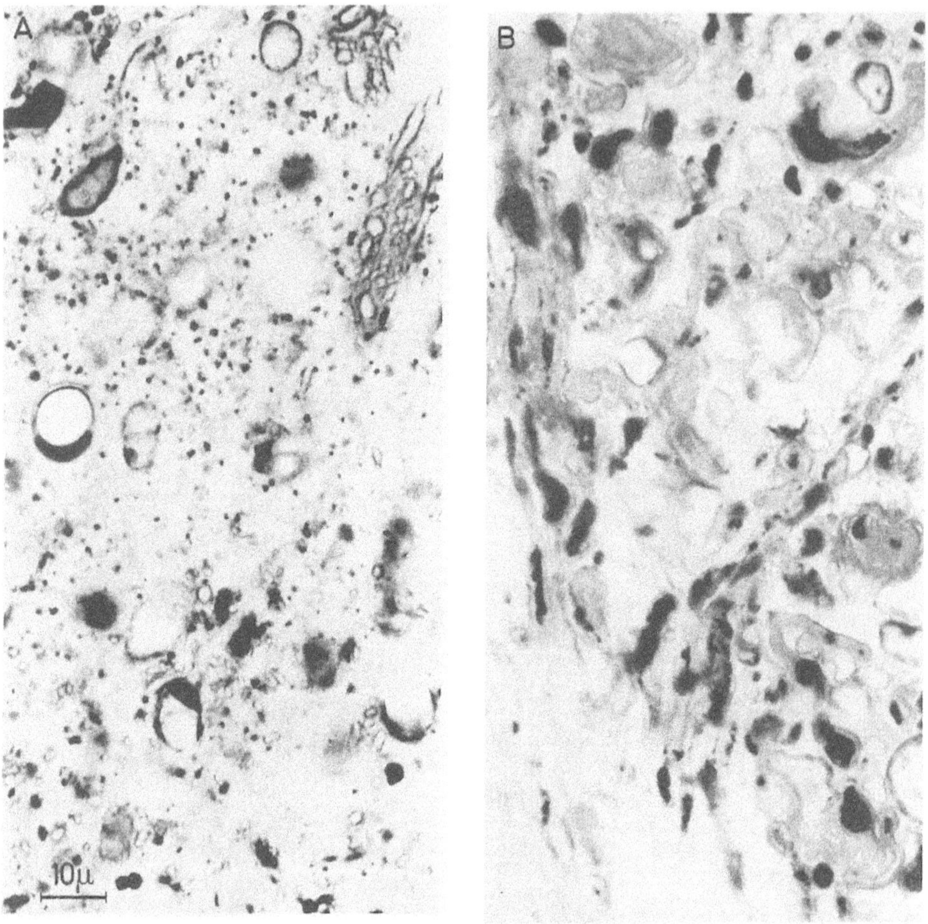

Fig. 5. A. Massive terminal degeneration in Rexed's lamina II (upper half of the figure) of the rat spinal cord following section of the corresponding ipsilateral dorsal root S_1 2 days before sacrifice. B. Degenerating dorsal root fibers penetrating the dorsal white column of a rat sacrificed 6 days after dorsal root section. Note the heavily argyrophilic axoplasm of the degenerating nerve fibers compared to the unstained axoplasm of the normal fibers. The semithin Epon-Araldite sections in A and B were impregnated overnight in a silver-pyridine solution prior to treatment in an ammoniacal silver solution and a reducer

and 3 were taken. Owing to better tissue quality and the absence of silver deposits, several of the electron-dense profiles appearing in this section are more easily identifiable as degenerating boutons than is the case in electron micrographs prepared from previously silver-impregnated sections.

The light micrograph shown in Fig. 5A illustrates silver-impregnated degenerating axons and axon terminals stained in a section taken from a block embedded in relatively soft Epon-Araldite. The section, which was impregnated overnight in a silver-pyridine solution, shows the distribution of dorsal root fibers in the substantia gelatinosa of a rat whose corresponding dorsal root (S_1) had been sectioned

2 days before sacrifice. The heavy accumulation of argyrophilic bodies suggests a massive termination of dorsal root fibers especially in layer II of Rexed (upper half of the picture). That dorsal root fibers in the rat terminate massively in layer II of the substantia gelatinosa was recently confirmed in a combined light- and electron-microscopic study (Heimer and Wall, 1968). Degenerating dorsal root fibers penetrating the dorsal white column of a rat sacrificed 6 days after dorsal root section are shown in Fig. 5B. Whereas the axoplasm of the fibers is heavily argyrophilic as a result of Wallerian degeneration, the myelin sheaths — regardless of whether they belong to degenerating axons or not — are only slightly argyrophilic and appear light brown in the light microscope.

Discussion

Several silver methods for semi-thin plastic-embedded sections — with or without previous extraction of the embedding medium — have been developed in order to facilitate the correlation of light and electron microscopic studies of different tissues, including normal components of the central nervous tissue (e.g. Berkowitz *et al.*, 1968). Recently the use of paraphenylenediamine has been suggested for the identification of degenerating axons and axon terminals in semi-thin Araldite sections (Holländer and Vaaland, 1968). The results presented here indicate that reduced silver methods, frequently applied in the light-microscopic tracing of degenerating nerve fibers in frozen sections and paraffin-embedded material, also can be used successfully for the selective impregnation of degenerating axons and axon terminals in plastic-embedded sections, without previous extraction of the plastic. Although recent findings by Walberg and Mugnaini (1969) suggest that the so-called "neurofilamentous" form of bouton degeneration may not be so easily demonstrated in plastic sections with the present method, it is possible that a modification of the procedure may reveal this relatively infrequent type of degeneration.

The method employed here is a modification of the original Nauta-Gygax method (1951) comparable to a recently described procedure for the impregnation of degenerating axons and axon terminals in frozen sections (Fink and Heimer, 1967). The results obtained in plastic-embedded tissue are similar in every respect to those obtained with the Fink-Heimer process in frozen sections. As in the latter (Heimer and Peters, 1968), metallic silver is deposited in greatest quantity and coarsest form in degenerating axoplasm — in the electron micrographs characterized by an increased electron density and by a deformation that becomes more marked with elapsing time of survival — and the regions of significant axon termination are characterized in the light microscope by a large number of more or less spherical black-impregnated structures usually measuring between 0.5 and 2 μ in diameter. That many of these bodies indeed represent degenerating boutons is evident from the electron microscopic study of silver-stained plastic sections (Fig. 3). Although increased electron density with concomitant loss of structural detail often prevent a classification of the degenerating profiles already on the second or third day of the degeneration cycle, some of the degenerating boutons are easily recognizable at this stage because of well preserved postsynaptic thickenings. The diffuse and apparently random deposition of fine dust-like silver grains throughout the tissue (Fig. 3) undoubtedly accounts for the brownish tint of the sections at the light-microscopic level. The myelin sheaths, whether normal or degenerating, are often slightly more argyrophilic than the rest of the normal tissue components (Fig. 3A), a phenomenon which causes normal axons to appear as ghost-like "railroad tracks" in longitudinal view (Fig. 2B), or as doughnuts in transverse section (Fig. 5). Although myelin sheaths occasionally exhibit a similar increase in argyrophilia in silver-impregnated frozen or paraffin sections, they are as a rule not apparent in the light-microscopic examination of such sections.

Serial Light and Electron Microscopic Mapping of Degenerating Axoplasm on the Same Block

A light-microscopic charting of the course and termination of a fiber system by means of silver methods is an essential prerequisite for the electron-microscopic examination of its synaptic details. Although adequately supplementary light- and

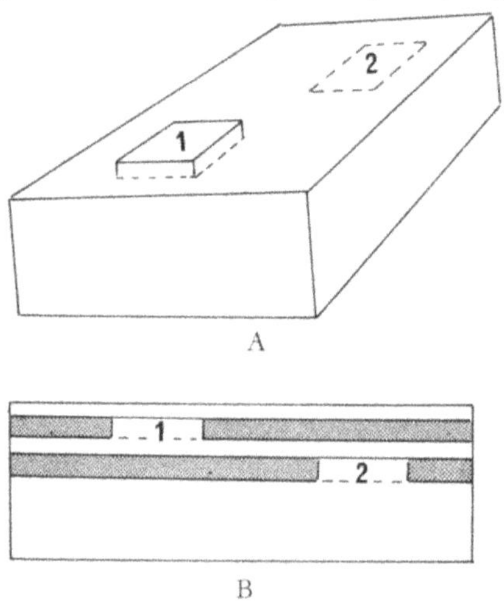

Fig. 6. Maximal use of the same block for correlative light and electron microscopic mapping of degeneration can be ascertained if regions of interest (A) are shaped into the form of islands ("mesa-pyramid") in successive trimming of the block with a glass-knife. Depending on the need for close correlation between light and electron-microscopy, semi-thin sections of the whole block face can alternate with ultra-thin sections cut from islands in different planes of the original block (B)

electron-microscopic data often can be obtained in separate experimental specimens, it is useful, and sometimes necessary — especially when dealing with more complex distribution patterns — to perform a serial light- and electron-microscopic analysis on one and the same experimental specimen. Following the initial light-microscopic identification and localization of the experimentally induced degeneration in the semi-thin section, the block can be trimmed down to the area of immediate interest prior to ultrathin sectioning for subsequent electron-microscopic analysis. Fig. 6 illustrates a procedure (McGee-Russell and Gosztonyi, 1967) allowing ultrathin sections to be obtained from different, circumscript regions of one and the same block.

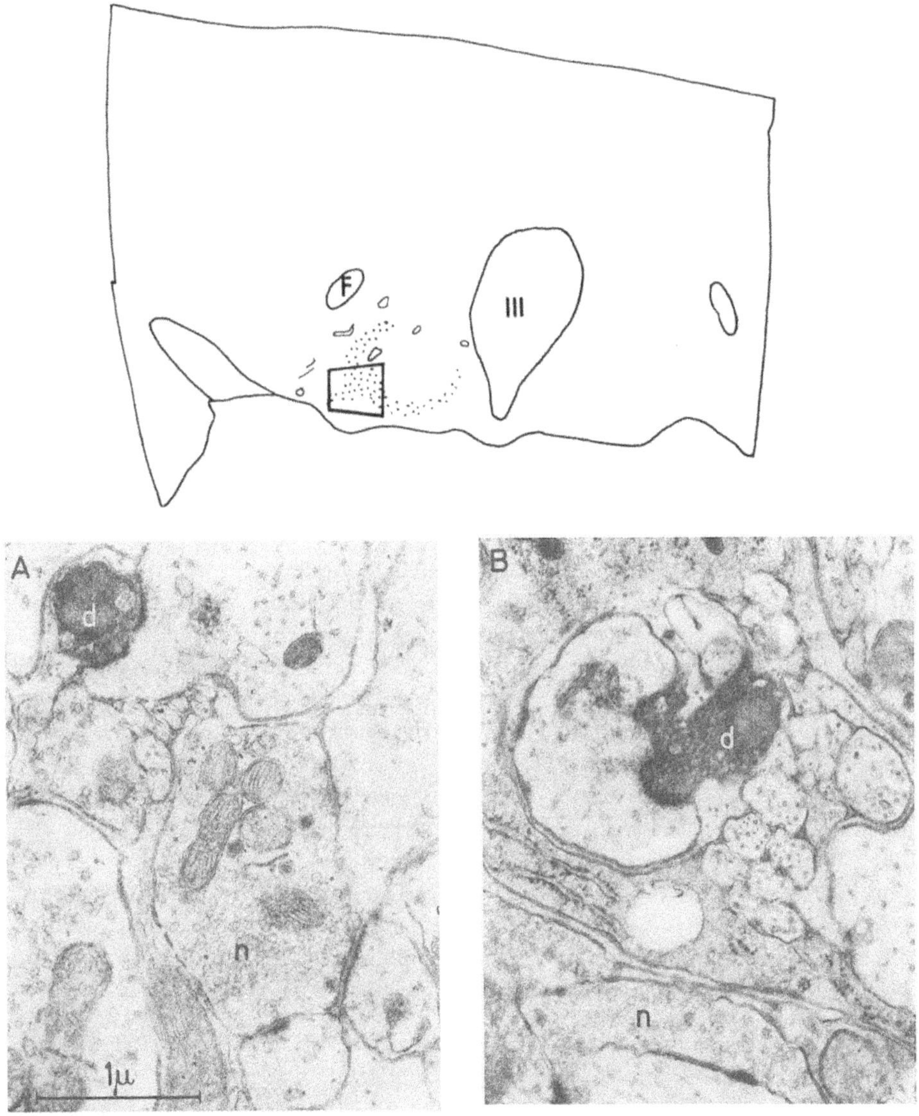

Fig. 7. Electron micrographs showing degenerating boutons (*d*) in synaptic contact with dendrites in the circum-ventromedial region 5 days after stria terminalis section. Upper drawing represents a silver impregnated semi-thin plastic embedded section used for light-microscopic charting of the field of terminal degeneration. The electron micrographs were made from nearby ultrathin sections cut after the block had been trimmed down to the small region outlined in heavy line in upper drawing. *n* normal bouton. [Drawing and photograph B reproduced from Brain Research **13**, 284—297 (1969) by courtesy of Elsevier, Amsterdam]

The use of this serial procedure for the purpose of orientation is exemplified by a case illustrated by Fig. 7. Findings in an experimental light-microscopic study of Wallerian degeneration in the rat had suggested that fibers belonging to the pre-commissural component of the stria terminalis terminate in the cell-poor zone surrounding the ventro-medial nucleus of the hypothalamus (Heimer and Nauta, 1969; see also Figs. 3 and 5 of the preceding chapter). To verify this apparent synaptic distribution of the stria terminalis, the region in question was examined electron-microscopically in a rat sacrificed 5 days after section of the stria. Brain slices of approximately 1 mm thickness were taken from the glutaraldehyde-formaldehyde-fixed brain, trimmed down to the border of the hypothalamic region, and embedded in an Epon-Araldite mixture following the standard post-fixation in osmic acid and dehydration. The field of terminal degeneration appearing in a silver-impregnated semi-thin Epon-Araldite section was plotted in an enlarged projection drawing of the section (Fig. 7, upper drawing), using blood vessels and other incidental profiles as landmarks. The Epon-Araldite block was thereafter trimmed down to the area of immediate interest (region outlined in heavy lines in the drawing) for ultra-thin sectioning. Electron micrographs from the corresponding region (Fig. 7A and B) confirmed the presence of degenerating boutons in contact with dendritic profiles in the circum-ventromedial region.

Addendum

After this article had been written, a paper highly relevant to the foregoing discussion came to the author's attention [Cohen, E. B., Pappas, G. D., Dark profiles in the apparently-normal central nervous system: a problem in the electron microscopic identification of early anterograde axonal degeneration. J. comp. Neurol., *136*, 375—396 (1969)]. In this important paper, Cohen and Pappas report having found numerous electron-dense boutons terminaux in the ventrobasal thalamic nucleus of unoperated cats, "...morphologically indistinguishable from those seen in... cats with dorsal column nucleus lesions of two and four days' duration", and suggest that this finding is associated with the artefactual phenomenon of the "dark neuron". Cohen and Pappas' findings again emphasize the need for a close correlation between light- and electron-microscopic observations in experimental material. Furthermore, although not mentioned explicitly in the foregoing account, precautionary measures such as continuous comparison of the operated with the unoperated side, and careful examination for encephalitic, vascular or traumatic pathology should be taken in each case of experimental lesion.

Neuronal Changes Central to the Site of Axon Transection. A Method for the Identification of Retrograde Changes in Perikarya, Dendrites and Axons by Silver Impregnation

G. Grant

Two basic pathological phenomena widely utilized in experimental neuroanatomy are the Wallerian degeneration and the retrograde cell reaction. The Wallerian degeneration makes it feasible to map the course and termination of fibre tracts, whereas the retrograde cell reaction enables the investigator to identify their cells of origin. There are, however, some problems within neuroanatomy that cannot be solved by the aid of these two phenomena. For example, in many regions of the central nervous system it is impossible to make a lesion restricted to a specific cell group without damaging fibres passing through or close to that group. The ensuing degeneration will therefore not represent exclusively the lesion of the cell group in question. The conclusions that can be drawn from an experiment of this type are therefore not always as satisfactory as one would wish.

A recent series of experiments performed in our laboratory have been concerned with the question whether perhaps in problems of this sort a profitable use could be made of the extreme severity of the retrograde cell reactions that follow axon transections in the *immature* nervous system (Brodal, 1939; La Velle and La Velle, 1958). Provided the animals used for the experiments were young enough, axonal transection could be expected to provoke a retrograde cell reaction so severe that the parent cell would die. It could furthermore be supposed that, as a consequence of this cell death, the axon projecting from the cell would degenerate and that it would be possible to identify this degeneration light-microscopically in sections impregnated according to the suppressive method of Nauta (1957). If our suppositions could be verified, a method would be available which might be useful for solving problems such as that mentioned above.

It should be noted here that our suppositions were directly contradictory to the negative statement in the Wallerian law, viz. that degenerative changes in the nerve fibre following axonal transection do not extend proximally beyond the origin of the nearest axon collateral (Cajal, 1928). Judged by the literature, this view is still widely accepted. In conflict with the Wallerian dogma, however, earlier investigators studying the effects of axonal transection had recognized in some fibre systems the occurrence of fibre degeneration proceeding *distalward* from the parent-cell body (see review by Beresford, 1965). This type of degeneration customarily has been called "retrograde fibre degeneration" or "ascending degeneration", but van Gehuchten (1903, 1906) who made the most extensive study of this type of degeneration suggested the term "indirect Wallerian degeneration" as more adequately descriptive of the phenomenon. None of these earlier investigators appears to have taken advantage of the general rule that the immature neurons of very young animals are more susceptible to the effects of axon trauma than are adult neurons.

The results of our early studies indicated that our assumptions had been correct. In addition to the finding of impregnated structures interpreted as degenerating axons, however, an unexpected discovery was made. In some preparations, neuronal

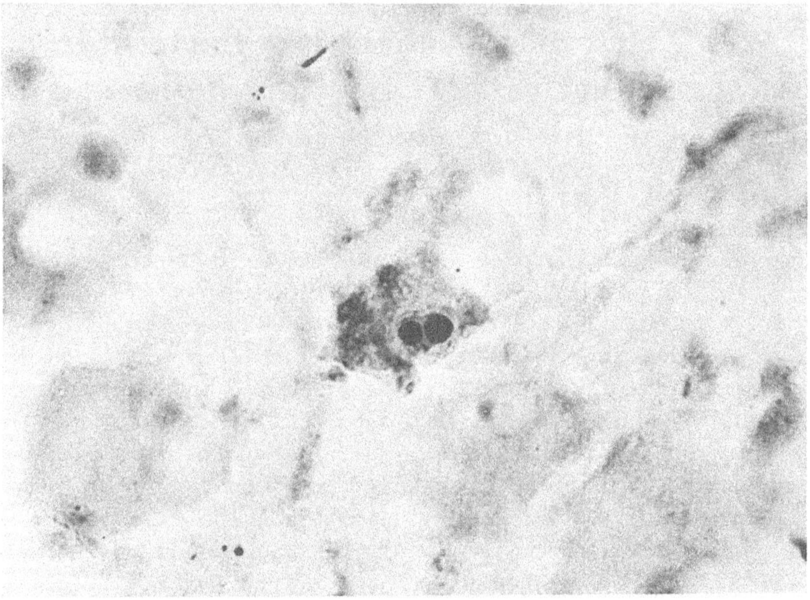

Fig. 1. Type I cell in the hypoglossal motor nucleus in a case of ipsilateral hypoglossal nerve transection. Note the two black granules within the nucleus and the absence of argyrophilia in the cytoplasm. Kitten 5 days old at operation, killed 3 days later. Immersion fixation. ×850

cell bodies were found to be heavily impregnated, and their likewise black impregnated dendrites exhibited a fragmentation suggestive of a degenerative process. This observation suggested the interesting possibility that silver methods might be used not only to follow axonal trajectories distal to the site of transection, but also to mark the parent cell body and to determine the orientation and the extent of the latter's dendritic processes. As will be discussed below, it has been possible to identify the phenomenon of retrograde dendrite degeneration also at the ultrastructural level.

Our systematic studies to date include observations made on hypoglossal motor neurons (Grant and Aldskogius, 1967), spinal motor neurons (Grant, 1968), olivo- and reticulocerebellar neurons (Grant; in preparation) and cervicothalamic neurons (Grant and Westman, 1968b). With the exception of the cervico-thalamic neurons which have been examined both in Nauta-stained light microscopical sections and in ordinary electron microscopical sections, these studies have been performed exclusively on Nauta-stained light microscopical sections. The cat has been used as the experimental animal in all these studies, but additional information has been obtained from experiments on rabbits.

With the procedure here outlined it has been possible to demonstrate in the Nauta-stained sections impregnated cells and dendrites as well as axons, probably with involvement of the myelin sheath. In electron micrographs of corresponding regions dendrites have been found which exhibit the same increase in electron density that has come to be recognized as a characteristic of degenerating axons (Grant and Westman, 1968a, b).

The impregnated *nerve cells* were found among neurons showing a retrograde cell reaction of the ordinary type, including chromatolysis and a peripheral displacement of the nucleus. In all instances so far the impregnated cells have been outnumbered

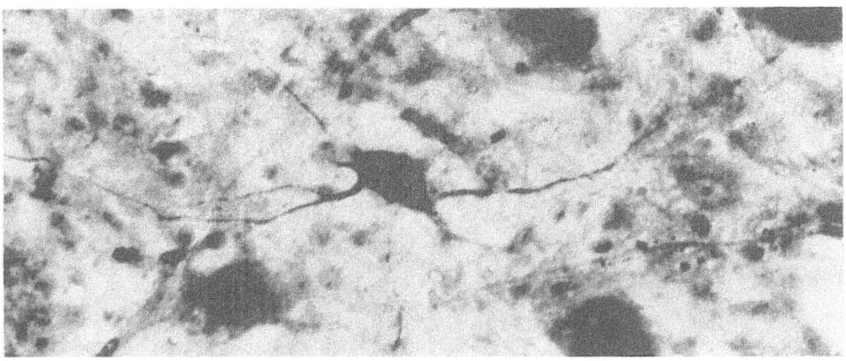

Fig. 2. Type II cell in the lumbar motor horn in a case of ipsilateral ventral root transection. Note the impregnation of the perikaryon and the cell processes. Note also the branching of the left dendrite and the beaded appearance of the distal portion of the right dendrite. Kitten 1 day old at operation, killed 7 days later. Immersion fixation. ×350. [From Exp. Brain Res. **6** (1968), by courtesy of the Publishers]

by the chromatolytic non-impregnated neurons. The impregnated cells have been classified in two types, type I and type II cells, respectively (Grant and Aldskogius, 1967). In type I cells the silver impregnation is confined to the nucleus, where it appears in the form of one or a few black globules (Fig. 1). The cytoplasm of such cells has a brownish colour but contains no black globules. Both cell body and the likewise brown stained dendrites appear markedly shrunken. In the hypoglossal neuron material, characteristic type I cells were found in very young kittens 3—6 days old at operation and killed as early as 3 days later. In type II cells a black, dense silver impregnation of the perikaryon is the characteristic feature (Fig. 2). The observation that some such cells shown an excentric area which is free from impregnation except for one or a few black globules indicates that the cell nucleus escapes complete impregnation. This suspicion has been strengthened by the observation in still other cells of a very small non-impregnated area close to the cell periphery, suggesting a tangentially sectioned non-impregnated nucleus. In the hypoglossal neuron material characteristic type II cells were found in sections from kittens 11—14 days old at operation and killed 7 days later, i.e. older than those in which type I cells were observed. Since type II cells are the only degenerative cells exhibiting black impregnated dendrites, they should be best suited for "marking" purposes. In recent studies in collaboration with Dr. Holländer of the Max Planck Institute for Psychiatry in Munich, ordinary Nauta stained sections were embedded in Araldite. In semi-thin sections taken from such preparations type I cells were found to have the silver impregnation confined to nucleolar-like structures, within the nucleus and the cytoplasm was free from silver impregnation. Type II cells were found to have the silver deposit confined to the cytoplasm. The cell nucleus of spinal motor neurons of type II was free from impregnation except for the same type of nucleolar-like structures that were found in type I cells and which were

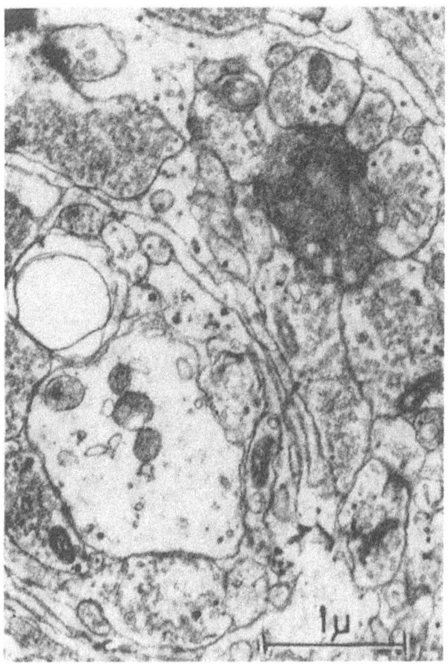

Fig. 3. Electron micrograph showing degenerating dendrite at upper right and normal dendrite at lower left. From the lateral cervical nucleus on the affected side in a case of transection of cervicothalamic axons. Note the normal-appearing boutons surrounding the degenerating dendrite. Kitten 3 days old at operation, killed 10 days later. Perfusion fixation. [From Exp. Brain Res. **7** (1969), by courtesy of the Publishers]

impregnated also in these type II cells. Whether the cell nucleus of type II cells from other regions also contains such nucleolar-like silver impregnated structures is at present not known.

The impregnated *dendrites* appearing in Nauta-stained sections (Grant, 1965; Grant and Aldskogius, 1967; Grant, 1968) exhibited a beaded appearance in all instances in which they could be traced for some distance away from the cell body (Fig. 2). In most cases, this distance did not exceed a few cell diameters; in only a few instances was it greater than five diameters. Occasionally, this distance has been long enough to include dendritic branches of the third order. It is possible that the use of sections thicker than the customary 20 μ would allow the tracing of a greater part of the degenerating dendritic trees.

Subsequent electron microscopic studies (Grant and Westman, 1968a, b) revealed the presence of electron-dense dendrites in the lateral cervical nucleus of the side

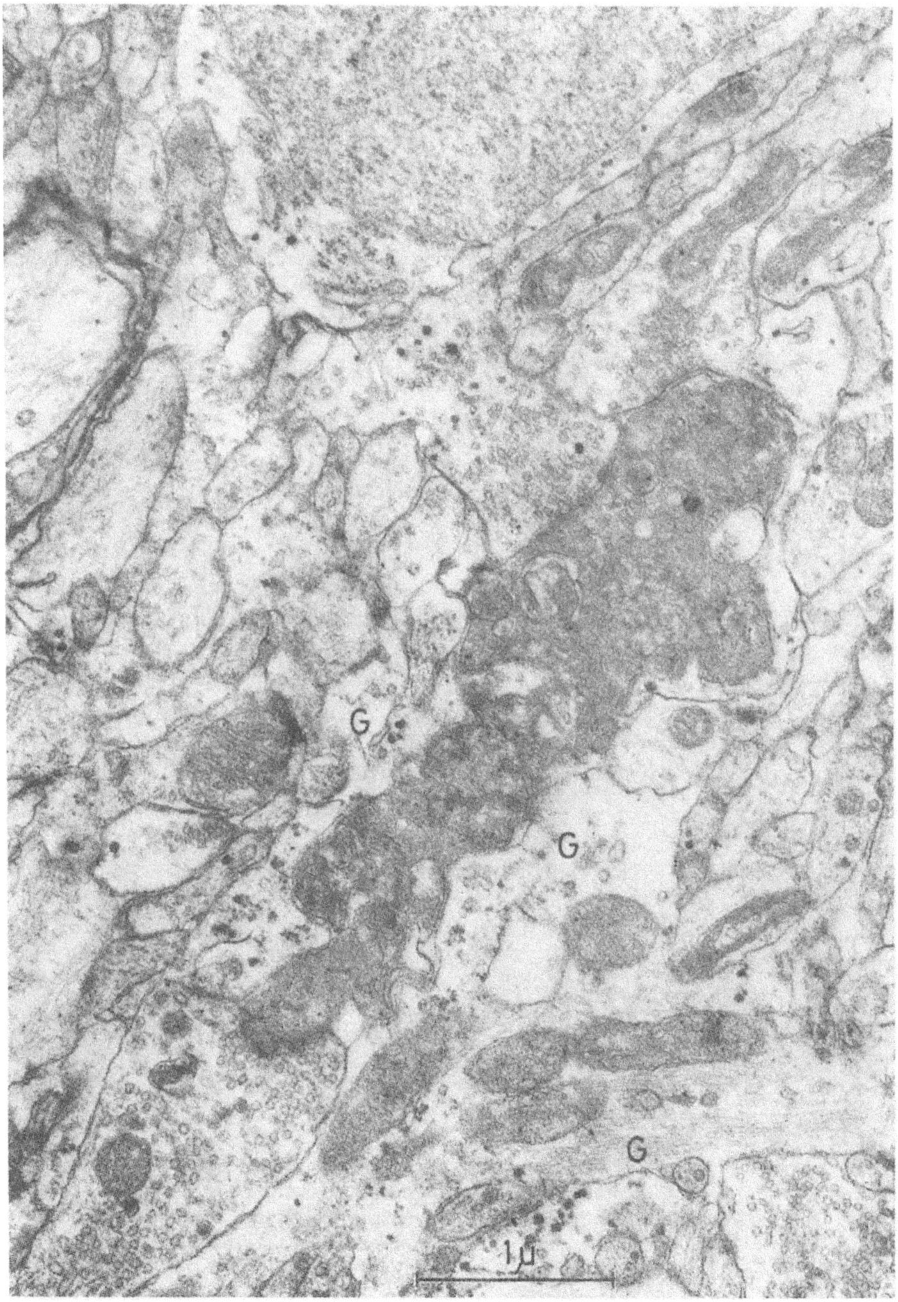

Fig. 4. Electron micrograph of a longitudinally sectioned degenerating dendrite. From the lateral cervical nucleus on the affected side in a case of transection of cervicothalamic axons. Note the fragmented mitochondria inside the dendrite. The dendrite is contacted by three boutons, the lower one containing several dense-core vesicles. The rest of the dendrite is contacted by astroglial cell processes (*G*). Kitten 3 days old at operation, killed 10 days later. Perfusion fixation. [From Exp. Brain Res. **7** (1969), by courtesy of the Publishers]

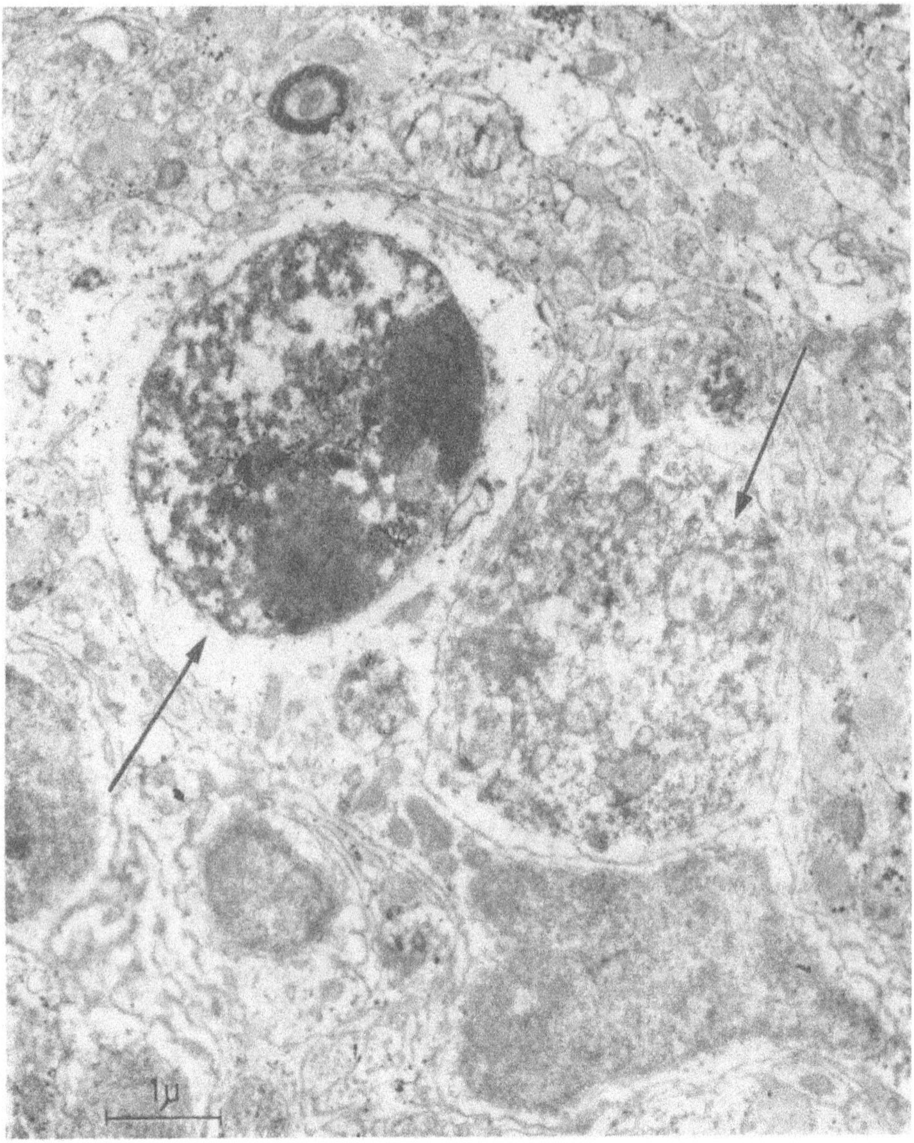

Fig. 5. Electron micrograph of phagocytic cell within the lateral cervical nucleus on the affected side in a case of transection of cervicothalamic axons. The cell nucleus is at lower right. The cell process projecting at upper left surrounds a large rounded inclusion (left arrow) that could be the phagocytosed nucleus of a nerve cell. To the right of this are other inclusions (right arrow) probably within another phagocytic cell. Kitten 9 days old at operation, killed 15 days later. Perfusion fixation. [From Exp. Brain Res. 7 (1969) by courtesy of the Publishers]

affected by axotomy (Figs. 3 and 4). In the animals sacrificed after short survival times, these evidently degenerating dendrites were found to be beset with apparently normal synaptic contacts. In cases of longer survival time a larger part of the dendritic surface appeared to be covered with astroglial cell processes, but even in these cases the axon terminals in contact with the dendrites had retained a normal appearance. In animals with long survival times degenerating structures were often found enclosed by phagocytic cells (Fig. 5).

The fibre impregnation of the *axons* in our Nauta-stained material generally exhibited a more granular appearance than is usually seen in true "direct" Wallerian degeneration (cf. Fig. 6). As in the latter process, however, pictures suggesting axon

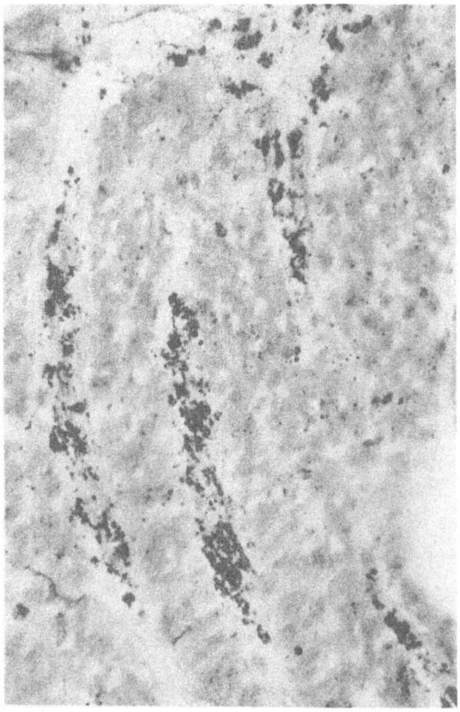

Fig. 6. Impregnated granules along intraspinal root fibres in the lumbar ventral funiculus in a case of ipsilateral ventral root transection. Kitten 2 days old at operation, killed days later. Perfusion fixation. ×400

fragmentation were observed (illustrated in Grant and Aldskogius, 1967). In some cases the impression was obtained that the myelin sheath was involved in the impregnation. An electron microscopic analysis of the mode of silver deposition occurring in retrograde fibre degeneration is in progress in our laboratory.

There can be little doubt that the fundamental phenomenon underlying the research strategy here discussed is cell death caused by an axonal lesion (cf. Grant and Aldskogius, 1967). Secondary to this necrotic process a fibre degeneration starting at the perikaryon proceeds along the axon in cellulifugal direction (Grant and Aldskogius, 1967). Since the cell death is the ultimate stage of a heavy retrograde cell reaction, the same factors which determine the retrograde cell reaction may be assumed to be critical also for this method of demonstrating degeneration of cell processes as well as of perikarya.

Our studies have shown that the age of the animal at operation is of importance (Grant and Aldskogius, 1967; Grant and Westman, 1968b). In our experiments involving the hypoglossal motor nucleus, for example, clear-cut neuron impregnation was observed in Nauta sections from kittens operated at ages ranging from 1 to 11 days and surviving 7 days postoperatively. By contrast, no such neuron impregnation was obtained in kittens operated at the age of 20 days and surviving the same period. And in no instances so far have we found neuron impregnation in full-grown animals. The possibility of provoking neuron impregnation in such animals will, however, be discussed below.

As expected, the postoperative survival period also proved to be of importance for the occurrence of neuron impregnation (Grant and Aldskogius, 1967; Grant, 1968 and Grant and Westman, 1968b). In the hypoglossal neuron study, kittens operated on at the age of 5 days showed no neuron impregnation after 1 or 2 days postoperative survival, whereas after 4 days neuron impregnation was obtained consistently.

The postoperative survival time has been shown to be essential also for another reason. Since the fibre impregnation starts at the perikaryon and then progresses along the axon in the distal direction (Grant and Aldskogius, 1967), the postoperative survival time should be extended beyond that necessary for the demonstration of cells and dendrites if the more distal parts of the neuron are to be demonstrated. In the cases of ventral root transection, for example, no good impregnation of intraspinal root fibres was obtained unless the animal had been allowed to survive surgery for 11 days, whereas cells and dendrites could be impregnated in animals that had been operated at the same age but had been sacrificed as early as the 7th postoperative day (see Figs. 4 and 2 in Grant, 1968). On the other hand, the degenerative argyrophilia of cells and dendrites seems to be a very short-lasting phenomenon (Grant and Aldskogius, 1967), probably as a result of the rapid rate of breakdown characteristic of immature cells (cf. La Velle and La Velle, 1958). Its demonstration might thus require a shorter postoperative survival than would be necessary for the demonstration of the axonal degeneration. This temporal differential may make it necessary to utilize more than one postoperative survival time if degenerative argyrophilia is to be demonstrated both in the cell body with its dendrites and in the axon.

The effects of other factors known to be of importance for the retrograde cell reaction, such as the distance between the cell body and the site of the axon trauma, the animal species used and the type of neuronal system studied (see Brodal, 1939), so far have not been studied systematically. Our finding of a greater amount of degeneration in the cases of ventral root transections than in those of sciatic nerve lesion might have been due to the more proximal site of the axon trauma. Before this interpretation is accepted, however, it is necessary to explore the importance of other possible factors, such as the presence of axon collaterals issued between the two transection sites, or to differences in the number of parent cell bodies in the spinal cord segments examined in the two experimental situations.

Our procedure in producing the axonal lesions has been somewhat different in the different studies. The hypoglossal nerve, the sciatic nerve and the ventral spinal roots were sectioned with the aid of fine scissors, and in transecting the nerves, but not the roots, a few millimeters of the peripheral stump were resected in order to prevent regeneration. In the cases of cerebellar lesions, a small sucker was used, and the transection of cervicothalamic axons in the midbrain was accomplished by the same technique. In spite of the variation of techniques, comparable signs of retrograde degeneration appeared in each of the neuronal systems studied. Needless to say, these findings do not exclude the possibility that the technical procedure followed in the production of axonal lesions is of some importance.

As regards the technique of fixation, we have used both perfusion (Grant, 1968; Grant and Westman, 1968b) and immersion (Grant and Aldskogius, 1967; Grant, 1968). Formaldehyde was used in all the studies, and the central nervous system was always dissected free immediately after exsanguination or perfusion of the animal. Perfusion was found superior to immersion since it seemed to reduce the occurrence in the Nauta-stained sections of dark neurons not related to the operation (see Grant, 1968). Retrograde degeneration, however, was demonstrable in the Nauta sections regardless of the type of fixation. In the electron microscopic studies perfusion with phosphate-buffered 4% formaldehyde was used.

The Laidlaw modification (Chambers *et al.*, 1956) of the Nauta method (1957) has been used throughout our studies for the demonstration of retrograde neuronal degeneration at the light microscopic level. Powell and Cowan (1964) used the same method in demonstrating retrograde changes in the anterior thalamic nuclei of adult rabbits which had been subjected to lesions of the cingulate cortex and hippocampus. There is, however, no reason to expect that such changes could not be demonstrated by any other method used for the impregnation of degenerating nerve fibres. On the contrary, there are reports of retrograde fibre degeneration demonstrated in adult animals with the Marchi method (Bregman, 1892; van Gehuchten, 1903; Glees *et al.*, 1951; and others) by the method of Nauta and Gygax (1954) (Guillery, 1959; Cowan *et al.*, 1961; Cragg, 1962; Powell and Cowan, 1964) and even with the methods of Glees (1946) (Glees *et al.*, 1951; Cowan *et al.*, 1961) and Bodian (Cowan *et al.*, 1961). It is possible, however, that each of these methods has its own requirements with respect to optimal survival periods (see, for example, Cowan *et al.*, 1961).

Throughout our light-microscopic studies we have used frozen sections of approximately 20 μ thickness. As pointed out previously (Grant, 1968), greater section thickness would undoubtedly have facilitated the tracing of impregnated degenerating fibres. On the other hand, the unavoidable occurrence of a brown "background" impregnation in Nauta material could be expected to impose certain limitations on optimal section thickness.

The identification of retrograde neuronal degeneration by the histological criteria here described thus allows the selective experimental marking of various types of neurons, especially in the immature nervous system. Even though the number of impregnated cells has been much lower than the number of chromatolytic neurons the procedure promises an applicability wider than that of the more commonly used retrograde chromatolysis. For example, it should permit not only the identification of the cells from which a particular axon system originates, but also the visualization of the dendrites of such cells. At the ultrastructural level, it promises the possibility of studying the relationships between "marked" dendrites and the various structures surrounding and contacting these dendrites. By combining the procedure with the transection of afferent fibre systems it might also be possible to obtain more detailed information concerning synaptic relationships. Its usefulness for the purpose of tracing the course of fibre tracts even in cases in which the direct destruction of the cells of origin is contra-indicated has already been demonstrated in studies of reticulo- and olivo-cerebellar connections following cerebellar lesions (Grant; in preparation). In addition, the procedure should allow the identification of collateral branches issued by degenerating axons (see Grant, 1968). The results of our recent studies on the primary vestibular fibres in young rabbits (Grant *et al.*, 1969) further indicate that the identification of retrograde neuronal degeneration in young animals by silver impregnation may be utilized also for the tracing of what we have called "transganglionic sensory degeneration". These observations suggest that it might be possible to determine the central representation of individual peripheral sensory nerve branches and thus, to overcome the traditional obstacle presented by the sensory ganglia.

There are some similarities between the present experimental method and the Golgi technique (Grant, 1968). The appearance of impregnated cells and dendrites affected by retrograde neuronal degeneration is in some respects comparable to that of normal Golgi-stained neurons. In both conditions, only a small percentage of the total cell population is impregnated. Moreover, in both techniques young animals are used.

This latter fact is undoubtedly a limiting factor, since certain morphological details may undergo changes during maturation. It is therefore of some importance that retrograde neuronal changes as here described may be brought about also in older animals (see Grant, 1968). It is possible that the elicitation of such changes in adult neurons requires more than the interruption of the cell's axon; at any rate, the observations reported by Glees *et al.* (1951) suggest that the appearance of marked retrograde changes in adult neurons is greatly facilitated by combining the axotomy with a de-afferentiation of the cell group in question. Stated in other words, retrograde changes as here described for immature neurons may in the adult nervous system require a combination of retrograde and transneuronal factors.

It is important at this point to consider a few common sources of histological misinterpretation. Foremost among these is the presence in silver-impregnated sections of dark-staining neurons exhibiting a somewhat shrunken appearance. In all likelihood these structures correspond to the so-called *dark neurons* observed in Nissl-stained material. Formerly often interpreted as evidence of a variety of neuropathological conditions, the dark neuron more recently has been shown to be an autopsy artefact (Cammermeyer, 1962). Dark neurons are particularly likely to occur in large numbers in brain biopsies and in brain material removed immediately post mortem and fixed by immersion. They are usually less numerous, but by no means uncommon in material fixed by routine perfusion (see Westrum and Lund, 1966). The appearance of dark neurons can be prevented altogether by the use of Cammermeyer's (1962) fixation technique, but this procedure is unfortunately incompatible with the Nauta silver method. In Nauta preparations in which the Laidlaw modification has been used, in the regions we have examined the dark neuron as a rule has been readily identified by its dark brown staining and often by a black, irregular perinuclear zone, characteristics quite distinct from the dense, black silver deposits filling the perikaryon and cell processes of the neuron undergoing retrograde degeneration.

Another important source of error lies in the frequent occurrence of small single silver granules dispersed among bundled fibres, apparently in association with myelin sheaths. Such granules have been encountered irrespective of the fixation procedure followed (Grant, 1968) and are especially prone to occur bilaterally in the spinal cord near the ventral root exit (Fig. 7) and in the dorsal funiculus of kittens of various ages. They can be suppressed by prolonging the permanganate pretreatment, but this frequently interferes with the optimal impregnation of other structures (Grant, 1968). Usually this artefact can be distinguished readily from the true degenerative granulation caused by axon transection by the relative sparseness and more random distribution of the granules (compare Figs. 7 and 6). The nature of these granules is obscure.

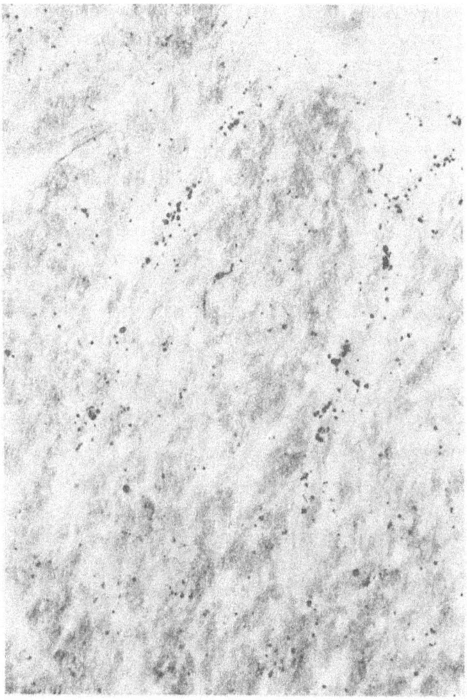

Fig. 7. Impregnated "artefact" granules along intraspinal root fibres in the lumbar ventral funiculus. Normal 9-day old kitten. Compare with Fig. 6. Perfusion fixation. ×400

Their similarity with the osmiophilic granules described by Key and Retzius (1873, 1876) and by Elzholz (1898) in peripheral nerves has been mentioned in a previous report (Grant, 1968). It appears possible that they represent breakdown products of myelin, a possibility suggested by Berthold and Skoglund's (1968a, b) observation of myelin sheath disintegration in normal feline peripheral nerves during postnatal development.

Discussion

Dr. Nauta: Can one expect to see retrograde changes such as the ones you observed in motor neurons of immature animals also after transection of the nerve in the *adult* cat? I have seen the same argyrophilic neurons in the thalamus of the hemide corticated adult rat, and yesterday Dr. Heimer showed a picture of the prepyriform cortex with such black-impregnated neurons in cases of olfactory bulb removal, likewise in the adult rat.

Dr. Grant: We have concentrated on young animals but we have also made some observations in the adult cat. So far, however, we have not found the same reaction in the adult.

It might well be that one could produce retrograde neuron degeneration in the adult animal by other means than we have used. One could transect the hypoglossal nerve axon closer to the cell body, for instance, at the exit from the brain stem instead of in the carotid triangle. Possible differences between different species and between different neuronal systems could also be utilized in attempts at provoking a retrograde neuron degeneration. For instance, some systems give rise to many collaterals while others do not.

Dr. Nauta: Even in one and the same system there may be differences in the reactions of individual neurons. In adult hemidecorticated rats, I have found intense argyrophilic reactions only in a rather modest percentage of the thalamic neurons. In any thalamic nucleus, such black-impregnated cells were always found among neurons that showed no more than the usual brown impregnation, despite the fact that they exhibited clear signs of retrograde chromatolysis. Your thought that the difference may lie in the number of collateral connections comes to mind naturally.

Dr. Heimer: May I follow your line of thought, Dr. Grant. We see the same kind of dark neurons in the prepyriform cortex, but the pattern of the olfactory system is complicated by a heavy two-way traffic system. After olfactory bulb lesions in the rat, we see these dark neurons only in a relatively restricted area of the olfactory cortex. Although we cannot rule out that this represents transneuronal degeneration, I find it more likely that it is a retrograde reaction of these neurons, especially because lesions restricted to this area result in the appearance of terminal degeneration in the olfactory bulb whereas lesions situated mainly outside this area in more caudal parts of the olfactory cortex do not seem to produce significant fibre degeneration within the olfactory bulb. The black neurons appear as early as a few days after olfactory bulb removal.

Dr. Nauta: The remarkable thing is that Nissl preparations at that stage and even a week later show little evidence of chromatolysis in these neurons, contrary to what one sees in the thalamus of the decorticate rat.

Dr. Cowan: Perhaps I may comment both on what Dr. Nauta has said and upon the reaction of the hypoglossal nerve in adult animals. Some years ago, Dr. Powell and I sectioned the hypoglossal nerve extracranially in a series of adult rabbits, specifically to look for retrograde, or indirect Wallerian, degeneration. We could find no evidence for this either in the nucleus or in its emerging rootlets. On the other hand, we did see occasional dark neurons and argyrophilic fragments (which we thought were due to the fragmentation of dendrites) in the dorsal nucleus of the vagus after interrupting the vagus in the neck. We have also seen a similar dense impregnation of neurons and fibre fragmentation in the anterior thalamic nuclei of rats after lesions of the cingulate cortex.

Dr. Myers: I have seen degeneration fairly often in the sensory component of the trigeminal nerve after making a variety of lesions in cats. This always seemed hard to explain as anything but a retrograde change such as you described. I would like to ask if anyone else may have seen this unexpected type of central degeneration of the trigeminal nerve after surgery in the head region?

Dr. Grant: That is of great interest indeed. As I mentioned briefly in my paper, we have made studies on the primary vestibular fibres in newborn rabbits. Following transection of the vestibular nerve branches peripheral to the vestibular ganglion, we have found terminal degeneration in the region of the lateral vestibular nucleus where Brodal and co-workers have demonstrated the termination of primary vestibular fibers. We have made similar findings with degeneration central to the sensory ganglia in the spinal cord of kittens in which the sciatic nerve has been cut. In both instances, the reasonable explanation is a degeneration of the central branch of the sensory neuron, secondary to a retrograde cell degeneration in the ganglion.

Dr. Peters: Do you find that the dark neurons of the type described by Cammermeyer have the same Nissl pattern as your dark neurons? Or is it different entirely?

Dr. Grant: I cannot answer that question since Nissl substance is not discernible in our argyrophilic neurons.

Dr. Guillery: I want to comment on the term "indirect Wallerian degeneration". It should not be applied to changes that occur in perikarya and dendrites. The use of this term to describe the axonal changes that occur between the cell body and the lesion is a little odd. Waller, after all, denied any changes central to the cut. The term "indirect Wallerian degeneration" thus attaches Waller's name to a phenomenon he denied.

Dr. Grant: To some extent, I have the same feeling. But still I think that the term "indirect Wallerian degeneration", which van Gehuchten originally proposed, is an appropriate term for characterizing the orthograde degeneration of the nerve fibre, proceeding from the parent cell body towards the site of the axonal lesion. Like the "true" or direct Wallerian degeneration, the degeneration proceeds in an orthograde direction. The term "retrograde fiber degeneration" which has been used by some authors for the same process is, therefore, misleading. Moreover, it is apt to cause confusion with the true retrograde fibre changes that take place close to the axonal lesion. Unlike the direct Wallerian degeneration the indirect Wallerian degeneration is the result of an indirect reaction. The axonal transection as such does not result in an orthograde fiber degeneration between the parent cell body and the site of the lesion. A heavy retrograde cell reaction leading to cell death is a necessary prerequisite.

Concerning nomenclature, I would like to suggest the following terms for the degenerative changes that can occur in neurons following axonal transections: retrograde neuron degeneration, as a general descriptive term; retrograde cell degeneration and retrograde dendrite degeneration as more specific terms; and finally, indirect Wallerian degeneration as a specific term of the axonal or fibre degeneration.

Originally we used the term "indirect Wallerian degeneration" not only for the degenerative fibre changes but also for the degenerative changes in nerve cell bodies and dendrites. But we feel now that it would be appropriate to have "retrograde" included in the terms for the changes of the cells and dendrites since the changes really are of a retrograde nature. The terms "retrograde cell degeneration" and "retrograde dendrite degeneration" can hardly be misinterpreted.

Dr. Fuxe: We have studied the effect of olfactory stimuli on the activity of the monoamine neurons and in these experiments we removed the olfactory bulbs. By chance we made interesting observation that the fluorescent properties of prepyriform cortex change as a result of this operation. The cells acquired a yellowish-brown auto-fluorescence 2—30 days after operation, which could not be observed either on the control side or in sham-operated animals. It seems to me that these changes might reflect a retrograde cell change.

Dr. Myers: When peripheral nerves are sectioned distally, efforts are made at regeneration leading to frequent re-innervation of denervated muscle. It has been difficult to reconcile this fact of regeneration with the retrograde degeneration and disappearance of neurons in the motor neuron pool in the spinal cord and lower brain stem with peripheral nerve section. Can you comment on the circumstances when successful peripheral regeneration will occur and when retrograde cell degeneration with loss of the nerve cell body will occur?

Dr. Grant: I could imagine that re-innervation in cases with disappearance of neurons could be explained by sprouting from neighboring neurons supplying the muscle.

Dr. Myers: Does section of the axis cylinder of any motor neuron close to the cell body inevitably result in retrograde dissolution of the neuron?

Dr. Grant: In very young animals this might well be the case. But in the adult, where the retrograde cell reaction is much less intense, one could not expect a dissolution of the neuron after axon transection generally. But the closer to the cell body the transection has been performed, the more probable is a retrograde dissolution of the neuron.

Electron Microscopy of Golgi Preparations for the Study of Neuronal Relations

Theodor W. Blackstad

Introduction

During the planning of this book the author was requested to prepare an account of his experiences (1965) with the combined use of Golgi impregnation and electron microscopy in neuroanatomical studies by the method of experimentally induced Wallerian degeneration. At the present time, however, it is no less useful to evaluate the application of a combined Golgi-and-EM procedure to the study of interneuronal connections in *normal* material. Therefore, and also because both approaches involve the same technical procedures, they will both be dealt with. The articles by Stell (1965a, b; 1967) and by the present author (1965) will form the main basis for the following account, for no further reports on this subject appear to have been published.

Despite the fact that the combined Golgi-and-EM strategy already has yielded some significant results — particularly in Stell's (1965b; 1967) elegant studies of the fish retina — it is only fair to state that it is at present in an early stage of experimentation and obviously in need of further development. What can be offered at the present time represents an idea or an attitude rather than a fully elaborated method ready for wide application. It is obvious that the practical procedures that have been tested so far will be supplanted by better ones. Anyone interested in this approach may find it profitable to modify rather than to adhere strictly to the practical procedures reported below. However, independent of future improvements, certain fundamental needs and principles will remain, and it is to these that the following discussion will be addressed.

It seems appropriate to consider first what could be obtained *in principle* by combining Golgi impregnation and electron microscopy, and then to relate these possibilities to existing needs.

Golgi impregnation of neural tissues serves to visualize single cells in a manner permitting classification of cells in terms of the ramification pattern of their processes. It may be necessary for combined Golgi-EM studies to demonstrate all, or only part of the processes for the classification, depending on the absence or availability of light-microscopic data. The long anterior horn cell dendrites, for example, will require more ample visualization than is needed for cerebellar or hippocampal granule cell dendrites. Some cells or parts of cells can be classified without Golgi impregnation: for example, the perikarya of Purkinje cells or of large motor neurons. However, in many other cases the Golgi method is indispensable, a fact too well known to require further documentation. One might think that the complete three-dimensional visualization of cells offered by the Golgi picture could be obtained also by reconstructions from serial electron micrographs. However, in practice the serial electron-microscopic reconstruction of whole neurons would involve a prohibitive amount of work (Karlsson, 1966).

Electron microscopy offers two or three major interdependent advantages in the study of the nervous system. The increased resolution permits visualization of processes too thin to be visible in the light microscope, and of organelles (plasma membranes, synaptic vesicles, microtubules, spine apparatuses, etc.) within structures that are large enough to be seen in the light microscope, but for which we have no fully adequate staining methods. Such organelles can thus acquire value as criteria for the identification of particular structures. Electron microscopy, furthermore, differs from light microscopic methods in the sense that it displays all tissue components together, thus allowing an evaluation of their interrelationships. However, electron microscopy usually fails to identify the continuity between a particular axon or dendrite and the distant parent cell body.

The *combination* of Golgi impregnation and electron microscopy permits one to combine the advantages of two approaches, and particularly, to study the environmental relations of Golgi impregnated structures by means of electron microscopy. It also allows the observer to characterize the ultrastructural properties of the impregnated structures themselves, irrespective of their surroundings.

In the following account only nerve cells will be considered, but glial cells may well be studied in the same manner.

With respect to the examination of *synaptic connections* with the Golgi-EM combination, two principal situations can be imagined: 1. Golgi impregnation of a postsynaptic cell, the presynaptic cell — either normal or "labelled" by Wallerian degeneration or other features — being a component of the non-impregnated tissue, and 2. Golgi impregnation of the presynaptic cell, particularly the axon and its synaptic terminals, the postsynaptic structure now being part of the unimpregnated tissue but possibly exhibiting features characteristic enough to permit its category to be identified. Such identifying characteristics might be provided, for example, by the retrograde dendritic changes shown by Grant (see elsewhere in this volume) or by retrograde chromatolytic cell changes.

A third situation, the impregnation of both the pre- and postsynaptic element of a synapse, will be too rare to have practical significance and to warrant further consideration here.

Turning to specific problems, there are several situations in which present methods for the study of connectivity are not satisfactory, and in which the principle of combining Golgi impregnation and electron microscopy would seem to hold some promise. Among such cases are 1. the classification of postsynaptic structures at sites of terminal bouton degeneration, and 2. the identification of the mode of termination of axons too short to be examined by the use of surgical methods. In these cases one would welcome techniques for the light-microscopic classification of postsynaptic and presynaptic elements, respectively, and yet have a tissue preservation adequate for an electron-microscopic examination of the same specimen.

Thus far, few attempts have been made to combine Golgi impregnation and electron microscopy. The experiments of Stell, as well as the author's own experiences with this double procedure are recorded below, with particular emphasis on difficulties, artifacts and failures, some avoidable, others unavoidable without drastic changes of procedures. This is done in an attempt to assist those who need this particular combination of techniques for some purpose, but who will have to invest efforts in developing the methods further. It is emphasized again that, although it is the author's belief that the combination of Golgi-impregnation and electron-microscopy holds considerable promise, the field of application of this research strategy is at present quite limited.

Procedures

Fixation

A *conditio sine qua non* is that the fixation method will allow both Golgi impregnation and a quality of tissue preservation reasonably adequate to the requirements of electron microscopy. The classical fixative for the Golgi method is potassium dichromate. Combined with sodium sulfate alone, as in Müller's liquid, it is unsuitable for a general preservation of ultrastructure, but it has been combined in Golgi studies with agents suitable as fixatives for electron microscopy, particularly osmium tetroxide (as in the rapid Golgi procedures), and formalin (as in the Kopsch modification). Furthermore, the chromation may be preceded by formalin fixation (Bubenaite modification and others). Both Stell and the author selected potassium dichromate-osmium tetroxide mixtures for chromation. In order to adhere strictly to the classical requirements for Golgi impregnation, Stell fixed fragments of retina by immersion in an icecold unbuffered mixture of 3.3% osmium tetroxide and 0.1 M (approximately 3%) potassium dichromate, at a pH of about 4. Despite this use of a procedure that would no longer be considered optimal today, Stell achieved a surprisingly good preservation of ultrastructural detail (Figs. 2, 4), no doubt because the organ under study was easily accessible and of small dimensions. However, when larger pieces of tissue, and particularly whole brain or spinal cord, are to be processed, it is necessary to fix by perfusion with a liquid containing a rapidly penetrating fixative and adjusted at a pH of 7.2—8. The present author used 10% formalin (4% formaldehyde) in 2% sucrose, buffered with m/15 phosphate. It is likely that any other formaldehyde or glutaraldehyde mixture that preserves ultrastructural detail will allow subsequent Golgi impregnation. Hypertonicity is advantageous in some respects and does not interfere with the Golgi-EM method. For a discussion of general problems of fixation for electron microscopy, the reader is referred to Peters' article in this volume, and to standard handbooks. However, considerable boldness is advisable in this as in other searches for new ways of study. Unorthodox procedures have proved in recent years to leave ultrastructure in a much better state of preservation than would have been thought possible a few years ago, before the factors of dehydration and polymerization had been brought under control. The experimenter, therefore, should not hesitate to try any modified fixation procedure that would seem to offer a particular advantage.

Impregnation

This procedure consists of two steps: chromation and treatment with the silver salt. The first of these steps may overlap and even be identical with fixation itself. It is obvious that this step should include treatment with potassium dichromate or any other soluble dichromate. The concentration of the dichromate solution, if used after initial aldehyde fixation, appears not to be critical. Both Stell and the author used concentrations in the range of 3—4%. By contrast, the time of chromation appears to be of importance and should probably amount to no less than one day. Stell used up to three four-day cycles of chromation and silver treatment and still obtained acceptable ultrastructural preservation. The present author chromated for 3—6 days. Furthermore, although low temperature is recommended for the initial fixation and possibly for the initial phases of chromation, final chromation is best done at room temperature and possibly even at higher temperature. This is inferred from the experience that, in light-microscopy, prolonged chromation promotes optimal preservation of myelin phospholipids. Actually, it seems probable that the long chromation at room temperature, although it may impair some cytoplasmic details, may preserve other components better than does the shorter standard stabilization of the tissue prior to dehydration and embedding (see intact myelin sheaths in Figs. 10, 16, 21 and 26).

Osmium tetroxide, commonly used — though not indispensable — for Golgi impregnation, is employed in nearly all procedures for electron microscopy either as a fixative or as a stabilizing or contrast-enhancing agent. It should therefore be included in a combined Golgi-EM procedure. Used as a primary fixative, it should have a concentration of at least 2%, regardless of whether fixation is achieved by immersion or perfusion. If used only after aldehyde fixation, its concentration may possibly be lower, as in classical light-microscopic Golgi procedures. However, in Stell's as well as in my own experience there is no stringent reason to reduce the osmium tetroxide content below 2—3%, for 1. higher percentages of osmium tetroxide do not interfere with Golgi impregnation, while 2. the dichromate component of the mixture possibly attenuates the effect of the osmium tetroxide, and 3. particularly strong osmium-induced contrast is sometimes desirable in ultrathin sections of Golgi material, especially when there are reasons to avoid further staining.

In summary, it is recommended that after fixation in aldehyde for $^1/_2$ to 3 hours, the tissue be chromated with a mixture containing 2—4% dichromate and 2—3% osmium tetroxide, buffered with KOH to pH 7.

Except for the exclusion of NaCl, the dichromate mixture here formulated is essentially identical with Dalton's fixative for electron microscopy. The maximal permissible duration of the treatment has not been established. After 3 days the author replaced the dichromate-osmium mixture by pure dichromate, for fear that a longer stay in the osmicated medium might cause excessive darkening of the tissue, but this fear may be unfounded. It is quite possible that chromation with or without osmium could be extended for weeks without untoward effect.

Penetration of osmium tetroxide into the deeper layers of tissue is very slow, even after aldehyde fixation. Dichromate likewise penetrates rather slowly. Naturally, these time factors must govern all decisions with respect to the duration of treatment of tissues in mixtures containing dichromates and osmium tetroxide. The effective period of treatment does not start the moment a tissue block is immersed in the fixative, but only when the structure of interest has been reached by the dichromate or osmium tetroxide. Perfusion of a brain with these substances in the course of initial fixation or after an initial aldehyde fixation obviously will shorten the latent period before effective chromation starts. The reader is referred to Morest and Morest (1966) for a more detailed consideration of this matter.

Silver nitrate is most often used in a strength of 0.75%. There are no apparent reasons to depart from this custom, even though weaker or stronger solutions may be equally effective. It appears possible, however, that too highly concentrated silver nitrate solutions might cause tissue shrinkage.

Re-cycling of dichromate-osmium and silver nitrate treatment, an established procedure in Golgi-impregnation (see Valverde's chapter in this book), was used by Stell (1965a, b; 1967). However favorable such alternating schedules may be from the point of view of Golgi workers, they may perhaps impair the preservation of ultrastructural detail and would therefore not seem advisable as routine methods when electron microscopy is contemplated.

Trimming and Embedding

Many different useful procedures of trimming and embedding are conceivable, but it is a fundamental requirement that the impregnated structure be clearly visible in the light-microscopic Golgi examination and re-appear undamaged in the block for electron microscopy.

Because the resinous embedding media used for electron microscopy are transparent and allow tissues to appear as well-cleared as in an ordinary light microscope clearing agent, the idea most naturally occurring is, to embed large blocks of tissue in plastic and cut thick sections for light microscopy. This approach was used by both Stell and the author (Figs. 1, 3, 5, 6). Stell used Epon and developed a method for sectioning before the medium had hardened completely. The Epon blocks were mounted on specimen supports, trimmed, and sectioned at 80 micra on a sliding microtome equipped with a heavy blade lubricated with an anhydrous mixture of 1 part of absolute ethanol and 1 part of amyl alcohol (moistening with non-polymerized Epon without the accelerator might be tried as an alternative). The sections were transferred serially to a Petri dish with absolute alcohol, flattened, freed of alcohol and allowed to dry slowly at room temperature. They were then mounted and cured in polymerizing Epon on plastic slides under plastic cover slips. The latter were later removed to facilitate examination under oil immersion. Areas of interest were dissected out and remounted on empty Epon bases (Figs. 1, 3). Stell stated that he remounted selected fragments of thick sections in any desired orientation on a transparent capsule-shaped support, using clear epoxy cement for bonding. It is obvious that many modifications can be invented for appropriate single or double remounting permitting sectioning in any plane that promises either greater convenience of interpretation, or a higher yield of the structures searched for.

In the procedure followed by the author, sections 50—200 micra in thickness were taken from a fully polymerized Araldite block. The tissue blocks had been progressively reduced in size during the steps of impregnation in silver nitrate, dehydration, and infiltration by non-polymerizing Araldite. Under a dissecting microscope, impregnated structures can be discerned to some extent with incident light even in larger non-transparent blocks, and may then be isolated by further trimming. Thick sections were cut from the polymerized blocks (made in gelatin capsules) by means of a razor-blade held in a small botanical hand-microtome. This procedure is by no means ideal since the blade tends to bend and remove wedge-shaped slices. The use of a microtome for hard tissues may be tried, although thick sections tend to break. On

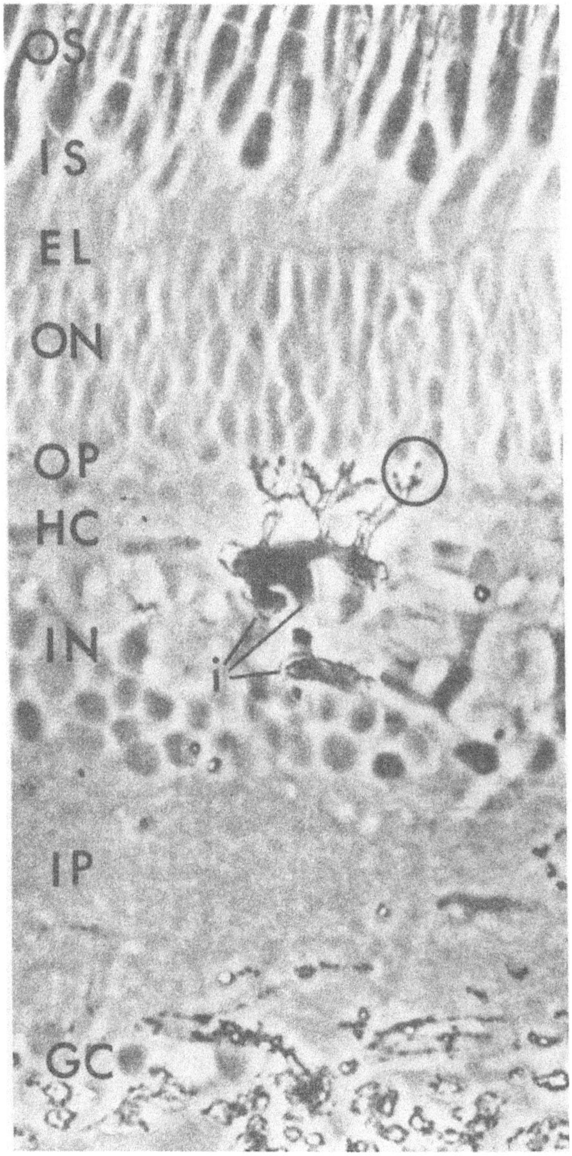

Fig. 1. Phase contrast micrograph of a 10 μ section from a Golgi-impregnated goldfish retina. A middle horizontal cell is seen (at i) with dendrites terminating at the rod synaptic endings. *OS* receptor outer segments; *IS* receptor inner segments; *EL* external limiting membrane; *ON* outer nuclear layer; *OP* outer plexiform layer; *HC* horizontal cell layers; *IN* inner nuclear layer; *IP* inner plexiform layer; *GC* ganglion cell layer. ×820. [Reproduced from Stell (1965b, Fig. 2, p. 395) by kind permission of the author and the Wistar Institute Press]

the basis of his experience the author would recommend to take sections from fully hardened blocks under a dissecting microscope with a razor knife, or razor blade, with or without the use of a hand-microtome.

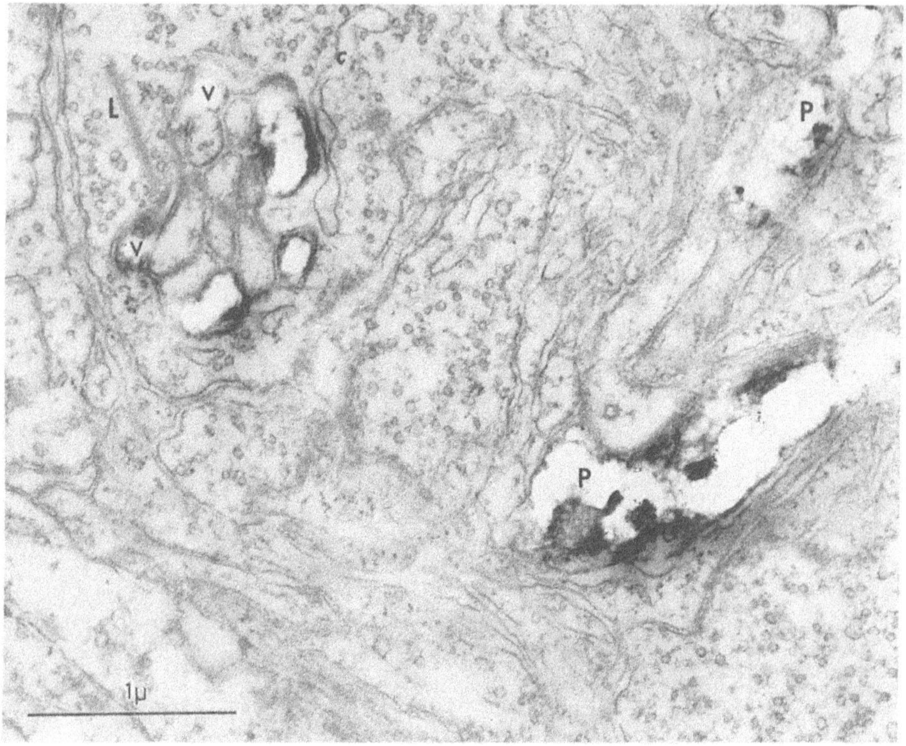

Fig. 2. Electron micrograph of goldfish retina (Stell, 1965a). Larger process of a Golgi-impregnated vitreal horizontal cell are seen at *P*, and fine processes (tips) occupy peripheral positions (*v*) among processes in the rod synaptic complex. *L* synaptic lamella; *c* cistern in the rod spherule. Sectioned on 5% acetone, lead-stained (Millonig, 1961). ×29,250

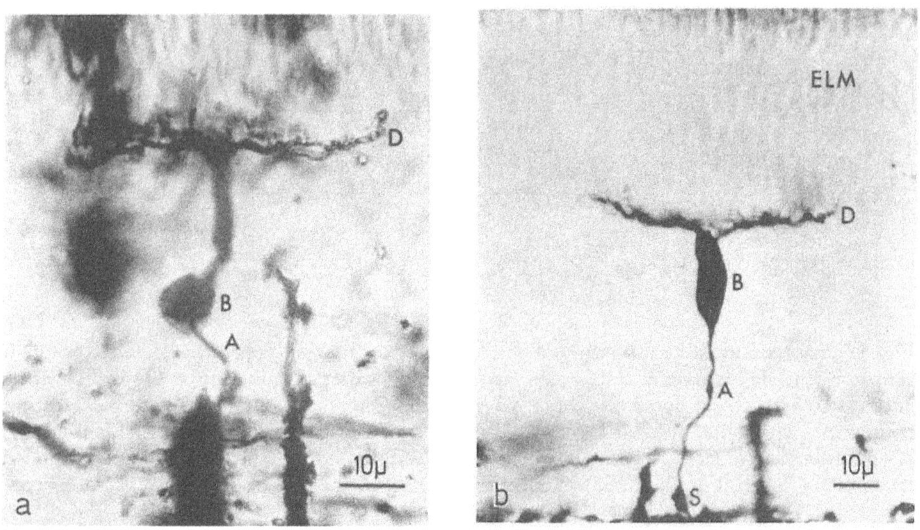

Fig. 3. Shown in *a* is a light micrograph of a large bipolar cell (from Stell, 1965a). Parts of the same cell are shown in Fig. 4. *A* axon; *B* cell body; *D* dendrites. Transilluminated 80 μ section. Marker indicates 10 μ. ×826. *b:* Light micrograph of another large bipolar cell (Stell, 1965a) in transilluminated block. *ELM* external limiting membrane; *D* dendritic arborization; *B* cell body; *A* axon; *S* large synaptic terminal. Marker indicates 10 μ. ×690

Remounting was achieved by placing the slice in a drop of polymerizing Araldite and pressing a flattened empty Araldite base down on it (see below: Araldite bases with embedded mirror). This is best done on a non-adhesive, flat and preferably shiny surface. Non-treated glass adheres too firmly to the polymerized Araldite. Metal foil rubbed with a little silicone oil can be used, but various types of plastic sheet may also

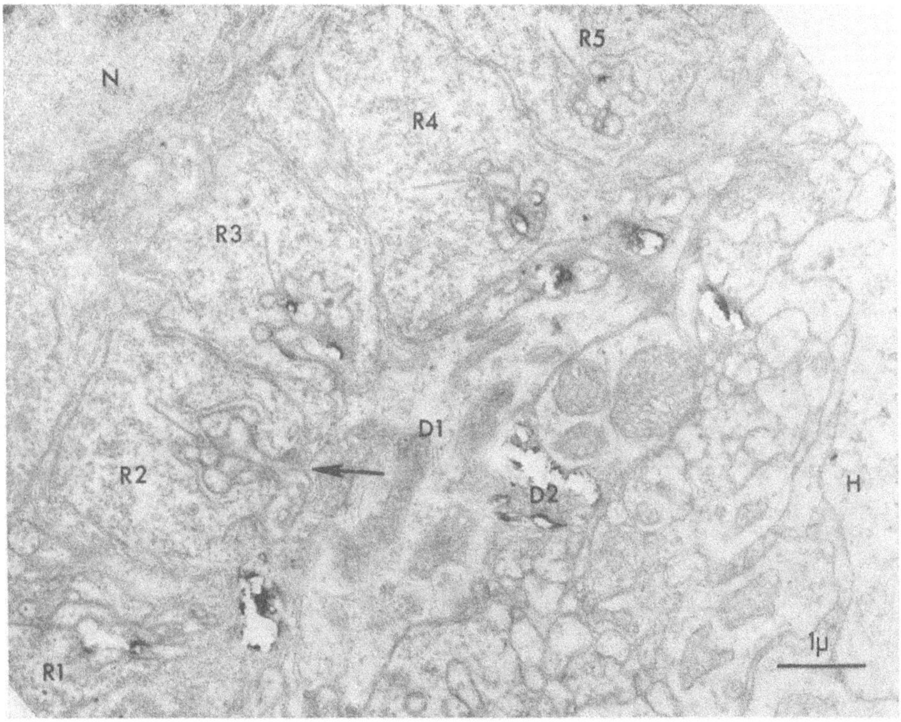

Fig. 4. Electron micrograph of dendritic field of bipolar cell shown in Fig. 3a (Stell, 1965a). Five rod spherules (*R*1—*R*5) are seen, all but one of which are related to impregnated bipolar dendritic terminals. One of the larger dendrites (*D*2) and several smaller ones of this cell are also seen. The one spherule not related to the impregnated cell (*R*2) appears to receive a branchlet (arrows) from an unimpregnated bipolar dendrite (*D*1). *H* horizontal cell of scleral row; *N* rod cell nucleus. Sectioned on 5% acetone, lead-stained (Millonig, 1961). Marker indicates 1 μ. ×12,320

be considered for the purpose. Because the tissue in the slice is already protected by polymerized plastic, rather high temperatures (60—90° C) can be used for polymerization at the stage of remounting. A test drop placed at the side of the specimen will show when polymerization is complete. This may take from one-half to several hours.

An alternative to the cutting of thick sections after polymerization would be, to cut as in conventional Golgi studies, viz., before or at the stage of dehydration. The sections can then be cleared by a suitable clearing agent for conventional classification and recording of cells, and subsequently be transferred to the final embedding plastic required for ultrathin sectioning. Numerous modifications could be imagined,

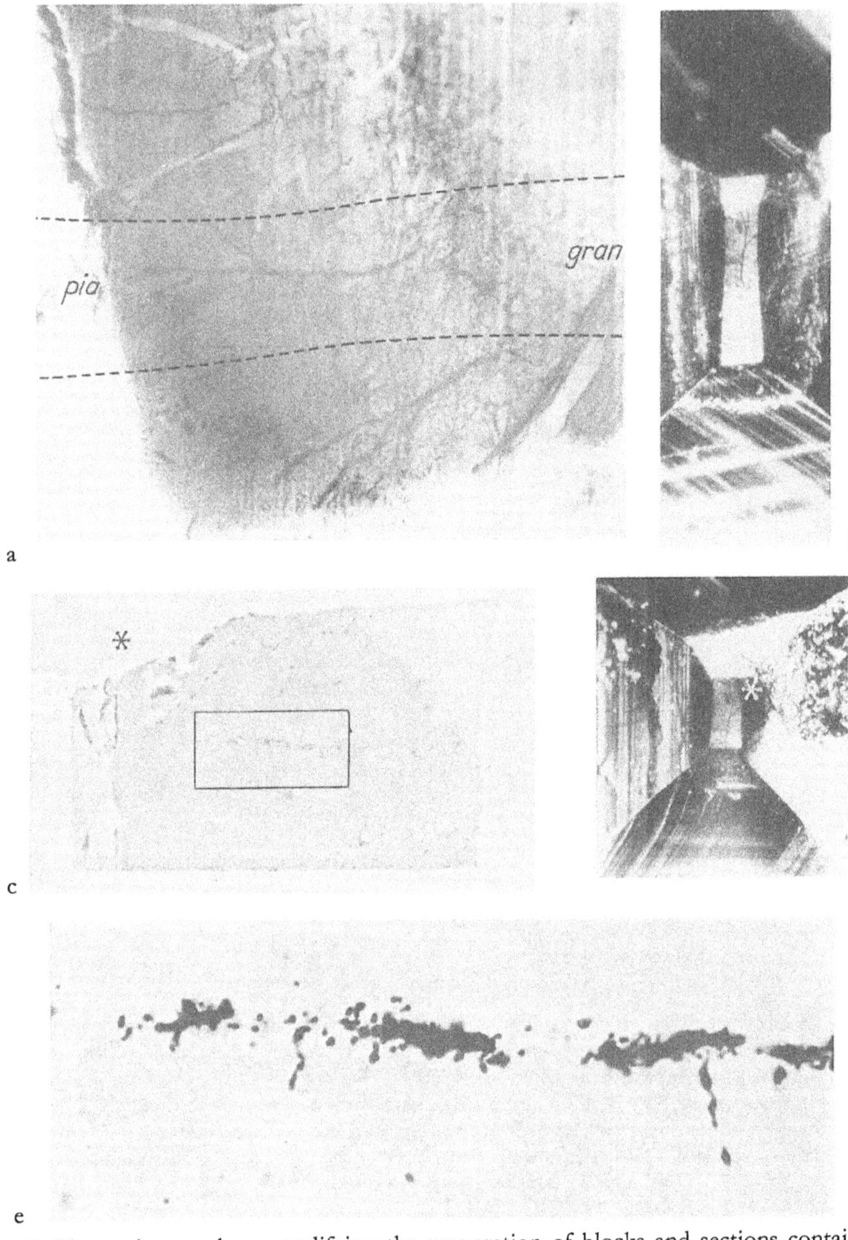

Fig. 5. Photomicrographs exemplifying the preparation of blocks and sections containing a specific portion of one identified nerve cell only. *a*: Major part of a slice removed from a polymerized block and here shown after remounting on an empty Araldite base. It contains the granular (*gran*) and molecular layers of the fascia dentata. Several dendrites of granular cells are seen. Broken lines delimit the area visible in *b*. ×248. *b*: The top of the same block (illuminated from below) after it had been trimmed down so as to contain only one impregnated granular cell dendrite (seen in *a*). ×64. *c*: The same block after further trimming. ×56. *d*: Section about 1 μ thick, from the same block. The main portion of the dendrite seen in *c* is visible within the rectangle. The asterisk indicates the corner marked accordingly in *c*. ×245. *e*: The area outlined with rectangle in *d*, shown at a higher magnification. Note abundant spines along dendrite, and the two axons crossing it at right angles. This thicker section was cut from the block between two groups of ultrathin sections, in one of which the same dendrite was present, as seen in the electron microscope. Oil immersion objective. ×1,525. [Reproduced from Blackstad (1965) by courtesy of Springer-Verlag, Berlin]

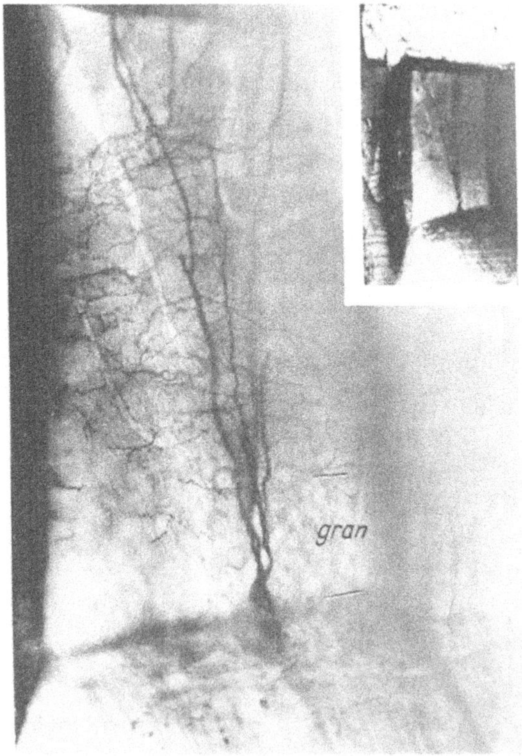

Fig. 6. Top of another block (a remounted thick section before trimming) from area dentata. One granule cell with several dendrites is visible. Thin axons are also seen. The granular cell layer has been labelled *gran*. *a*: ×245, *b*: ×57. [Reproduced from Blackstad (1965), by courtesy of Springer-Verlag, Berlin]

but in principle it would seem best to introduce as few steps as possible. Non-polymerizing plastic might be chosen as the clearing agent[1].

The technique of cutting thick sections in the conventional way would appear to be time-saving as well as optimal for the purpose of wide survey. However, it remains to be seen whether this procedure may not cause mechanical damage visible only at the ultrastructural level.

It is easy to imagine suitable procedures other than those mentioned here. One would be, to facilitate sectioning by embedding in celloidin or a similar soft material that could be dissolved or penetrated by the resins needed for ultrathin sectioning. Several procedural details of this category might be suggested by the technical traditions and the equipment of the particular laboratory.

[1] Holländer (1969) has developed a method for embedding sections of Golgi impregnated material in Araldite. On one sheet of cellulose acetate is placed a second sheet with an opening cut out in the middle, to which the dehydrated and cleared section is transferred with some polymerizing Araldite. A third sheet is placed on top and kept in place by a weight. After curing, the cellulose acetate sheets are easily stripped off and the slice with associated plastic is ready for light-microscopic examination and further trimming for electron microscopy.

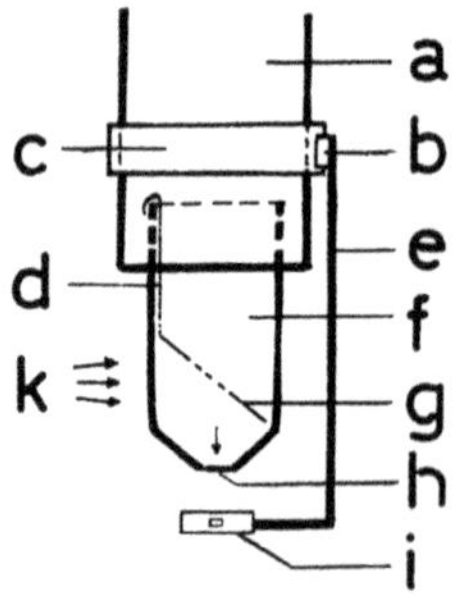

Fig. 7. Diagram of arrangement for control of blocks during ultramicrotomy. For explanation of symbols, see text. [Reproduced from Blackstad (1969), by courtesy of UCLA Forum in Medical Sciences, Los Angeles]

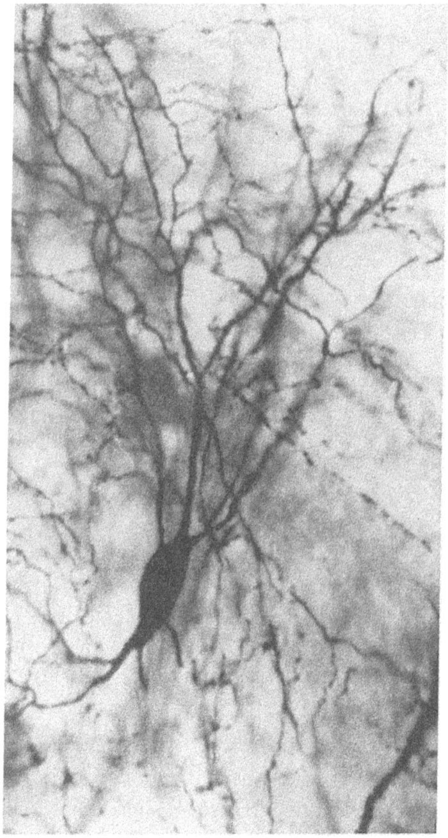

Fig. 8. Neuron with short axon from molecular layer of fascia dentata. The cell body and axon ramification are located approximately equidistant from the pial surface and the granule cell layer. The dendrites are virtually without spines, and without the coarse "varicosities" ("beading") often seen in poorly fixed material but occasionally also in well-fixed tissue. Fixation: formalin perfusion and dichromate-osmium tetroxide treatment as described by Blackstad (1965). A few measurements on this and other similar cells are mentioned in Blackstad (1969). Compare with Fig. 9. × approx. 725

In the 50—300 μ thick sections of impregnated tissue general histological features are often seen fairly well in the light microscope. The osmium treatment gives a faint but useful contrast which permits visualization of cell layers, myelin sheaths, blood vessels and other structures (Figs. 5a, 6). Phase contrast microscopy may bring out

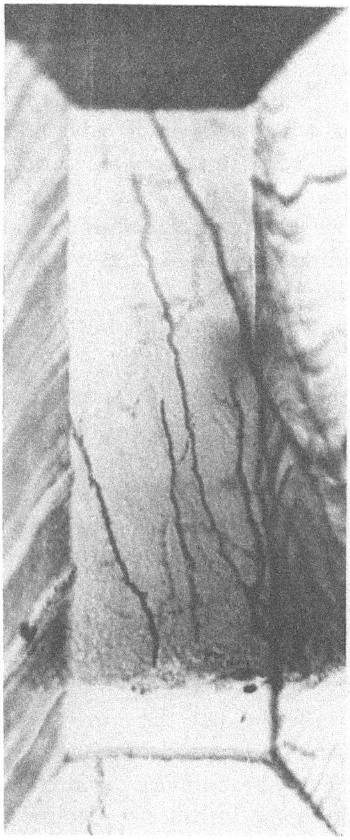

Fig. 9. Top of transilluminated Araldite block with dendrites of cell similar to the one shown in Fig. 8, after removal of most of the cell. Two non-spinous dendrites are seen in the middle, flanked by granule cell dendrites with spines. Part of the longest spineless dendrite was studied in serial ultrathin sections and used for a pilot analysis of the environment of spine-free dendrites (Blackstad, 1969). Compare Fig. 8. × approx. 940

the picture even more clearly, and is particularly useful for examining 0.5—2 μ control sections (Fig. 1). Such sections may also be stained with conventional methods for semi-thin sections, but staining solutions dissolving the Golgi precipitate should, of course, be avoided. Semi-thin sections allow observations of considerably greater detail than can be made in the thick slices, and there is every reason to collect as much information as possible from such material before electron microscopy. Conversely, once acquainted with the ultrastructural details of a tissue one may develop an increased competence for interpreting semi-thin sections and profitably exploit the much easier light-microscopic method of study.

Ultrathin Sectioning

The ultrathin sectioning of Golgi material is no different from that of any other tissue, but the usual scarcity of impregnated structures per unit volume makes it particularly important to select only structures of interest for electron microscopy, in order to avoid time-consuming work with "empty" tissue. Single impregnated cells, and even parts of impregnated cells, should be trimmed free of irrelevant tissue. As long as the 50—200 μ thick sections are unmounted, trimming with fragments of razor blade held with forceps is easily done under a dissecting microscope at a magnification of 25—40. Final trimming is hardly advisable at this stage because the piece of tissue may become too small to be handled during remounting (see above) and is more conveniently postponed until after the section has been remounted. The transparent base with the tissue may be studied with the light microscope much like a slide (Figs. 5b, c, 6). The impregnated structures contrast conspicuously with the almost transparent surrounding tissue. The depth of the structure of interest below the block surface can be estimated by use of the micrometer knob.

For the purpose of final trimming after the block has been fastened in the chuck of the ultramicrotome, a slight modification of the usual procedure has been found useful (see Fig. 7). Its particular feature is to permit further trimming, even after ultrathin sectioning has been started, without removing the block or the chuck from the ultra-microtome. The empty plastic base used for remounting (f in Fig. 7) has had a small slanting mirror (g) fashioned of aluminum foil (d and g) embedded in it prior to polymerization. It reflects light from a light source (k) through the remounted slice (h) with the effect that one can examine the Golgi-impregnated tissue in transmitted light even after the block is mounted in the microtome chuck. Second, a small mirror (i) about 2×3 mm large and made by evaporating aluminum onto a coverslip, is fastened to the microtome chuck (a) by means of a detachable bracket (c), the arm (e) holding the mirror being hinged (at b) so as to permit positioning of the mirror in front of the block during intermissions in the ultrathin sectioning. The tissue may be examined through this mirror by means of the usual microscope on the microtome in order to determine whether the neuronal processes, cell bodies, or other structures of interest have been reached by the microtome knife and to what stage the sectioning of these structures has proceeded. This procedure is useful in determining at which stage of sectioning sections should be picked up on grids, and thus helps to prevent loss of significant material. The limit of resolution is determined by the optical system of the stereomicroscope used, and may be improved by placing an aperture between the mirror and the microscope. The microtome light can be used as the source of transilluminating light, but it is preferable to install a separate, stronger light source. A photographic camera may, of course, be fitted to the optical equipment.

Stell (1965a, b; 1967) controlled the trimming and sectioning of Golgi-impregnated remounts by transilluminating the mounted specimen by means of a split condensor, i.e., with the upper part of the light microscope condensor removed, in order to eliminate some of the optical haze due to internal reflection of a focussed beam in the epoxy mounting block.

The need for such checking procedures varies with the sparsity of the impregnated structures to be examined in the electron microscope. They are highly useful in cases in which the object of interest is a single dendrite (as in Fig. 9), but hardly worthwhile for the sectioning of a rich axonal plexus.

Serial ultrathin sections of Golgi material are not only relatively easy to obtain but are highly recommended because the outlines of impregnated structures are more reliably distinguished from random artifacts in a sequence of sections. As a rule, only short series of ultrathin sections are needed for this purpose. Other reasons for the use of serial sections include all those valid for other forms of biological material and therefore need not be enumerated here.

The Trough Liquid

In the initial experience of both Stell (1964) and the author, the precipitate which had been so clearly visible in the light microscope had largely vanished from ultrathin sections that had been floated for a few minutes on water or on a weak solution of acetone or alcohol (Figs. 2, 4, 15). It was found, however, that this loss of impregnation can be prevented by adding potassium dichromate to the trough liquid or saturating it with silver dichromate (Figs. 10, 11, 14, 21, 24). Actually, in the latter condition it is not unusual to see signs of further crystallization on the impregnated structures (Fig. 18). Evaporation of trough liquid causes crystals to appear along the knife edge and the sides of the trough. These undesirable effects can be reduced by diluting the solution slightly by adding 1 part of water to 3—5 parts of saturated silver dichromate solution, depending on temperature, speed of cutting, etc. Sometimes one may be satisfied with less than the full amount of precipitate in the impregnated structures, in order to see their outlines better. In that case one can select a still lower concentration of silver dichromate and create the effect desired in each case. In some instances, even complete absence of precipitate might be accepted or preferred. Stell (1965b) held that "the spaces remaining after the precipitate dissolves are as reliable for identifying impregnated cells as the precipitate itself, as shown by their consistent location in the tissue, by the presence of an enclosing plasma membrane and by serial sections (Figs. 2, 4). Since dissolving of the precipitate also results in clearer sections, the only real disadvantage is that the edges of the holes show the usual tendency to retract under the electron beam. For routine purposes, we allow the stain to dissolve, although occasional sectioning on dichromate may be necessary where interpretation is questionable." It is probably the best practice to start examining sections with well-preserved precipitate and later proceed with the less stark pictures provided by sections from which the precipitate has been dissolved.

The use of potassium dichromate instead of silver dichromate in the trough must reduce the tendency of silver dichromate crystals to form on the sections. Stell used a 0.1 M solution, i.e. about 3% potassium dichromate. Whether silver dichromate or potassium dichromate is used, it is advisable to dip the grid in distilled water a split second after it has been raised from the trough, in order to rinse off the trough liquid before it begins to evaporate. A minimum of time should be allotted to this submersion in order to avoid loss of precipitate. Immediate subsequent blotting of the grid is recommended for the same reason.

According to Stell (1965b), the Golgi precipitate is preserved by floating the sections on organic solvents.

Staining

Staining of the ultrathin sections may interfere with the preservation of the Golgi precipitate in the sections, and for some purposes unstained sections will have sufficient contrast (Figs. 10, 11, 14—18, 21—25). The author found lead citrate by itself ineffective but obtained a definite increase in contrast with uranyl acetate and lead citrate used in the regular way. However, this double staining procedure tends to dissolve the precipitate (cp. Fig. 19) and therefore requires modification, perhaps by a suitable choice of solvents and by adjustment of staining times and temperatures. Stell (1965b) found that Golgi precipitate, once retained after sectioning, is not removed by lead staining. It is likely that the best understanding of the tissue will result from a comparison of stained and unstained sections.

Sandwiching

Sections of Golgi material, even when mounted on carbon films, are liable to rupture at the sites of impregnated structures during electron microscopy. Even from the outset, after ultramicrotomy, breaks exist in the hard crystalline mass of silver dichromate (see Fig. 5 of Stell, 1965b), and these fractures widen in an unacceptable and highly disturbing way during microscopy, apparently as a result of local overheating. Similar effects observed in attempts to sandwich sections with carbon in the coating plant were probably likewise due to heat. Sandwiching with formvar, however, will eliminate this difficulty almost completely, and allows a high beam current to be used (the carrying film being carbon or carbon combined with formvar). The formvar film should be picked up with the grids using the slide (not paper) method (see e.g. Reimer, 1967, page 324), which allows the moisture to be blotted off immediately. This is important, since otherwise dissolution occurs, followed by precipitation outside the Golgi-impregnated structures (Figs. 16, 17). In all likelihood, the tendency of the section to rupture at the holes left after leaching-out of the precipitate observed by Stell, will likewise be abolished by sandwiching.

Observations

General Appearance

The electron-microscopic appearance of Golgi impregnation has already been presented earlier in this chapter. Stell's and my own observations are almost identical. That excessive precipitates form in some places and transcend cell borders is well-known from light microscopy. However, it also has become clear that in places where such excessive accumulation is lacking the precipitate respects the lipoprotein membranes of the tissue. Electron microscopy shows that the dark, fully impregnated dendrites and boutons contain the precipitate within the plasma membrane, but the precipitate usually spares intracellular structures surrounded by lipoprotein membranes such as mitochondria, nucleus, etc. Although proper Golgi-impregnation thus corresponds to an intracellular precipitate, in neurons as well as in glial cells, precipitate occurs extracellularly in many places between plasma membranes (see Figs. 12, 18, 20, 23a). Larger areas in which such extracellular precipitates are prevalent (Fig. 20) correspond to brownish, spoiled regions in the light microscope. Sometimes the extracellular precipitate may outline cells for some distance relatively selectively as a light brown border. Fig. 12 shows Purkinje cells so outlined. Although this mode of impregnation represents an unreliable definition of single cells, it may possibly be exploited for special purposes.

In some cases, very unusual results are obtained. Fig. 24, for example, shows a nearly complete impregnation of the endoplasmic reticulum of a nerve cell.

Although the nature of the Golgi precipitate and the basis for its formation have been much debated, there is little reason to doubt that the precipitate consists of silver dichromate. Only silver ions and dichromate ions are present initially, and further treatment involves no reaction that might change the silver dichromate. The view is supported by the experience that the Golgi-precipitate is dissolved more readily by distilled water than by a solution of silver or potassium dichromate. Once formed, however, the precipitate can be changed. A reducing bath such as used in photographic development probably will lead to metallic silver. Attempts to identify the precipitate with selected-area electron-diffraction (unpublished) did not show silver dichromate but reduced compounds, which was to be expected because of the bombardment with the negative electrons.

The precipitate no doubt is formed by a progressing crystallization and the impregnation of a given tissue area is either complete or totally absent. However, an occasional cell (or any other membrane-bound volume) may be encountered in which impregnation appears not to have been completed, part of the cell being black and another part entirely without precipitate. This corresponds to the light-microscopic experience that impregnation is not always entirely complete, a phenomenon probably less frequently encountered after triple impregnation. It is therefore evident that caution is needed in the evaluation of electron microscopic pictures of Golgi sections. The limits of a metallic precipitate should not be automatically equated with a cell border, just as a cell border should not be overlooked when covered by excessive precipitate.

Examples of Application; General Comments

Normal Material

In Stell's investigations on fish retina, the Golgi-EM method was exploited for visualization of *postsynaptic* dendritic structures[2]. In Stell's (1965a, b) study, horizontal cells of the retina were identified by the shape and location of their perikarya and their typical dendritic pattern (see Fig. 1 of the present paper). In subsequently prepared electron micrographs the tiny dendritic tips could be traced to the characteristic synapse between receptor and bipolar cell (Fig. 2). Stell's unpublished thesis (Stell, 1965a) also contains pictures of impregnated bipolar cells (Fig. 3a, b) and electron micrographs showing their dendritic tips located centrally in the synaptic complex just mentioned (Fig. 4). More illustrations are presented in one of Stell's (1967) later papers. These findings represent excellent examples of the potential of the Golgi-EM method. In this case, it was possible to demonstrate the specific nature of peripheral and central parts of a synaptic complex even though impregnation destroyed some of the finer detail, a deficiency that could easily be compensated for by electron-microscopy of conventional normal material in which, for instance,

[2] For the sake of simplicity, the processes of the retinal horizontal cells are here termed dendrites. See Stell (1967) for a discussion of their possible nature.

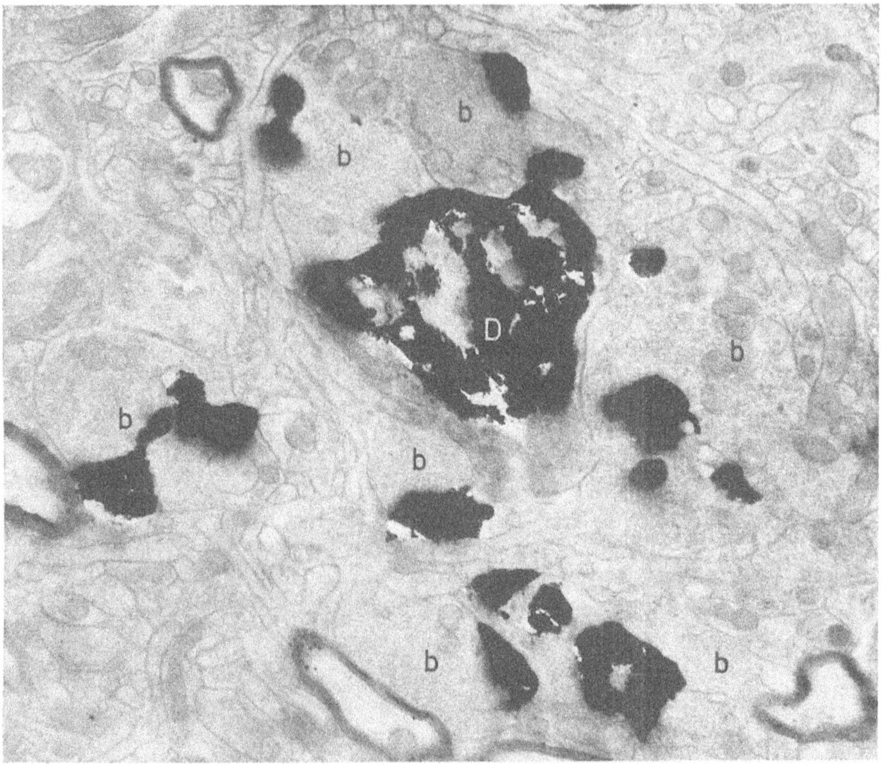

Fig. 10. Impregnated dendrite of a hippocampal (CA3) pyramidal cell surrounded by unimpregnated giant boutons of mossy fibers. Impregnated excrescences arising from the dendrite interdigitate with negative reliefs of the giant boutons. This picture was taken from the hilus of the fascia dentata, but identical morphology exists in the layer of mossy fibers proper. ×12,000

postsynaptic densities can be recognized. Only very tedious three-dimensional reconstruction could provide the same information. Stell emphasized the suitability of the Golgi-EM combination when the impregnated structure is easy to recognize. He foresaw a more difficult application of the principle in cases in which all components are small, of similar outline, or less regularly arranged.

The present author likewise identified dendrites by light microscopy and then proceeded to electron microscopy for further identification of their environment. In the regio inferior (CA3) of the hippocampus, giant pyramidal cells were impregnated and the interdigitation of the latter's ramified dendritic spines (excrescences) with the characteristic giant boutons of the mossy fibers was identified (Fig. 10; see also Fig. 13b). Further observations were made on the pyramidal cell dendrites. Their pattern of ramification was observed in the stratum radiatum of the regio superior (CA1) and in the stratum oriens of the regio inferior (CA3). In the molecular layer of the fascia dentata the dendrites of the granular cells were studied (Figs. 5 and 6); in this region some neurons other than granule cells were isolated (Fig. 8) the dendrites of which possess virtually no spines. Brief mention has been made (Blackstad, 1969) of a pilot attempt to measure the proportions of structures

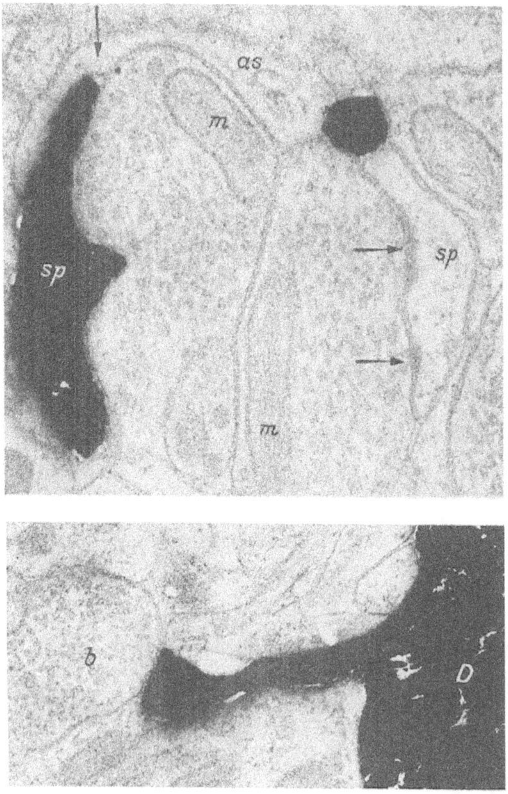

Fig. 11. Frame *a* shows two dendritic spines in the molecular layer of the fascia dentata. One of these forms part of an impregnated granule cell. A normal bouton occupies the concavity of each spine and a type 1 synaptic contact (of Gray) associated with this shape of spine is visible at the unimpregnated spine (dense postsynaptic substance at horizontal arrows). Note that Golgi precipitate is confined strictly to the spine (at vertical arrow and elsewhere). ×36,000. *b*: A portion of a granule cell dendrite (*D*) with an associated spine. Note flattened outline toward the normal bouton, suggesting a type 1 synaptic junction at this site rather than a random non-synaptic contact. The Golgi precipitate has spared the plasma membrane of the bouton as well as the larger part of the intercellular cleft in front of the spine head. ×24,000. [Reproduced from Blackstad (1965) by courtesy of Springer Verlag, Berlin]

in contact with such smooth dendrites from serial sections of selected single dendrites (Fig. 9).

Up until the present time, work on impregnated structures from the regions just mentioned has led to no more than a limited amount of new synaptological information, but the potential value of the strategy for various forms of analysis is hardly disputable. For example, further studies by the combined Golgi-and-EM method may be expected to provide significant information with respect to possible selective relationships of previously identified dendrites to any one or more of the currently recognized classes of axon terminals.

Golgi-impregnated *axonal structures* can also be identified. The short axons arising from the granule cells of the fascia dentata, the hippocampal mossy fibers, and

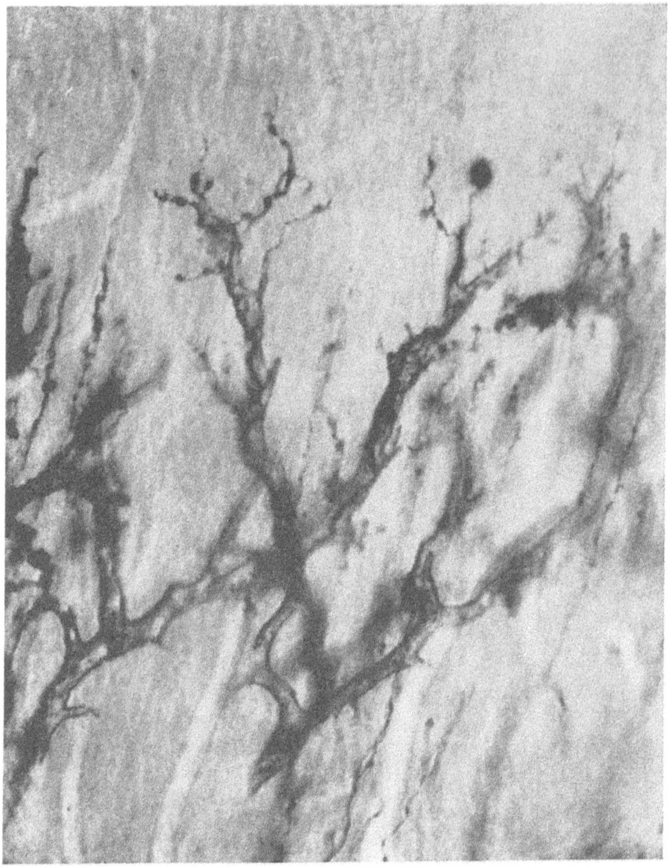

Fig. 12. Photograph of an approximately 50 μ thick section from a Golgi-impregnated block of cerebellum. At middle levels of the molecular layer a few branches of Purkinje cell dendrites are seen outlined by a precipitate. Subsequent EM analysis showed this precipitate to be extracellular (Fig. 18). Furthermore, climbing fibers can be identified along the outlined levels of the dendrites, as well as more superficially. A few of the climbing fibers were examined on the screen of the microscope and were found to adhere closely to Purkinje cell dendrites. × approx. 500

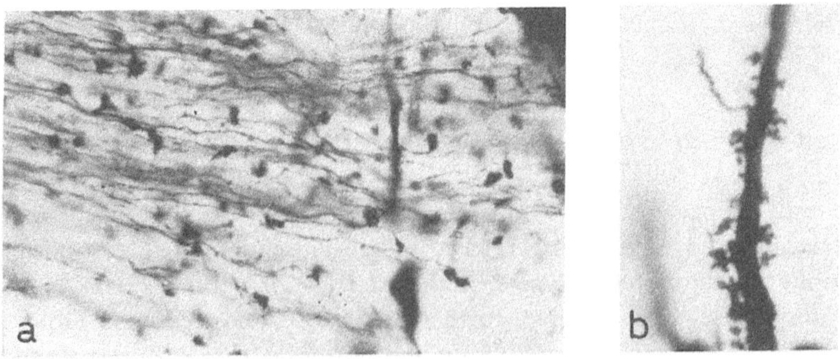

Fig. 13. Conventional Golgi sections for light microscopy from the layer of mossy fibers of the guinea pig. *a*: Mossy fibers and boutons. ×378. *b*: Giant pyramid-like cell dendrite with ramified spines (excrescences) from the hilus of the fascia dentata. Identical excrescences are seen in the layer of mossy fibers. ×973

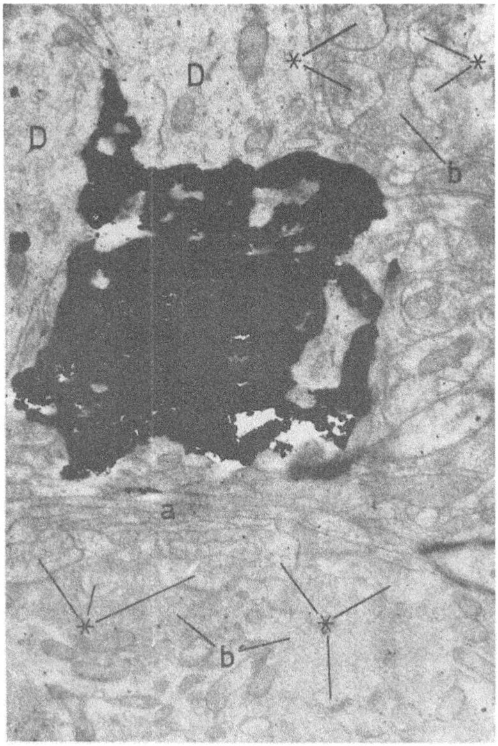

Fig. 14. Impregnated giant bouton of a mossy fiber in the hippocampus. Asterisks indicate unimpregnated dendritic excrescences interdigitating with neighboring unimpregnated giant boutons (b). *D* dendritic shafts of pyramidal cells; *a* thin axons with the appearance of mossy fibers. ×12,500

their large boutons were found to be easily demonstrable by light- and electron-microscopy after Golgi impregnation (Figs. 13, 14); their appearance in the impregnated state corresponded closely to the descriptions of these structures in non-impregnated EM material (Hamlyn, 1961, 1962; Blackstad and Kjaerheim, 1961). The grooves and channels in these large boutons, filled by the excrescences of the CA3 pyramidal cells, were seen as non-impregnated areas[3]. The case of the mossy fibers represents a first example of the practical possibility of mapping, at the ultrastructural level, the relations of short, impregnated axons in the central nervous system.

Stell reported never to have observed spread of Golgi precipitate across a synaptic junction. The present author likewise has seen little evidence of such spread and believes that the tendency to be bridged by Golgi precipitate is no greater for synaptic clefts than it is for any other inter- or intracellular space.

Cerebellar structures appear easy to select for electron microscopy when impregnated. In the author's preliminary studies of the cerebellum Golgi-impregnated climbing fibers (Fig. 12), mossy fibers, and Golgi cell axons and their synaptic terminals were identified and studied in the electron microscope. Purkinje cell

[3] This is a particularly striking example of absence of transsynaptic impregnation.

dendrites are fairly easy to recognize even without impregnation, and their intimate relationship to impregnated climbing fibers was readily verified in these studies by the combined Golgi-and-EM strategy. In impregnated blocks, the glomeruli of the granular layer can be recognized easily and, since their components are no more difficult to identify electron-microscopically in Golgi-impregnated tissue than in conventional material, the relationships of impregnated Golgi cell axons and boutons within these complexes should be an easy object for detailed analysis.

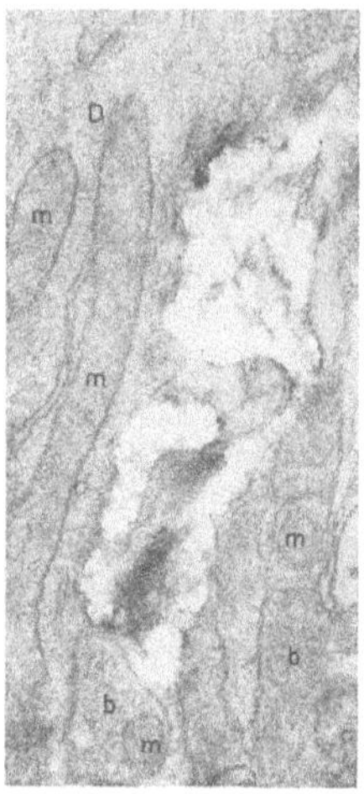

Fig. 15. Part of an ultrathin section that had been floated on water in the trough of the ultramicrotome. The light area represents a giant bouton from which the precipitate has been dissolved. Before ultramicrotomy, mossy fibers and giant boutons were seen in the light microscope to contain dark Golgi precipitate. *D* dendrite; *b* boutons; *m* mitochondria. ×15,000

Only a few applications of the combined Golgi-and-EM method have been mentioned in the foregoing account. There are, no doubt, numerous similar problems with respect to other areas of the central nervous system. Cells with short axons exist in most brain regions and are currently receiving considerable attention (Brazier, 1969). Visualization of long axonal systems can, of course, also be obtained, but for this category the method does not offer any clear-cut advantages over other methods of study, particularly because of the inherent difficulty of origin-identification.

Applications of the Golgi-EM strategy to normal tissue other than such guided by synaptological interests could also be conceived. One is the measuring of volume or surface of brain structures too tortuous to be traced in conventional material in single or serial sections. A majority of profiles derived from such structures must, among the wealth of tissue components of similar appearance, remain unidentified in electron-micrographs unless they can be identified, by virtue of impregnation, with structures recognized in the preceding light-microscopic examination of the specimen. Determination of surface and volume of cortical dendrites, and of the proportion of these parameters attributable to spines, represents one type of problem that might be solved with the combined Golgi-and-EM analysis of one and the same block of tissue.

It might be added here that Golgi impregnation combined with electron-microscopy may have applications outside the nervous system. The capacity of the Golgi method to impregnate bile capillaries, for example, is well known, and the method may well be found useful for demonstrating other fine clefts or channels in animal tissue. One could imagine that such structures as the T-tubules of striated muscle could be accessible to this form of analysis. Golgi impregnation of endoplasmic reticulum, including the perinuclear space, was shown in Fig. 24; in the case in question the Golgi apparatus had remained unimpregnated. Examples of Golgi impregnation of fine extracellular spaces in nervous tissue were mentioned above.

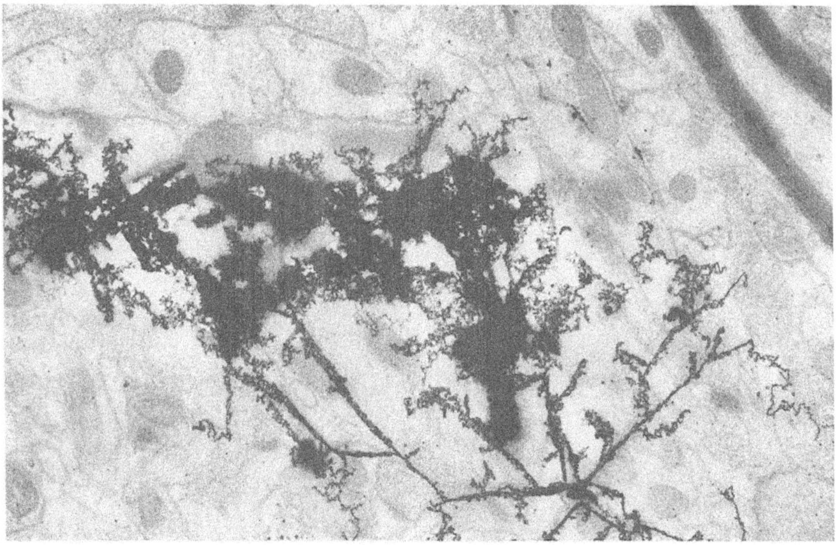

Fig. 16. Part of section of stratum oriens of hippocampus (CA3) with Golgi-impregnated basal dendrite of pyramidal cell. Precipitate partially dissolved and recrystallized on the section as a result of contact with moisture during sandwiching. The grid and the second film was picked up with paper and allowed to dry slowly. ×18,000

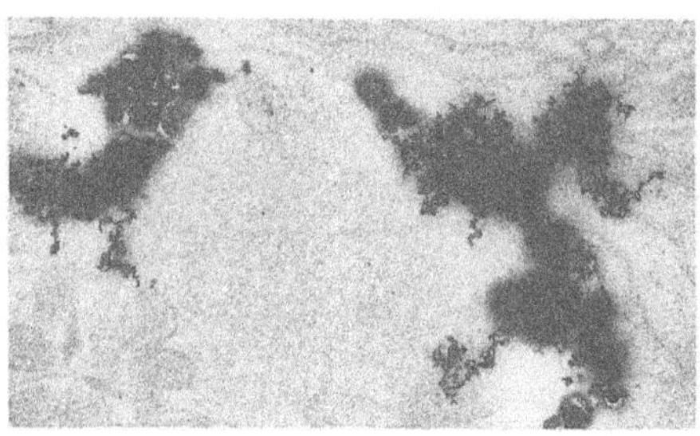

Fig. 17. Ramified spines (excrescences) from the hilus of the fascia dentata. Part of a giant bouton appears in the center. Partial dissolution and recrystallization of the Golgi precipitate as in Fig. 16. ×18,000

Material with Degenerating Axon Terminals

There exists in neuroanatomical research a serious need for unequivocal identification of the postsynaptic structures associated with axon terminals found degenerating as a result of experimental lesions. This need was the author's primary motivation to attempt a combination of Golgi impregnation and electron microscopy. It was hoped that Golgi impregnation would occur in some at least of the cells with which a transected group of fibers establish contact. As reported in 1965, and illustrated in the present paper, this aim was accomplished to a certain extent. Commissural fibers distributed to the zone of basal dendrites of hippocampal (CA3) pyramidal cells (Figs. 21, 22) formed the main object of the pilot experiments, while further observations involved the inner zone of the molecular layer of the dentate area (Fig. 23).

Degenerating boutons proved to be easily distinguishable in electron micrographs of Golgi-impregnated blocks. Many of these terminals of commissural fibers were found in direct contact with dendrites of pyramidal cells. In some cases the mode of contact between a degenerating bouton and a spine is characteristic enough to allow its identification as a true synaptic contact of a type already familiar from observations in experimental unimpregnated tissue. Astroglial envelopment of the structures engaged in such contacts may support the identification of the contact as a synaptic

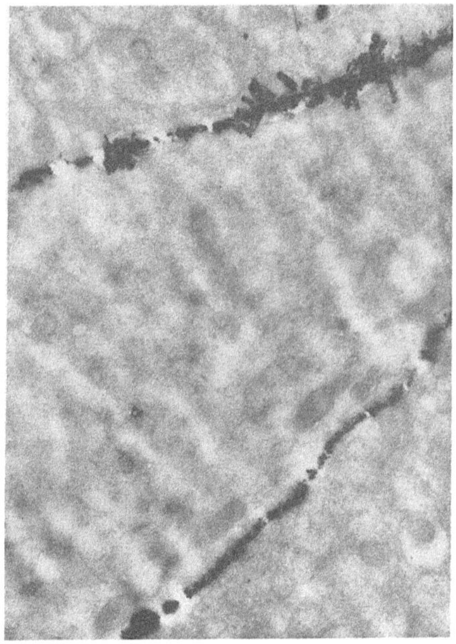

Fig. 18. Extracellular precipitate along Purkinje cell dendrite (compare Fig. 12). The section was cut with silver dichromate in the trough of the microtome and crystalline growth of precipitate on the section has occurred. × approx. 12,000

one (Fig. 21b, c). There are undoubtedly other regions in the nervous system where the synaptic nature of an immediate contact between a degenerating bouton and a Golgi-impregnated dendritic profile can be inferred from certain unmistakable form characteristics. Dendritic excrescences such as those seen in the layer of hippocampal mossy fibers would be one example, calyciform spines of the fascia dentata another. However, the user of the combined Golgi-EM method would soon discover that the configuration at many contact sites is not characteristic enough to preclude the possibility of random, i.e. non-synaptic, contact.

The foregoing remarks point up the fundamental deficiency inherent in the method as it stands at the present time, namely, that the postsynaptic densities in dendrites, synaptic vesicles in axons, or other structural details necessary for the morphological definition of a synaptic junction, are masked by the dark precipitate. The author therefore believes that, except for those fortunate situations in which such details are not essential for the identification of a synapse (cp. Figs. 2, 4, 10, 11, 14, 21b, e), it is as yet premature to recommend the combined Golgi-and-EM method for general use in attempts to establish the identity of the neurons engaged

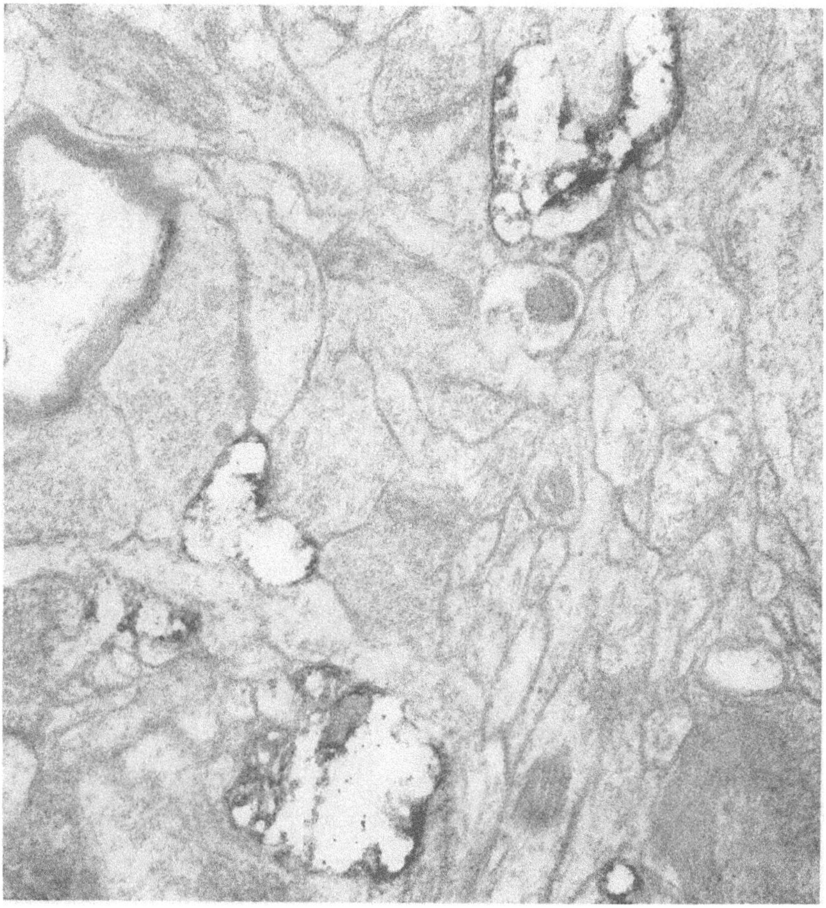

Fig. 19. Neuropil from the hippocampal region. Section floated on water in the trough of the microtome. Stained with 10% uranyl acetate in 50% ethanol for 10 minutes, and undiluted Reynolds' lead citrate for 4 min. ×39,000

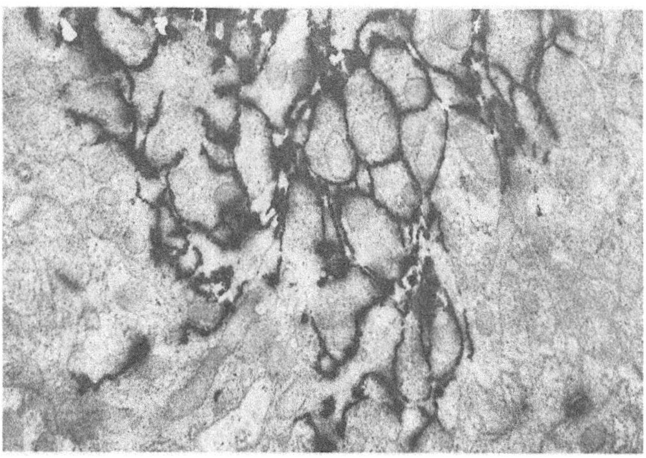

Fig. 20. Extracellular precipitate in neuropil from the hippocampal region. This tissue looked evenly brownish in the light microscope. × approx. 20,000

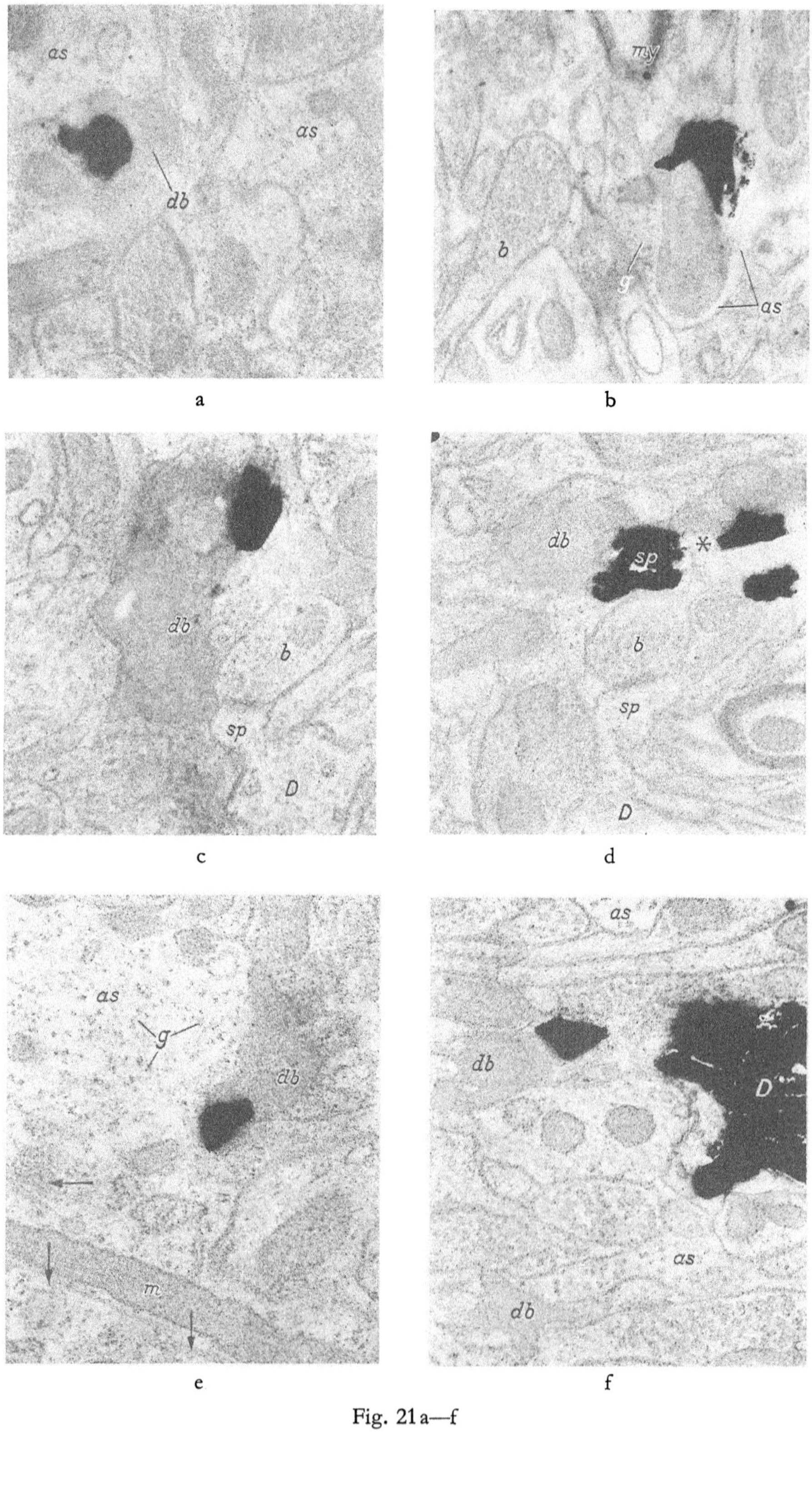

Fig. 21 a—f

in a synaptic connection. This difficulty notwithstanding, it may be worthwhile to explore in each case the possibility of applying the combined Golgi-EM procedure. The mapping of axonal distribution areas with reduced silver methods is an unavoidable prerequisite for such work. In some cases the mere number of contacts involving degenerating boutons may give some information, namely, if that number is either identical with, or significantly lower than the number of axon terminals found in the area in normal, conventionally prepared material. It goes without

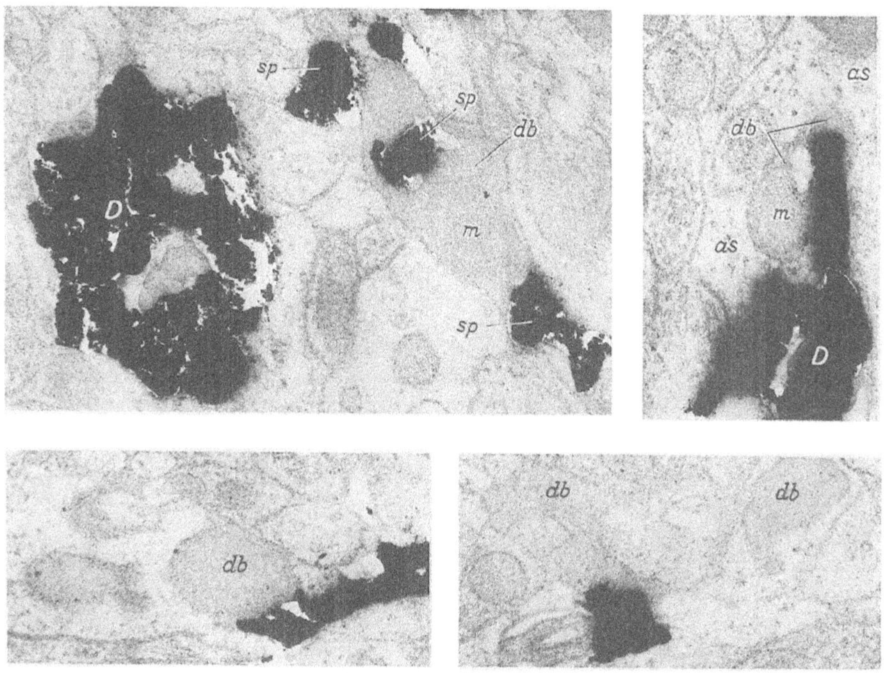

a b c d

Fig. 22. Electron micrographs from the same block as Fig. 21. Note, in *a*: invagination of a spine into the degenerating bouton; two additional spine heads are in contact with the degenerating bouton. The unimpregnated areas in the parent dendrite probably are mitochondria. *b*: Another instance of invagination of the spine into the degenerating bouton; part of the parent dendrite is included. An astroglial process surrounds the bouton-spine junction. In *c* a degenerated bouton is applied to the distal, slightly excavated part of a spine. *a—d*, ×27,600. [Reproduced from Blackstad (1965) by courtesy of Springer Verlag, Berlin]

Fig. 21. Electron micrographs from stratum oriens of regio inferior hippocampi 4 days after transsection of the commissural afferents. The sections were made from a block containing a number of impregnated pyramidal cells, and trimmed so as finally to contain only the stratum oriens with impregnated basal dendrites. The spine head visible in each micrograph thus belongs to a basal dendrite of a hippocampal pyramidal cell. Each spine is in contact with a degenerating bouton. Note particularly, in *a*: invagination of spine into degenerating bouton; in *b*: the same phenomenon as well as the presence of an astroglial process (g) around the contact; in *c* and *d*: the additional presence of an unimpregnated spine with a type 1 contact (serial sections showed the spine stalk at the site marked by an asterisk); in *e*: astroglial cytoplasm around the spine and bouton, and in *f*: the flattened plane of apposition of spine and degenerating bouton. *a—e*, ×27,600. [Reproduced from Blackstad (1965) by permission of Springer Verlag, Berlin]

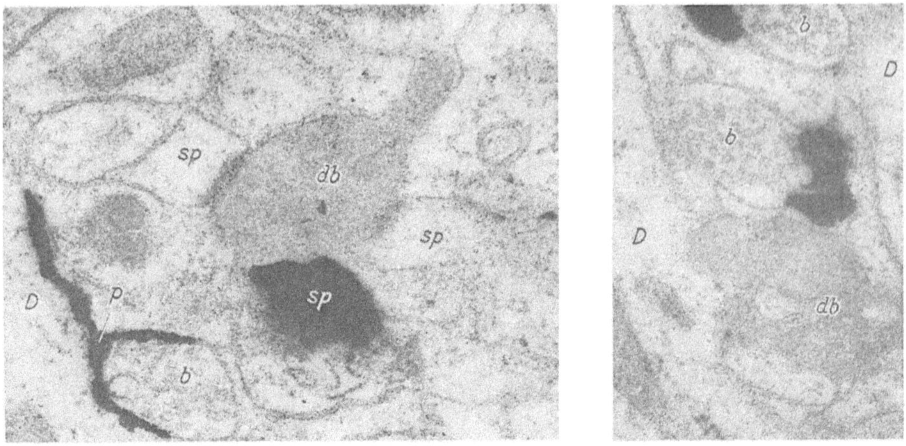

a b

Fig. 23. Electron micrographs from the inner part of the molecular layer of the area dentata, removed for fixation 4 days after transection of the commissural afferents. In each figure spine head (*sp*) belonging to a granule cell dendrite is seen in contact with a degenerating bouton (*db*). In *a* the bouton is also in contact with two unimpregnated spine heads. Precipitate mainly in extracellular clefts is present in *a* at *p*. *D* dendrite; *b*: normal boutons. *a*—*b*, ×35,000. [Reproduced from Blackstad (1965) by permission of Springer Verlag, Berlin]

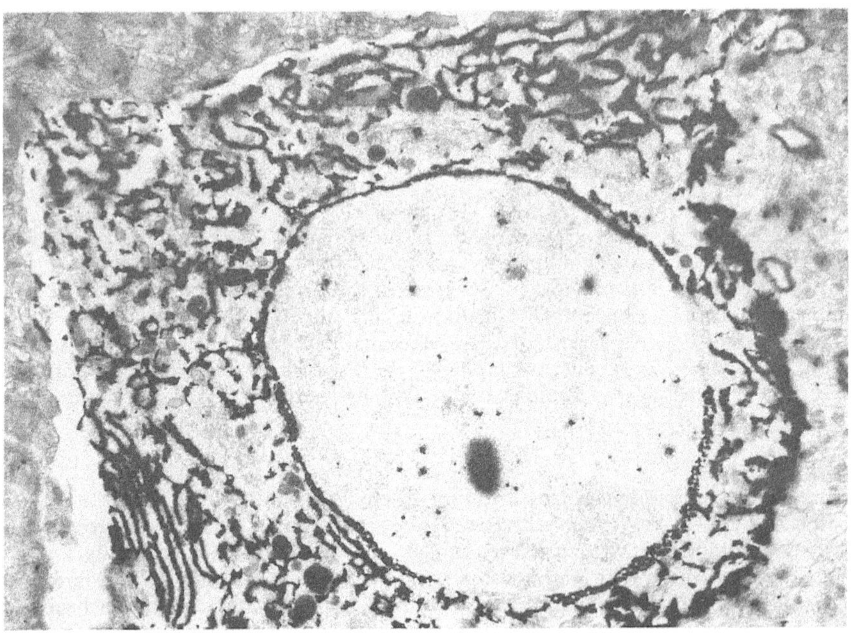

Fig. 24. Neuronal perikaryon from the hilus of the fascia dentata. Impregnation has occurred in the endoplasmic reticulum and the perinuclear space. White areas associated with the latter may be nuclear pores. ×16,500

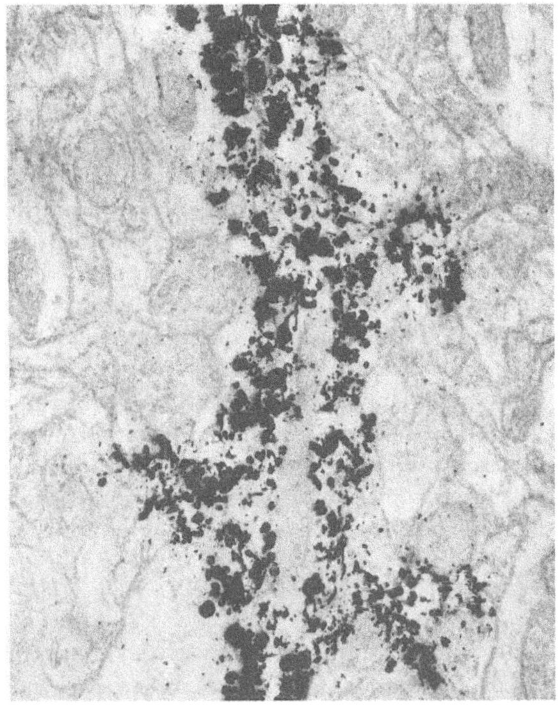

Fig. 25. Dendrite from a Golgi-impregnated block treated with 1% sodium sulfide (pH about 8.7) before dehydration and embedding in Araldite. Diffusion of silver sulfide (replacing silver dichromate) has occurred. × 21,000

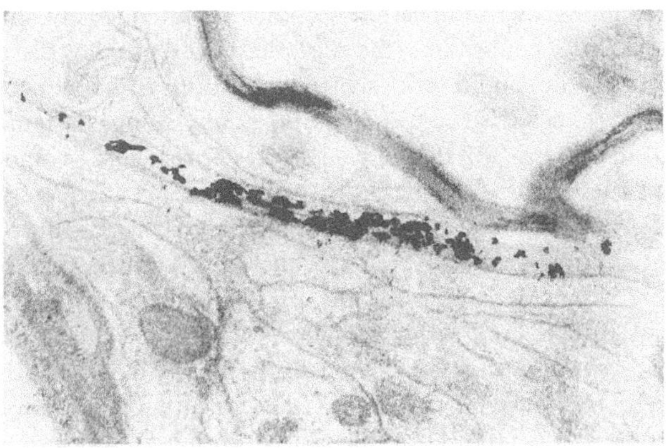

Fig. 26. Partial dissolution of Golgi precipitate in an axon by treatment of the block with 2% ammonium nitrate before dehydration and embedding. ×15,000

saying that severance of afferents should be as massive as possible, to increase the chance of encountering tissue elements in contact with degenerating structures. Preferably, the pathway under study should be totally transected. The use of adult animals and short survival times will help to avoid misinterpretations due to transneuronal processes or "indirect Wallerian" degeneration. Many combinations of experimental and quantitative studies can be imagined, but will not be dealt with here.

To become more widely applicable, the combined procedure of Golgi impregnation of the postsynaptic structure and a Wallerian labelling of degenerating axons needs some drastic improvement of the impregnation procedure in order to become more fruitful. Either the dichromate and silver must be replaced with some other combination of ions resulting in a less electron-opaque crystalline substance visible in the light microscope[4], or a better method must be developed to remove the silver dichromate, totally or partially, after the impregnated cell has been characterized. The present procedure of dissolving the substance by floating the ultrathin sections on the trough fluid results in more or less empty holes that are impossible to interpret[5]. Fig. 26 shows part of a block that was "de-impregnated" with ammonium nitrate before final embedding. The result obtained in this case suggests that Golgi impregnation does not necessarily destroy ultrastructural details, and that the latter can be re-visualized by appropriate partial or complete dissolution of the precipitate. Should such re-visualization become a routine possibility, a very wide applicability of the combined Golgi-and-EM strategy could be expected.

Summary

This chapter deals with electron microscopy of Golgi preparations. It includes the general purpose of the procedure, some practical aspects, and considerations of applications and future possibilities. The silver dichromate procedures are so far the methods of choice, and must be applied after proper fixation that preserves the ultrastructural detail. Thick (30—200 μ) survey sections are needed, to permit light-microscopic study of impregnated structures prior to electron microscopy. The sectioning and staining of the ultrathin sections are attended by special problems, because the impregnating substance (the precipitate) will grow or dissolve depending on the composition of the liquids with which the sections come into contact. Electron microscopy of impregnated structures requires sandwiching, preferably with formvar. In general, the Golgi-EM method offers an approach to problems concerning the relationships of structures too large or tortuous to be adequately surveyable in single or serial ultrathin sections. For example, it permits detailed studies of the relations of dendrites or axons to the surrounding structures. The combination of Wallerian degeneration (a marker for the presynaptic component of the synapse under investigation) and Golgi impregnation (to show the postsynaptic structure) may be useful, provided the shape of the synaptic structure is sufficiently characteristic to be recognized. However, if this is not the case, the obscuration of the postsynaptic membrane specializations by the Golgi precipitate seriously interferes with the identification of true synaptic contacts. The same difficulty is encountered in the case of impregnation of the presynaptic structure. Further development of the procedure is needed to amplify its applicability.

---

[4] Stabilization of lipoprotein membranes with dichromate, prior to impregnation proper (formation of crystalline precipitate), might still prove to be necessary or advantageous.

[5] A replacement of dichromate with sulfide in the unimbedded block resulted in diffusion (Fig. 25) and loss of resolution. It is quite possible that replacement by other ions will have similar adverse effects.

Acknowledgments

The author gratefully acknowledges the technical assistance of Mrs. J. L. Vaaland, Miss I. Madsen, Mr. K. Jörgensen and Mr. B. Krunderup. Special thanks are due Dr. W. K. Stell (present address: Laboratory of Neurophysiology (9/B-10), National Institutes of Neurological Diseases and Stroke, National Institutes of Health, Bethesda, Md. 20014, U.S.A.) for permission to include material from his thesis and other unpublished data in this account.

Discussion

Dr. Peters: I wonder whether anyone has looked at methylene blue preparations in the electron microscope. Is methylene blue electron-opaque?

Dr. Blackstad: I would be very skeptical with regard to methylene blue because one would have to immerse fresh tissue in the staining liquid for a long time. Cajal (1952) gave a frightening description of the artifacts that can result, including progressive disappearance of dendritic spines.

Dr. Peters: So, the only other hope might be the technique developed recently for "spearing" neurons and getting the dye-stuff in directly. Perhaps one could inject electron-opaque material into individual neurons.

Dr. Blackstad: I doubt that it would help us much. You deposit something in the cell body, but I think it would be much more difficult to fill the whole dendritic tree.

Dr. Sidman: I want to rise to the defense of my colleagues in the Department of Neurobiology. The dyes which they are injecting with intracellular micropipettes do diffuse very rapidly through all parts of the cell body and the dendrites. At present they detect the dyes by fluorescence, not by electron opacity. The method allows one to record from a given cell and then to localize it and define its size and shape (Stretton, A. O. W. and Kravitz, E. A., 1968).

Dr. Ramón-Moliner: First of all, I would like to congratulate Dr. Blackstad for this excellent work. I would also like to retract a statement that I made yesterday when I said that in Golgi material the impregnation was massive and appears to fill the whole cytoplasm with silver chromate or with mercury. While looking at some of your electron micrographs, I got the impression that, at least in some dendrites, the Golgi impregnation remains confined to a peripheral, submembranous zone and leaves a clear space inside. Would you agree to that?

Dr. Blackstad: Yes. Stell and I have had exactly identical experiences on this point. The precipitate usually fills the cell processes completely, but it may spare parts of a cell, e.g., stop at mitochondrial membranes, leaving the mitochondria unimpregnated. Cell nuclei are also usually unimpregnated. The precipitation process may be stopped, resulting in incomplete impregnation, and then, of course, you see a portion of a cell completely impregnated and another part of the cell actually unimpregnated. Where impregnation takes place it is complete, but another region of the cell may be without impregnation.

Dr. Valverde: Dr. Blackstad, do you mean that you have observed partial impregnation of a cell?

Dr. Blackstad: Yes.

Dr. Valverde: Was this with the Golgi triple impregnation procedure?

Dr. Blackstad: Stell used triple impregnation in some of his experiments. I impregnated only once, and we both saw instances of incomplete filling of the cells. It must be added here that the experienced investigator will see which cells and which areas are useful, and which ones are not useful.

Dr. Valverde: This means that the saying that "the Golgi impregnations is an all-or-none phenomenon" requires qualification. I do not believe I have ever seen this phenomenon of partial impregnation after triple impregnation.

Dr. Blackstad: I would tend, with my most limited experience, to agree with you. I think the impregnation is much more complete in the processes than in the cell body, and when it fails, it is mostly in large areas inside the cell body.

Dr. Valverde: There is another thing I think is very interesting. In the picture you showed of the climbing fiber and Purkinje dendrite, you mentioned that the staining of the Purkinje dendrite was only peripheral. Could you elaborate on that?

Dr. Blackstad: It was no staining of the Purkinje cell at all, but rather an outlining of it by extracellular precipitate. It only occurred in this single case.

Dr. Valverde: I have seen cases of a pale outlining of calyces of Held, and wonder if that may have been likewise a "shadowing" effect.

Dr. Nauta: Dr. Blackstad, if I understood you correctly, the preparations you showed us were all processed by the rapid Golgi method. Have you tried any Golgi-Cox?

Dr. Blackstad: No, I have not. One reason for that was that mercury is a precipitating agent for proteins and would, therefore, hardly do as an initial fixative in electron microscopy. But now that it has been shown that it can be used after initial aldehyde fixation, it might be worth a try.

Dr. Nauta: Could anything be said about the electron density of mercury salts as compared to silver? Would mercury be better or worse for your purpose?

Dr. Blackstad: I have no idea! Whether it would be more easily subject to sublimation, I don't know.

Dr. Nauta: I was only thinking about its properties vis-à-vis the electron beam. Dr. Ramón-Moliner told us that there is a more or less transparent form of mercury prior to alkalinization. Could this possibly be also less than completely opaque to the electron beam?

Dr. Ramón-Moliner: I think that we may encounter another difficulty with the Golgi-Cox method because, in the finished preparation, mercury appears to be present not only in the opaque structures, but also in the background, that is, in the transparent parts of the section. I suspect that there is non-reduced mercury left in the non-stained regions because they have a very high refraction index. Presumably, this would affect the electron-transparency of the background. On the other hand, it would be very interesting to know what happens in Golgi-Cox stained material prior to the action of the alkali. We know that there is a deposit of certain complex compounds of mercury that becomes black only after the alkali acts on them. It would be interesting to find out, by means of electron microscopy, if mercury in its bound, non-reduced form displays a more even distribution than that displayed by the granules of silver chromate in the Golgi technique. In addition, the possibility remains that some specific affinity of mercury for membranes or cytoplasmic organelles may be demonstrated.

Anterograde and Retrograde Transneuronal Degeneration in the Central and Peripheral Nervous System

W. M. Cowan

Introduction

Although it has been known for more than half a century that certain groups of neurons may atrophy, or degenerate, following the death of the cells upon which they project or after the interruption of their afferents, it is not generally appreciated how widespread such transneuronal changes may be. As recently as 1956 Torvik, in a study of the effects of lesions in the brainstem and cerebral hemisphere upon the inferior olive and pontine nuclei, could state: "In adult animals it seems questionable whether transneuronal changes appear at all in other areas than the lateral geniculate body". However, even at that time there was a considerably body of evidence, both experimental and clinico-pathological, to indicate that both types of transneuronal degeneration — that following deafferentation and that secondary to retrograde cell degeneration — may occur in a number of neuronal systems. And since the publication of Torvik's paper there have been a number of additional studies which have served to emphasize the widespread occurrence of transneuronal effects both in the central and peripheral nervous systems. The etiological factors responsible for the atrophy or degeneration of neurons which have suffered no direct injury are still largely unknown, but nevertheless the time seems opportune to review the now extensive literature on this subject.

The intention of the present paper is to draw attention to the fact that transneuronal changes may occur in many regions of the central nervous system, and in some parts of the peripheral nervous system, and to emphasize how important it is in the interpretation of cellular degeneration in the nervous system always to consider the possibility of transneuronal effects. In addition, it may be helpful in pointing out that transneuronal degeneration may be used as an additional technique for the elucidation of neural connections in certain systems. Hopefully, it may also stimulate interest in these forms of neuronal degeneration and, in particular, to draw attention to the need for more critical studies of the cytochemistry and functional properties of deafferented neurons and of neurons which have been deprived of their normal projection field.

Terminology

Before considering the evidence for the distribution of transneuronal changes, a few remarks are necessary on the terminology current in the literature and that which will be used in this paper.

The terms "transneuronal" and "trans-synaptic" which are sometimes distinguished in the literature will be treated here as though they were synonomous. In the strictest sense, *trans-synaptic* should perhaps refer only to the changes which occur across a synapse (i.e., in the postsynaptic process secondary to some change in the presynaptic component) and the term *transneuronal* should then be used more generally to cover the changes which involve the entire second order neuron. However, little is to be gained by this distinction, and in what follows the term "transneuronal" will be used to cover all types of atrophic or degenerative change which may occur in a neuron, other than those in which there has been some direct injury to the nerve cell or its processes.

Similarly, in the literature, the terms "degeneration" and "atrophy" are sometimes used synonomously, and at other times are sharply distinguished. Following common pathological usage it might be more appropriate to restrict the term "degeneration" to those situations in which there is actual neuronal death and neuronophagia, and to use the term "atrophy" for all other situations where the reaction of the cell falls short of its death (De Long and Sidman, 1962). Since in many situations the initial transneuronal effect may express itself as an atrophy (in the sense that most of the cells simply appear to be shrunken) and later lead to a true degeneration (with a significant degree of cell loss) it is difficult to be completely consistent in the usage of these terms and in most of what follows the two terms will be used interchangeably.

It is important to distinguish between what are strictly atrophic conditions and those in which there is a retardation or cessation of growth on the one hand, and a failure of maintenance of differentiated neurons or neuroblasts on the other. This distinction, which has not always been appreciated, is of considerable significance in neuroembryological studies and must be considered in all cases in which transneuronal changes are found in fetal or immature animals.

It is also important to define the *direction* of the degenerative change; the terms "anterograde" and "retrograde" are those most generally used to characterize the two common forms of transneuronal effect. By an *anterograde transneuronal degeneration* is meant a transneuronal atrophy or degeneration which is secondary to deafferentation; the atrophy seen in the cells of the mammalian lateral geniculate nucleus following interruption of the optic nerve is of this type. The term *retrograde transneuronal degeneration* refers to those degenerative conditions which are secondary to a retrograde cell reaction of the type which follows axon-section. For example, in the mammalian visual system (which is one site in which such changes are well-documented) the degeneration of the retinal ganglion cells after lesions of the visual cortex is a retrograde transneuronal degeneration since it occurs as a secondary phenomenon consequent upon the retrograde cell degeneration in the lateral geniculate nucleus. It is often necessary to further qualify the terms anterograde and retrograde transneuronal degeneration as it is known that both changes may be sequential, in the sense that more distantly related groups of neurons may also show degenerative changes. Two examples may illustrate this point. After eye enucleation anterograde transneuronal changes may occur not only in the lateral geniculate nucleus but also in the visual cortex, and clearly these two effects should be distinguished. Similarly, severe retrograde transneuronal changes may occur in the ventral tegmental nucleus of the midbrain after lesions of the limbic cortex which result in an initial retrograde transneuronal degeneration in the medial mammillary nucleus. Such changes as occur in the lateral geniculate nucleus after eye enucleation will be referred to as a *primary anterograde transneuronal degeneration* and those found at the next synaptic level (in this case the visual cortex) may be termed a *secondary anterograde transneuronal atrophy*. More distant effects, where these occur, can then simply be referred to as a *tertiary* or *quarternary anterograde transneuronal degeneration*. In the same way the type of degeneration seen in the medial mammillary nucleus after lesions of the limbic cortex may conveniently be referred to as a *primary retrograde transneuronal degeneration* and those, like the changes in the ventral tegmental nucleus, as a *secondary retrograde transneuronal degeneration*.

While there are several reasons for believing that the anterograde and retrograde transneuronal degeneration share a number of features in common it is convenient to consider the two phenomena separately.

Anterograde Transneuronal Degeneration

Before discussing some of the cytological aspects of this form of transneuronal degeneration and the factors which may be implicated in it, the evidence for its widespread distribution in the central and peripheral nervous system will be reviewed.

1. The Mammalian Visual System

a) The Dorsal Nucleus of the Lateral Geniculate Body

The changes seen in the dorsal nucleus of the mammalian lateral geniculate body after section of the optic nerve represent the classic example of anterograde transneuronal degeneration, and, as has been indicated above, it is still widely believed that this is the only site in which such changes have been unequivocally demonstrated. The literature on transneuronal degeneration in the lateral geniculate is considerable and as it has been reviewed on several occasions (see for example, Minkowski, 1920; Le Gros Clark, 1932; Tsang, 1937; Cook, Walker and Barr, 1951; Matthews, Cowan and Powell, 1960; Glees, 1961) it need not be considered here from an historical point of view. However, there are several features of the degeneration in the geniculate which characterize this form of transneuronal effect and for this reason it is necessary to deal with a number of the more recent papers on this subject at some length.

One of the most important of these studies, judged by the influence it has had on subsequent work, is that of Cook, Walker and Barr (1951) who attempted for the first time to quantify the severity of the cellular atrophy and to systematically investigate the effect of different postoperative survival periods. Since most of their observations were based upon experiments in adult cats and on the neurons in lamina A of the lateral geniculate nucleus (which receives fibers from the contralateral eye) we shall only consider this aspect of their work. Briefly, they found that there was no statistically significant change in the mean cross-sectional area of the cells in lamina A following eye removal at 14 and 32-days, but from the 63rd through the 313th postoperative day the cells in this lamina showed a reduction in mean cross-sectional area ranging between 16.1 and 30.2%. Over the same period there was also a significant reduction in the cross-sectional area of the nucleus (of between 8.6 and 18.4%) and in the 3 animals with the longest survival periods (210, 241 and 313 days, respectively) the profile of the nucleolus also showed a statistically significant reduction in area amounting to between 16.7 and 26.1%. In the 7 female cats used in their experiments there appeared to be no alteration in the position of the nucleolar satellite at any stage during the transneuronal atrophy. Essentially similar, although less marked, changes were seen in the rabbit experiments but these were not consistently present until about 5 to 6 months after eye enucleation. Cytologically the most striking change in the cells undergoing transneuronal degeneration was their progressive shrinkage and loss of cytoplasmic basophilia; this was quite unlike the rapid dissolution of Nissl substance and early neuronal swelling seen in the typical chromatolytic reaction which follows axon section. Only in the animals with the longest survival periods were there additional signs of degeneration in the form of cytoplasmic vacuolation.

A similar study of transneuronal degeneration in the monkey lateral geniculate nucleus was undertaken 10 years later by Matthews, Cowan and Powell (1960). This study was prompted by the finding that whereas in the more common laboratory animals the changes in the lateral geniculate nucleus after eye removal are rather slight and occur comparatively slowly, in Primates the degenerative process is a fulminating one, and may occur as early as 7 days after eye enucleation in the monkey (Glees and Le Gros Clark, 1941). In order to confirm this, and to provide additional data on the time course of the degeneration in the geniculate, a detailed quantitative study of the changes seen in each of its six laminae was undertaken in animals which were allowed to survive for periods ranging from 4 days to 4 months after eye removal. Some of the more significant findings in this work may be summarized as follows: (i) The atrophy in the geniculate cells occurred in 2 phases: an early and rather rapid phase lasting approximately 2 weeks during which the cells were reduced in cross-sectional area by as much as 25%, and a later, slowly developing phase, which progressed throughout the period of the study, but even after 4 months the cell shrinkage amounted to little more than 40%. (ii) The degeneration occurred even earlier than had been previously reported: clear atrophic changes could be recognized in one case as early as 4 days postoperatively and by the eighth day the cells in nearly all laminae were significantly shrunken. (iii) There were significant differences in the rate of degeneration in the different laminae. Thus the cells in the laminae which receive afferents from the contralateral eye (the crossed laminae) showed the earliest reaction, being somewhat shrunken, and decidedly pale-staining, by the fourth day postoperatively, while the neurons in the uncrossed laminae (connected to the ipsilateral eye) showed little sign of atrophy until some days later. However, by the eighth postoperative day the atrophy in the uncrossed laminae clearly exceeded that in the crossed. (iv) The early difference in the reaction of the cells in the crossed and uncrossed laminae did not continue into the later phase of the degeneration; by the end of the second month after eye removal the changes in the two sets of laminae were essentially the same. (v) The two large-celled laminae (laminae 1 and 2) were found to degenerate more slowly than the small-celled laminae (laminae 3—6) but within the crossed or uncrossed groups (laminae 1, 4 and 6, and laminae 2, 3 and 5, respectively), the overall pattern of degeneration appeared to be the same. (vi) Despite the early and rather rapid progression of the degenerative change there was very little evidence of actual *cell loss* in the affected laminae. At the longest survival period studied (129 days) the loss of neurons (measured in a selected area of each lamina of the geniculate) amounted in no case to more than 7.8% and was statistically significant in only one lamina.

The absence of any significant degree of cell loss in the geniculate even 4 months after eye removal was in striking contrast to the earlier findings of Goldby (1957) who had reported that only about half the neurons in the affected laminae of the lateral geniculate had survived in a human case which he examined almost 40 years after eye removal. However, despite the severity of the cell loss in this case the mean cross-sectional area of the cells in the first two laminae was reduced by only about 32% and those in laminae 3 to 6 by approximately 48%.

The problem of cell loss in the Primate lateral geniculate nucleus after eye removal was later re-examined by Matthews (1964) who carefully studied both lateral geniculate nuclei of a monkey one year after eye enucleation. In this animal the degree of cell shrinkage and pallor was such that at low magnifications, the cells in the affected laminae were "quite inconspicuous" (c.f. Polyak, 1957). However, estimates of the percentage cell loss showed that this amounted to only between 4% (in lamina 2) and 22% (in lamina 1). The small-celled laminae on the side of the eye enucleation showed a cell loss of just over 16% and those on the side contralateral to the enucleation were reduced by just over 14%. The severity of the cell shrinkage was much more striking: the mean cross-sectional areas of the cells in the first two laminae were reduced by 54.3 and 59.9%, and most of the neurons in the small-celled laminae were reduced to about one-third of their normal cross-sectional area. The nuclear areas were similarly reduced between 39.4 and 62.3% and there was a reduction in the area of the nucleoli of all laminae of about 45%.

Another feature of interest in Matthews' study was a series of observations on the interaction of anterograde transneuronal atrophy in the lateral geniculate (produced by eye enucleation) and retrograde degeneration in the same cells after lesions of the visual cortex. While this phenomenon needs to be investigated more fully, it appears that when retrograde degeneration is combined with deafferentation for the same period, the changes in the neurons are qualitatively similar in character to that produced by either form of degeneration alone but the severity of the atrophy may differ to some extent. For example, the combined degeneration is more severe than transneuronal degeneration alone even at the shortest interval studied (4 days), but between the 6th and 12th days postoperatively, neurons undergoing both retrograde and transneuronal degeneration appear to be less severely affected than those undergoing retrograde degeneration alone.

Kupfer (1965) has also presented quantitative data on transneuronal degeneration in the human lateral geniculate nucleus from a series of cases examined at intervals between 1 month and 73 years after unilateral enucleation or after unilateral macular lesions. During the first two years after enucleation there appeared to be no cell loss, but after longer survival periods there was a progressive loss of neurons ranging from 20 to 50%. In addition there appeared to be a clear difference in the behavior of the cells in the crossed and uncrossed laminae, the changes in the uncrossed laminae being the more severe in five of the seven cases studied. However, in only two cases was the difference in the severity of the degeneration in laminae 3 and 5, on the one hand, and in laminae 4 and 6 on the other, statistically significant. No difference was found in the reaction of the cells in the magnocellular and parvocellular laminae at least up to

9 years after enucleation; after longer survival periods there was some evidence to suggest that the small-celled laminae became progressively more severely affected than laminae 1 and 2.

In a later chemical and cytochemical study of transneuronal degeneration in the monkey, Kupfer and Downer (1967) observed a slight *increase* in the RNA concentration of the lateral geniculate nucleus during the first two days after optic tract section. However, by the fifth day postoperatively the RNA level in the affected nucleus had fallen to less than 50% of that on the control side and declined from that time onwards, so that by the twenty-third day after tract section the RNA concentration in the affected geniculate was only 30% of the normal value. Accompanying these changes in the level of RNA in the geniculates undergoing transneuronal degeneration they found comparable changes in the reduced-NAD: lipoamide oxidoreductase activity, in glucose oxidation and in the rate of incorporation of labeled leucine into protein. The significance of these chemical studies is not clear at this time, but they are mentioned here because they show quite clearly that important changes may occur in the Primate lateral geniculate nucleus as early as the first 24 to 48 hours after deafferentation.

A number of workers have tried to determine the effects of light deprivation upon the cells in the lateral geniculate nucleus by rearing animals in the dark and in monochromatic light or by suturing the lids of one or both eyes (see for example, Goodman, 1932; Le Gros Clark, 1943; Chow, 1955; Terry, Roland and Race, 1962). For the most part the findings in these studies are somewhat inconclusive and many are flatly contradictory. But, in 1963 Wiesel and Hubel convincingly demonstrated that visual deprivation (produced either by unilateral eyelid suture or by the use of a translucent contact lens) in young kittens could result in significant differences in the size of the cells in the related laminae of the lateral geniculate body. For example, in two kittens which were monocularly deprived from birth by eyelid closure for about three months, the mean cross-sectional area of the cells in lamina A of the contralateral geniculate nucleus was reduced by about 40% and that in lamina B by about 25%. The cells in lamina A_1 of the ipsilateral geniculate were reduced by approximately the same amount as those in lamina A of the contralateral side. Despite these rather severe morphological changes in the lateral geniculate nuclei the operative procedure had no detectable effect upon the retina, the optic nerves, or other parts of the visual system. Monocular lid closure in kittens at the age of two months also resulted in a reduction in size of the cells in the related geniculate laminae as did the introduction of a translucent contact lens at the normal time of eye opening but in neither instance were the changes as marked as after earlier eyelid suture. Eyelid closure for three months in an adult cat had no effect upon the geniculate cells. These observations indicate that there is a rather clear relationship between the severity of the cellular changes in the geniculate and the age at which visual deprivation was begun and, from certain other experiments which they reported, this appears to be true also for the duration and completeness of the light deprivation. Despite the morphological similarity of the changes seen in the lateral geniculate nucleus after early visual deprivation and the anterograde transneuronal degeneration seen after eye enucleation in adult animals the two phenomena are probably quite different. Wiesel and Hubel were careful to point out that there are significant postnatal growth changes in the neurons of the lateral geniculate nucleus and that early eyelid closure may result in an arrest or retardation of growth rather than a true degenerative process in the visually deprived neurons.

However, in a later study Wiesel and Hubel (1965) found that a morphologically similar, although less severe, neuronal shrinkage could occur in young cats subjected to eyelid closure at 3 months, at which age the geniculate cells appear to have reached their maximum size. The implication of this is that in young animals visual deprivation may cause a true neuronal atrophy but, of course, it does not argue against the view that eyelid closure at an earlier stage may act primarily by retarding neuronal growth and maturation. Another finding of importance in their second study was the observation that there is no significant recovery of the cells in the lateral geniculate nucleus in kittens which had been monocularly deprived from birth for about 3 months, even though the eyelids were reopened for recovery periods of up to 14 months.

Since so little has been done on the functional properties of deafferented neurons it is interesting to note that Wiesel and Hubel (1963) found that geniculate cells which received their input from a previously sutured eye generally behaved quite normally and had normal receptive fields. In some cases the overall activity of the cells appeared to be diminished and a few of the cells were found to have receptive fields with abnormally large centers. In view of the striking morphological changes seen in these cells their relatively normal functional characteristics may seem somewhat surprising. But, since the receptive fields of geniculate neurons are largely determined by their retinal input it is perhaps more significant to consider the "output" from these cells as measured by the discharge characteristics of neurons in the visual cortex. This was examined in their later study in which they found that animals which had been subjected to eyelid suture for the first 3 months of life were, behaviorially, virtually blind, and, judging from the unusual properties of single cortical cells, there appeared to be a rather "severe deterioration of innate cortical connections" from which there was no appreciable functional recovery after reopening the eyelids. A few cells showed some response to stimulation of the originally deprived eye but most of these responses were obviously abnormal. This suggests that the period of visual deprivation has resulted in a significant alteration in the pattern of connections in the visual cortex (probably because of the altered input from the affected geniculate neurons) and that connections once lost are probably incapable of being re-established.

The effect of visual deprivation on the lateral geniculate nucleus has also been examined cytochemically by Kupfer and Palmer (1964) and compared to the degenerative changes due to unilateral enucleation done at the same age. They concluded that the primary effect of visual deprivation produced by eyelid closure is to retard the growth of the geniculate neurons without altering their normal metabolism. Thus, even though the deprived cells were reduced in cross-sectional area by as much as 30—40% as compared to their controls, they were not significantly smaller than neurons in the corresponding laminae of normal kittens shortly after birth (i.e., at the age at which the eyelid closure was usually performed). Further, there were no discernible changes in the activity of several enzyme systems (TPN-diaphorase, DPN-diaphorase, cholinesterase, and the dehydrogenases of 6-phosphogluconate, succinate and lactate-DPN) in the deprived laminae. This absence of cytochemical change after eyelid suture is in marked contrast to the rapid alteration in the metabolic activity after early eye enucleation. As early as 5 days after enucleation there was a significant reduction in the activity of all the enzyme systems tested, and this reduction progressed with increasing postoperative survival.

It is also interesting to note that in Kupfer and Palmer's study there was an appreciable shrinkage in the denervated geniculate neurons by the fifth day after eye enucleation, and that at this early stage there was a distinct difference in the reaction of the cells in the crossed and uncrossed laminae, the former being clearly affected at this stage, while the cells in the uncrossed laminae were essentially normal. This difference in the reaction of the crossed and uncrossed laminae was also apparent after eyelid closure; in two kittens operated upon shortly after birth the crossed laminae showed a clear reduction in the cross-sectional area of the cells as compared to the controls as early as 10 days after eyelid suture. More will be said about the significance of the age of the animal at the time of deafferentation later, but these findings, when taken in conjunction with those of Cook, Walker and Barr (1951) and Cohn (1956) provide a singularly clear example of the greater susceptibility of immature neurons to anterograde transneuronal degeneration.

In reviewing the literature on transneuronal degeneration in the lateral geniculate nucleus special attention has been paid to the studies on cats and Primates because of their general interest. However, it should be noted that comparable changes have been seen in the lateral geniculate nucleus of a wide variety of mammals, including phalangers (Packer, 1941), sheep (Nichterlein and Goldby, 1944), goats (Minkowski, 1920), tree shrews (Glickstein, 1967), rats (Tsang, 1937), and guinea pigs (Hess, 1957b). It is also worth pointing out that transneuronal degeneration in the lateral geniculate has proved to be a most useful tool, not only in disclosing the significance of laminar patterns in cases where this is not normally apparent from the study of Nissl or fiber preparations (c.f. Glickstein, 1967), but especially in the elucidation of the organization of the retinal projection upon the lateral geniculate, a problem which had proved to be extremely difficult to resolve with most fiber degeneration methods (Le Gros Clark and Penman, 1934). The work of Le Gros Clark and Penman in which relatively small lesions were placed in different parts of the retina and the distribution of the resulting transneuronal changes in the lateral geniculate nucleus was carefully mapped, is a classic example of the usefulness of anterograde transneuronal changes for determining connections in the central nervous system.

b) The Superior Colliculus

The transneuronal changes seen in the lateral geniculate are so striking that to a large extent they have obscured the fact that similar changes may also occur in other parts of the visual system. And although the evidence for transneuronal degeneration in these other areas is generally less conclusive, there are a number of studies which indicate that a distinct atrophy may occur in the superior colliculus after unilateral eye enucleation.

One of the most thorough studies of transneuronal degeneration in the colliculus is that of Tsang (1937) which also includes an excellent review of the early experimental and clinicopathological literature on transneuronal changes after eye removal and in cases of congenital anophthalmia. Tsang examined the changes in volume of the rat superior colliculus after eye enucleation at different ages and after different postoperative survival periods. Volumetric measurements of this kind do not clearly indicate the severity of the changes in the collicular neurons; nevertheless, it is clear from Tsang's experiments that the reduction in volume of the colliculus is not simply due to the degeneration of retino-collicular fibers, and it is noteworthy that the hypoplasia of the superior colliculus was consistently more marked than that in the lateral geniculate nucleus. The atrophy in the colliculus was most marked in a group of animals which had been enucleated at birth and was appreciably less severe in those operated upon at 13 days of age and very slight in adult animals. Similarly, the most severe changes were seen in the animals which were allowed to survive postoperatively for the longest period (7 months). As Tsang's work was the first attempt to quantify the severity of the transneuronal atrophy it represents a most important landmark in this area but unfortunately it has never received the attention it deserves.

Some years later Hess (1957b) showed that unilateral destruction of the eye in fetal guinea pigs (under 45 days of age) could lead to a severe neuronal degeneration in the superior colliculus as early as four days after the operation. Following the same operative procedure in older animals (50 days or older) comparable changes were not found in the colliculus until nine days postoperatively. The changes in fetal animals are clearly more comparable to those seen in the neuroembryological studies discussed below than to the transneuronal effects described by Tsang (1937), but it is worth noting that in the fetal guinea pigs there was no detectable neuronal loss in the lateral geniculate nucleus so that at this stage the cells in the superior colliculus appear to be much more susceptible to transneuronal influences than those in the geniculate.

In a similar study of young mice subjected to eye enucleation or eyelid closure shortly after birth, Terry, Roland and Race (1962) found clear atrophic changes in the superior colliculus (but not in the lateral geniculate nucleus) 3 weeks after eye removal. No clear changes were seen in any part of the visual system after closure of the eyelids up to 9 weeks postoperatively.

In the same year De Long and Sidman (1962) published the results of a careful radioautographic study of the postnatal histogenesis in the mouse superior colliculus and of the effects of eye removal shortly after birth. The earliest indication of a change in the colliculus was found 7 days after eye removal and the resulting hypoplasia progressed slowly over the remaining period of their study (until the 50th postnatal day). From volumetric measurements and estimates of the number of labeled cells they concluded that eye removal has two primary effects: it produces a rapid transneuronal degeneration of differentiated neurons and also a failure of many neuroglia to proliferate. Although very few degenerating neurons were seen in the colliculus they felt that this was because the deafferentation resulted in an unusually rapid disintegration of the immature neurons so that comparatively few dying cells would be evident at any one stage. Comparing their findings with those of other workers they concluded that if deafferentation occurs before normal connections are established, development is generally unaffected until about the stage at which the connections are normally formed, but it then leads to a very rapid, and frequently a complete, cellular degeneration of the deafferented neurons. On the other hand deafferentation at a later stage usually results only in a neuronal atrophy characterized by cell shrinkage rather than neuronal death.

It is not at all clear why such striking changes should be observed in the superior colliculus in rodents whereas in carnivores and Primates optic nerve section appears to have no significant effect upon the cells in the colliculus. It is probable that in the "higher" mammals there is a significant change in the ratio of retino-collicular to retino-geniculate fibers (Polyak, 1957) but whether this is the significant factor or whether it is due to some other difference in the retinal projection to, or intrinsic organization of, the colliculus, is quite unknown. Indeed, until very recently the nature of the retinal projection to the superior colliculus was still unsettled; and although it is now known that many, if not all, of the fibers which pass from the optic tract to the superior colliculus in cats and rats are collaterals of the retino-geniculate fibers (Hayashi, Sumitomo and Iwama, 1967; Sefton, 1968), it is still not known if this is true of other species.

The superior colliculus, or its homologue, the optic tectum, has been the subject of a number of experimental neuroembryological studies. Generally these have involved either the removal of the optic vesicle and optic cup, or sectioning of the optic nerve, with or without rotation of the eye. A review of the extensive literature on this topic is clearly out of place in the present context but it should be pointed out that several authors have described cellular changes in the optic tectum after early removal of the eye in fish (Leghissa, 1951; Pflugfelder, 1952), in amphibians (Larsell, 1931; Kollros, 1953; McMurray, 1954; Terry and Gordon, 1960) and in birds (Filogamo, 1950; Leghissa, 1959; Cowan, Martin and Wenger, 1968). There is some question as to whether the nature of the changes in these different vertebrates is the same (for example, Kollros, and Terry and Gordon, are of the opinion that the hypoplasia seen in the amphibian optic tectum after eye removal is due exclusively to a reduction in mitotic activity, whereas most other workers feel that the changes are essentially due to a failure in maintenance of differentiated neurons) but together these studies clearly indicate that the eye exerts an important trophic influence upon the developing optic tectum. The many studies of the changes in the superior colliculus and other parts of the visual system in cases of congenital anoph-thalmia (see for example, Monakow, 1889; Cosnettatos, 1931; Chase, 1945; Moscowitz and Noback, 1962; Duckworth and Cooper, 1966) point to the same conclusion.

c) The Ventral Nucleus of the Lateral Geniculate Body

Although the ventral nucleus of the mammalian lateral geniculate body receives a substantial input from the retina and, in lower mammals at least, is clearly an important visual center, it has received nothing like the attention of the other components of the visual system. There are a few reports in the literature of transneuronal changes in this nucleus after eye removal, and of these the most significant is that of Tsang (1937). As part of his wider study of the effects of eye removal in rats at different ages, Tsang described an appreciable reduction in the volume of this nucleus which was due, in part to the loss of retinal fibers, and in part to a neuronal atrophy within the nucleus. Qualitatively the degeneration in the ventral lateral geniculate nucleus appeared to be as severe as that seen in the dorsal nucleus, but this does not seem to be the case in most other mammals that have been studied.

d) The Visual Cortex

Many of the earlier workers who studied the effects of eye removal, especially in young animals, reported that there were distinct histological changes in the visual cortex (see Tsang, 1937, for references). Much of this literature is rather difficult to evaluate since it is not well illustrated and no attempt was made to quantify the changes described. But during the last few years there have been several studies, using mainly the Golgi method, which have shown that transneuronal degenerative changes may occur in the visual cortex both after lesions of the lateral geniculate nucleus and after eye removal. Thus, Globus and Scheibel (1966, 1967) and Valverde (1967) have found that there is a statistically significant loss of dendritic spines on the apical dendrites of certain pyramidal cells in the striate area after eye enucleation in rabbits and mice. The loss of spines is generally confined to a specific segment of the dendritic tree (in what is assumed to be the site of termination of the geniculo-cortical fibers), but even in this segment an appreciable number of normal spines persists. There are apparently no detectable changes in the cell perikarya so that Nissl preparations give no indication of this secondary anterograde transneuronal effect.

Whether the loss of dendritic spines is consequent upon the degeneration of the presynaptic terminals (as must occur after lesions of the lateral geniculate body), or is secondary to some kind of "functional dissue" of the afferent fibers (as might be the case after eye enucleation) is not clear. Nor is it known whether the apparent loss of spines is due to a true degeneration of these dendritic appendages or simply to their resistance to impregnation with the Golgi method. Colonnier (1964) has suggested from an electron microscopical study of terminal degeneration after undercutting of the cerebral cortex, that dendritic spines may degenerate, and be removed by glial action, along with the degeneration of the presynaptic terminals. This finding, which has been widely quoted in support of the view that deafferentation may lead to a loss of spines, has not yet been confirmed by other workers and for this reason it is necessary to consider the alternative possibility that the degeneration of a presynaptic terminal may in some way render the dendritic spines resistant to silver impregnation (Powell, 1967; and see below).

More recently Valverde and Esteban (1968) have claimed that there may be a further loss of dendritic spines from the pyramidal cells in the peristriate cortex on the same side as the eye enucleation. Counts of the number of dendritic spines in the upper part of layer III of the peristriate cortex (i.e., in the region where most of the fibers from the contralateral visual cortex terminate) show a statistically significant reduction in 24 and 48-day-old mice subjected to unilateral eye enucleation at birth. The comparable cells in the peristriate cortex on the side contralateral to the enucleation were unaffected. This finding is of particular interest because it appears to be the first clear example of a *tertiary* anterograde transneuronal degeneration. That is to say, the changes in the peristriate cortex have occurred after a primary anterograde transneuronal degeneration in the lateral geniculate nucleus on the side contralateral to the eye removal, and after the secondary transneuronal degeneration in the visual cortex on that side.

The difficulty which exists in the interpretation of secondary and tertiary anterograde degenerative changes such as these is that at present there is no good quantitative study of the primary effects of eye enucleation in the lateral geniculate nuclei of young mice or rabbits. If, for example, eye enucleation is not followed by an appreciable degree of cell loss in the geniculate it would be necessary to postulate that there is some degree of degeneration in the terminals of the thalamo-cortical fibers although the bodies of the geniculate cells are preserved (although possibly in a shrunken form). Alternatively, if the thalamo-cortical fibers and their terminals do not degenerate then presumably the loss of dendritic spines in the visual cortex and the contralateral peristriate cortex is due to some form of functional disuse. Clearly this whole problem requires a good deal of further work before these difficulties can be resolved. A systematic search for secondary and tertiary transneuronal changes in the Primate visual and peristriate cortex after early eye enucleation might be particularly valuable at this time in view of the more drastic primary transneuronal changes in the lateral geniculate nucleus. It is also of some importance for studies of transneuronal degeneration in general to determine whether or not the shrinkage of the neuronal soma which is so characteristic of this type of degenerative change is accompanied by an active degeneration in the axons of the affected cells.

2. The Auditory Relay Nuclei

We shall now consider the evidence for anterograde transneuronal degeneration in the auditory relay nuclei since in several respects it parallels the evidence presented above for the lateral geniculate nucleus, and because this problem has been the subject of two particularly important experimental studies. The first of these is a study of the effects of the destruction of the cochlea (including the spiral ganglion in which the first order neurons of the auditory system are located) upon the primary and secondary auditory relay nuclei in the brainstem of adult cats (Powell and Erulkar, 1962). The second is an earlier study (Levi-Montalcini, 1949) of the effects of early removal of the otocyst on the development of the acoustico-vestibular centers in chick embryos.

The study by Powell and Erulkar (1962) was designed to parallel that of Cook, Walker and Barr (1951) on the lateral geniculate and the data from the two studies is directly comparable. In addition to sampling cell sizes in the three subdivisions of the ventral cochlear nucleus, and in the dorsal cochlear nucleus, quantitative data were presented for two of the secondary auditory relay nuclei (the lateral superior olive and the medial trapezoid nucleus). The principal findings of this study were: (i) After deafferentation, the cells in the three parts of the ventral cochlear nucleus undergo a transneuronal atrophy which has much the same time course as that in the lateral geniculate nucleus, i.e., there is comparatively little change after one month, but from the second month onwards there is a statistically significant atrophy in all three subnuclei amounting to between 15.5% and 36.3% for the cell perikarya, between 7.2% and 26% for the nuclei, and between 9.9% and 25.1% for the nucleoli. (ii) As in the geniculate, after an initial delay of some weeks the severity of the degeneration reaches a maximum and appears to be stable from about the 60th day until almost one year. (iii) There are no detectable changes in the cells of the dorsal cochlear nucleus even though it receives a substantial input from the cochlea (Powell and Cowan, 1962). (iv) Despite the differences in the mode of termination of the cochlear fibers in the ventral cochlear nucleus (the endings in the anteroventral nucleus are of the calyciform type [Held, 1893; Cajal, 1909] whereas those in the posteroventral nucleus are of the more common bouton variety) the time course of the transneuronal atrophy in the different subdivisions of the ventral cochlear nucleus is essentially the same. (v) Similarly, although there are appreciable differences in the size of the cells in the different nuclei (the cells in the posteroventral nucleus are considerably larger than those in the medioventral nucleus) the appearance, the severity and the time course of the transneuronal degeneration is not demonstrably different.

A further finding of interest in this study is that destruction of the cochlea leads not only to a primary anterograde transneuronal degeneration in the ventral cochlear nucleus, but also to a *secondary* transneuronal degeneration in the more distant relay nuclei in the auditory system as was first reported by Baginsky (1886) after lesions of the auditory nerve in newborn rabbits. From the detailed quantitative data presented by Powell and Erulkar it is clear that both the time course and the severity of the transneuronal degeneration in such secondary relay nuclei as the lateral superior olive were essentially the same as that in the ventral cochlear nucleus — the cell perikarya, nuclei and nucleoli shrinking to about 25% of their normal size. The perikarya of the cells in another secondary nucleus which was examined (the medial trapezoid nucleus) showed a similar reduction in cross-sectional area, but for reasons which are not yet apparent, there seemed to be no change in the area of the nuclei or the nucleoli of these cells at any stage during the degenerative process. No quantitative estimates were made of cell or nuclear size in the nuclei of the lateral lemniscus or in the ipsilateral preolivary nuclei, but qualitatively these also appeared to have undergone a secondary transneuronal degeneration. In fact, the only secondary relay nucleus which did not show cellular changes, on either side, was the medial superior olive, and it is not without significance that the cells in this nucleus are known to receive an approximately equal number of afferents from the cochlear nuclei of both sides (Stotler, 1953; Powell and Cowan, 1962; Goldberg and Brown, 1968).

As it is a matter of some importance for the interpretation of the degeneration in the secondary relay nuclei, it should be added that Powell and Erulkar (1962) could find no evidence of cell loss in the ventral cochlear nucleus even after a postoperative survival period of 319 days. There was also no reduction in the number of cells in the lateral superior olive which had undergone a secondary transneuronal degeneration. As was pointed out above, the finding of a *secondary* anterograde transneuronal degeneration in the absence of cell loss in the primary nuclei raises a number of problems concerning the nature of the transneuronal effect at the synapses in the secondary relay. Furthermore, the similarity in the time-course and severity of the degeneration in the primary and secondary auditory relay nuclei in the cat and that in the lateral geniculate of this animal raises the question whether or not there are morphological or functional limits upon the degenerative process in adult animals. That is to say, if transneuronal changes occur in different systems in a given species, do they always occur after the same delay; do they follow the same time-course, and are they limited in the severity they can attain? If so, then it might be possible to use these criteria for detecting transneuronal changes provided the necessary quantitative data could be obtained.

So far there does not appear to have been a comparable study to this in any other species. It would be of particular interest to know whether anterograde transneuronal changes occur in the ventral cochlear nucleus, and in the secondary auditory relay nuclei of Primates and, if so, whether these are of comparable severity to those observed in the lateral geniculate nucleus and also if they occur with the same rapidity.

Levi-Montalcini's study (1949) is of considerable importance because it represents one of the best documented examples of an anterograde transneuronal effect in a developing neural system. It appears that early extirpation of the otocyst in chick embryos does not affect cellular proliferation or normal differentiation in the primary auditory relay nuclei (the so-called *nucleus angularis* and the *nucleus magnocellularis*), but sometime after neuroblast differentiation has been completed both nuclei show degenerative changes. In the nucleus angularis the degeneration is so severe that by the time of hatching more than 80% of the cells have disappeared; the nucleus magnocellularis, on the other hand, is less severely affected, the cell loss at the time of hatching amounting to only about 30%. In a third nucleus (the *nucleus laminaris*), which probably receives most of its afferents from the two primary relay nuclei, there are no signs of degeneration, at least up to the 17th day of incubation. According to Levi-Montalcini the difference in the reaction of the two primary auditory relay nuclei is probably explicable on the basis of the completeness of the deafferentation: the most severe changes occur in the nucleus angularis because it is deprived of almost all its afferents by the removal of the otocyst, and the survival of a fairly large number of cells in the nucleus magnocellularis is probably due to the fact that it receives a significant proportion of its afferents from other sources.

In another group of experiments reported in this paper it was found that early extirpation of several different parts of the brain above the level of the medulla did not affect the normal development of the cochlear or vestibular centers, at least until the time of hatching. This important observation suggests that intracentral afferents are less significant for the maintenance of the primary relay nuclei than the extrinsic afferents from the cochlear and vestibular division of the eighth nerve.

3. The Vestibular Nuclei

As yet there does not appear to have been a systematic study of transneuronal degeneration in the vestibular nuclei, but in both Levi-Montalcini's (1949) and Powell and Erulkar's (1962) papers reference was made to these nuclei. In Levi-Montalcini's experiments the absence of the vestibular root fibers seemed to be without effect in all but one of the vestibular nuclei. The single exception was the *nucleus tangentialis* which receives the largest fibers in the vestibular nerve (Cajal's "collosal fibers"). In this nucleus the absence of the ingrowing root fibers resulted in a failure of normal neuronal differentiation with the result that at about the time of hatching the nucleus was considerably reduced in size and had failed to migrate from its early medial position in the brainstem into its definitive location near the point of entrance of the collosal fibers.

In one of Powell and Erulkar's (1962) experiments the vestibular division of the eighth nerve was incidentally damaged and they reported in passing that the cells of the medial and inferior vestibular nuclei in this brain were shrunken and pale-staining. In the lateral vestibular nucleus some of the cells also appeared to be shrunken but the larger (Deiter's) neurons were unaffected and there was no appreciable change in the cells of the superior vestibular nucleus. Since no quantitative observations were made in this experiment, and as there appears to be no other relevant information in the literature, it is difficult to compare these changes in the vestibular system with those in other primary relay nuclei. A careful study of the effects of deafferentation upon the vestibular nuclei would be valuable at this time since so much is now known about their structure and function (Brodal, Pompeiano and Walberg, 1962).

4. The Mammalian Olfactory System

There are several incidental observations in the literature which indicate that anterograde transneuronal changes may occur in the mammalian olfactory bulb and prepiriform cortex (Le Gros Clark, 1957), and recently a number of studies have appeared dealing specifically with this problem. Of these the most detailed is the paper by Matthews and Powell (1962). In this study the olfactory epithelium was surgically removed in a series of young rabbits and the resulting changes in the olfactory bulb and lateral olfactory tract were studied in Nissl, Bodian and Golgi-Cox preparations. Six and twelve days postoperatively the only change was the loss of olfactory nerve fibers on the surface of the olfactory bulb but by the sixteenth day clear degenerative changes were seen in the mitral and tufted cells. The changes in these cells became more severe after longer survival periods, and in addition, changes became apparent in the periglomerular and granule cells. Thus, after a survival period of 130 days the glomerular layer was reduced in cross-sectional area by approximately 45%, the outer plexiform layer by about 43% and the granule cell and internal plexiform layers by almost 36%. Measurements of the perikarya of the mitral and tufted cells at this stage showed them to be reduced by almost 50 and 40% respectively. In Golgi-Cox preparations after comparable survival periods very few mitral, tufted or periglomerular cells appear to be impregnated, but the granule cell layer, although reduced in area, showed an approximately normal number of impregnated neurons. In the region of the mitral cell layer there was a conspicuous number of rather unusual cells which were considered to be unusually impregnanted neuroglial elements. Since, in Nissl and reduced silver preparations the mitral and tufted cells were stained quite normally (although reduced in size) Matthews and Powell were inclined to interpret their absence in Golgi preparations as being due to a failure of impregnation, possibly brought about by an alteration in the metabolic activity of the cells, consequent upon their deafferentation (see also Powell, 1967).

The significance of this study on the olfactory bulb was threefold. First, it clearly demonstrated that transneuronal atrophy can occur in the olfactory system, and that in young rabbits at least, it may occur quite rapidly after deafferentation. Secondly, it showed that the transneuronal effect is not confined to the cells which are directly denervated but may also involve the second and third order neurons which are separated by one or more synapses from them. Thirdly, it was the first demonstration that transneuronal degeneration may result in such a significant alteration in the affected cells as to render them resistant to impregnation by the Golgi method.

Most of the findings in Matthews and Powell's paper refer to the principal olfactory bulb, but comparable changes were seen in the accessory olfactory bulb when the sensory epithelium of the vomero-nasal organ was removed; for example, in the experiment cited above with a postoperative survival of 130 days, the mitral cells in the accessory bulb were reduced in cross-sectional area by more than 38%. It may also be noted that the vomero-nasal organ was not affected by the atrophic rhinitis which commonly affects the olfactory epithelium of older rabbits and frequently leads to atrophic changes in the main olfactory bulb (Le Gros Clark and Warwick, 1946).

Transneuronal changes in the olfactory system are by no means limited to the olfactory bulb; in fact there is a good deal more literature on the effects of interruption of the lateral olfactory tract and extirpation of the olfactory bulb upon the prepiriform cortex than on the primary changes in the bulb itself. The first clear example of transneuronal degeneration in the prepiriform cortex was presented as long ago as 1918. In that year Winkler reported a shrinkage of the pyramidal cells in the prepiriform cortex of the rabbit within one month after interruption of the olfactory peduncle. This finding was confirmed some years later by Allison (1953), who has also described a cellular atrophy in the prepiriform area of a human brain in which the olfactory tract had been accidentally interrupted during a prefrontal leucotomy two years previously (Allison, 1954). Later work has shown that the degenerative change involves not only the pyramidal cell perikarya but also their dendritic processes. For example, Jones and Thomas (1956, 1962) have shown that removal of the olfactory bulb in rats may lead in time to a significant reduction in the number of secondary and tertiary dendritic branches of the pyramidal cells, and White and Westrum (1964) have similarly reported a loss of the spines from the dendrites of these cells after deafferentation.

Since the widely quoted view that the primary site at which the transneuronal degenerative process may act in cortical cells is the dendritic spine, stems partly from this work and from that on the visual cortex, it may be relevant at this point to consider the evidence for this hypothesis. Apart from the circumstantial evidence that in certain areas which have been deafferented there may be a statistically significant reduction in the number of dendritic spines, the only other finding which points in this direction comes from Colonnier's (1964) electron microscopical study of the effects of undercutting the visual cortex. Within a few days of deafferenting cortical cells in this way, Colonnier found that certain of the dendritic spines which were in contact with degenerating axon terminals appeared to be engulfed by a reactive glial cell process. On this basis he rather tentatively suggested that the spines might be ingested by the glial cells along with the degenerating presynaptic terminals and that this might be indicative of a trans-synaptic effect. There is a good deal of ancillary evidence to suggest that the attachment of the pre- and postsynaptic elements at a synapse is rather strong (e.g., it may resist the homogenization and centrifugation of brain tissue needed for the preparation of synaptosomes [Gray and Whittaker, 1960]), and it is conceivable that the ingestion of a degenerating presynaptic terminal by a glial cell may be accompanied by a "nipping-off" of the postsynaptic process. However, it should be borne in mind that Colonnier's findings were based upon cortical undercutting and it is impossible to exclude that in those cases in which an apparent engulfment of dendritic spines was seen the cortical neurons were undergoing a retrograde rather than a transneuronal degeneration, and as the critical findings were not based on serial thin sections the interpretation of the engulfment of the presynaptic process is itself open to question. In this same context it is important to note that in a later electron microscopical study Westrum (1966) observed that after removal of the olfactory bulb in rats many of the postsynaptic processes in the prepiriform cortex were clearly in contact with glycogen containing glial cell profiles. This suggests that the more common pattern after deafferentation is for the degenerating presynaptic processes to be phagocytosed by reactive glial cells and for the postsynaptic elements to be left intact, at least during the first few weeks after denervation. The long-term effects of denervation upon the postsynaptic processes remain to be determined in most areas. McMahan (1967) has found that there may be a disruption of the postsynaptic membrane thickenings, some weeks after deafferentation of cells in the rat lateral geniculate nucleus but as yet this observation has not been confirmed (see below).

5. Spinal Cord and Somatic Afferent System

There is an extensive literature dating back to the latter part of the last century on the effects of interrupting dorsal roots, and of lesions of the motor cortex, upon the cells of the spinal cord. As early as 1898 Warrington reported finding chromatolytic cells in the gray matter of the cord after dorsal root section and after transection of the cord at higher levels. Several years later Barron (1933) also described chromatolyzed neurons in the spinal cords of rats after lesions involving the pyramidal tracts and Foerster, Gagel and Sheehan (1933) found changes in the large cells of the posterior and anterior horns of the cord after interruption of the dorsal roots in monkeys and in a number of human cases. Foerster and Gagel (1934) also claimed to find similar changes in the gracile nucleus after transection of the posterior columns in the lumbar region. Taken by themselves these studies would seem to be quite conclusive, but unfortunately there is almost as complete a body of evidence to the effect that dorsal root section and pyramidal tract lesions do not result in unequivocal transneuronal changes in the cord (see for example Schimert, 1939; Hare and Hinsey, 1940). Cook, Walker and Barr (1951), who have reviewed much of the earlier literature on the effects of deafferentation upon the cells of the spinal cord, reexamined this problem in a series of cats in which they had placed unilateral lesions in the motor cortex, or in which some of the lumbar and upper sacral dorsal roots had been unilaterally divided. They concluded this part of their study with the statement that "examination of the spinal cord material failed to reveal any cell changes which could, with certainty, be attributed to degeneration of either corticospinal or dorsal root fibers".

The only clear example of transneuronal degeneration in the somatic afferent system is the finding of severe cell shrinkage and neuronophagia in the principal and spinal nuclei of the trigeminal nerve following alcohol injection into the human Gasserian ganglion (Penman and Smith, 1950). In the spinal nucleus about two-thirds of the medium-sized neurons showed a moderate degree of chromatolysis and many of the smaller cells appeared shrunken and had dark-staining nuclei. In the main sensory nucleus there was a definite reduction in the number of neurons and many of the remaining cells were dark-staining and shrunken. The mesencephalic and motor nuclei were unaffected.

In view of the very rapid and severe degeneration which occurs in the monkey lateral geniculate nucleus, it might be thought that interruption of the medial lemniscus and spinothalamic tracts would result in similar changes in the ventral posterior nucleus of the thalamus. In 1957 Bowsher reported in an abstract that he had found transneuronal changes in a small number of cells in this nucleus after lesions of the spinothalamic tract in a series of human brains but as this work has not been published in full, and since the finding does not seem to have been based on quantitative data, it is difficult to evaluate the significance of this observation. Moreover, Glees and Le Gros Clark (1941) have reported that they were unable to identify transneuronal changes in this nucleus in a single monkey in which the medial fillet was divided.

6. The Pontine and Inferior Olivary Nuclei

The reasons for considering these two nuclear groups together is that they form the subject of an important paper on transneuronal effects produced by unilateral lesions of the forebrain, cerebral peduncles and brainstem by Torvik (1956). For this work, which represents one of the most carefully documented studies of transneuronal degeneration in young animals, lesions were placed in young kittens between 8 and 20 days of age and the resulting changes observed over a period of between 3 and 8 days. Following the appropriate lesions (in the brainstem in the case of the inferior olive, and in the cerebral cortex and cerebral peduncles in the case of the pontine nuclei) marked cellular degeneration occurred in both areas as early as 4 or 5 days after surgery. Since the distribution of the degeneration varied predictably with the site of the lesion, it was clear that the changes were due to the selective deafferentation of different topographical portions of the inferior olive and pontine nuclei.

The changes in the inferior olive closely resembled the retrograde degeneration seen in this structure after lesions of the cerebellar cortex. The earliest changes were in the form of cell shrinkage and were observed $3^1/_2$ days after operation; over the next 48 hours the cells proceeded to degenerate in one of two ways: Most commonly the degeneration passed from a stage of cellular pyknosis to a complete and rapid disintegration of the neuron, but in some cells the degeneration was characterized by a brief phase of chromatolysis resembling the "primäre Reizung" of Nissl. Accompanying this rapid neuronal death there was an acute glial reaction which was generally more severe than that seen after cerebellar lesions. The changes in the pontine nuclei were less intense than those seen in the olive and also occurred slightly later, reaching their maximum at about 6 or 7 days postoperatively. Again, both types of cellular degeneration were present in the pontine nuclei but by far the more common reaction was the progressive shrinkage and ultimate dissolution of the neurons.

There are earlier accounts of transneuronal degeneration in the pontine nuclei (v. Monakow, 1882; Masuda, 1914; Beck, 1950) but in none of these were the changes so dramatic or so clearly described as in Torvik's paper and, in addition, Torvik's was the first description of transneuronal degeneration in the inferior olive. This is of particular interest for the understanding of transneuronal degeneration because the severe changes in the inferior olive occurred *bilaterally* after unilateral brainstem lesions, which implies that under some circumstances transneuronal degeneration may occur in a structure when at the most only 50% of its afferents have been removed. As the completeness of deafferentation is generally considered to be one of the more important factors determining whether or not transneuronal changes occur any given situation, this observation is clearly very critical.

The severity of the cellular reaction in the inferior olive and pontine nuclei is also unusual. Cell loss is not commonly a feature of anterograde transneuronal degeneration even after long survival periods and it seems likely that in Torvik's experiments the use of very young animals was responsible for the early and rapid disintegration of the deafferented cells. From a cytological point of view the changes more closely resemble those found after axon section or during retrograde transneuronal degeneration in young animals (see below).

7. Other Sites in the Central Nervous System

There are several other areas in the brain in which it has been suggested that anterograde transneuronal degeneration may occur. Many of these suggestions have not been confirmed by later work and so will not be discussed here. However, there are a few systems in which a *prima facie* case may be made for postulating that transneuronal effects are present and these will be discussed in some detail because they show how important it is to consider the possible effects of deafferentation in nearly every case of cellular degeneration in the central nervous system.

a) The Reticular Nucleus of the Thalamus

Several studies of thalamo-cortical relations have included reports of cellular degeneration in the thalamic reticular nucleus (Walker, 1938; Rose and Woolsey, 1943; Rose, 1952; Chow, 1952), and it has been suggested that the nucleus projects in an ordered manner upon all parts of the cerebral cortex. However, there are certain features of the cellular reaction in this nucleus which suggest that it may not be a retrograde degeneration. For example, whereas most thalamic nuclei undergo a rapid and virtually complete retrograde cellular degeneration within 2 or 3 weeks after even quite small cortical lesions, in the reticular nucleus distinct cellular changes are only found after relatively large lesions and occur much more slowly, so that even after rather long survival periods a substantial number, if not most, of the cells persist (Rose, 1952).

Recently Carman, Cowan and Powell (1964a) have found that there is a precise topographical projection of cerebral cortex *upon* the reticular nucleus and as this parallels the regional degeneration described by Rose (1952) they have tentatively suggested that the cellular changes seen in the nucleus after cortical lesions may be transneuronal rather than retrograde in character. Clearly the fact that the nucleus receives a substantial input from the cerebral cortex does not of itself establish that the cellular changes are transneuronal (a number of alternative explanations have been put forward to account for the cellular changes [Rose, 1952; Chow, 1952]), but the fact that the cells are deafferentated after a cortical lesion is highly significant and certainly a transneuronal effect cannot be excluded. Furthermore, the finding in Golgi preparations that the axons of the cells in the reticular nucleus project caudally rather than into the internal capsule (Cajal, 1911; Scheibel and Scheibel, 1966) strongly supports this view.

b) The Claustrum

A similar argument has been put forward to account for the cellular degeneration seen in the claustrum after lesions of the cerebral cortex. Narkiewicz (1964) has described a systematic pattern of cellular degeneration in the claustrum after localized cortical lesions and has interpreted this as evidence for an organized projection of the claustrum to most parts of the neocortex. However, there is evidence from Nauta studies that most, if not all, parts of the cerebral cortex project in an orderly manner *upon* the claustrum (Carman, Cowan and Powell, 1964b). This again raises the possibility that the cellular changes found in the claustrum may be transneuronal rather than retrograde in character and certainly the appearance of the cellular changes is consonant with this view.

c) The Intralaminar Nuclei of the Thalamus

The nature of the relationship between the intralaminar nuclei of the thalamus and the cerebral hemisphere has been a source of concern to neuroanatomists for many years, and despite a great deal of experimental work the problem still remains unresolved (see Powell and Cowan, 1967). In the present context the only relevant experimental finding is that after cortical lesions there is some degree of cellular atrophy in the intralaminar nuclei generally in the form of cell shrinkage but there may also be some slight cell loss (see for example, Walker, 1938; Nashold, Hanbery and Olszewski, 1955; Murray, 1966). Recently Powell and Cowan (1967) have suggested that these changes may be due to the deafferentation of the intralaminar nuclei by cortical lesions as there is now a good deal of evidence that the intralaminar nuclei receive a substantial, and organized, projection from the cerebral cortex (cf., Auer, 1956; Petras, 1964; Mehler, 1966; Powell and Cowan, 1967; Sakai, 1967). Also, the cellular reaction in these nuclei after cortical lesions is quite different from that after lesions of the internal capsule and striatum which indubitably interrupt the axons of the intralaminar cells. A good deal more work is required before this problem can be resolved but, as Powell and Cowan (1967) have pointed out, transneuronal factors must always "be seriously considered as a possible explanation, or at least as a contributing factor, in all cases of neuronal degeneration when there is evidence that the causative lesion may have involved an appreciable number of afferents to the affected area, and especially where the distribution of the cellular degeneration agrees with the topical organization of the afferents".

8. The Peripheral Nervous System

In this section only the changes seen in certain of the sympathetic and parasympathetic ganglia will be considered; the morphological and functional changes found in striated muscle after denervation will not be discussed as these have been the subject of an excellent recent review (Guth, 1968).

The only studies of transneuronal degeneration in the autonomic nervous system comparable to those discussed above are those of Hamlyn (1954) and Levi-Montalcini (1947). Hamlyn, who has also reviewed some of the controversial earlier literature on this topic, examined the changes in the superior cervical ganglion of adult rabbits following the interruption of the sympathetic trunk. The earliest changes in the postganglionic neurons after deafferentation were seen 40 days postoperatively; at this stage the cells appeared to have lost some of their cytoplasmic basophilia and to be slightly shrunken. By the 100th day postoperatively the cell shrinkage appeared to have reached a maximum (about 17%) and remained at this level through the 150th day. At no stage during the degenerative process did the cells exhibit the characteristic features of a chromolytic reaction, and even after the longest postoperative survival period there appeared to be no cell loss. Hamlyn's results differ from the earlier work of Sternschein (1920) and de Castro (1923), both of whom claimed to find chromolytic changes in sympathetic ganglion cells following preganglionic nerve section, and they also contradict Gibson's claim (1940) that there are *no changes* in the perikarya of postganglionic cells after preganglionic nerve section.

The second problem which has received particular attention is the effect of early extirpation of the midbrain upon the development of the ciliary ganglion in the chick (Levi-Montalcini, 1947). This procedure, if carried out between the 32nd and 38th hours of incubation, had no effect upon the initial proliferation, differentiation or migration of the cells which give rise to the ciliary ganglion. The cells appear to grow quite normally until about the 10th day of incubation at which stage they begin to show regressive changes. These changes are so severe that by the time of hatching the ganglion seems to have undergone an almost complete cell loss. The implication of this is that the presence of a normal afferent input to the ganglion at about the 10th day of incubation is essential for the further growth and maintenance of the postganglionic cells.

There does not appear to have been a comparable study of the effects of early deafferentation upon the development of other sympathetic or parasympathetic ganglia and it is important to emphasize that the extremely severe degeneration seen in the developing ciliary ganglion is not necessarily typical. It is also important to point out these changes are not at all comparable to those descibed by Hamlyn (1954) in the superior cervical ganglion; whereas the latter are clearly a transneuronal atrophy in the generally accepted sense the degeneration in the ciliary ganglion is to be thought of as a failure of maintenance of differentiated, but immature, neuroblasts, due to the absence of their normal afferent input.

The ganglia of the autonomic nervous system are one of the few sites in which it has been possible to study the effects of deafferentation electrophysiologically and with the electron microscope. One of the most critical studies of this problem is that of Hunt and Nelson (1965) in which the IXth and Xth sympathetic ganglia were denervated in frogs, and the subsequent changes studied over a period of about 40 days. For approximately the first 24 to 36 hours after preganglionic nerve section transmission through the ganglion was not markedly affected but concurrently with the degeneration of the terminals of the preganglionic fibers (approximately 48 hours after nerve section) synaptic transmission failed although the deafferented cells continued to respond to antidromic activation throughout the period investigated. At no stage were there morphologically detectable changes in the postganglionic neurons but the membrane resistance of the denervated cells was found to be markedly increased.

The absence of morphological changes in the postsynaptic membrane specializations following denervation is somewhat surprising and at present there is some disagreement in the literature on this subject. For example, Taxi (1965) has described a clear fragmentation of the specialized subsynaptic formation found in frog sympathetic ganglion cells, within five days after denervation. On the other hand, Sotelo (1968) claims that neither this subsynaptic formation nor the more common postsynaptic membrane specializations are affected, at least up to 12 days after denervation.

This problem clearly calls for further work and it would be of particular interest to study the fate of the synaptic complex in the cells of the superior cervical ganglion which show clear signs of transneuronal degeneration (Hamlyn, 1954).

Some General Observations on Anterograde Transneuronal Degeneration

Before reviewing the work on retrograde transneuronal degeneration, it may be helpful to summarize some of the principal findings on the anterograde transneuronal degenerations. However, it is important to emphasize at the outset that there are still so many lacunae in our knowledge that it is extremely difficult to make valid generalizations about either the morphological features of the degeneration or about the etiological factors which may underlie them.

Perhaps the most definite conclusion that can be drawn from the evidence currently available is that the effects of deafferentation vary inversely with the age of the animal at the time the deafferentation is carried out. That is to say, all forms of anterograde transneuronal degeneration seem to be more severe and to occur more rapidly in immature animals. In the special case where the "deafferentation" is carried out before normal synaptic connections are established, the neurons which are deprived of their afferent input may fail to differentiate fully, but almost invariably undergo a complete degeneration. The reasons for the greater susceptibility of young or immature neurons to deafferentation are not clear and conversely it is not known why cells become relatively resistant to deafferentation with increasing age. Some possibilities which have been suggested are: (i) Immature cells are in a metabolically unstable state and for this reason more susceptible to any form of noxious influence; as the cells mature they become progressively more stable and hence more resistant to indirect injury. (ii) Immature cells may be more dependent upon some substance(s) derived from their afferents, and as the cells grow older they become increasingly self-sufficient and independent of such trophic influences. (iii) It is possible that immature cells which have not yet received all their afferents and whose axons have not yet established all their synaptic connections are unable to withstand the effects of denervation at this stage in their development. The later formation of additional synaptic connections would then confer upon the cells a certain resistance to deafferentation. In a sense most of these explanations are tautological and provide us with little real understanding of the factors responsible for the greater severity of transneuronal degeneration in young animals. This point will be considered later in connection with the retrograde transneuronal degenerations and here it need only be added that there is a good deal of evidence that immature neurons appear to be more sensitive to all forms of injury *(direct or indirect)* than fully developed cells.

The second generalization which can be made with some confidence is that there are significant differences in the reactivity of neurons in adult animals of different species to deafferentation of the same nuclear groups. The lateral geniculate nucleus is perhaps the best example of this; eye enucleation in Primates leads within a few days to a severe atrophy of most of the cells in the related part of the geniculate whereas in cats and rabbits there is little indication of the effects of deafferentation until about 2 and 6 months after eye removal, respectively, and in rats it appears that deafferentation may lead to no detectable change in the geniculate cells. What the nature of this "species difference" may be is again quite unknown. It has been suggested that there may be significant differences between species in either the number, and possibly the nature, of the retinal and nonretinal afferents to the lateral geniculate nucleus. Thus it was thought at one time that in monkeys each geniculate cell received only a single axosomatic ending (from the optic nerve), whereas in the cat and rabbit each cell was in contact with a number of axosomatic and axodendritic boutons from the retina and other sources (Glees and Le Gros Clark, 1941; Glees, 1941, 1942). However, it is now known that in Primates as well as in the subprimates there is at least one other source of afferents to the lateral geniculate nucleus (from the striate cortex) and there is every reason to think that the quantitative differences in the synapses which have been described in silver preparations are quite misleading. There may well be other more subtle qualitative differences in the retinal endings within the geniculate (Colonnier and Guillery, 1964; Szentagothai, 1963; Peters and Palay, 1966; McMahan, 1967) and the differences in their reaction to section of the optic nerve may be one important clue to this problem.

A third factor of some significance is the location of the deafferentated cells. Although there is some evidence that cells in certain systems may respond to deafferentation after the same delay and to approximately the same extent (cf. the changes described above in the auditory relay nuclei and the lateral geniculate nucleus of the cat, and the atrophy in the rabbit lateral geniculate nucleus and superior cervical ganglion), this is clearly not always the case. Again, to cite an example from the visual system, in Primates and carnivores eye enucleation results in clear-cut transneuronal changes in the lateral geniculate nucleus but has comparatively little effect upon the cells in the superior colliculus, but the reverse appears to be true in the immature rodents.

A fourth factor which has been stressed in the literature is the completeness of the deafferentation. There are a few situations in which the severity of the transneuronal degeneration appears to be rather directly related to the number of afferents removed (e.g., the auditory relay nuclei in the chick), and in the special cases of visual deprivation, to the completeness of the derpivation (cf., the effects of eyelid suture and the insertion of a translucent corneal lens in young kittens). This hypothesis has frequently been invoked to explain the absence of clear transneuronal changes in many systems which have unquestionably been deafferented, but there is in fact very little direct evidence to support it. In fact, there are many observations which would seem to contradict it. For example, the cells in the superior cervical ganglion of the rabbit show only a mild degree of cell shrinkage after complete denervation, whereas the cells of the inferior olive in young kittens may degenerate although deprived of only half their afferents from the brainstem.

The mode of termination of the afferents does not appear to be a very significant factor. If it were, the cells in the anteroventral cochlear nucleus which receive large, calyciform endings from the eighth nerve might be expected to be more severely affected by destruction of the cochlea than the cells in the posteroventral nucleus which receive only the more usual form of terminal bouton. However, this is not the case (Powell and Erulkar, 1962) and at present there is no good evidence that the morphological character of the presynaptic terminal has any direct relationship to the nature or severity of the transneuronal changes after denervation.

Lastly, it is true to say that in every situation which has been studied the severity of the transneuronal changes has been shown to be a function of the duration of the postoperative survival period; that is to say, the longer the survival period the more severe is the atrophy in the deafferented cells. This does not imply that the degenerative process proceeds uniformly from the time of deafferentation; in fact, there is a good deal of evidence that there is a latent period of a few days or longer before the effects of deafferentation make themselves evident, at least morphologically. In the Primate lateral geniculate nucleus the delay is as short as 4 days but in the rabbit lateral geniculate it may be as long as several months. As a rule once it begins the degeneration proceeds quite rapidly to a near maximum condition and progresses only slightly beyond that point even after very long survival periods. Why neurons should show little or no morphological evidence of denervation until some considerable time after the degeneration and removal of their afferents is not at all clear. It is known, for example, that in cats the optic nerve terminals degenerate within a few days after eye enucleation and yet it is not until almost 60 days later that the earliest signs of degeneration become evident in the cells of the lateral geniculate nucleus. It is difficult to postulate a mechanism which would account for such a long delayed reaction in the postsynaptic neurons, and it seems likely that there may be significant functional, chemical, and possibly ultrastructural, changes in the deafferented cells some time before the overt atrophy of the cell perikarya. Further functional and cytochemical studies like those of Hunt and Nelson (1965) and Kupfer and Palmer (1964) are clearly needed to clarify this issue.

It is generally agreed that in adult animals the only consistent cytological change during anterograde transneuronal degeneration is a progressive reduction in the mean cross-sectional area of the cell soma, nucleus and nucleolus. The progressive cell shrinkage is such that the affected neurons when seen in isolation do not appear grossly abnormal. It is only when the deafferented cells are compared to normal neurons in the same system that the transneuronal effect becomes evident and it is generally only in systems like the lateral geniculate nucleus in which the affected and normal neurons are in close approximation that the degeneration is readily recognizable. Since most studies have been based upon Nissl-stained material, it is understandable that these should be the most frequently discussed changes, and it is likely that if more use were made of reduced silver or Golgi preparations greater emphasis would be placed upon the atrophy of the dendrites and their appendages. In fact, it now appears that during deafferentation there is a generalized reduction of the surface area of the postsynaptic cell and Powell and Erulkar (1962) have put forward the interesting suggestion that the size of a neuron's perikaryon and dendritic tree may be directly related to (and determined by) the number of functional synaptic contacts upon it.

In some situations *cell loss* appears to occur after relatively short survival periods (e.g., the principal and spinal nuclei of the trigeminal nerve) but this is not usually a feature of anterograde transneuronal degeneration in adult animals except after very long survival periods (usually some years in Primates). In young and immature animals on the other hand, there may be as rapid and as complete a cell loss after denervation as after axon section (see Torvik, 1956).

There have been too few studies done of the fine structural changes during transneuronal degeneration to permit of any clear conclusions at this time but in most of the sites which have been studied the postsynaptic processes appear to be relatively unaffected at least during the first few weeks after denervation. There may be some change in the postsynaptic membrane specialization but this has been found in only a few studies and has been disputed in others. As yet there has been no electron microscopical study of the more fulminating types of transneuronal degeneration such as that seen in the inferior olive of young kittens where one might expect to see early and marked degenerative changes in the dying cells.

From the few physiological studies of deafferented neurons which have been carried out it appears that there is no striking change in their ability to conduct impulses, and to date the only significant finding is that the cell's membrane resistance may be appreciably increased (Hunt and Nelson, 1965). There is also some reason to think that in the case of visually deprived kittens the pattern of connections established by the axons of deprived geniculate neurons in the visual cortex may be abnormal although the response characteristics of the geniculate cells themselves do not seem to be grossly affected (Wiesel and Hubel, 1963).

There has been a good deal of speculation about the nature and the mechanism of the transneuronal effects which follow deafferentation. Some workers have postulated that it is due to a hypersensitivity reaction in the denervated neurons, others that it is due to some toxic (or otherwise noxious) influence of the degenerating axon terminals upon the postsynaptic processes or to the effect of prolonged functional disuse. Unfortunately, there is almost no direct evidence which can be cited in support of any of these views and much of the indirect evidence which is commonly advanced is either disputable, or at least, open to alternative interpretations. For example, the hypothesis that transneuronal atrophy is the expression of a hypersensitivity reaction seems to have been put forward solely on the basis that "the 'law of denervation' as formulated by Canon appears to be generally applicable" and "the cytological appearance of deafferented cells ... is not incompatible with the presence of hypersensitivity" (Cook, Walker and Barr, 1951). Similarly, Beresford's (1966) suggestion that the large degenerating retinal endings found in the monkey lateral geniculate nucleus after eye removal are "parasitic" upon the geniculate cells, and are responsible for the early transneuronal atrophy by taking "an unusually great amount of amino acids from an adjoining glial cell which usually supplies the host neurone with the same materials", is based upon a view of the relationships between neurons and glia which can hardly be substantiated at this time. The "disuse hypothesis" has rather more to commend it, and it certainly provides the simplest explanation for a number of experimental findings such as the effects of visual deprivation and the occurrence of secondary transneuronal changes in the absence of cell loss at the site of the primary atrophy. However, this hypothesis provides no real indication as to the nature of the transneuronal effect. A fourth hypothesis which is widely held is that transneuronal effects are due to the removal of some (as yet undefined) trophic influence exerted by the presynaptic processes upon the neurons with which they are in contact. Since the rôle of trophic influences has recently been reviewed at length by Guth (1968) the subject need not be discussed here. However, it is perhaps worth pointing out that even at the neuromuscular junction where this problem has been most intensively investigated our understanding of the mechanisms involved is still grossly inadequate, and in the central nervous system the study of this subject has barely advanced beyond the descriptive phase.

Retrograde Transneuronal Degeneration

Between 1870 and 1884 Gudden reported the results of a series of experiments in which extensive areas of the cerebral cortex were ablated in young and newborn rabbits. In most of these animals (which had been allowed to survive for periods of up to 1 year) there was a severe degeneration of the principal nuclei of the dorsal thalamus and, in addition, there was a marked atrophy of the posteroventral part of the medial mammillary nucleus and in what is now known as the ventral tegmental nucleus. Although at that time it was impossible to adequately interpret these findings (mainly because the anatomical relationships between the structures involved had not yet been determined, but also because of the severe distortion of the brain which occurred after such lengthy postoperative survival periods), it is now clear that Gudden's work included the first description of a primary and secondary retrograde transneuronal degeneration. As a result of the involvement of the limbic area by the cortical lesions the anterior thalamic nuclei had undergone a typical retrograde cell degeneration; secondary to this the cells in the posteroventral part of the medial mammillary nucleus underwent a *primary retrograde transneuronal degeneration*. The degeneration in the medial mammillary nucleus, in turn, resulted in a *secondary retrograde transneuronal degeneration* in the ventral tegmental nucleus.

Similar reports of retrograde transneuronal degeneration in the medial mammillary nucleus, and in the optic nerve and retina after involvement of the visual cortex, were presented by several workers in the latter part of the 19th century (Ganser, 1882; v. Monakow, 1882, 1889), and in the early part of the present century (Münzer and Wiener, 1902; Nissl, 1913; Winkler, 1918). Unfortunately, much of this early work was either neglected or vigorously disputed (possibly because of the threat it seemed to pose to the establishment of the "neuron doctrine"), and it was not until recently that secondary changes, after retrograde cell degeneration, have begun to receive the attention they deserve. Much of the credit for this is due to a group of neuro-ophthalmologists who recognized the that late occurrence of optic atrophy in patients who had suffered damage to the visual cortex may be the result of a secondary degeneration of the retinal ganglion cells consequent upon the degeneration of the lateral geniculate nucleus (Walsh, 1947; Haddock and Berlin, 1950).

In 1954, Cowan and Powell showed that the occurrence of retrograde transneuronal degeneration could be used as an experimental tool for determining rather precisely the interrelations between the different subdivisions of the medial mammillary nucleus and the three fields of the limbic cortex, and since that time there have been a number of experimental studies of retrograde transneuronal degeneration in this system and in the visual pathway. Nevertheless, compared to the anterograde transneuronal degenerations, secondary retrograde effects are still not widely recognized.

In addition to the medial mammillary and ventral tegmental nuclei and the ganglion cells of the retina, retrograde transneuronal degeneration has been described in the pyramidal cells of the motor cortex, the inferior olive, the mossy fibers of the cerebellum, and the avian homologue of the Edinger-Westphal nucleus. Since the first two systems have received the greatest attention they will be considered in more detail than the others.

1. The Medial Mammillary and Ventral Tegmental Nuclei

The early experimental finding of degeneration in these nuclei after lesions involving the limbic cortex in young animals has been confirmed in a number of recent studies (Cowan and Powell, 1954; Peacock and Combs, 1965; Bleier, Bard and Woods, 1966; Bleier, 1969; Cowan and Hart, 1969).

While studying the connections of the midline and intralaminar nuclei in rabbits, Cowan and Powell (1955) incidentally observed degeneration in the medial mammillary nucleus in several young animals in which the lesion had involved the projection fibers of the anterior thalamic nuclei. In another study (Cowan and Powell, 1954) use was made of this observation to determine the relationship between the different subdivisions of the mammillary nucleus and the three cytoarchitectonic fields of the limbic cortex defined by Rose and Woolsey (1948). The details of the morphological relationships between the mamillo-thalamic system and the limbic cortex need not be considered here but some of the findings in this study on the nature and time course of the primary retrograde transneuronal degeneration in the medial mammillary nucleus are pertinent. Cowan and Powell (1954) found that distinct changes could be seen in the medial mammillary nucleus as early as 17 days after a lesion of the limbic cortex in 3-week-old rabbits; the earliest changes were in the form of an overall shrinkage of the cells, a loss of Nissl material, and an early and quite severe gliosis. With longer survival periods, or when younger animals were used, the degenerative changes were appreciably more severe and could lead to a complete cell loss in the affected parts of the mammillary nucleus. In young adult animals, some neuronal shrinkage was observed 24 days after operation, but even after 68 days there was comparatively little cell loss. On this basis it was suggested that the severity of the retrograde transneuronal degeneration in the medial mammillary nucleus after lesions of the limbic cortex is an inverse function of the age of the animal at the time of operation, and varies directly with the duration of the postoperative survival period. Since the distribution of the degeneration in the medial mammillary nucleus bore a consistent relationship to the site of the cortical lesion and to the distribution of the retrograde cell degeneration in the anterior thalamic nuclei, attention was drawn to the possibility of using retrograde transneuronal degeneration as an experimental method for determining neuronal connections in certain systems.

A similar, but more detailed, study of retrograde transneuronal degeneration in the medial mammillary nucleus of young rabbits has recently been undertaken by Bleier (1969). In addition to clarifying some of the interconnections between the different subdivisions of the mammillary nucleus and the anterior thalamic nuclei and limbic cortex, Bleier has considered the effects of the age of the animal at the time of operation and the duration of the postoperative survival period in considerably more detail, and has expanded upon an observation reported in her earlier study with Bard and Woods (1966), namely, that severe cellular degeneration can occur in the ventral tegmental nucleus after lesions of the limbic cortex. The latter finding, although not completely new, is of particular interest since it is the first clearly documented example of a secondary retrograde transneuronal degeneration in recent years. From Bleier's description (1969) of the changes in the ventral tegmental nucleus it is clear that the severity of the secondary retrograde transneuronal degeneration is directly proportional to the severity of the primary transneuronal changes in the posterior and basal parts of the medial mammillary nucleus and, like the primary degeneration, varies with the age of the animal at the time of operation and the duration of the postoperative period. In newborn rabbits the degeneration in the ventral tegmental nucleus is very severe even after a survival period of only 1 week and appears to reach its maximum within 3 weeks, so that it is no more severe in animals which were sacrificed after 60 days than in one which had died 22 days after operation. In a group of animals which were allowed to survive postoperatively for 2 months there was a mild degree of cell loss and a moderate atrophy of the surviving cells in those operated upon at 7 and 15 days of age, but in an animal which was 20 days old at operation only occasional cells in the nucleus showed signs of atrophy. In one animal which was more than 4 months old at operation there were no signs of secondary transneuronal degeneration in the ventral tegmental nucleus, even though most of the posterior limbic cortex had been ablated. That is to say, in newborn animals almost the entire ventral tegmental nucleus is dependent upon the integrity of the medial mammillary nucleus, by 3 weeks of age a percentage of the cells (mainly in the dorsal part of the nucleus) has become independent of the mammillo-thalamic system and from that age onward the nucleus as a whole becomes progressively independent of the mammillary complex. Bleier favors the view that the increasing stability of the cells in the ventral tegmental nucleus is due to the development of collateral connections with structures other than the medial mammillary nucleus. In support of this she mentions finding a number of abnormally large cells persisting in the severely affected mammillary nuclei, which she regards as cells that were not involved by the initial retrograde transneuronal degeneration and grew to their unusual size by the sprouting of new collaterals. By comparing the changes in the ventral tegmental nucleus in animals with lesions in different parts of the limbic cortex, Bleier was able to show that the nucleus was only clearly involved in experiments in which there was some damage to the cingulate area of Rose and Woolsey (1948).

Cowan and Hart (1969) have carried out a quantitative study of the primary and secondary retrograde transneuronal degeneration in the ventral tegmental nucleus of young rats following lesions of the anterior thalamic nuclei and limbic cortex, respectively. The principal findings of this study are that after lesions of the anterior thalamus in 5-day-old rats which were allowed to survive for 3 weeks postoperatively, there was a cell loss in the ventral tegmental nucleus amounting to more than 60%. In older animals (10 and 15 days of age at operation) the cell loss was of the order of 45% and 20%, respectively; in 28-day-old animals with the same survival period there was no detectable cell loss. In addition to the marked cell loss in the nucleus, most of the surviving cells were markedly shrunken and pale-staining. Again the severity of the cell shrinkage varied with the age of the animal at the time of operation; thus, in the animals operated upon at 10 days of age the mean cross-sectional area of the surviving cells was reduced by approximately 30%, in 15-day-old animals it was reduced by just over 25%, and in the 20-day-old animals by just under 10%. The nuclei of the persisting cells were also reduced in cross-sectional areas, but always to a lesser degree.

The effect of the duration of the postoperative survival period upon the severity of the secondary retrograde transneuronal degeneration in the ventral tegmental nucleus was studied in a series of rats in which limbic lesions were placed on the fifth day postnatally. These animals were allowed to survive for periods ranging from 1 to 7 weeks and, as was expected, the degree of cell loss and the severity of the shrinkage in the surviving cells increased progressively with increasing postoperative survival. For example, there was no detectable cell loss in the ventral tegmental nucleus after a survival period of one week, but by the end of the second week more than 20% of the cells had disappeared and by the end of the seventh week there was a cell loss amounting to approximately 35%. This study provides the first quantitative evidence for the effect of the age of the animal at the time of operation and of the duration of the postoperative survival period on the severity of retrograde transneuronal degeneration, and fully confirms the qualitative impression that the changes are always more severe in young animals and after longer survival periods. In addition, it has helped to clarify the relationship between the ventral tegmental nucleus and the different subdivisions of the medial mammillary nucleus, a problem which had proved to be rather difficult to resolve by the more direct procedure of placing lesions in the mammillary body (Cowan, Guillery and Powell, 1964).

Retrograde transneuronal changes also have been observed in the medial mammillary nucleus of young monkeys subjected to hemidecortication (Peacock and Combs, 1965), and in the medial mammillary and ventral tegmental nuclei of cats after lesions involving the limbic cortex and anterior thalamic nuclei, respectively (Bleier, Bard and Woods, 1966).

2. The Ganglion Cells of the Retina

What can now be interpreted as a retrograde transneuronal degeneration was first described in the visual system by Ganser (1882) who reported a reduction in the optic nerve fiber layer of the retina several months after an extensive lesion of the cerebral cortex in a young kitten, and by von Monakow (1889) who made essentially the same observation in a series of young puppies. Since these early studies there have been a number of incidental reports of degeneration of the retina after lesions of the optic radiations or visual cortex (see, for example, Klüver, 1937), but no systematic anatomical study of this problem was undertaken until 1963 when Van Buren described the appearance of the ganglion cell and inner nuclear layers of the retina after lesions of the visual cortex and of the optic nerve in a group of monkeys and in a number of human cases. For a detailed account of the histological changes in the retina in the human material reference should be made to Van Buren's (1963a) monograph; here we shall only consider the experimental evidence reported in his later paper (Van Buren, 1963b). In the retinae of two monkeys examined about 20 months after surgical interruption of the optic chiasm there was an extensive loss of cells in the ganglion cell layer of the retinae medial to the fovea, and although there were no evident changes in the internal plexiform layer, the inner nuclear layer showed clear evidence of cystic degeneration. The external plexiform layer and the receptor cells were apparently unaffected. In a single monkey examined 48 months after an occipital lobectomy the optic tract was found to be reduced in cross-sectional area by more than 50% and there was a distinct loss of ganglion cells in the homonymous halves of the two retinae; in this case there were no detectable changes in the deeper retinal layers.

These experimental findings together with the observations in the human cases (which were essentially the same) provide an adequate explanation for the earlier clinical reports of optic atrophy following lesions of the visual cortex or visual radiations. As this material has been reviewed elsewhere (Walsh, 1947; Van Buren, 1963a), reference need only be made to a single example presented by Haddock and Berlin (1950). In this case a severe bilateral optic nerve atrophy was found $3^1/_2$ years after extensive damage to both occipital lobes, which Haddock and Berlin considered to be due to a secondary degeneration of the retinal ganglion cells following an earlier retrograde cell degeneration in the lateral geniculate nucleus.

As yet there has been no systematic study of the effects of age, of the duration of the postoperative survival period or of possible species differences on retrograde transneuronal degeneration in the ganglion cells or in the cells of the inner nuclear layer of the retina. Such a study would seem opportune at this time as in many respects the visual system represents an ideal site in which to investigate the cellular changes which occur during retrograde transneuronal degeneration and also the changes in the axon terminals of the cells undergoing the secondary degeneration.

3. The Avian Homologue of the Edinger-Westphal Nucleus

The only other well-documented example of a retrograde transneuronal degeneration is the finding of a severe cell loss in the accessory oculomotor nucleus (which is generally thought to be the avian homologue of the Edinger-Westphal nucleus, i.e., the nucleus of origin of the preganglionic fibers to the ciliary ganglion) in young chicks some weeks after eye removal (Cowan and Wenger, 1968). In the same study the effects of early removal of the optic vesicle upon the development of accessory oculomotor nucleus were investigated, and the results of this part of the study will be considered in some detail because they represent the first description of what may be regarded as a form of retrograde transneuronal degeneration in a developing system.

The ciliary ganglion being formed from the neural crest is not directly affected by early removal of the optic vesicle (Levi-Montalcini and Amprino, 1947; Hammond and Yntema, 1958). However, by the 7th or 8th day of incubation clear degenerative changes can be seen in the ciliary ganglion on the side of the optic vesicle removal and these progress so rapidly that by about the 12th day of incubation the ganglion shows an almost complete cell loss (Amprino, 1943; Cowan and Wenger, 1968). Secondary to the degeneration in the ciliary ganglion clear degenerative changes are observed in the accessory oculomotor nucleus. This degeneration begins at about the 10th day of incubation and proceeds over the next 8 to 10 days to an atrophy of virtually all the cells in the nucleus. The degeneration in the accessory oculomotor nucleus is considerably more marked after early removal of the optic vesicle than after eye enucleation in newly hatched chicks but whether this is due to the formation of collateral connections during the later stages of development or to some other factor is not known.

The primary degeneration of the cells in the ciliary ganglion in these experiments, although commonly referred to in the neuroembryological literature as a retrograde degeneration, is clearly an example of the failure of maintenance of differentiated neuroblasts in the absence of their projection field (cf Hamburger and Levi-Montalcini, 1950; Hess, 1957a). The changes in the accessory oculomotor nucleus on the other hand appear to be a true transneuronal degeneration since there is evidence that the axons from the nucleus reach the ciliary ganglion several days before the cells begin to show degenerative changes. It is not known whether comparable changes occur in other systems during embryonic development; this seems likely, but it should be emphasized that the synapses in the avian ciliary ganglion are somewhat unusual and the degenerative changes in the nucleus of origin of its preganglionic fibers may be quite unique.

4. The Motor Cortex

In 1905 Campbell reported finding a degeneration of the pyramidal cells in the precentral gyrus in the brains of 5 patients in whom amputations of the leg, arm or hand had been carried out some years previously. The simplest explanation for this observation is that the changes in the cortex were due to a retrograde transneuronal degeneration secondary to the retrograde reaction of the cells in the anterior horn of the spinal cord whose axons had been divided by the amputation. It is interesting to note that the localization of the cortical degeneration corresponded rather closely with what was known of the functional organization of the motor area, and Campbell's findings may be regarded as the first clear use of retrograde transneuronal degeneration to determine connections within the central nervous system. Unfortunately, these observations do not appear to have been followed up and Campbell's findings have generally been overlooked or neglected.

5. The Inferior Olive

Greenfield (1958) has suggested that the degeneration seen in the inferior olive in cases of heat stroke may be a form of retrograde transneuronal degeneration since it appears to occur secondarily to the destruction of the Purkinje cells of the cerebellum. However, it is difficult in pathological material of this kind to exclude that the terminals of the olivary cells are themselves not involved by the primary lesion and, for the present, judgment as to the nature of the changes in the olive in this condition should be reserved. But it is interesting to note that in a recent electron microscopical study Herndon (1968) described clear degenerative changes in the mossy fiber terminals in the cerebellar cortex a few days after the destruction of the granule cells produced by the administration of a sulphur-containing heterocyclic hydrocarbon, Thiophen. This observation is of considerable interest since it is the first unequivocal demonstration of a change in a presynaptic axon terminal consequent upon the degeneration of the postsynaptic element with which it is in contact.

Some General Conclusions on Retrograde Transneuronal Degeneration

The limited number of studies on retrograde transneuronal degeneration makes it rather difficult to generalize about its distribution or about the necessary conditions for its occurrence. However, there are a few features which appear to be common to those systems in which it is best known and which have been most carefully investigated, and these may provide some clue to the etiological factors which may be implicated in this form of transneuronal degeneration.

Perhaps the most interesting feature common to the three areas in which retrograde transneuronal degeneration has been unequivocally domonstrated is that they are all components of what may be termed "neuronally-closed" systems, in the sense that they are located in neuronal pathways which have rather limited connections with other parts of the nervous system. The medial mammillary and ventral tegmental nuclei, for example, are closely interrelated by way of the mammillo-tegmental tract and the mammillary peduncle (Cowan, Guillery and Powell, 1964) and, as far as is known, their only other major connections are with the anterior thalamic nuclei and the hippocampus which themselves are closely related by way of the limbic cortex and the fornix (see Raisman, Cowan and Powell, 1966). Similarly, there is a remarkably close topographic relationship between the retina, the lateral geniculate nucleus and the visual cortex, and recent work has shown that there is a similarly organized reciprocal connection from the visual cortex to the lateral geniculate nucleus (see Guillery, 1967a; Garey, Jones and Powell, 1968). Apart from its retinal and cortical connections the lateral geniculate nucleus is not known to receive an input from any other source. The retinal ganglion cells probably receive no other afferents than their input from the bipolar cells but they do, of course, project to other areas in addition to the lateral geniculate nucleus. The accessory oculomotor nucleus in the avian brain is probably the only source of preganglionic fibers to the ciliary ganglion and although its afferent connections are not known it is likely (by analogy with the mammalian Edinger-Westphal nucleus) that its major input is indirectly from the retina.

The second feature which seems to be significant is the very severe degeneration which occurs in the first order neurons. Generally this retrograde degeneration takes the form of a virtually complete cell loss which effectively deprives the second order neurons of their entire projection field. To what extent the restricted efferent projection of the first order neurons is responsible for the severity of the retrograde degeneration is not known for certain but it would be surprising if this were not an important factor, and the fact that in most cases the critical lesion also deprives these cells of one of their principal sources of afferents is no doubt also significant. In the special case of the secondary retrograde transneuronal degeneration in the ventral tegmental nucleus the initial retrograde cell degeneration in the anterior thalamic nuclei and the primary retrograde transneuronal degeneration in the medial mammillary nucleus are equally severe, and here, certainly, the death of the cells in the medial mammillary nucleus would deprive the neurons in the ventral tegmental nucleus of a considerable number of their afferents.

Closely related to this second point is the finding that retrograde transneuronal changes are always more severe in young animals and, within certain limits, the degeneration becomes progressively more marked as the duration of the postoperative survival period is increased. The significance of the age of the animal at the time of the causative lesions has already been discussed under the anterograde transneuronal degenerations and it is likely that the same factors are responsible for the greater sensitivity of young and immature neurons to the removal (by degeneration) of their projection field. The cellular changes in retrograde transneuronal degeneration seem at first sight to be consistently more severe than in the anterograde degenerations. However, this is probably more apparent than real, because the majority of the studies of retrograde transneuronal degeneration have been carried out on very young animals and for this reason are not comparable to most of the work on the effects of deafferentation. If one compares Torvik's (1956) findings on anterograde degeneration in the inferior olive and pontine nuclei of young kittens with the retrograde transneuronal changes seen in the medial mammillary and ventral tegmental nuclei, it is apparent that the maximum severity of the two degenerations is essentially the same. The fact that cell loss occurs comparatively early in retrograde transneuronal degeneration in young animals obviously sets a limit upon the effect of the duration of the postoperative survival period, but in older animals, when the cellular reaction is less marked, the severity of the degeneration clearly increases with time.

If, as Bleier (1969) has suggested, the increasing resistance of older animals to retrograde transneuronal effects is due to the continuing development of axon collaterals and alternative synaptic connections, it would seem that one of the most significant factors determining whether or not cells are susceptible to this form of degeneration is the extent of their axonal ramifications. In this context it is interesting to note that in the two best-documented examples of this degeneration there are approximately the same number of neurons in the nuclei undergoing the primary and secondary degenerations (i.e., in the anterior thalamic and medial mammillary nuclei, and in the lateral geniculate nucleus and the retinal ganglion cell layer).

As yet, little or no attention has been paid to the sequence of changes which occur in the axons of cells undergoing retrograde transneuronal degeneration or to changes in the synapses upon cells during acute retrograde degeneration. The few light microscopical studies of this problem have yielded rather equivocal results (Barnard, 1940; Barr, 1940; Schadewald, 1941, 1942) and to date the only electron microscopical observations of early retrograde transneuronal changes in synaptic endings are those of Herndon (1968) on the mossy fibers of the cerebellum. Unfortunately, it is not known in this case whether the cells of origin of these affected synapses undergo a transneuronal change and it would be extremely interesting to examine the synapses in the anterior thalamic nuclei or the internal plexiform layer of the retina after lesions of the limbic cortex and optic nerve, respectively. There have similarly been no cytochemical or functional studies of cells undergoing retrograde transneuronal degeneration and for this reason speculation about the causative factors involved in this secondary degeneration are clearly premature at this time.

Discussion

Dr. Voneida: There are so many questions that arise after a paper like this that one can choose only a few among them. Your picture of retrograde changes in the preganglionic sympathetic cells of the chick brought to mind the interesting observation that following chiasm section in cats and also in primates there is an extreme dilation of the pupils. Caulfield, Robinson and I (1968) have recently reported a longer-term study of this dilation. With time, the dilation recedes and reflex pupilloconstriction reappears, although it never reaches normal levels. The explanation of this has been a real dilemma. I am wondering if the stage of extreme dilation may not correspond to a very early stage of retrograde cell change, with gradual recovery later on.

Unidentified Speaker: The common explanation for this sort of phenomenon is that it is a form of denervation hypersensitivity, and this is also one of the explanation given for the transneuronal cell changes. When you try to find real evidence in support of your theory, you will find it completely lacking.

Dr. Voneida: As a second comment, I want to say that the information contained in Dr. Cowan's paper could be of great importance to behavioral physiologists. It seems quite possible to me that the behavioral effects of brain operations may result not only from the primary lesion, but also from more or less remote retrograde and transneuronal changes. This aspect of the brain-behavior relationship has been too long neglected.

Dr. Van Buren: I was waiting with particular anticipation for Dr. Cowan's presentation. In our studies of the cerebral thalamic relationships in man we are dealing with periods of degeneration rarely under six months, and often as long as 15—20 years. This circumstance brings up difficulties in interpretation. We have, for example, seen cases in which cerebral damage was associated with pronounced cell changes in such thalamic cell groups as the nucleus limitans, the reticular nucleus, the intralaminar nuclei, and I believe also in the periventricular midline regions. (These last-mentioned cell groups I have found difficult to delineate in man.) Dr. Cowan's account gave me the feeling that what I have seen in these outlying cell territories may not have been imaginary, even though it is to a large extent in conflict with the classical papers. Dr. Cowan's findings make me wonder how many synaptic leaps away from the cortex our cell changes were taking place.

Dr. Cowan: There are reasons for thinking that some of the cellular changes seen in the thalamic reticular nucleus, and in certain of the intralaminar nuclei, are transneuronal in nature and result from the inevitable deafferentation which these nuclei suffer after large cortical lesions. This interpretation is, however, still open to question: for example, it is possible that the axons originating from these nuclei have cortical collaterals although the evidence for these is rather equivocal. In the case of the intralaminar nuclei, axonal degeneration studies have shown that their efferent projection is primarily to the corpus striatum rather than to the cortex. But the point I should like to stress here is that whenever we have reason to believe that a large number of afferents to a nucleus has been removed, one has to consider the prossibility that cellular changes seen in the nucleus may be transneuronal rather than retrograde in character. This is particularly important in thalamocortical relations since cortical lesions not only damage the axon terminals of the related thalamic nuclei, but may also deprive the nuclei of a substantial part of their afferent supply.

Dr. Ramón-Moliner: You mentioned the work of Matthews and Powell, and that of Jones and Thomas, who reported either a decrease in the stainability of the dendrites following transneuronal degeneration, or possibly an actual decrease in their degree of ramification. I think that, in this context, one should also mention the opposite findings by Kuypers and Türk who reported an apparent increase in the richness of the dendritic ramifications in the nuclei of the dorsal funiculus following section of the contralateral medial lemniscus at mesencephalic levels. They also reported an increase in the number of dendritic spines, but their photographs suggest that the spiny processes which they observed may not have been spines of the common type.

Dr. Cowan: I think the real interest of both findings is that deafferentation may so radically interfere with the internal economy of a neuron as to affect its impregnability with the Golgi stain.

Dr. Petras: Are you able to distinguish in silver preparations orthograde from retrograde degeneration? For example, if you lesion the cortex and look at the thalamic nucleus, could you identify argyrophilc particles as representing one type or the other?

Dr. Cowan: Dr. Powell at Oxford and I tried to do this a few years ago (1964). There are certain morphological criteria which are helpful in distinguishing retrograde from anterograde axonal degeneration in silver preparations, but I think we are being forced more and more to rely upon electron microscopical findings. If we really want to know whether cortico-thalamic fibers are present, we need to look with the electron microscope for degenerating synapses after cortical lesions. In those cases where the pathways to and from the nucleus are different, the problem is a good deal simpler.

Dr. Guillery: May I comment on this briefly? There is a danger of misinterpretation even if you find degenerating synaptic terminals in the thalamus with the electron microscope, for the reason that the thalamic neurons appear to have a tremendous number of local branches. I don't think you could argue that electron microscopic evidence of terminal degeneration would be any more conclusive than are the pictures of degeneration provided by our silver material. We still would not know whether the degeneration in the thalamus is retrograde or Wallerian.

Dr. Cowan: Yes, I agree. At present, of course, we do not know what may happen to axon collaterals during retrograde cell degeneration, especially at the stage before detectable changes occur in the cell.

Although this is not related to the subject of my paper, perhaps I may mention that there are a number of morphological features which characterize retrograde fiber degeneration: for example, this degeneration appears initially in the form of rather fine dust-like fragmentation throughout the area showing retrograde cell degeneration or chromatolysis. At a later stage one may see very coarse silver deposits which appear to be due to the fragmentation of the axons of the degenerated cells. In the thalamus, to cite one example, such coarse fragments can be traced into the thalamo-cortical radiations and they always appear some time after the appearance of cell changes in Nissl preparations.

Dr. Blackstad: Was not Dr. Guillery a bit pessimistic? (Laughter!) Anterograde axon degeneration starts very early after the neuron is damaged. In many mammalian systems, it is very clear-cut after two days, and it begins to become discernible even before that. Would a retrograde reaction make its appearance so early?

Dr. Guillery: In some systems you can see a retrograde cell change in three days (Chow and Dewson, 1966). In studying a system which has a lot of local collaterals, the interpretation of axon degeneration seen with the electron microscope should be very carefully correlated with the changes seen in Nissl preparations. If you are looking at a system where the first cell change is seen at fourteen days, you can safely regard axonal changes that occur after three or four days as anterograde. In a system where you see earlier cell changes, you should be concerned in case some of the degenerating boutons are collaterals of cells undergoing retrograde change.

Dr. Karten: In regard to the point Dr. Guillery is raising, I think we certainly should take into consideration the recent findings of Kreutzberg who has found a marked glial proliferation in the facial motor nucleus as early as 24-72 hours after severing the corresponding facial nerve.

Dr. Myers: In my own studies with the Nauta technique, I too have been concerned with the distinction between Wallerian and retrograde axon changes. I convinced myself fairly early that the Nauta technique only stained Wallerian degeneration, at least with the survival times used. In the monkey's visual system, the geniculo-calcarine fibers are sharply separated from the cortico-thalamic fibers in the white matter of the hemisphere. The geniculo-calcarine course in the external, the latter in the internal sagittal stratum. In all cases of lesion in the visual cortex, I found Nauta fiber degeneration exclusively distributed in the internal sagittal stratum. Likewise, thalamic lesions produced degeneration only in the external sagittal stratum. This evidence was lacking in the circumstance of complete segregation of fibers, for Nauta staining of retrograde fiber changes, at least when looked at during the first two weeks after the lesion. Evidence was also lacking for retrograde dendrite changes in the thalamus that could be compared to Dr. Grant's findings in immature animals. All considered, I felt rather confident that the axon degeneration traced from cortex lesions to the thalamus in our studies represented cortico-thalamic fibers.

Dr. Guillery: I think all of us who have described cortico-thalamic fibers would like to share that confidence. (Laughter!).

Dr. Petras: Dr. Myers' comment raises the question whether retrograde changes such as the ones described in very young animals by Dr. Grant ever appear in silver-stained material from adult animals. Or would they perhaps take longer to develop in the adult nervous system? Is it possible that both Dr. Grant's and your own studies may have given such clear-cut evidence because they were done in very young animals?

Dr. Cowan: The work of Dr. Powell and myself on retrograde fiber degeneration, to which I referred earlier, was done on adult rabbits and I believe that the slide Dr. Nauta showed of retrograde changes in the thalamus was from an adult rat. So I don't think the age of the animal is particularly critical. The disturbing thing is that retrograde cell changes may be evident very much earlier than most of us think, in some cases as early as 3—4 days following the lesion.

Dr. Konigsmark: I would like to return to a point that Dr. Van Buren made, namely the application of retrograde and transneuronal degeneration methods to the human brain. We found in the ventral cochlear nucleus shrunken neurons much like those shown by Dr. Cowan. I think the study of transneuronal and retrograde changes could have a very wide application in the study of anatomical connections in the human brain with natural lesions in various locales.

Autoradiographic Methods and Principles for Study of the Nervous System with Thymidine-H^3

RICHARD L. SIDMAN

Introduction

The history of autoradiography as a method is closely interwoven with the idea of radioactivity itself. Niepce de St. Victor (1867) and Henri Becquerel (1896) made the first autoradiograms when they showed that uranium salts altered the emulsion in an adjacent photographic plate, though they had formulated the experimental design on the premise that light mediated the response. In 1898 Marie Curie reinterpreted Becquerel's observations and added the measurements that established the concept of radioactivity (Romer, 1955). Subsequent applications to biological problems followed basically the same technical approach, but histological (microscopic) autoradiography became practical only after several more decades, with the development of methods for maintaining accurate alignment of a thin slice of tissue and a photographic film through the many steps of staining and development, so that a radioactive source could be localized in the tissue by direct microscopy (e.g., Bélanger and Leblond, 1946). The subsequent development of tritium as a tracer, of a range of fine-grained precision emulsions, and of refinements in the methods for applying them to tissue specimens has been reviewed by Caro (1964, 1966), Bélanger (1965), Salpeter (1966), Stevens (1966), and Rogers (1967).

The outstanding methodological problems of autoradiography today are those of histochemistry in general: identification of the substance to be localized, accurate preservation of its position in the tissue during fixation and the subsequent steps of the histological process, and precise localization. Radioactive atoms are relatively easy to detect, since their emitted β-particles produce a latent image in the emulsion, but there may be great difficulty in identification of the chemical species which contain the radioactive atom after its reworking by the metabolic machinery of a living specimen. Preservation of the position of the labeled metabolite in the tissue is particularly difficult in the case of water-soluble or lipid-soluble substances and has been given too little attention in general. If it is assumed that position of the metabolite has been maintained, the radioactive source can be localized by light microscopy with a resolution of about 2 μ in the case of C^{14} (Robertson *et al.*, 1959) and about 0.3 μ for H^3 (D. K. Hill, 1962). By electron microscopy, a resolution with H^3 of about 0.1 μ is attainable (Caro, 1962; Moses, 1964; Bachmann and Salpeter, 1965). These histochemical issues largely determine the types of questions that can be studied profitably by autoradiography.

Applications of autoradiography to the nervous system specifically have come mainly within the past decade, particularly in relation to the tracing of the life history of cells after they have incorporated labeled thymidine into deoxyribonucleic acid (DNA). During the previous decade it had become clear that exogenous thymidine served almost exclusively as a precursor of DNA (Reichard and Estborn, 1951; Friedkin *et al.*, 1956) so that the problem of defining the labeled metabolites, alluded to above, was largely circumvented in this instance. Next, Hughes and his colleagues (1958) synthesized tritium-labeled thymidine of high specific activity as a potential chemotherapeutic agent against cancer and thus incidentally gave impetus to the autoradiographic study of cell turnover in a variety of tissues. Within another year, Sauer (1959) and Sauer and Walker (1959) in the chick and Sidman and Miale (1959) and Sidman *et al.* (1959a and b) in the mouse independently had recognized the value of the method for analysis of cell behavior in the embryonic nervous system. Leblond and his colleagues promptly applied it to analysis of cell turnover in the brains of postnatal animals (Amano *et al.*, 1959; Leblond *et al.*, 1959).

This review will deal mainly with the methodological principles and limitations pertaining to the light microscopic study of neurogenesis with tritiated thymidine, though many of the principles apply to the use of other metabolites as well. The form of the presentation will follow the design of an *in vivo* experiment, from initial injection of labeled metabolite to interpretation of data.

Technological Considerations

Tritiated thymidine is most commonly provided commercially with the label in the methyl group, at a Specific Activity of 0.36 c/mmole to above 10 c/mmole and at a concentration of 1 mc/ml. Since it withstands 120° C for 1 hour without measurable decomposition or loss of H^3, it can be sterilized by autoclaving (Crowter *et al.*, 1960). Self radiation causes decomposition at a rate of 0.5 to 1.0% per month. The rate is greater at $-20°$ than at 0° C, though it is much reduced at $-196°$ C (Apelgot *et al.*, 1964). Therefore storage in a refrigerator is preferable to the usual deep freeze.

Intravenous injection is the preferred route of administration of thymidine-H^3. Incorporation into a nuclear fraction insoluble in most histological fixatives, largely DNA, presumably proceeds at a maximum rate initially, when the circulating concentration of substrate is highest, and reaches completion within 40 to 60 minutes (Quastler and Sherman, 1959; Koburg and Maurer, 1962). In the mouse 95% of a single injected dose has been cleared from the circulation within 25 minutes (Staroscik *et al.*, 1964), though phosphorylated derivatives can remain available for 16 to 32 hours (Moffat and Pelc, 1966). After 4 intravenous injections of 4 μc/g at 6 hour intervals in adult mice, circulating thymidine-H^3 is reduced to less than 0.005 μc/ml 24 hours after the last injection, although total tritium counts in plasma are still considerable (Konigsmark and Sidman, 1963).

Pelc and Appleton (1965) examined the distribution of water-soluble metabolites of thymidine-H^3 after a single injection in adult mice and demonstrated that almost all cells show nuclear and cytoplasmic labeling during the first hour. Counts were nearly equal to background at 24 hours except over cells with labeled DNA. Clearly thymidine-H^3 is available to many more cells than incorporate it into DNA and is turned over rapidly except in those cells which retain virtually all of it for DNA synthesis. Since thymidine-H^3 is available for so short a time, the route of injection influences the tissue distribution and the absolute grain counts; the counts are more uniformly distributed after intravenous than after intraperitoneal injection. With reference to the nervous system, they found that after intraperitoneal injection the counts were lower over brain cells than over cells of any other organ examined. The effectiveness of intraperitoneal and intramuscular routes has been compared by Skougaard and Stewart (1967).

A number of techniques have been described for the intravenous injection of small laboratory animals such as adult mice (e.g., Feder, 1958; Ronai, 1966; Billings, 1967; Mylrea and Albrecht, 1967; Champlin and McGill, 1967; Virolainen, 1968). Pre-weanling mice are injected intraperitoneally. Thymidine-H^3 reaches mouse embryos effectively following intravenous or intraperitoneal injection of the mother from the 12th day of gestation onward (the day a plug is found is counted as day 0). The labeling of the embryos is distinctly weaker, though sufficient for most purposes, on the 10th and 11th days of gestation; either the contacts between maternal circulation and fetus are less adequate or the pool of DNA precursors is larger in cells of the younger embryos. On the 9th day of gestation, thymidine-H^3 may be

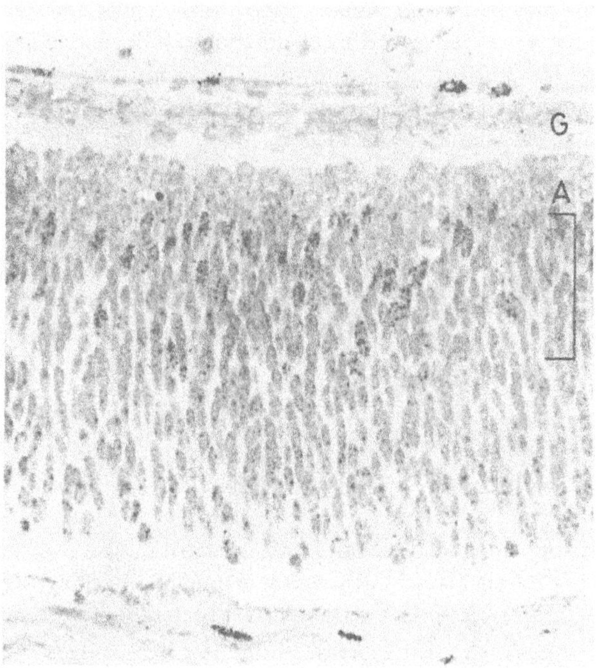

Fig. 1. Retina of a 2-day-old mouse injected subcutaneously with thymidine-H^3 (5 μc/g body weight) and killed 3 hours later. The vitreal surface, at the top, is lined with vascular endothelial cells and fibroblasts, several of which are heavily labeled. The same is true of choroidal cells at the bottom of the picture. Within the retina, ganglion cells (*G*) and amacrine cells (*A*) are unlabeled. Nuclei of bipolar and photoreceptor precursor cells lying just external to the amacrine cells are labeled (bracket), but more weakly than the non-neural cells. Some labeled retinal nuclei are moving toward the external retinal surface, where they will divide. Tissue was fixed in 10% acrolein, embedded in glycol methacrylate, sectioned at 2.5 μ, exposed autoradiographically for 3 weeks, and stained with toluidine blue. 275×

injected directly into the embryo through the uterine wall with a 30 gauge hypodermic needle (Pierce, 1967).

We usually inject thymidine-H^3 at a dose of 5 μc/g body weight, whatever the route. This dose is 2 to 10 times higher than the ones commonly used in the study

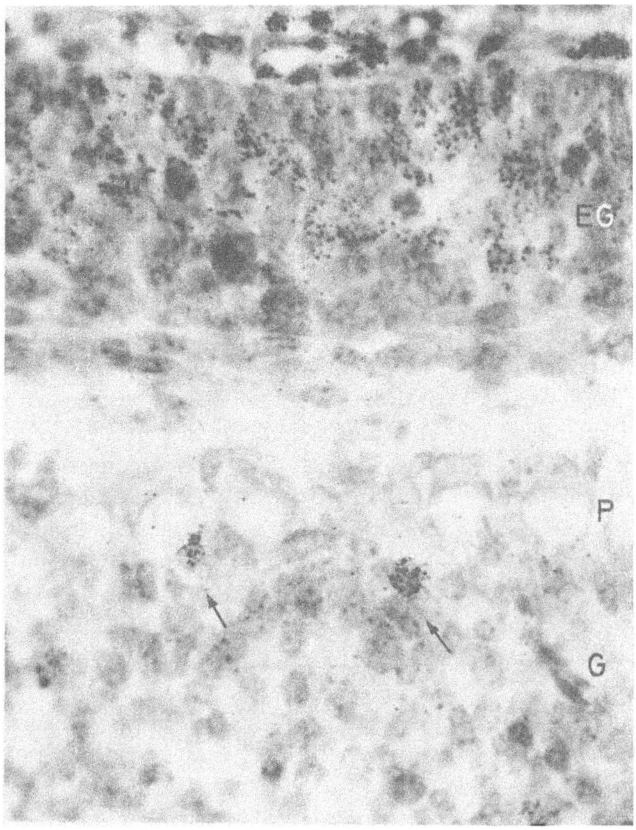

Fig. 2. Cerebellar cortex of an 8-day-old mouse injected intraperitoneally with thymidine-H^3 (5 μc/g body weight) and killed one hour later. The pia-arachnoid, containing labeled cells runs across the top. The outer two thirds of the external granular layer (*EG*) contains numerous labeled neuron precursors. The large Purkinje cells (*P*) are unlabeled, but heavily labeled glial cells (arrows) lie close to them in the outer part of the granular layer (*G*). Most of the glial precursors are more heavily labeled than most of the neuron precursors. Tissue was fixed in 10% acrolein, embedded in polyester wax, sectioned at 8 μ, exposed autoradiographically for 3 weeks, and stained with toluidine blue. 570×

of other organs than the nervous system, but the reason is that the brain cells typically show lower grain counts. Fig. 1 illustrates the contrast in labeling intensity between retinal neuron precursors and extraretinal cells in an eye of a 2-day old mouse killed 3 hours after intraperitoneal injection of thymidine-H^3. The grain counts over the nuclei of 50 neuron precursors showed a mean of 7.7 grains with a standard error of ± 0.6, compared to counts over choroidal and vitreal fibroblasts and endothelial cells in the same sections of 31.7 ± 2.2. The difference may reflect in part local differences in rate of exchange of thymidine-H^3 between plasma and cells, the importance of which was emphasized by Pelc and Appleton (1965), but proliferating cells in the avascular cornea and lens typically show higher counts than retinal or brain neuron precursors. Differences in rate of DNA synthesis would influence the counts, but our impression is that glial precursors often label more heavily than nearby neuron precursors (Fig. 2), even though they probably do not synthesize DNA faster (see pages 9—10 below). Differences in concentration of DNA in nuclei of different sizes cannot account for the data, for small neuron precursors with high concentrations of DNA often are labeled more weakly than cells with larger nuclei (Figs. 1 and 2). By elimination, the size of the precursor pools, the relative activities of synthetic versus degradative enzymes, or blood-brain barrier phenomena must influence the extent of incorporation of exogenous thymidine into DNA.

Thymidine-H^3 is incorporated by cells in developing nervous tissues in culture (Sidman, 1963) and in supravital preparations of embryonic mouse or human tissues (Rakic and Sidman, 1968a).

If the precursor pools and enzyme activities do influence uptake of thymidine-H^3, one might anticipate that the incorporation by some cells might be subthreshold. A striking instance was found by Adelstein *et al.* (1964), who showed that dividing cells of some rodent species fail to become labeled after exposure to the usual concentrations of thymidine-H^3. The cause is high activity of a pyrimidine nucleoside degrading enzyme (Adelstein and Lyman, 1968). On examination of the nervous system in neonatal animals of the same species, we have found generalized cytoplasmic and nuclear labeling at first variable from region to region (Figs. 3A and 3B), followed by persistent labeling of some nuclei (Fig. 3C). The general lesson is that cells may replicate DNA of the conventional base compositions without incorporating exogenous thymidine on the standard time schedule, and it remains to be learned whether any cell populations in species such as mouse or man behave similarly at any stage of life.

The relatively high dose of thymidine-H^3 used in studies of neurogenesis *in vivo* raises the issue of toxicity. Doses of 1 to 2 μc/g body weight cause little radiation effect (Cronkite, 1959; Mendelsohn, 1960; Smith *et al.*, 1962), but spermatozoa were damaged at 5 μc/g (Johnson and Cronkite, 1959; Bateman and Chandler, 1962) or 50 μc/g (Kisieleski *et al.*, 1964). Other toxic effects were recorded in whole animal experiments by Grisham (1960), Post and Hoffman (1961), Baserga *et al.* (1962, 1966), and Johnson and Cronkite (1967). By contrast, Altman (1966) recorded no toxic consequences after injecting 10 μc/g in long term experiments in young rats, and Fujita (1967) gave an injection of 10 μc/g followed by 1 μc/g every 5 or 12 hours for up to 20 days to young mice, with apparent impunity. Pierce (1967 and unpublished observations) has injected 0.01 mc of thymidine-H^3 in 0.01 ml directly into individual 9-day mouse embryos and has observed no malformations or impairment of survival in several litters; DNA in various classes of neurons became effectively labeled in these mice. The most extensive study on the effects of chronic exposure to tritium during embryonic life was carried out by Fliedner *et al.* (1968). They injected thymidine-H^3 at a dose of 0.4 μc/g body weight intraperitoneally into pregnant rats every 6 hours from the 9th day of gestation to birth and measured the subsequent growth of the offspring, their breeding performances, and the general health of the F2 generation. No abnormalities in these measurements were detected, no malformations, and no increase in frequency of chromosomal abnormalities. Thus,

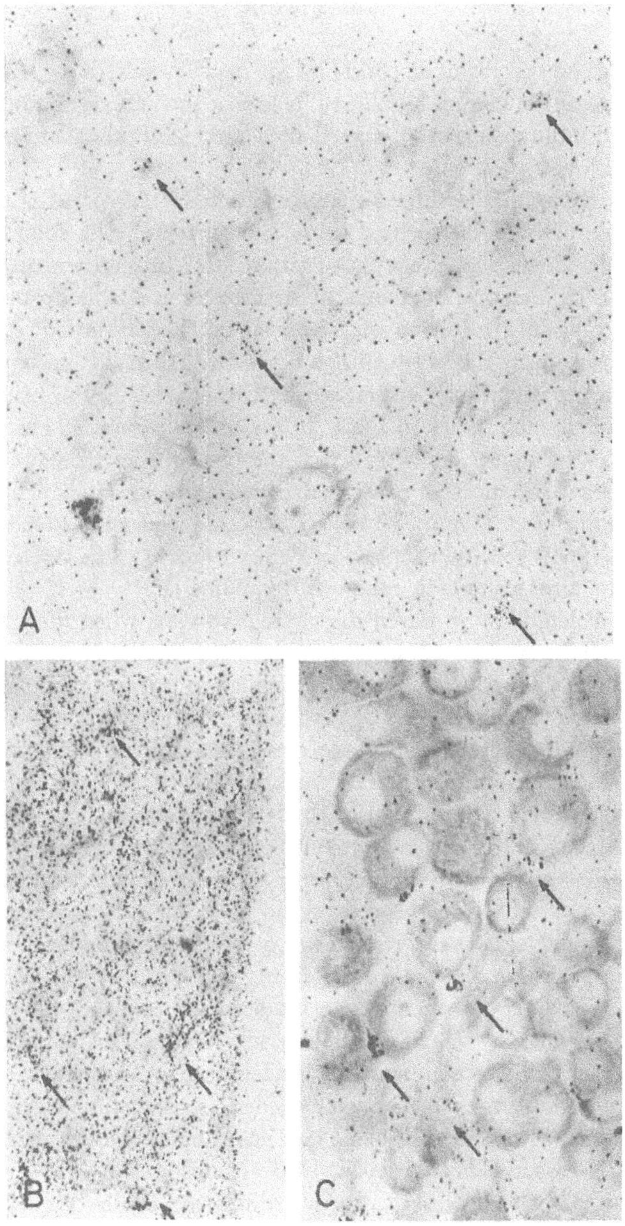

Fig. 3. A. Ventral horn of spinal cord in a 4-day-old ground squirrel, Citellus tridecemlineatus, injected intraperitoneally with thymidine-H^3 (5 μc/g body weight) and killed one hour later. The silver grains are diffusely distributed over neuron cytoplasm and neuropil, and appear to be more concentrated over the nuclei of a few small cells (arrows). B. Dorsal root ganglion at the same spinal level in the same animal as in Fig. 3A. Labeling is diffuse and much more intense in general, but is particularly concentrated in nuclei of a few small satellite Schwann cells (arrows). C. Dorsal root ganglion of a littermate ground squirrel, also injected on postnatal day 4, but killed on day 16. Most of the persistent label is concentrated in nuclei of occasional satellite Schwann cells (arrows). Tissues were fixed in 10% acrolein, embedded in polyester wax, sectioned at 8 μ, exposed autoradiographically for 24 weeks, and stained with toluidine blue. ×275

in several studies no specific toxic effects on the nervous system have been recorded, but the doses in current use are high enough so that such effects should be sought as more subtle and more quantitative aspects of neurogenesis are analyzed (e.g., Hicks and d'Amato, 1968).

While it is commonly assumed that thymidine-H^3 serves as a pulse label in mammals on the basis of the incorporation and excretion data summarized above, there is ample evidence for either reutilization after breakdown of labeled DNA from dying cells or retention of thymidine derivatives in a labile form normally lost during the histological processing of the specimen (Robinson and Brecher, 1963). The salvage mechanism may be utilized normally, especially in sequestered sites where cell turnover is extensive (discussed by Cleaver, 1967). For example, in bone marrow 35% of thymine bases in DNA of erythroid precursors are used for local DNA synthesis after nuclei are extruded at the orthochromatic stage (Feinendegen *et al.*, 1966). Other instances of presumed reutilization are documented by Bryant (1962), Diderholm *et al.* (1962), M. Hill (1962), and Galassi (1967). In the careful study of Steel (1966), cells of a rapidly growing tumor became labeled between the 4th and 8th days after a single injection of thymidine-H^3.

Moffat and Pelc (1966) proposed the alternative view when they prepared autoradiograms of cryostat sections so as to retain water-soluble labeled metabolites. They found nuclei to be more heavily labeled than in conventional paraffin sections at 2 to 19 hours after a single dose of thymidine-H^3 and interpreted this to indicate retention of thymidine or its phosphate esters in a soluble form which does not diffuse through the nuclear membrane. They plucked hairs from the mice just prior to the single injection and found that cells of the hair bulb became labeled only after an interval of 16 to 32 hours. Their conclusion was that plucking of hairs blocked DNA synthesis for about a day by inhibition of the phosphorylating enzymes or of DNA polymerase, after which the cells could use soluble metabolites that had ben present all the time. Neither reutilization nor delayed utilization has been demonstrated in the nervous system, but the issue is clearly important, especially with regard to the assigning of cellular "birthdays", to be discussed on pages 10—13 below.

DNA is a very large molecule that is easily rendered insoluble by a range of fixatives. The agent of choice in many studies has been an acidified or neutral alcohol which allows nuclear images to be crisply stained, is compatible with the autoradiographic emulsions (Ficq, 1959; Rogers, 1967), and permits differential extraction of DNA and RNA (Baserga and Kisielski, 1963; Jacobson *et al.*, 1968). However, the nervous system is not well preserved by such fixatives. We have preferred to fix by perfusion or immersion with 10% acrolein in 0.85% saline, a good method for morphological purposes and for retention of the molecular shapes of nucleic acids (e.g., Feder and Wolf, 1965). Other aldehyde fixatives commonly used in neurological studies, such as formalin or glutaraldehyde, are suitable. A note of caution is offered by Peters and Ashley (1967), who point out that labeled free amino acids may be bound to tissue components after exposure to certain aldehyde and other fixatives and can be wrongly interpreted as labeled protein. Labeled thymidine or its metabolites may be similarly retained, for cytoplasmic labeling is sometimes observed after use of acrolein or glutaraldehyde fixatives. Other explanations are possible, such as incorporation of label into mitochondrial DNA or breakdown of methyl-H^3 thymidine by bacteria contaminating the stock solution, followed by synthesis of labeled amino acids and eventually of cytoplasmic protein (Cleaver, 1967). Another note of caution is that fixatives containing mercury or chromium salts may drastically alter the sensitivity of autoradiographic emulsions.

Choice of embedding material is not critical for autoradiography, provided that sections of good quality can be cut. Clusters of silver grains are likely to appear wherever there are discontinuities in the tissue, such as along the luminal edges of blood vessels or along edges of shrinkage artifacts or tears. These "false positives" are rarely confusing in studies with tritiated thymidine, where one's attention is focused on cell nuclei, but they pose a problem with other labeled metabolites, particularly if frozen sections must be used. For anatomically-oriented autoradio-

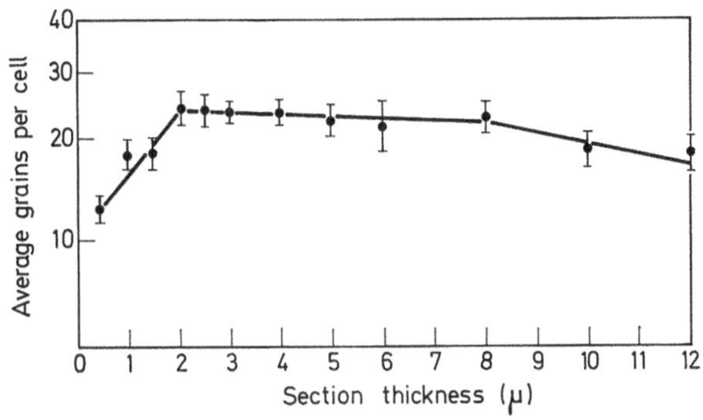

Fig. 4. A semi-logarithmic plot showing changes in grain counts per cell as a function of section thickness. Data for each point were obtained by counting 50 cells in each of 4 sections through the external granular layer of the cerebellar cortex of a 2-day-old mouse that had been injected with thymidine-H³ one hour before death. Tissue was fixed in 10% acrolein and embedded in glycol methacrylate. Six sections were cut at each thickness through one block of tissue. Each value is given as a range encompassing ±1 Standard Error from the Mean

graphic studies we embed specimens in polyester wax (Sidman *et al.*, 1961) and cut serial sections at 6 or 8 μ; for quantitative studies we embed small specimens in glycol methacrylate (Ashley and Feder, 1966) and cut sections at 2.5 μ.

Section thickness is an important variable. Resolution in autoradiograms is controlled largely by the strength of the β-particle and the thickness and closeness of apposition of the section and the emulsion (Doniach and Pelc, 1950; Caro, 1962; Rogers, 1967). Sensitivity in tritium-detecting autoradiograms also is influenced markedly by section thickness (Fig. 4). The slope of the left limb of the curve and the position of the inflection indicates that 3 μ sections should give more reproducible grain counts than thinner ones. The decreased sensitivity of ultrathin sections (especially when coupled with thin emulsions) is quickly recognized by everyone who undertakes electron microscopic autoradiography, and is one of the major limitations of that method (see discussion by Rogers, 1967).

Staining methods merit brief comment. Of the ones most widely used in neurohistology, some are most suitably applied prior to application of the radiosensitive emulsion and some after (table). We routinely stain our autoradiograms with 0.05% toluidine blue in 0.02 M sodium benzoate-benzoic acid buffer, pH 4.4. A grainy black precipitate of undefined origin occasionally appears and is prevented by immersing slides for 5 minutes before staining them in a solution prepared by adding one volume of concentrated ammonium hydroxide containing 28 to 30% ammonia to 19 volumes of water, followed by a rinse in water for 10 minutes. After the autoradiograms are stained with toluidine blue, they are dehydrated with 95% ethanol, followed by 100% ethanol and xylol. Sections are mounted in Permount. Certain neurological stains decrease the sensitivity of the emulsion so drastically as to be incompatible with autoradiography, for example, the Cajal gold chloride-sublimate method for astrocytes or del Rio Hortega's silver carbonate stain for oligodendroglia and microglia. A suitable barrier to be interposed between such stained sections and the emulsion remains to be devised. Golgi-stained sections, in their usual form, are too thick for autoradiography, though a great need exists for correlation of Golgi and autoradiographic data.

Table. *Time of application of neurological stains during autoradiography*

Stain prior to application of emulsion	Stain after development of emulsion
Feulgen	Toluidine blue, cresyl violet, and related dyes
Periodic acid-Schiff	Counterstain for periodic acid-Schiff
Reduced silver methods (Palmgren, Bodian, Holmes)	Luxol fast blue Sudan dyes Hematoxylin and eosin "Triple acid" connective tissue stains

The fixative and stain may modify the sensitivity of the emulsion in less drastic ways than those mentioned above. Such effects can be recognized by careful comparison of grain counts over tissue and off the tissue, in both labeled and unlabeled specimens exposed to the same technological treatment. For example, the counts over tissue might be hardly above background and yet be quite meaningful in an instance where carefully processed unlabeled controls show tissue counts well below background values. Such conditions are rarely met in the case of incorporation of thymidine-H^3 into nuclear DNA, but are most likely to be encountered during studies of a diffusely distributed metabolite that requires unconventional histological processing for its retention.

The many types of radiosensitive emulsions and the methods of applying them have been reviewed in detail (Caro, 1964; Pelc *et al.*, 1965; Stevens, 1966; Rogers, 1967). For light microscopic autoradiography we have used the dipping method of Kopriwa and Leblond (1962) and NTB-2 or NTB-3 emulsion diluted 1:1 with distilled water (NTB-2 emulsion is preferred). De-waxed or plastic sections are dipped from the dry state, while mounted frozen sections are immersed in distilled water at room temperature for 1 hour before they are dipped. We include standard radioactive and nonradioactive slides in every run, and test each new batch of emulsion on receipt; over a period of several years we have rejected about one bottle of emulsion in 10 because of excessive background fog. We assemble sets of microscope slides in plastic slide holders (Lipshaw Mfg. Co.) prior to dipping them, and after dipping we stand them in an oven at 28° C and 50 to 80% humidity (achieved by placing two large pans of water in the oven) for 30 minutes so that the emulsion can drain to a coating of uniform thickness. Then the slides are packaged in plastic slide boxes, sealed with black photographic tape and stored in a Dry Ice chest at $-70°$ to $-75°$ C for the desired exposure period of several days to several months. Some radiosensitive emulsions are subject to latent image fading, which is thought to be reduced by the low temperature and CO_2 atmosphere; avoidance of fading may be achieved also under other conditions of exposure, varying with the emulsion (see, for example, the discussion following Pelc *et al.*, 1965). The sensitivity of at least one emulsion is reduced about 30% at $-75°$ C, compared to $+4°$ C (Pelc *et al.*, 1965). The conditions for optimum "signal to noise" ratio should be worked out by each investigator for his own material.

Interpretation of Data on Thymidine-H^3_I Incorporation

Grain Counts

If all labeled cells in a given specimen behaved alike, the grain counts would be expected to show a Poisson distribution because of the natural randomness of radioactive decay. Animal cells exposed to thymidine-H^3 almost always display grain counts with greater variation around the mean, so that additional factors must influence the data. Several of these have been mentioned already: thickness of section, emulsion, and the space between them; concentration of precursor that reaches the cells across the vascular and extracellular spaces, and rate of change in precursor concentration; rates of degradation of precursor; size of the pool of endogenous metabolites. Another critical factor is the density of the tissue components which the β-particle traverses in its route between source and emulsion. Maurer and Primbsch (1964) have calculated that "infinite thickness" for tritium is 0.05 g/cm^2, or a path of less than 3 μ through nucleoplasm, 2 μ through cytoplasm, and 0.6 μ through nucleolus. The densities of retinal rod outer segments (and presumably of myelin, which has a similar structure), of mitochondria (measured in the ellipsoid zone of photoreceptor cells), and of glycogen (measured in the paraboloid zone of cone photoreceptors) are such (Sidman, 1957) that "infinite thickness" for tritium would be close to 1 μ. The importance of the inevitable tissue inhomogeneities was emphasized by Pelc and Welton (1967), who calculated that about 6 times more β-particles might emerge toward the emulsion from a squashed and spread tritium-labeled nucleus compared to the same labeled nucleus in interphase in a 4 μ section.

Additional factors increase variability of grain counts. Nuclear size of diploid cells varies greatly in the nervous system, so that the concentrations of labeled DNA, and the resultant grain counts, may be very different for two cells with identical percentages of labeled DNA. Also, the absolute grain counts influence the form taken by the data. For example, in a histogram with a mean grain count per nucleus of about 30, one can ignore cells showing 1 to 5 grains in calculating variance, but if the mean is 10 to 15 grains, it is less easy to separate "true" from "background" counts (Cleaver, 1967). (While it is conventional to ignore cells with fewer than 4 to 5 grains per nucleus, such cells are usually labeled far above background and should receive more attention.) Finally, the accuracy of replication should be measured by each investigator engaged in quantitative studies. Representative data were recorded by Cleaver and Holford (1965), who obtained a standard error of about 10% when replicate slides were processed together, and about 15% when they were processed separately.

Kinetics of Cell Turnover

The single greatest application of autoradiography in the past decade has been to measure the phases of the generation cycle in cells that had been exposed to thymidine-H^3 (summarized in Stohlman, 1959; Lamerton and Fry, 1963; Thrasher, 1966; Cleaver, 1967). We will concentrate on applications to the nervous system. (See also the detailed review by Gracheva, 1967.)

The terminology of Howard and Pelc (1953) was the first to acknowledge that DNA replication precedes cell division, and it remains the one in widest use to

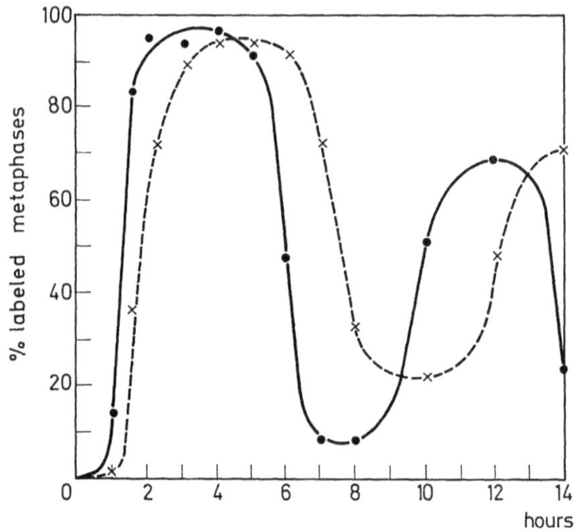

Fig. 5. A plot of the per cent of metaphase figures that are labeled as a function of time after exposure of 10-day (solid line) and 11-day (dashed line) mouse embryos to a single dose of thymidine-H^3. From such curves the timing of the phases of the cell generation cycle can be estimated. (Redrawn from Kauffman, 1968)

designate the phases of the cell generation cycle: S, period of DNA synthesis; G2, the gap between DNA synthesis and mitosis; M, period of mitosis; and G1, the gap between mitosis and DNA synthesis. The most reliable method for measuring the phases of the cycle, short of direct observation in tissue culture combined with autoradiography (e.g., Sisken, 1964), is to expose cells to a pulse dose of thymidine-H^3, sample them at approximately hourly intervals, and record the percentage of mitotic figures that are labeled (Quastler and Sherman, 1959). The advantages are that this method does not depend on the grain counts and does not use the labeling index (percentage of cells labeled immediately after the pulse period), both of which are subject to uncontrollable errors. However, the method is cumbersome, particularly with timed mammalian embryos or diseased animals, which are difficult to obtain.

Curves obtained for the thoracic spinal cord of 10 and 11 day mouse embryos are presented in Fig. 5 (Kauffman, 1968). The time between injection of thymidine-H^3 and appearance of the first labeled mitotic figure theoretically is a measurement of G2; a more practical value to record is the time to the appearance of 50% labeled mitotic figures, equal to $G2+{}^1/_2M$ (1.5 hours for the 10 day embryo), and the interval between corresponding points on two successive cycles provides a measure of the total cycle (8.5 hours for the 10 day embryo). Other values are obtainable by subtraction. The duration of the total cycle, and of G1, S, and G2 was longer in the 11 day than in the 10 day embryo, thought S was about the same (Kauffman, 1968; see also Atlas and Bond, 1965).

A lengthening of the cycle with increasing embryonic age had been found earlier by a different autoradiographic principle (Fujita, 1962). Working with the chick embryo, where the injected thymidine-H^3 remains continuously available, he measured the percentage of cells labeled 45 minutes after injection, and the time when the

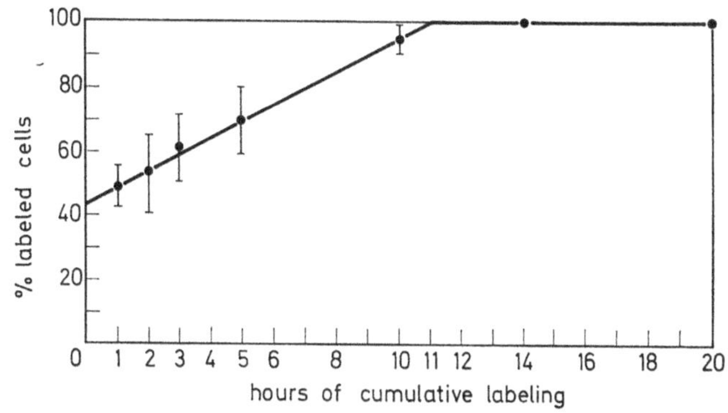

Fig. 6. A plot of the per cent of labeled cells as a function of time, under conditions in which thymidine-H^3 is almost continuously available (cumulative labeling method). See text for calculation of the timing of phases of the cell generation cycle. (Redrawn from Fujita, 1968)

mitotic cells became labeled, the time when all cells of the germinal zone became labeled non-homogeneously, and the time all cells of the germinal zone acquired a homogeneous label. From the measurements he calculated the individual phases of the generation cycle. These and other analytical studies are summarized in Fujita (1966).

Fujita (1967) has applied the cumulative labeling technique in a quantitative analysis of cell proliferation and migration in the external granular layer of the postnatal mouse cerebellum (Fig. 6). When thymidine-H^3 was injected every few hours, a linear increase in the percentage of labeled external granule cells was observed and reached 100% at 11 hours. That time interval represents G2+M+G1, or the total cycle minus S. The labeling index (43%) is proportional to S/total cycle, if one assumes complete asynchrony of the cell population (see, however, Dawson and Field, 1964). From the two equations he calculated the total cycle and S as 19 and 8 hours, respectively. The mitotic index was measured to be 2.6% and is proportional to M/total cycle; from this, M was calculated to be 0.5 hour. G2 was measured as the time for the first labeled metaphase to appear and was found to be 2 hours. G1, by subtraction, was 8.5 hours.

The continuous labeling method as worked out in other cell systems is discussed by Wimber (1963) and Puck (1964).

Most of the proliferative populations in the nervous system are non-homogeneous and the analysis of cell kinetics is more difficult. Asbury (1967) applied the pulse label method to the study of Schwann cell proliferation in the sciatic nerve of the 2-day-old mouse. When the percentage of labeled mitoses was plotted as a function of time after the single injection of thymidine-H^3, the descending limb of the curve never came down to the 50% level, probably because of variability from cell to cell in the durations of S, M, and perhaps G2. Despite this, he was able to estimate S as the interval between the initial rise in the curve (2 hours after injection) and the post-plateau point at which the first unlabeled mitotic figure was found (10 hours after

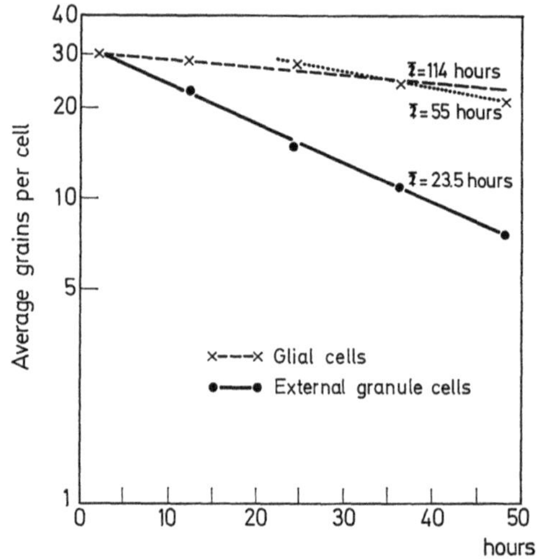

Fig. 7. A semi-logarithmic plot showing the rate of decrease in average grains per cell in the cerebellar cortex as a function of time after a single injection of thymidine-H^3 subcutaneously in a series of 2-day-old mice. Grain counts decrease faster, i.e., the cell generation cycle is shorter, in the neuron precursor population of the external granular layer than in the glial population of cerebellum and adjacent brainstem. Specimens were fixed in 10% acrolein, embedded in glycol methacrylate and sectioned at 2.5 μ. Each point represents the average count of 200 cells

injection). As a further complication, it was suspected that some daughter cells ceased dividing after each cell cycle. Evidence supporting this was obtained when histograms of grain count distributions showed that about 10% of the initially labeled cells retained a heavy label up to 96 hours, the last time point examined. This datum was used as a correction factor in calculating the total generation time from plots of grain count halving times. A value of 24 hours was recorded for the generation time. The estimates of S and the generation time, along with the labeling index, in turn were used to calculate the percentage of Schwann cells proliferating in the sciatic nerve on postnatal day two, according to the equation:

$$\frac{\text{S}}{\text{generation time}} = \frac{\text{labeling index}}{\text{percent of cells proliferating}}$$

A value of 27% was obtained. Or, almost three-quarters of the Schwann cells were not proliferating, and perhaps instead were maturing and making myelin sheaths around axons.

Another example of the usefulness of grain counts following a pulse dose of thymidine-H^3 is illustrated in Fig. 7, based on study of the brainstems of the same mice whose sciatic nerves were analyzed by Asbury (1967). It had not proved feasible to study the kinetics of central glial cells by the accurate method of plotting labeled mitotic figures against time after injection because very few mitotic figures were recognized, labeled or not, except among the external granule cells of the cerebellar cortex (neuron precursors) and in scattered vascular endothelial cells. (See, however, Hommes and Leblond, 1967.) Whatever the mechanics of glial cell division, the grain count halving data (Fig. 7) indicated a generation time of 114 hours (if a straight line was fitted to all the points) or 55 hours (if it was fitted to the last three points). The latter value is probably more accurate, since measurement of the halving time should begin when half the labeled cells have completed their first division (Fried, 1968). Another uncertainty relates to the homogeneity of the glial cell population that became labeled. In any case, glial cell proliferation appears to be quite different from proliferation of the small neuron precursors of the cerebellar cortex that were analyzed in the same sections and were found to proliferate with a generation time of 23.5 hours (Fig. 7).

Another approach to analysis of cell kinetics is via double labeling with thymidine-H^3 and thymidine-C^{14} (Hilscher and Maurer, 1962; Wimber and Quastler, 1963; Davies and Wimber, 1963; Baserga and Lisco, 1963; Bresciani, 1964; Pilgrim and Maurer, 1965; Thrasher, 1966; Wimber and Lamerton, 1966; Song and Tabachnick, 1968; Lala, 1968). This method involves the use of fewer animals to obtain the kinetic data, but has not been applied yet to the nervous system.

The quantitative data reviewed so far have shown that the cells of particular germinal zones in the brain of the immature mouse or chick behave as a single population, and that their generation cycle slows down in the course of development. Other zones contain less homogeneous populations but can be analyzed nonetheless, as shown in the case of the Schwann cells of sciatic nerve. Glia of the central nervous system appear to proliferate differently from neuron precursors.

As more of the analytical facts on cell proliferation are obtained, the problem of control mechanisms will become increasingly germane. On principle one accepts that genetic mechanisms govern cell proliferation, but it is likely that controls may be mediated via nearby cells and even by more distant environmental influences. Only a few autoradiographic studies have been addressed to this issue. DeLong and Sidman (1962) showed that 46% of the glial cells destined to incorporate thymidine-H^3 on postnatal day 3 failed to become labeled in one side of the superior colliculus after the opposite eye had been removed on the day of birth. This decrease in glial proliferation preceded the loss of neurons, which was another consequence of the enucleation. Possible relationships in both directions between postnatal cell proliferation in the nervous system and the way an organism interacts with its environment have been considered speculatively by Altman and his colleagues (Altman and Das, 1964; Altman, 1966).

Determination of Cell "Birthdays"

Until now we have been discussing predominantly methods that yield quantitative data on cell turnover. The student of development also needs autoradiographic data that relate more explicitly to the anatomical organization of the nervous system. Among the most intriguing features of the nervous system are its great range of specialized cell types, the long time span over which its cells arise, the spatial separation between the site of origin of cells and their positions at maturity, and the long post-mitotic life of neurons and of many glial cells. Autoradiography with thymidine-H^3 gives unique information on the times of cell origin and on the migrations that occur during the early and critical stages of differentiation. Preparation of serial sections and the mapping of cell positions in relatively large areas of tissue are called for in such studies, and a given autoradiogram rarely meets both the kinetic and the neuroanatomical objectives. The technical problem of plotting the position and other properties of the labeled cells at magnifications of 200× or more, while at the same time retaining the general topographic features of the specimen as recognized at much lower magnifications, is a common one in neuroanatomical work and is best met with the aid of a good drawing tube (Angevine, 1965) or a pantograph apparatus (e.g., Karten, 1968).

When one plans to decipher the site and time of origin, the routes and rates of migration, and the specialized differentiation of particular types of nerve cells, it proves expedient to begin by determining the "birthdays". This term is defined empirically on the basis of autoradiographic data as the last day on which nuclear DNA is replicated in that cell line. Neurons ordinarily do not divide after the developmental period, so that once the label is incorporated by the parent cell and distributed to the daughters, the neuron can be described as "branded for life" (Angevine, 1965).

It would be expected, then, that if an animal were injected with thymidine-H^3 once during the developmental period and killed as an adult, the heavily labeled cells would be the ones that had been generated on the day of injection, and weakly labeled cells would be progeny of further divisions subsequent to that day. In general this is valid, but injections should be made at a systematic series of developmental stages, all animals to be killed at adulthood, and two criteria should be met to establish birthdays with confidence. First, the analysis should account for all the major cell types of region under study, as has been accomplished for several regions of the mouse nervous system: cerebellum (Miale and Sidman, 1961; Pierce and Sweet, 1967), retina (Sidman, 1961), hippocampal formation (Angevine, 1965), inner ear (Ruben, 1967), cochlear nuclei (Pierce, 1967), and olfactory bulb (Hinds, 1969a). Second, a coherent, changing relationship between lightly and heavily labeled cells should be found as one compares adult specimens that had been exposed to thymidine-H^3 at different developmental stages. A given type of cell would appear to be unlabeled when the thymidine-H^3 had been injected several cell generations prior to its birthday, weakly labeled in specimens that had been injected within a few generations of the birthday, then heavily labeled (defining the birthday), and unlabeled when the injection is made at any subsequent time. The sweep of these phases across each major cell population of the mouse hippocampal formation was vividly presented by Angevine (1965).

In species where the pulse labeling method is difficult to apply, such as the chick *in ovo*, the cumulative labeling method allows determination of cell birthdays (Fujita and Horii, 1963; Fujita, 1964). It serves also as a cumbersome — but very reliable — method in mammals, as exemplified in Fujita's (1967) study of the cerebellar cortex in the mouse. He gave multiple daily injections of thymidine-H^3 during the first two postnatal weeks on a schedule that initiated the injections in some animals on day

1, some on day 2, and so on. Any cell in the cerebellar cortex that was unlabeled at the end of the cumulative period must have arisen prior to the onset of the injections.

Several difficulties interfere with the establishment of cell birthdays. The various geometric factors reviewed on p. 1—8 influence the grain counts, quite apart from the number of cell replications, so that interpretation of "lightly labeled" versus "heavily labeled" cells is uncertain at times. Cells engaged in their final DNA replication but exposed to thymidine-H^3 during only part of their S phase will show misleadingly low grain counts, as will cells exposed for one reason or another (see pages 3—6 above) to an unduly low effective concentration of labeled precursor. Changes in nuclear size arising during differentiation may cause differences in grain counts among cells that initially were labeled alike, and similarities among cells that were initially different or that had multiplied a different number of times after becoming labeled. Finally, if the exposure time is too short or too long, the weakly and heavily labeled populations will merge into one.

Most of these factors decrease the grain counts, so that the nervous system of a mature animal exposed to thymidine-H^3 at a suitable developmental stage may contain cells that had become "branded" on their birthdays, but are labeled lightly nonetheless. The incidence of such cells is easy to determine in a specimen that had been exposed to thymidine-H^3 on the final day of generation, recognized as the last on which injection leads to any labeling of the particular cell class. All radioactive cells in such a specimen had become labeled on their birthdays by definition, regardless of grain count. The frequency of weakly labeled "birthday" cells in specimens exposed to thymidine-H^3 earlier in the generation period is harder to establish. A crude calculation was made of the percentage of cells formed on each day of the generation period in the subcommissural organ (Rakic and Sidman, 1968) and yielded an estimate that was quite at variance with the assessment obtained by the more conventional procedure of recording only the heavily labeled population.

The precision of the assignments of birthdays would be diminished if the label in DNA is reutilized after cell death or if labeled precursors remain available for prolonged periods (see p. 4—5 above). Cell death does occur in the developing nervous system (Hamburger and Levi-Montalcini, 1949; Glücksmann, 1951; Källén, 1965; Saunders and Fallon, 1967; Hughes, 1968). On the basis of some assumptions that are plausible but difficult to prove, Hughes (1961) estimated that in the lateral ventral horns of the S8—S10 segments of the spinal cord in *Xenopus laevis* tadpoles, about 8 or 9 large motor cells degenerated for every one that survived. In the same material, two spinal ganglion neurons were thought to degenerate for each surviving one (Prestige, 1965).

Cell death merits autoradiographic study, both with reference to possible reutilization of DNA components, and as a problem in its own right. Once cells become labeled at the time they are generated, it is feasible to measure the time interval to any subsequent event. Thus, Nordlander and Edwards (1968) injected thymidine-H^3 into developing insects, recorded the subsequent appearance of labeled pyknotic nuclei in the nervous system, and established that many nerve cells survived only 3 or 4 days after they were generated. A similar observation in the early postnatal mouse retina is recorded in Fig. 8. In a somewhat different example, DeLong and Sidman (1962) counted labeled neurons in the superior colliculus after removal of one eye at birth and found degeneration of 40% of the neurons on the deafferented side, most of them disappearing very rapidly and without obvious accompanying signs of pathology between the 4th and 7th days after enucleation.

Reference was made to the possibility that labeled soluble precursors of DNA might persist for more than a day after injection of thymidine-H^3 (Moffat and Pelc, 1966) and cause errors in the determination of cell birthdays, but no direct studies on the nervous system have been carried out. It is quite doubtful that such metabolites

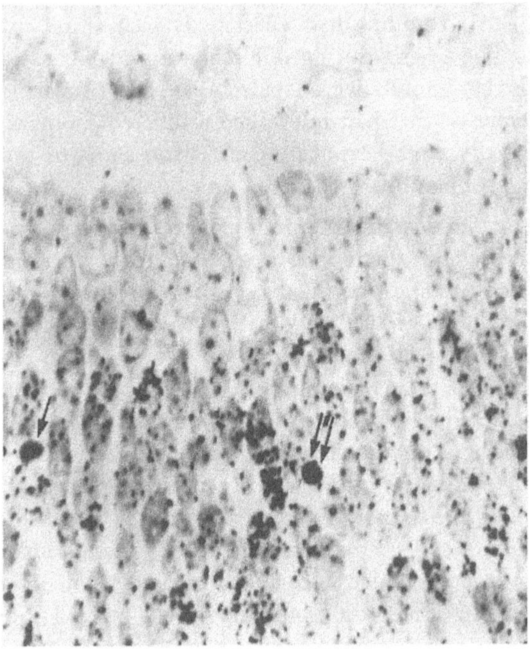

Fig. 8. Retina of a 3-day-old mouse injected subcutaneously with thymidine-H^3 (5 μc/g body weight) and killed 20 hours later. Occasional pyknotic nuclei are scattered through the neural retina (arrow) and some are labeled (double arrow). The life span of such cells could be calculated by plotting the per cent labeled pyknotic nuclei as a function of time after a single injection. The tissue was processed as described for Fig. 1. ×920

are available for DNA synthesis in the embryonic nervous system in hours immediately beyond the first hour after injection, for if they were, the "to-and-fro" nuclear displacement in the primitive ependymal zone would be obscured (see the following section). However, incorporation of persistent labeled precursor into DNA at later times cannot be ruled out.

Fig. 9. A. Cerebellar cortex of a mouse that was injected subcutaneously with thymidine-H^3 (4 μc/g body weight) on postnatal day 7 and was killed at 707 days of age. Numerous well-labeled granule cell neurons are seen, at all levels of the granular layer. B. Lateral vestibular nucleus of the same mouse. Many glial cells have remained heavily labeled for 2 years here and through most parts of the brain. Tissue was fixed in 10% acrolein, embedded in polyester wax, sectioned at 8 μ, exposed autoradiographically for 3 weeks, and stained with toluidine blue. ×275

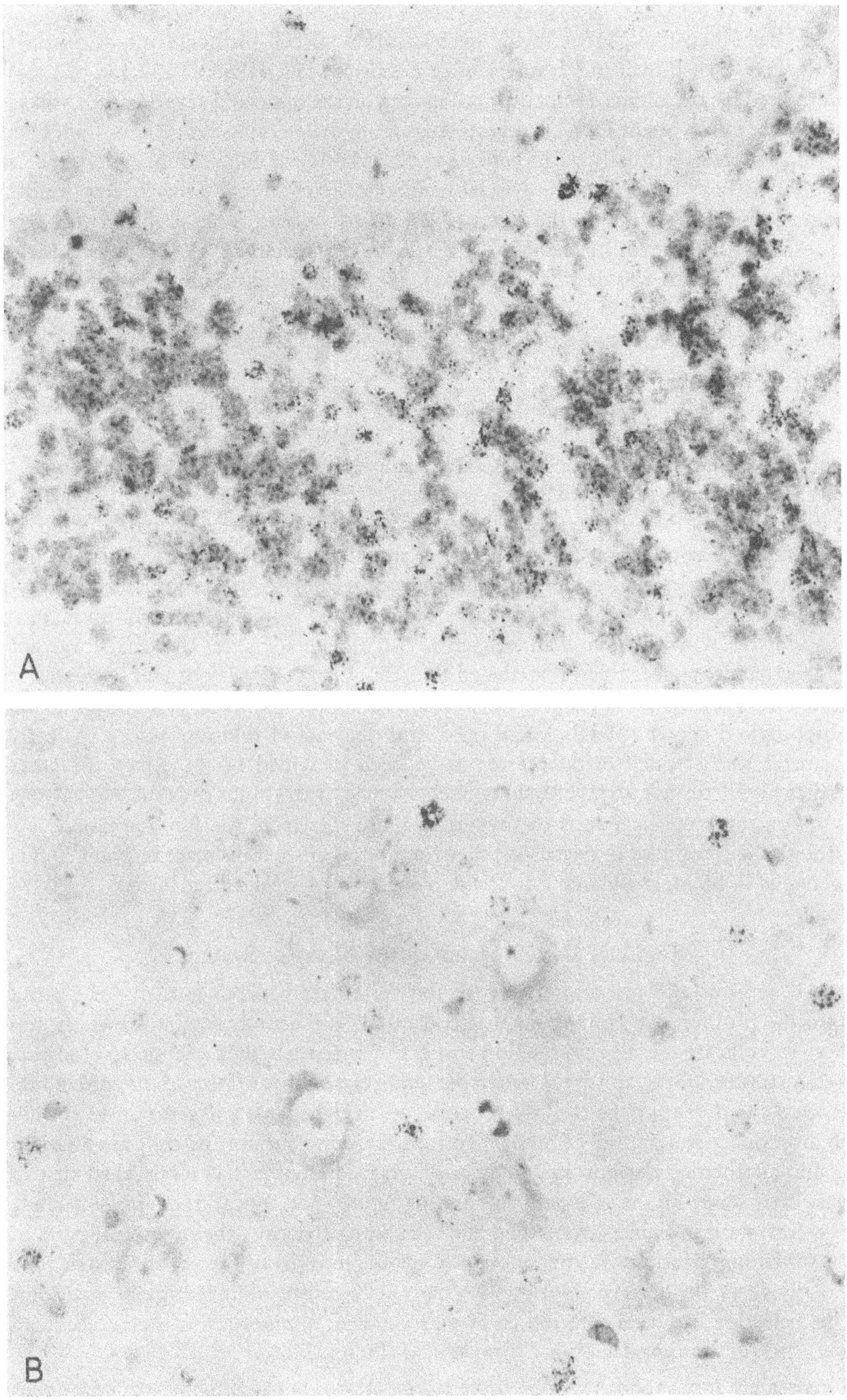

Fig. 9 A and B

A "false negative" error in the determination of cell birthdays might result if DNA were turned over independent of cell division. Pelc (1958, 1963a and b, 1964) interpreted his thymidine-H^3 incorporation data on the brain and several other organs to indicate two types of DNA synthesis apart from that serving for cell division. One of these types is virtually as intense as the synthesis for DNA duplication and is thought perhaps to serve some special metabolic function; the other is responsible for nuclear grain counts on the order of 2% of those over S phase nuclei and may represent a DNA repair mechanism. The half life of "metabolic DNA" in brain was estimated at 115 to 800 days (Pelc, 1963a). We have found that many neurons and glial cells remain heavily labeled for 700 days in the mouse brain (Fig. 9), but we do not have reliable quantitative data, either on grain counts or cell numbers as a function of age.

Another challenge to the "birthday" concept is found in the cytophotometric measurements of Lapham and others, which indicate that certain classes of large neurons, including Purkinje cells, hippocampal pyramidal cells, and autonomic ganglion cells, are tetraploid (see references in Herman and Lapham, 1968). The change from diploid to tetraploid state probably occurs in the neonatal period in man and in certain laboratory species (Lapham, 1968). A somewhat worrisome, but not critical, observation is that the change appears not to be accompanied by thymidine-H^3 incorporation. If the facts are verified, this would be the first known instance of DNA synthesis in man, cat, or rat for which exogenous thymidine was not an effective precursor. Even if a labeled exogenous precursor is not readily found, it should be possible to approach the problem autoradiographically by labeling the initial diploid content of DNA and measuring the rate of dilution during the early postnatal weeks. Such evidence may have been provided in the course of other studies by Fliedner *et al.* (1968), who labeled 100% of the brain cells present at birth by giving multiple injections of thymidine-H^3 to rats from the 9th embryonic day through the remainder of gestation; they found a significant decrease in grain counts over cells in brain, cord, and autonomic plexuses by 4 weeks after birth.

Regional Development in the Nervous System

All neurons and macroglial cells of the vertebrate nervous system are derived from the proliferative epithelium that constitutes the neural plate and early neural tube. Its cells are initially cuboidal in shape, but very early they assume an elongate, pseudostratified form, and are oriented perpendicular to the surfaces of the epithelium. Toward the outer surface the cell processes in time become splayed out, to form a narrow zone virtually free of nuclei. Toward the inner surface the cells are attached to one another by desmosomes. Autoradiographic analysis has established that at these early stages all the cells are proliferating rapidly and appear to comprise a single class (reviewed in Fujita, 1966). The nuclei in S phase occupy the outermost parts of the epithelium; during G2 they move toward the luminal surface of the neural tube and divide at the surface. During G1 the nuclei return to the original positions. This behavior has been referred to as a "to-and-fro" migration of nuclei (Sidman *et al.*, 1959a), or as an "elevator" movement (Fujita, 1966).

After an appropriate number of cell generations, nuclei of some daughter cells migrate away from the original zone and pass between the outer processes of the pseudostratified cells. At this stage one must apply names to the concentric zones that have developed from the initially simple epithelium. The innermost zone, representing the original tissue, has been assigned many names, including *primitive ependymal layer* and *matrix layer*. The outermost zone, relatively free of cell nuclei, is called the *marginal layer*. The cells which leave the primitive ependyma form the intermediate *mantle layer*. This concentric pattern foreshadows the arrangement of ependyma, gray matter, and white matter in the mature spinal cord and brainstem.

The autoradiographic evidence indicates that most of the early cells to enter the mantle layer are postmitotic and are destined to differentiate into neurons (Sidman, 1961; Miale and Sidman, 1961; Fujita, 1963, 1964; Martin and Langman, 1965). After a variable time, however, another population of cells appears at the junction of the primitive ependymal and mantle zones in most parts of the neuraxis and forms a new *subependymal layer*. These cells are distinguished from the others by incorporating thymidine-H^3 but not showing the "to-and-fro" nuclear displacement. They are derived from the primitive ependymal layer, but probably have a longer generation cycle than their predecessors. Subependymal cells seem to differentiate into both neurons and glia, though no data are available as to whether or not a single subependymal "stem cell" gives rise to both classes.

The subependymal layer becomes quite prominent in the walls of the lateral ventricles from late gestation to adulthood (Smart, 1961) and serves as a source of glial cells postnatally (Smart and Leblond, 1961; Altman, 1966). The subependymal cells migrate outward into and through the adjacent nerve fiber tracts and continue to multiply as they go. Some of the progeny continue to multiply in the cortical gray matter (Hommes and Leblond, 1967), but many remain as permanent postmitotic cells (Fig. 9).

In the cerebrum and cerebellum, new waves of migration occur and markedly distort the basic tissue pattern. In these regions, some of the postmitotic cells passing outward from the primitive ependymal layer do not stop in the mantle layer, but continue outward and form a new zone of gray matter just short of the distalmost part of the marginal layer. In the cerebellum, these cells are the Purkinje cell population. Subsequently the Golgi II neurons arrive by the same route. In the cerebral cortex the somas of each wave of cells to reach the cortex take up positions distal to their predecessors, with some exceptions, so that the cortex is formed in an "inside-out" sequence (Angevine and Sidman, 1961; Berry and Rogers, 1965, 1966; Hicks and D'Amato, 1968).

Another striking pattern of cell migration occurs in the cerebellar cortex (Miale and Sidman, 1961). On the 13th embryonic day in the mouse, proliferating cells leave the lateral part of the primitive ependymal layer that lines the roof of the 4th ventricle and begin to migrate along the external surface of the cerebellar anlage. These small round cells multiply as they migrate, and after about 48 hours they have come to coat the entire cerebellar surface as the *external granular layer*. Their rate of further proliferation exceeds the rate of surface growth of the cerebellum and continues through the second postnatal week. At its maximum extent, the external granular layer is about 8 to 10 cells thick and covers an organ with a very large folded surface. Beginning on about postnatal day 4 some of the daughter cells cease dividing and migrate inward to become predominantly neurons of the cerebellar molecular layer during the first postnatal week and predominantly granule cell neurons during the second week (Miale and Sidman, 1961; Fujita, 1967). The granule cells migrate past the Purkinje cell somas to attain their final positions. The rates of cell proliferation and migration have been plotted by Fujita *et al.* (1966) and Fujita (1967).

Other extensive migrations that have been analyzed autoradiographically are those from the primary and secondary rhombic lips into the brainstem (Pierce, 1965, 1966), from the ganglionic eminence of the telencephalon to the pulvinar region of the diencephalon (Sidman and Rakic 1968; Rakic and Sidman, 1969), into the optic tectum of the chick embryo (Fujita, 1964) and the olfactory bulb of the mouse (Hinds, 1969b), and the extraordinary migrations of neural crest cells (Weston, 1963). Neuroblasts migrate toward the developing rat cortex at a rate of 15 to 30 μ/hour (Hicks and d'Amato, 1968). In the larval frog, cells proliferate at the periphery of the retina, near the ciliary body, and migrate horizontally through the inner nuclear layer toward the fundic region of the eye, the velocity decreasing steadily from 16 to 9 μ/hour over the course of 4 days (Hollyfield, 1968). Lastly, a given proliferative region is commonly programmed to produce several classes of cells in succession, but the control mechanisms are completely unknown.

In certain cases it appears likely that the migrations serve to bring into contact classes of neurons that are destined to acquire synaptic contacts (Sidman, 1961; Angevine and Sidman, 1961; Miale and Sidman, 1961). Such a mechanism would represent only a special case of a more general mode of intercellular recognition in the developing nervous system. Some support for this idea has been obtained from study of the *reeler* mouse, a mutant in which cells of the developing cerebellar cortex, cerebral isocortex, and hippocampal formation arise and migrate normally at first, but fail to acquire their correct positions in the "home" territory and fail to make proper synaptic connections; the regions of the nervous system most severely affected by the mutation are those in which cells appear to find their way by crossing the positions of predecessor cells (Sidman, 1968).

A few final, general comments on neurogenesis as studied with thymidine-H^3 are pertinent. It appears that the large neurons of a given region generally arise earlier than the small ones. In the mouse brainstem, most large and medium-sized neurons have arisen by embryonic day 15, and in the cerebral cortex by day 17. Small nerve cells, by contrast, continue to arise well into the postnatal period, as emphasized by Altman (1966a and b). The circuitry of the brain for some reason appears to call for much greater numbers of small neurons than large ones, and the logistics are easier to arrange if small cells form later, and if they form on surfaces capable of comfortable expansion. For example, external granule cells are present on the surface of the mouse cerebellum for about 3 weeks, but more than 80% of the granule cell neurons are generated in the final few days (Fujita, 1967). A comparable population of cells coats the human cerebellum for many months after birth. Astrocytes also appear to form over a very long time span, some attaining birthdays early in development, coincident with nearby neurons (Angevine, 1968, and personal communication), but the majority arising postnatally. As noted earlier, the generation cycle is longer for glial cells than for neuron precursors, so that even if both cell types were programmed to undergo a comparable number of divisions, the glia would form predominantly in the early postnatal period. It should be emphasized that glial cells are present in the developing central nervous system almost as early as neuroblasts;

the analysis of "birthdays" in a sense obscures this fact, for the majority of the glial cells in the embryonic brain continue to multiply, whereas few of the neuroblasts do so. Lastly, a given proliferative region is commonly programmed to produce several classes of cells in succession, but the control mechanisms are completely unknown.

Discussion

Dr. Frontera: First of all, I would like to congratulate Dr. Sidman with this first-rate work. The approach taken by Dr. Sidman has an important bearing on the future of Comparative Neuroanatomy, and I believe that it may provide answers to many of the questions that were raised at a recent conference in New York with respect to the identification of inter-phyletic homologies. I would like to ask if it is possible with this method to use the knowledge of the so-called birthday of a cell or cell group in such a manner that you go back and in a neat series of experiments study its development and differentiation in more detail.

Dr. Sidman: Once you have demonstrated in the adult animal that a given population is labeled then often you can go back. If, for example, Purkinje cells arise on embryonic day 12 in the mouse, I can inject thymidine H^3 on day 12, examine at any subsequent stage, and identify the labeled cells in the cerebellum as Purkinje cells, even though they had none of their characteristic morphological of topographical features as yet. This works if no other population occupies similar positions in the cerebellum at that time.

Dr. Konigsmark: Would you care to comment on the relative value of keeping autoradiograms in carbon dioxide or nitrogen?

Dr. Sidman: When you buy film for your camera, a little note on the package may say, "do not keep exposed film in the camera any longer than necessary". The latent image produced by light (or radioactivity) will fade with the passage of time because the altered silver halide crystals revert to the original form. Oxygen is a major factor in this, and temperature is another. We develop our autoradiograms in a Dry Ice chest, at —75° C temperature, in a CO_2 atmosphere. Others develop autoradiograms under different conditions, such as refrigerator temperature with a drying agent in the boxes containing the slides. The procedure of choice may depend on what brand and type of emulsion one uses. Apart from fading of the latent image, it is not practical to expose an autoradiogram for more than 6 months or so because in order to double the number of grains you have to double the exposure time.

Dr. Konigsmark: You showed an autoradiogram with a few silver grains in the spinal cord of a ground squirrel. I am wondering how you differentiate these grains from the artifact of background activity?

Dr. Sidman: The background of the tissue was negligible. As a control for the emulsion, we always process two control slides with each batch of autoradiograms; one is a specimen with a known radioactivity and the other a slide with no radioactivity. In the case you mentioned, the controls were as they should have been. Of course, all we demonstrate is tritium; I do not know what molecular form the radioactivity was in.

Dr. Konigsmark: Would you care to comment about whether the external granule cell layer of the cerebellum also gives rise to glia?

Dr. Sidman: Dr. Fujita has suggested that at the end of the proliferative period the external cell layer turns from making neurons to making glial cells. I do not have any critical data on this point. To modify the question a bit, I think that in general, one should not accept the easy conclusion that a given proliferating population first produces neurons and later glia. The autoradiogram is misleading when used as a "birthday method" for the study of gliogenesis. If I give thymidine to the embryo and kill the animal as an adult, I see relatively few labeled glial cells. But that does not mean that the embryonic brain lacks glial cells. It has glial cells virtually throughout the period of time in which it has neuroblasts. Rather, the result means that most glial cells of the embryo, unlike the majority of the neuroblasts, continue to proliferate, so that when the animal is killed as an adult the tritium has become diluted below threshold. The notion that the young brain is full of neurons and has no glia until much later is quite wrong.

Dr. Cowan: I have two questions. The first concerns the validity of determining the "birth-date" of cells, as it were, negatively. In the case of chick embryos in which it is difficult, if not impossible, to pulse-label, do you think it is valid to simply inject labeled thymidine on different days and then define the time of formation of a population of cells as the last day the cells could be labelled?

Dr. Sidman: Yes, I do. Fujita has used the continuous labeling, negative birthday method in chicks and mice successfully.

Dr. Cowan: The other question concerns the mechanism of release of daughter cells from the neural epithelium as they differentiate to form neuroblasts. Do you have any ideas as to how this might happen and how it may be recognized?

Dr. Sidman: No, I do not believe anyone has looked into that important issue as yet.

Dr. Myers: Could you comment on programmed cell death during embryogenesis and whether there is anything about the character of cells, as shown by your techniques, that would pressage cell death?

Dr. Sidman: No. I am supposed to be a neuropathologist, but I do not know how to recognize that an embryonic cell is destined to die.

Dr. Petras: Could you clarify for me what you said about the appearance of the dendritic tree, both in the granule cell and in the Purkinje cells in reeler mice? Are there some granule cells that take a position ventral to the Purkinje cells, and then develop their own dendritic tree, and others that stay superficial and are neuroblastic in appearance? I am not sure I understood what you said.

Dr. Sidman: The Purkinje cells of the reeler mouse migrate outward from their site of origin but do not attain positions quite as close to the surface as they should. They may not recognize exactly where they should go. Most of the granule cells migrate inward from their site of genesis on the external surface of the cerebellum, but they end up in a position superficial to the Purkinje cells. Now, a Purkinje cell may send its dendritic tree among the granule cells, and in fact some of the dendritic branches will reach the molecular layer and contribute to what little molecular layer there is. Granule cells do synapse upon those Purkinje cells. The molecular layer itself contains parallel fibers, Purkinje spines, and so on. The deeper-lying Purkinje cells, on the other hand, are totally disoriented, the volume of dendritic material is reduced, the spines are virtually absent, and few or no granule cell axons contact them. Possibly most or all of the surviving granule cells contact Purkinje cells in reeler mice, but distorted Purkinje cells appear to survive without receiving a granule cell input.

Dr. Fox: There are cells in the molecular layer that are difficult to identify. One cannot tell whether they are stellate cells or basket cells. Larramendi thinks they are granule cells. I would like to ask you, Dr. Sidman, do you have any evidence that some granule cells in the normal adult never make it and stay in the molecular layer?

Dr. Sidman: No, I do not have any evidence one way or the other.

Dr. Voneida: Do you have any data as to whether those cells which migrate over very long distances retain connections with the structures from which they arose?

Dr. Sidman: Findings in Golgi material suggest that they do not retain such early connections. Barring the possibility that there is some extraordinary long process which fails to stain, I would have to say that migrating cells in general do not retain their original connections. However, one can speculate that they may have a genetic history in common with cells which migrate from the same source and go in another direction or with cells which are located in distant areas but are similarly specified.

Dr. Voneida: That is what I was getting at.

Dr. Sidman: But I do not think they have to maintain any physical contiguity.

Dr. Voneida: This would perhaps be an important means for telling us how cells which make connections over very long distances know where they are going.

Fluorescence Microscopy in Neuroanatomy

Kjell Fuxe, Tomas Hökfelt, Gösta Jonsson and Urban Ungerstedt

Introduction

The ever-increasing use of fluorescence methods in bio-medical research clearly bespeaks a steadily increasing demand for more sensitive methods allowing the identification of biologically active substances present in tissues in low — and often extremely low — concentration. Beside improvements in chemical methodology, the development of more adequate and commercially available equipment has played a major role in promoting the use of fluorescence methods.

Up to the present time, only relatively few histofluorescence methods have found an application in neuroanatomical research. In addition to the method of Falck and Hillarp for the demonstration of biogenic monoamines (DA[1], NA, A and 5-HT) with which this account will deal almost exclusively, the fluorescence instrumentarium now available to neuroanatomists includes techniques for histamine (Juhlin and Shelley, 1966; Ehinger and Thunberg, 1967; Håkansson and Owman, 1967), tryptamine (Björklund, Falck and Håkansson, 1968), lipofuchsin granules (see Sjöstrand, 1944; Koenig, 1963; Reichel, 1968) and an immunofluorescence technique for the visualization of myelin (see Rauch and Raffel, 1964). Recently, also immunofluorescence techniques for the demonstration of DA-β-hydroxylase and chromogranin, two proteins present in NA neurons, have been introduceed by Geffen, Livett and Rush (1969), opening up a new field in neurohistology. Nuclei acids can be demonstrated with the well-known acridine-orange technique (see Pearse, 1960).

As mentioned above this review will deal with the Falck-Hillarp technique with which it has been possible to visualize and map out previously unknown neuron systems containing biologically active monoamines. The technique has been extensively used not only in purely morphological research but also in approaching problems of a dynamic nature such as the synthesis, storage, transport, uptake, release and catabolism of neuronal and extraneuronal monoamines, and the action of certain drugs. It is beyond the scope of this publication to review the vast literature reporting the results obtained by the method. Instead, it has been the authors' intent to present relevant data with respect to the basic chemistry, specificity, sensitivity, limitations, and quantitative aspects, as well as the details of the practical procedure, with an account of the capacities and some applications of the method.

---

[1] Abbreviations used: A = adrenaline, CA = catecholamine (s), DA = dopamine, 6-OH-DA = 6-hydroxydopamine, DOPA = 3,4-dihydroxyphenylalanine, 5-HT = 5-hydroxytryptamine, 6-HT = 6-hydroxytryptamine, 5,6-diHT = 5,6-dihydroxytryptamine, 5-HTP = 5-hydroxytryptophan, NA = noradrenaline.

Historical Remarks

Until a few years ago, the cellular localization of the biogenic monoamines remained almost completely unknown, although these substances had been identified in significant amounts in various organs and tissues. Several earlier attempts at histological localization were hampered by the low sensitivity and questionable specificity of the methods used. Nonetheless, despite their imperfections, these older methods have been of great value in the study of cells containing large amounts of CA and 5-HT (*e.g.* adrenal medullary cells and enterochromaffin cells).

The oldest of these earlier methods is the classical *chromaffin reaction* in the adrenal medulla published by Henle as early as 1865 (for references see Pearse 1960; Coupland, 1965). Henle's method was followed by the potassium iodate method for selective staining of NA (Hillarp and Hökfelt, 1953, 1955), the *argentaffin reaction* for the demonstration of enterochromaffin cells (several modifications are used; see Pearse, 1960), and the fluorescence methods of Erös (1932, see also Barter and Pearse 1953, 1955) and Eränkö (1955). Erös and Eränkö were the first to use formaldehyde condensation and fluorescence microscopy for the visualization of NA and 5-HT, but their techniques did not possess the sensitivity required for a successful demonstration of these substances in neurons.

In more recent years, techniques were developed that allowed the demonstration of, and differentiation between, NA and A in adrenal medullary cells both at the ultrastructural and light microscopic level (Wood and Barnett, 1964; Coupland, Pyper and Hopwood, 1964; Tramezzani, Chiocchio and Wasserman, 1964). A similar technique has been used for electronmicroscopic localization of 5-HT in enterochromaffin cells (Wood, 1965). By means of a potassium-permanganate fixation-procedure it recently has been possible to demonstrate biogenic monoamines at the ultrastructural level in a more general manner (Richardson, 1966; Hökfelt, 1968).

Unique possibilities were opened up when Falck and Hillarp and co-workers succeeded in developing a fluorescence-histochemical method sensitive enough to demonstrate monoamines stored in neurons both in the peripheral and central nervous system (Falck, Hillarp, Thieme and Torp, 1962; Falck, 1962; Carlsson, Falck and Hillarp, 1962). This method is based on the principle that biologically active monoamines condense with formaldehyde to yield strongly fluorescent products. The procedure developed is quite simple: pieces of tissues or organs to be studied are freeze-dried or, if prepared as whole-mounts, air-dried, and then exposed to gaseous formaldehyde vapour the optimal humidity of which varies with the nature of the tissue studied. During this treatment the biogenic monoamines are converted to highly fluorescent compounds which can be localized by means of fluorescence microscopy. The practical histochemical procedure will be described in detail elsewhere in this review.

Chemistry

The method of Falck and Hillarp has been extensively studied from the methodological point of view, in particular by the use of model systems (see Corrodi and Jonsson, 1967; Jonsson, 1967c). It must be emphasized that a knowledge of the chemical and histochemical background is of the utmost importance for the evaluation and interpretation of the results obtained with the method. The fundamental reaction involves a condensation of a β-arylethylamine with formaldehyde to yield a tetrahydro-derivative (a *Pictet-Spengler* reaction, see Whaley and Govindachari, 1951), which in a second, protein-promoted, reaction is dehydrogenated to its corresponding 3,4-dihydrocompounds. The products thus formed from CA are 6,7-dihydroxy-3,4-dihydroisoquinolines, and intermediates in this reaction are the non-fluorescent 6,7-dihydroxy-1,2,3,4-tetrahydroisoquinolines (see Fig. 1). The formed compounds (*3, 6, 10*) are in a pH-dependent equilibrium (maximum between pH 6 and 10) with their tautomeric quinoidal forms (*3a, 6a, 10a*) which absorb strongly between 360 and

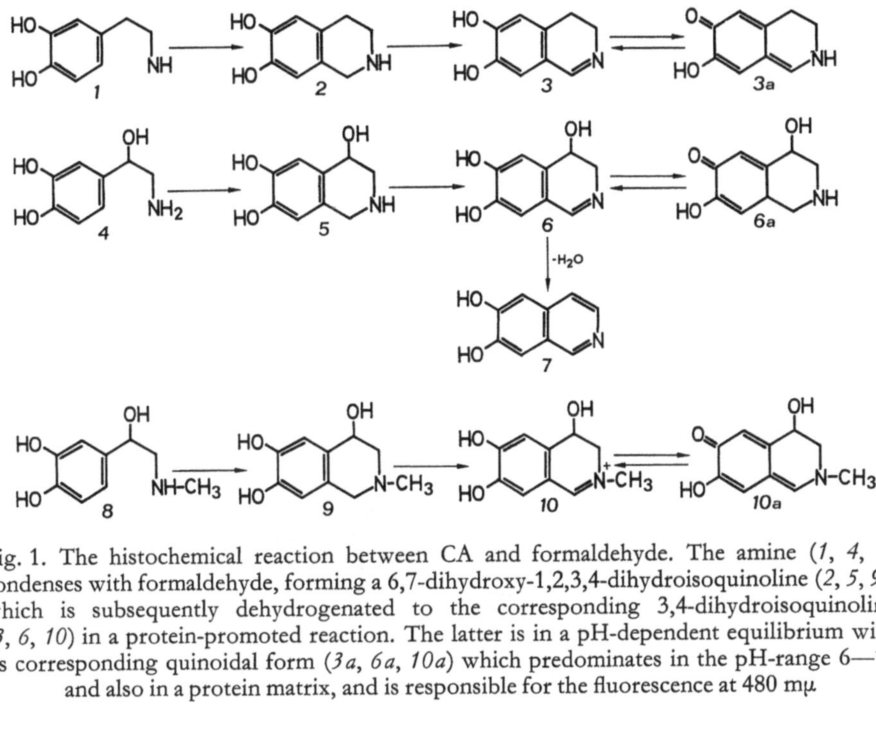

Fig. 1. The histochemical reaction between CA and formaldehyde. The amine (*1, 4, 8*) condenses with formaldehyde, forming a 6,7-dihydroxy-1,2,3,4-dihydroisoquinoline (*2, 5, 9*), which is subsequently dehydrogenated to the corresponding 3,4-dihydroisoquinoline (*3, 6, 10*) in a protein-promoted reaction. The latter is in a pH-dependent equilibrium with its corresponding quinoidal form (*3a, 6a, 10a*) which predominates in the pH-range 6—10 and also in a protein matrix, and is responsible for the fluorescence at 480 mμ

420 mμ and are thus the source of a strong fluorescence which in a protein layer has its peak of excitation/emission at 410/480 mμ, see Fig. 2. By comparing the spectral characteristics of various synthetic 6,7-substituted 3,4-dihydroisoquinolines it has been shown that the quinoidal form is a 6-quinone structure (Jonsson, 1966). The quinoidal and the protonated non-quinoidal forms have different spectral characteristics, a phenomenon which can be used for testing specificity (see below). The first step, the so-called *Pictet-Spengler* reaction, proceeds as readily in solution as in the solid state, and in the case of primary CA (DA and NA) both quantitatively and with a high velocity at temperatures down to +20° C. The protein-promoted dehydrogenation reaction also proceeds easily, but under the mild conditions used for histochemistry (50° C—80° C for 1 hour) it requires the presence of a protein with a high catalyzing power. Most of the native proteins possess this property which, incidentally, appears to be attributable to certain amino acids, for only glycine and alanine and also dipeptides containing these amino acids have been found to be excellent catalyzers (Corrodi and Hillarp, 1963, 1964; Jonsson, 1967b). All of the tetrahydroisoquinolines tested — including the non-substituted derivative — are readily dehydrogenated to

their corresponding 3,4-dihydroisoquinolines, indicating that this reaction is of a general type and thus not linked to any specific substitution in the isoquinoline nucleus. This dehydrogenation reaction produced by formaldehyde represents a previously unknown phenomenon in chemistry, and its exact nature has so far escaped elucidation (see Jonsson, 1967c). Both the reaction of ring closure and the subsequent dehydrogenation step can occur in solution but, as stated above, the second step requires the presence of a protein or at least certain amino acids (Jonsson, 1967b). The reactions can also take place in tissues when performed in solution (Eränkö and Räisänen, 1966; Eneström and Svalander, 1967; Laties, Lund and Jacobowitz,

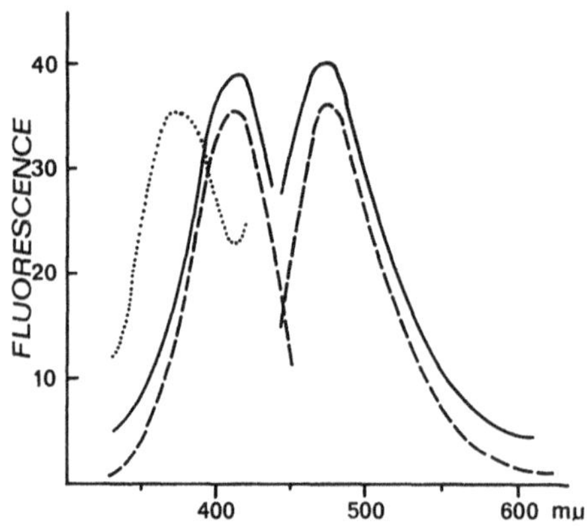

Fig. 2. Excitation and emission spectra obtained from adrenergic nerve terminals in rat iris (—) and from a model protein layer containing NA (---) showing typical formaldehyde-induced CA-fluorescence (excitation/emission maxima: 410/480 mμ) An excitation spectrum from adrenergic nerve terminals in rat iris (·····) after exposure to hydrogen chloride (maximum at about 360 mμ)

1967), but it is difficult to avoid diffusion of the amines and/or their fluorescent compounds and to obtain quantitative yields of the fluorophors under these conditions.

The secondary catecholamine, adrenaline (A) also reacts with formaldehyde to yield N-methyl-4,6,7-trihydroxy-3,4,dihydroisoquinoline (IX), but since the dehydrogenation step involves the formation of a quarternary nitrogen, a higher incubation temperature and longer times are needed for good yields (Corrodi and Hillarp, 1963; Jonsson, 1967b). This difference in reaction between primary and secondary CA can be used to some extent for their differentiation (Falck, Häggendal and Owman, 1963; Norberg, Ritzén and Ungerstedt, 1966). The secondary catecholamines, N-isopropyldopamine and isoprenaline, yield practically the same fluorescence as does A after identical treatments (Corrodi, Malmfors and Sachs, 1966).

The amino acid DOPA, precursor in the NA biosynthesis, gives a formaldehyde-induced fluorescence the spectral characteristics of which are practically identical with that of primary CA. The structure of the fluorophor in this instance probably is 3-carboxy-6,7-dihydroxy-3,4-dihydroisoquinoline (Jonsson, 1967c).

Under certain conditions CA can give rise to a more yellowish fluorescence caused by a batochromic shift in emission maximum from 480 mμ to 500—550 mμ (Caspersson, Hillarp and Ritzén, 1966; Jonsson, 1967b). This is due to a side reaction, which has been found to occur when large amounts of CA are present (e.g. in adrenal medullary cells) and in model protein layers when the CA/protein quotient is very high. Primary CA more easily give this side reaction, but it is possible to prevent it by using milder formaldehyde gas reaction (less humid gas, shorter incubation time or lower temperature). It has not been possible to establish accurately the molecular structure of the formed fluorophor, but the reaction probably proceeds via the tetrahydro- to 3,4-dihydroisoquinoline to the final fluorescent compound or compounds. The 4- and 7-hydroxygroups of the phenylethylamine and isoquinoline molecule respectively seem to be involved in the reaction and the end product may be the result of a polymerization and/or oxidation (see Jonsson, 1967b). This spectral shift has so far not been found to interfere significantly in the histochemical localization of CA in adrenergic neurons, and since it can be prevented by varying the reaction conditions it can be used as a specificity test whenever it occurs.

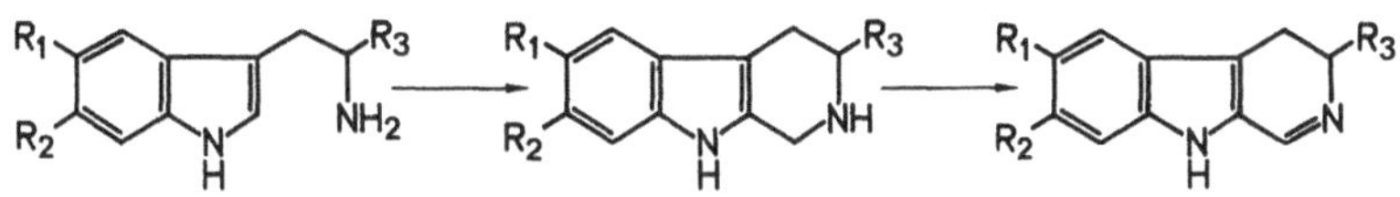

Fig. 3. The histochemical reaction of a β-indolylethylamine (5-HT: $R_1 = OH$, $R_2 = R_3 = H$; α-methyl-5-HT: $R_1 = OH$, $R_2 = H$ $R_3 = CH_3$; 6-HT: $R_1 = R_3 = H$, $R_2 = OH$; 5,6-diHT $= R_1 = R_2 = OH$ $R_3 = H$; Tryptamine: $R_1 = R_2 = R_3 = H$) with formaldehyde. The amine (*11*) first condenses with formaldehyde yielding a 1,2,3,4-tetrahydro-β-carboline (*12*) which is dehydrogenated in a protein-promoted reaction to its corresponding fluorescent 3,4-dihydro-β-carboline (*13*)

The *m*-tyramines (*e.g.* *m*-tyramine, α-methyl-*m*-tyramine/*m*-hydroxy-amphetamine/, *m*-hydroxy-norephedrine/metaraminol, Aramine/, *m*-tyrosine and α-methyl-*m*-tyrosine) enclosed in a dried protein layer and exposed to formaldehyde gas readily enter a *Pictet-Spengler* reaction forming 6-hydroxy-1,2,3,4-tetrahydroisoquinolines which subsequently are dehydrogenated to corresponding 3,4-dihydroisoquinolines. The latter are in equilibrium with their tautomeric quinone structures which are intensely fluorescent. These fluorescent compounds have, however, two emission maxima: one at 420 mμ and another at 510 mμ (Corrodi and Jonsson, 1966). It has been shown by use of microspectrofluorometry that the 420 mμ-peak becomes dominant in neurons after injection of metaraminol or α-methyl-*m*-tyrosine (Jonsson and Ritzén, 1966), whereas under the same conditions the 510 mμ peak dominates in certain non-neural cell types such as cardiac muscle fibers (Sachs, 1966), glia cells (Dahlström and Fuxe, 1964) and the pericytes of brain capillaries (Hamberger, 1967). This difference in emission spectra of 6-hydroxy-3,4-dihydroisoquinolines when present in adrenergic neurons or non-nervous structures has remained unexplained so far, but it seems reasonable to assume that the 510 mμ-peak may be due to an association of 6-hydroxy-3,4-dihydroisoquinoline molecules.

The tryptamines such as T, 5-HT, 5-HTP, 6-diHT and 6-HT, if enclosed in a dried protein matrix, are readily converted by formaldehyde gas to 1,2,3,4-tetrahydro-β-carbolines which subsequently are dehydrogenated to their corresponding fluorescent 3,4-dihydroderivatives (see Fig. 3). In the case of 5-HT and 5-HTP the excitation maximum lies in the same wavelength range as that of CA but the emission peak of 525 mμ which characterizes these substances (Fig. 4) differs from that of

CA. The fluorescence is very sensitive to ultraviolet light (Fig. 5). The reaction is in principle similar to that taking place in the case of CA, but tryptamines require more energetic reaction conditions if quantitative yields of the fluorophor are to be obtained (Corrodi and Jonsson, 1965b; Fuxe and Jonsson, 1967).

The tryptamines, *e.g.* 5-HT and 5-HTP and their 3,4-dihydro-β-carbolines to some extent also can undergo a side reaction when more energetic reaction conditions are used (see Ritzén, 1967b). This ,however, has not been found to interfere with the histochemical demonstration of the β-indolylethylamines. By means of spectral analysis this side reaction can be used for testing specificity.

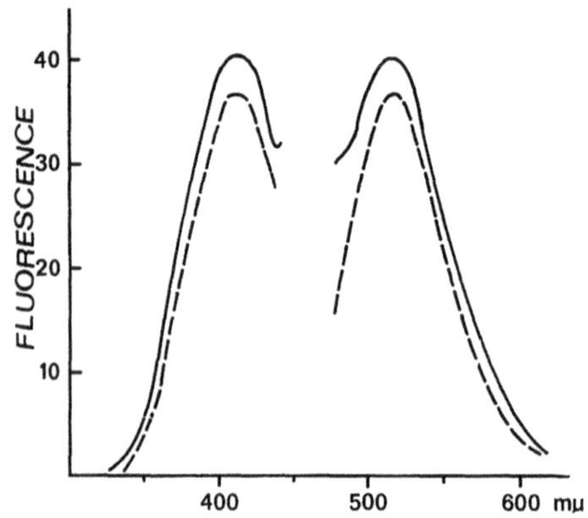

Fig. 4

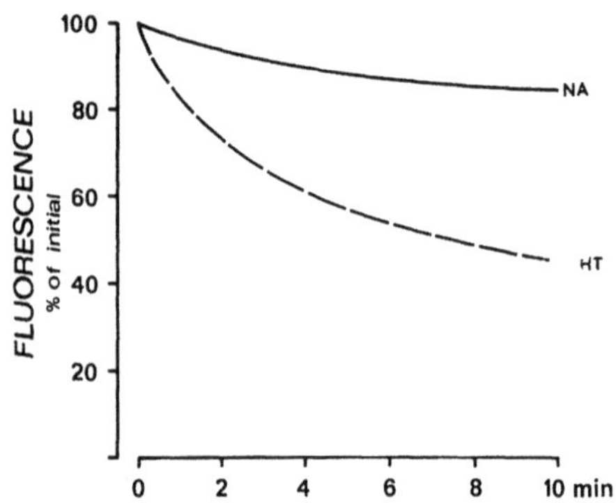

Fig. 5

Fig. 4. Excitation and emission spectra obtained from a mast cell (—) and from a model protein layer containing 5-HT (---) exhibiting typical formaldehyde-induced 5-HT-fluorescence (excitation/emission maxima 410/525 mμ)

Fig. 5. Effect of irradiation at maximal excitation wavelengths on fluorescence intensity of NA (—) and 5-HT (---) in dried protein layers exposed to formaldehyde gas

Extensive studies have shown that the mechanism of the *Pictet-Spengler* reaction consists in an electrophilic attack on the benzene and indole nucleus, and is facilitated by an increased electron density of the carbon atom at the site where the ring closure takes place. This condition is amply fulfilled by 3-hydroxylated β-phenylethylamines, *e.g.* NA and DA, and also by β-indolylethylamines *e.g.* 5-HT and 5-HTP (see Kovács and Fodor, 1951; Whaley and Govindachari, 1951). These substances meet the structural requirements for a compound to react with formaldehyde and to enter the primary condensation reaction during the histochemical procedure (see Jonsson, 1967c). In the β-phenylethylamine molecule it is the 3-hydroxygroup and in the indole nucleus the heterocyclic nitrogen which facilitates the ring closure, due to their great activation. The protein-promoted dehydrogenation step is a general reaction for 1,2,3,4-tetrahydroisoquinolines and β-carbolines. The reaction also requires a reactive carbonyl compound such as formaldehyde.

The amine must be either primary or secondary; tertiary amines and amides do not give rise to any fluorescence. In complete agreement with these theoretical aspects it has been found that only 3-hydroxy-or 3,4-dihydroxyphenylethylamines and their β-hydroxylated and α-methylated analogues (and their corresponding amino acids *e.g.* DOPA, α-methyl-DOPA and 3,4-dihydroxyphenylserine/DOPS/) react with formaldehyde under the conditions used in the histochemical procedure to form fluorescent 3,4-dihydroisoquinolines (Jonsson, 1967c). From the histochemical point of view it is of greatest interest that the biogenic compounds closely related to CA and 5-HT — *e.g.* phenylalanine, tyrosine, tyramine, tryptophan, 5-hydroxyindoleacetic acid and the deaminated and 3-0-methylated metabolites of CA — do not react at all or too little to be of any significance for the histochemical localization of biogenic monoamines (see Fig. 6). The aromatic amino acids, however, might contribute to some of the background fluorescence exhibited by tissue proteins. Melatonin (N-acetyl-5-methoxy-tryptamine), which is non-basic, does not react with formaldehyde and model experiments have failed to show any fluorescence attributable to this compound (Corrodi and Jonsson, 1965b).

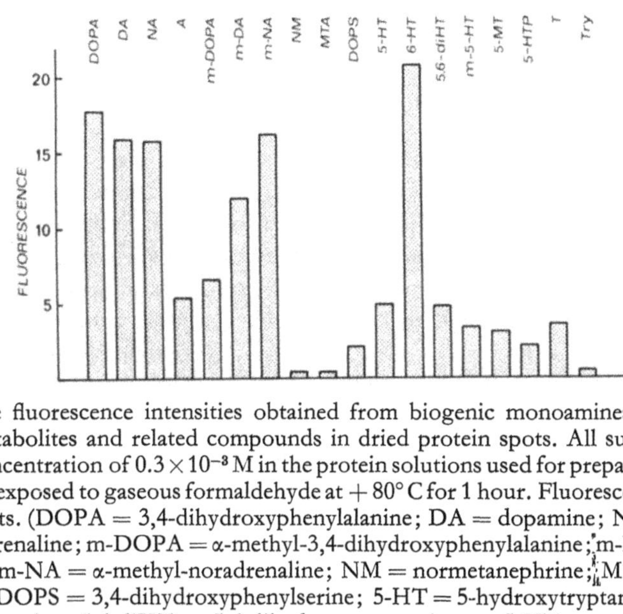

Fig. 6 Relative fluorescence intensities obtained from biogenic monoamines, monoamine analogues, metabolites and related compounds in dried protein spots. All substances were present at a concentration of 0.3×10^{-3} M in the protein solutions used for preparing the spots: the spots were exposed to gaseous formaldehyde at + 80° C for 1 hour. Fluorescence expressed in arbitrary units. (DOPA = 3,4-dihydroxyphenylalanine; DA = dopamine; NA = noradrenaline; A = adrenaline; m-DOPA = α-methyl-3,4-dihydroxyphenylalanine; m-DA = α-methyl-dopamine; m-NA = α-methyl-noradrenaline; NM = normetanephrine; MTA = 3-methoxytyramine; DOPS = 3,4-dihydroxyphenylserine; 5-HT = 5-hydroxytryptamine; 6-HT = 6-hydroxytryptamine; 5,6-diHT = 5,6-dihydroxytryptamine; m-5-HT = α-methyl-5-hydroxytryptamine; 5-MT = 5-methoxytryptamine; 5-HTP = 5-hydroxytryptophan; T = tryptamine; Try = tryptophan)

It should be mentioned in this context that it has been reported that also histamine can be visualized with a formaldehyde vapour technique (Lagunoff, Phillips and Benditt, 1961; Orden, Wugman and Giarman, 1965), but the fluorescence yield is very low when compared to primary CA, and in any case histamine-fluorescence shows other spectral characteristics and therefore will not interfere seriously with the localization of biogenic monoamines.

The fluorophors formed from the biogenic monoamines appear not to be bound to the surrounding proteins by *e.g.* methylene bridges, but are instead enclosed in the protein network formed by the tanning action of formaldehyde. The degree of binding (enclosure) can be reinforced by more energetic formaldehyde treatment which may provide possibilities for the improvement of autoradiography of these types of substances (see Ritzén, Hammarström and Ullberg, 1965; Hammarström, Ritzén and

Ullberg, 1966; Fuxe, Hökfelt, Ritzén and Ungerstedt, 1968). The use of formaldehyde gas fixation seems in general to be a valuable technique in histochemistry (Lagunoff, Phillips and Benditt, 1961; Falck and Owman, 1965).

Specificity

Considering the known molecular requirements for a compound to give rise to fluorescence, and the general experience obtained in extensive studies performed since the method was first published, the chemical specificity must be very high for CA and 5-HT and their immediate precursors (DOPA and 5-HTP). It is, however, of utmost importance to check the specificity of the fluorescence observed, i.e. to ascertain that the fluorescence is indeed due to the presence of one of the reactive monoamines. This is especially important in the study of tissues and organs not previously examined, or in the case of changes in the normal fluorescence picture following the administration of drugs of unknown effect, changes that may result from native or formaldehyde-induced fluorescence of the drug. The specificity can now safely be checked with several tests. The general background fluorescence of tissue proteins and various autofluorescent tissue structures are the main factors that will interfere with the microscopical demonstration of biogenic monoamines (Sjöstrand, 1944; Koenig, 1963; Sainte-Marie, 1965; see also Fig. 7). Collagen and elastic fibers are examples of autofluorescent structures (see Corrodi, Hillarp, and Jonsson, 1964).

By virtue of the physico-chemical properties of the various fluorophors of monoamines, it has been possible to introduce a number of histochemical critera of specific fluorescence. At least the first of the criteria listed below must be met if an observed fluorescence is to be regarded as specific.

Histochemical Criteria

For testing specificity of fluorescence, a tissue structure can be analyzed with respect to the following characteristics.

1. Reaction Conditions for Development of Fluorescence

The simplest and most often overlooked test of specificity naturally consists in omitting the formaldehyde gas treatment. The specific formaldehyde-induced fluorescence of reactive monoamines (3-hydroxylated β-phenylethylamines and β-(3-indolyl)-ethylamines), however, only develops under certain conditions. The most important reaction variables are the humidity of the formaldehyde gas (generated from paraformaldehyde), incubation time and temperature. The formaldehyde gas treatment must be performed in a well-closed vessel containing about 5 g of paraformaldehyde per liter, but the amount is not critical. The standard reaction time is 1 hour at $+80°$ C. During this treatment more or less quantitative yields are obtained for primary CA, their α-methylated analogues, DOPA and *m*-tyramines (provided an optimal humidity of the formaldehyde gas). Secondary CA, *e.g.* A, and indolylethylamines, *e.g.* 5-HT and 5-HTP, require more energetic conditions for the development of maximal fluorescence (higher humidity of the formaldehyde gas, higher temperature and longer incubation time). These variables must, however, be controlled rigorously, since more energetic reaction conditions generally will cause an increased background fluorescence from tissue proteins, promote diffusion of the fluorescent compounds, and may even cause a decrease of monoamine fluorescence, probably as a result of destruction of the formed 3,4-dihydroisoquinolines and β-carbolines. Such reaction conditions will also promote side reactions, resulting in spectral changes. For all these reasons it may be advantageous, especially in the case of 5-HT and 5-HTP, to use a modified formaldehyde gas treatment consisting of a primary treatment with a formaldehyde gas having a humidity low enough not to cause any noticeable diffusion, followed by exposure of the tissue specimen to a more humid formaldehyde gas which will increase the yield of fluorescent compounds (Fuxe and Jonsson, 1967).

The histochemical reaction requires a certain amount of water to proceed, and paraformaldehyde is used as the main water source (Hamberger, Malmfors, and Sachs, 1965; Hamberger, 1967). The humidity of the formaldehyde gas is *very critical*, since if it is too dry, the observed fluorescence will be weak, although distinctly localized. Increases in water content of the gas will gradually raise the fluorescence intensity to an optimum, but too humid formaldehyde gas will cause diffusion of the monoamine and/or their reaction products. Therefore, the formaldehyde gas should have a certain optimal humidity to allow the development of maximal fluorescence, and it must be remembered that even the slightest diffusion makes it impossible to demonstrate the localization of fluorescent substances in structures as delicate as the finer nerve terminals. A great advance has been the introduction of paraformaldehyde standardized with respect to its water contents (Hamberger, Malmfors and Sachs, 1965; Hamberger, 1967).

The exact role played by the water in the histochemical reaction is not completely understood, but it certainly increases the reactivity of the formaldehyde gas. The effect of the water in the gas might also partly be due to the fact that it causes a diffusion of the amines or their fluorophors from the intraneuronal storage granules to the extragranular space, thereby decreasing or abolishing the fluorescence quenching that might otherwise result from high local concentrations of the fluorescent products (see Jonsson, 1969). It has also been found that CA-fluorescence produced by dry formaldehyde gas can be increased by exposure to ammonia vapour, since ammonia favours the existence of CA's strongly fluorescent quinoidal form, implying that the effect of water may in part also be due to increase of pH in the tissue environment (Corrodi, Jonsson and Malmfors, 1966).

2. Spectral Characteristics

The peak of excitation/emission of the formaldehyde-induced CA-fluorescence is 410/480 mμ and that of 5-HT-fluorescence 410/525 mμ (see Figs. 2 and 4). *m*-Tyramines can have two emission maxima, 420 and 510 respectively and maximal excitation at 360 and 385 mμ corresponding to non-quinoidal and quinoidal forms. The non-substituted indolylethylamine derivative T has its peak of excitation at 370 mμ and its peak of emission at 490 mμ. The spectral analysis can be most safely performed by using microspectrophotofluorometry.

3. Sodium Borohydride Reduction

This is a chemical specificity test that has the advantage of being directly applicable to tissue sections. It is performed in alcoholic solutions. During this treatment the fluorescent compounds of the reactive monoamines (3,4-dihydroisoquinolines and

β-carbolines) are rapidly and almost quantitatively reduced to their corresponding non-fluorescent 1,2,3,4-tetrahydroderivatives. These can, however, be converted to their fluorescent 3,4-dihydro compounds upon renewed formaldehyde gas treatment (Figs. 1 and 3, see also Corrodi, Hillarp and Jonsson, 1964). This is a very simple and reliable test of high specificity, but the important points is that the actual proof for specificity is the regeneration of fluorescence after reduction, hence sufficient controls must always be used to exclude solvent extraction of the fluorescent compound as a possible explanation for failure to regenerate the fluorescence.

4. *Sensitivity to Irradiation with Ultraviolet Light*

The formaldehyde-induced monoamine fluorescence is sensitive to irradiation with ultraviolet light with resultant photodecomposition of the fluorophors (see Fig. 5). The sensitivity is often different for different compounds (Jonsson, 1967c; Jonsson and Sandler, 1969). The rate of photodecomposition is preferably determined with a microfluorometer.

5. *Quenching of the Fluorescence by Water*

The 6,7-dihydroxy-3,4-dihydroisoquinolines and the 6-hydroxy-3,4-dihydro-β-carbolines exhibit a stronger fluorescence in the solid state (*e.g.* in a dried protein matrix) than in solution (Corrodi and Jonsson, 1966; Ritzén, 1966). In the case of CA this difference is 10 times and in the case of 5-HT 80 times (Ritzén, 1967c). When present in a dried protein layer, both in models and tissues, the fluorophors show a much weaker intensity if surrounded by a water-containing medium (Falck, Hillarp, Thieme and Torp, 1962; Corrodi, Hillarp and Jonsson, 1964; Corrodi and Jonsson, 1965; Ritzén, 1967b). The fluorescence reappears almost completely after removal of the water-medium, provided that the fluorescent compounds have not been extracted during the procedure.

Histochemical Differentiation between the Various Monoamines

Since many of the reactive monoamines show almost identical spectral characteristics, there have been several attempts to develop methods for their differentiation. Using microspectrofluorometry a safe differentiation between several compounds can be made:

1. CA and DOPA; peak of emission at 480 mμ.
2. *m*-Hydroxyphenylethylamines; peak of emission at 420 and 510 mμ.
3. 5-HT and 5-HTP; peak of emission at 525 mμ.
4. T and Tryptophan; peak of emission at 490 mμ.

The main differentiating problems in monoamine fluorescence histochemistry are between the various CA and between CA and 5-HT.

Differentiation between the Various CA

Owing to the almost identical spectral characteristics of the fluorescent compounds of CA and DOPA, these substances cannot be directly distinguished from one another. Primary CA can be differentiated from secondary CA on the basis of pifferences in the kinetics of their respective reactions with formaldehyde (Corrodi and Hillarp, 1963; Falck, Häggendal and Owman, 1963). Although the differences in fluorescence intensity can be estimated by the unaided eye, far more accurate comparisons are made by microfluorometric measurements (Norberg, Ritzén and Ungerstedt, 1966). Even then the results require care interpretation. Baumgarten, Ehinger and Falck (1967) have published another differentiation method which is based on the fact that primary CA more easily show spectral changes than do secondary CA, owing to a side reaction of the formed 3,4-dihydroisoquinolines. This method cannot, however, be considered as a general method for differentiation between primary and secondary CA, since it is based on the premise that the concentrations of monoamine in the structures under investigation are in the same order of magnitude. Other methods for differentiating between NA and A have also been reported: a borohydride-periodic acid treatment (Mukherji, Ray and Sen, 1966) and a modification of the trihydroxy-indole method (Angelakos and King, 1967).

Differentiation between the primary CA (NA and DA) can be made by spectral analyses after treatment with hydrochloric acid. This method rests on the findings 1. that the fluorescent product formed from NA, 4,6,7-trihydroxy-3,4-dihydroisoquinoline, has a labile hydroxy group in position 4, which is easily split off by HCl-treatment and 2. that the resultant fully aromatic 6,7-dihydroxyisoquinoline has spectral characteristics different from those of the 6,7-dihydroxy-3,4-dihydroisoquinolines (Corrodi and Jonsson, 1965a; Björklund, Ehinger and Falck, 1968).

Differentiation between CA and 5-HT

a) Difference in Fluorescence Colour

The most reliable differentiation is made by spectral determination of the emission-peak difference between both types of compounds. This difference can also be used for differentiation directly in the fluorescence microscope, since the colour of the emitted light differs, being green to yellow-green for CA and yellow for 5-HT (using a cut-off filter strongly absorbing below 490 mμ). It must be borne in mind, however, that a yellowish formaldehyde-induced fluorescence is not always attributable to the presence of 5-HT or 5-HTP, for CA and especially primary CA when present in high concentration can give rise to a fluorescence colour very similar to that of 5-HT and characterized by a maximal emission at about 500—550 mμ (Caspersson, Hillarp and Ritzén, 1966; Jonsson, 1967b). This side reaction can be prevented by using less energetic reaction conditions, less humid formaldehyde gas and shorter incubation time. The specificity and differentiation can thus be made by using different reaction conditions, preferably in combination with objective registration methods.

Another pitfall to be mentioned in this context is, that in certain cases the fluorescence colour can appear as yellow while objective measurements nonetheless indicate a peak of emission at 480 mμ typical for CA-fluorescence (Norberg, Ritzén and Ungerstedt, 1966; Jonsson, 1967b). This somewhat puzzling phenomenon is probably due to sensitivity differences of the human eye at various parts of the spectrum and at various intensities (see Ritzén, 1967c). When high concentrations of primary CA are present, the formaldehyde-induced fluorescence colour appears more yellowish, despite the fact that the peak of emission is at 480 mμ. This may be explained by the fact that maximum sensitivity of the human eye is shifted towards higher wavelengths with an increasing intensity of the light. The phenomenon has been observed in the small, intensely fluorescent (SIF) cells of the superior cervical ganglion and adrenal medullary cells of rats (Norberg, Ritzén and Ungerstedt, 1966; Jonsson, 1967b).

Considering all these findings, it is quite obvious that in cases of even the slightest doubt about the fluorescence colour and other characteristics, spectral analysis should be carried out.

b) Rate of Photodecomposition

Irradiation of the fluorescent products of CA and 5-HT with ultraviolet light will cause a decrease in fluorescence intensity attributable to photodecomposition. This decrease is much faster for the 3,4-dihydro-β-carbolines than for the 6,7-dihydroxy-3,4-dihydroisoquinolines (see Fig. 5). This property can be used for differentiation, but it must be regarded as no more than complementary to other differentiation tests (see Dahlström and Fuxe, 1964; Caspersson, Hillarp and Ritzén, 1966; Jonsson, 1967a, 1967b).

c) Exposure to Hydrogen Chloride

Treatment of CA-formaldehyde products with hydrogen chloride causes a transition of the fluorescent tautomeric quinoidal forms (3a, 6a, 10a) to their corresponding protonated nonquinoidal forms (3, 6, 10). This transition is revealed by a hypsochromic shift of the excitation maximum from 410 to 360 mμ. A bathochromic shift of the emission maximum from 480 to 510 mμ also occurs, but this change is not as easily accomplished as the hypsochromic shift. These spectral changes can be used for differentiation between CA and 5-HT, since they are specific for the 3,4-dihydroisoquinolines formed from CA and DOPA.

Use of Pharmacological Approaches for Evaluating Specificity and Differentiation

Problems of specificity, and hence of differential identification of fluorescent compounds, can be approached also by pharmacological-biochemical analysis. This often-used strategy has been the subject of detailed studies and reviews (see Carlsson, Falck and Hillarp, 1962; Falck, 1962; Dahlström and Fuxe, 1964; Norberg and Hamberger, 1964; Malmfors, 1965) which have emphasized the great value of the pharmaco-analytical techniques. Nonetheless, since several histochemical tests based on a variety of principles are now available, the major application of the method has shifted from specificity-testing to the study of the effects of drugs on the monoamine fluorescence. The method requires critical interpretation, as well as a thorough acquaintance on the part of the investigator with the chemical and histochemical aspects of the method. A prime reason for caution in interpretation arises from the fact that under certain conditions, *e.g.* after the administration of precursors, amine analogues, or drugs interfering with the monoamine metabolism, compounds meeting the structural requirements for fluorescence reactions may accumulate in the tissues. Several problems therefore cannot be adequately elucidated without the combined use of pharmacological-biochemical and histochemical investigations.

Sensitivity and Quantitative Aspects

It has not been possible to quantitate accurately the sensivity of the method owing to local differences in concentration of the amine in the same neuron and between different neurons. In general, the formaldehyde vapour technique can be considered very sensitive in comparison with other histochemical- and cytochemical methods, and superior to all other techniques applied to this field of research. A 10 μ thick protein layer containing 5 μg of primary CA/g protein (bovine serum albumin) exhibits a fluorescence intensity detectably higher than the control. The limiting factor seems to be the background fluorescence exhibited by the tissue proteins. The concentrations of biogenic monoamines in tissues, as determined by chemical assay, generally lie in the range of 0.3—1.5 μg/g wet weight, depending mainly on the nerve density in the tissue. The amines must thus be concentrated in tissue stores to be detectable; if diffusely spread out in the tissue they could hardly be detected.

From studies on adrenergic neurons in the superior cervical ganglion of the cat, Norberg and Hamberger (1964) calculated that quantities below 1 pg (10^{-6} μg) evenly distributed throughout the perinuclear cytoplasm (volume of about 10 to $20\times10^3\mu^3$) of a nerve cell body still may be easily detectable in the fluorescence microscope (10 μ thick section). The amount of NA in these cell-bodies, which exhibit a moderate to strong fluorescence, has been calculated to be in the order of 0.5×10^{-6} μg, a value corresponding to 10 μg/g wet weight (Norberg and Hamberger, 1964; Dahlström and Häggendal, 1966). DA has been shown to be present in similar concentrations in cell bodies of the substantia nigra (Andén, Fuxe, Hamberger and Hökfelt, 1966). Dahlström, Häggendahl and Hökfelt (1966) have calculated that the strongly fluorescent varicosities of adrenergic nerves of rat iris and vas deferens contain an average of 5×10^{-9} g, corresponding to a concentration in the varicosity cytoplasm of about 1—4×10^3 μg/g wet weight. The concentration of CA in varicosities of the nigro-neostriatal DA-neuron system has been found to be of the same order (Andén, Fuxe, Hamberger and Hökfelt, 1966). In studies on drug-induced depletion of the adrenergic transmitter and its subsequent reappearance or refillment of the emptied monoamine stores by *in vitro*-incubation with ^{3}H-NA, the transmitter could be detected even if the tissue stores were depleted by more than 90—95 percent (Fuxe and Sedvall, 1964; Andén, Fuxe, Hamberger and Hökfelt, 1966; Jonsson, 1969; see also Olson, Hamberger, Jonsson and Malmfors, 1968). These figures are, however, somewhat difficult to interpret exactly, since the neuron systems in question might differ from each other in their reaction to drugs (Malmfors, 1965).

A question often raised in this context concerns the extent to which the fluorescence intensity really reflects the amount of monoamine present in the neuron. In the case of primary CA, combined chemical assay and histochemical analysis have so far generally shown a good correlation, especially in the case of distribution of monoamine neurons (see *i.a.* Häggendal and Malmfors, 1965; Corrodi and Malmfors, 1966; Andén, Fuxe, Hamberger and Hökfelt, 1966) although the general experience has been that relatively large changes in monoamine levels are required to allow the detection of differences in fluorescence intensity. For example, an increase above normal fluorescence intensities generally cannot be observed unless the increase is a drastic one. A more striking discrepancy, however, has been observed between the biochemically determined levels of 5-HT in the brain (especially in the telencephalon and diencephalon) and the number of 5-HT containing nerve terminals seen in the fluorescence microscope (see Fuxe, 1965b, 1965c). The difficulties involved in the demonstration of 5-HT are probably due in part to the fact that 5-HT requires more energetic reaction conditions for its quantitative conversion into the fluorescent 6-hydroxy-3,4-dihydro-β-carboline than are required for the conversion of the primary CA to their 6,7-dihydroxy-3,4-dihydroisoquinolines. In addition, the former compounds have a lower fluorescence yield (see Fig. 6) and are extremely sensitive to irradiation with ultraviolet light (see Fig. 5). The method for demonstrating 5-HT has been improved by using a modified formaldehyde gas treatment (Fuxe and Jonsson, 1967), but still cannot be considered as sensitive as that for primary CA. Of greatest interest is, however, that an isomer of 5-HT, namely 6-HT, has been found to give rise to a formaldehyde-induced fluorescence which on a molar basis is comparable in magnitude to that of primary CA. The use of this compound, recently introduced in monoamine fluorescence histochemistry, has yielded promising results with respect to localization of 5-HT neurons in the central nervous system (Jonsson, Fuxe, Hamberger and Hökfelt, 1969).

The theoretical-quantitative aspects of the method of Falck and Hillarp has been investigated by Ritzén using model protein microdroplets and microfluorimetric techniques (see Ritzén, 1966; 1967a, b). The studies on the fluorescence — concentration relationship of formaldehyde-treated CA (NA and DA) have shown that the fluorescence intensity increases in proportion with concentration up to a concentration value of about 4.5×10^{-2} M (moles per liter dried protein) corresponding to approximately 8×10^{3} μg/g wet weight of a cell. Above this value a considerable concentration-dependent quenching occurs. Corresponding figures for formaldehyde-5-HT fluorescence are 9×10^{-2} M and 16×10^{3} μg/g wet weight of the cell. Microfluorimetry of formaldehyde-treated mast cells (containing 5-HT) and sympathetic nerve cell bodies (containing NA) have shown that in these systems the fluorescence intensity correlates well with the chemically assayed monoamine-levels. The calculated concentration of NA in adrenergic nerve terminals (1—4×10^{3} μg/g wet weight) is below the quenching value, but since the transmitter is stored mainly in submicroscopic vesicles the local concentration must be considerably higher and thus quenching will occur. This seems in fact to be the case, judging from the results of a combined microfluorimetric and ^{3}H-NA uptake study of rat iris (Jonsson, 1969; see also Olson, Hamberger, Jonsson and Malmfors, 1968). It was found that the fluorescence intensity of NA increased proportional to the concentration of the substance up to about 30—40% of the endogenous level but above this value a concentration-dependent quenching occurred (*c.f.* van Orden, Bensch, Langer and Trendelenburg, 1967). The Quenching-Value for NA stored in the granules is somewhat lower, compared to the situation when NA is extragranularly distributed and in the former case the relative fluorescence yield as expected was lower. In the linear part of the fluorescence — concentration relationship very reliable semiquantitative estimations of the fluorescence intensity proved possible. From this study it can be concluded that, whereas changes from normal fluorescence levels with certainty reflect true changes in amine concentration, the converse is not necessarily true for considerable changes in amine concentration may escape detection in the fluorescence microscope when the amine-concentration is so high that quenching occurs. Since it is not possible to extrapolate quenching-values from one tissue to another, it is necessary in doubtful cases to correlate chemical assay and microfluorimetric data. Other microfluorimetric studies performed in this field are those of van Orden, Bloom, Barrnett and Giarman (1966) and Avakian and Gillespie (1968).

Some authors have suggested that the present technique has a serious limitation in that it only demonstrates amines located in the amine storage granules (Csillik, 1964; Csillik and Erulkar, 1964; Angelakos and King, 1965, 1966, 1967; Gillis, Schneider, van Orden and Giarman, 1966). On the contrary, however, extensive uptake studies have shown that the method demonstrates both granular and extra-granular monoamines, as well as monoamines taken up extraneuronally (see i.a. Malmfors, 1935; Hamberger, 1967; Jonsson, 1969). The qualitative spectral characteristics are the same for all of these storage forms, but the relative fluorescence yield is higher when the amine is stored outside granules (Jonsson, 1969). It must be emphasized, however, that the individual intraneuronal monoamine-storage granules cannot be demonstrated with this technique (Jonsson and Sachs, 1969).

A Detailed Description of the Histochemical Fluorescence Method for CA and 5-HT

The general procedure usually involves the following steps, listed in the order in which they are performed: (1) freeze-drying, (2) reaction with formaldehyde gas, (3) embedding, (4) sectioning and mounting. Instead of freeze-drying, cryostate sectioning can also be used, and in certain cases it can be of great advantage to use stretch-preparations (see below). Reference is here made to Eränkö's (1967) review.

1. The basic principle of this step involves rapid freezing of the pieces of tissues or organs to be studied and subsequent dessication *in vacuo* until most of the tissue water has been removed. The tissue specimen is dissected out as soon as possible after sacrifice of the animal and is then frozen by rapid dipping into liquid propane cooled by liquid nitrogen. The biogenic monoamines are not destroyed by this procedure and will remain at their cellular sites in the freeze-dried tissues (Corrodi, Jonsson and Malmfors, 1966). The tissue pieces can be rather large (e.g. whole frontal sections of the rat forebrain) provided the thickness does not exceed 3 mm. For identification purposes the specimen should be transferred before freezing to a small piece of fairly thick, rigid paper (thin papers usually are warped in the propane and as a consequence the specimen will crack considerably more than usual). It is almost impossible to prevent all cracking of brain pieces during the freezing, but the pieces can be kept together by covering the slices with a nylon net (Ungerstedt, to be published). The time in propane is not critical and can be extended to several hours.

The importance of rapid freezing is that it prevents the formation of tissue-disrupting ice crystals. If the pieces cannot be dried immediately, they can be stored on dry ice or in liquid nitrogen for several weeks without significant loss of quality. On transferring the specimens to the metal plate for drying *in vacuo*, thawing must be avoided by previous cooling of the plate in liquid nitrogen.

In the drying-procedure, large freeze-driers were initially used, with dry ice and liquid nitrogen in the cold trap (see Falck, 1962; Dahlström and Fuxe, 1964c; Falck and Owman, 1965). Thieme (1966) later designed a small fast-acting freeze-drier in which phosphopentaoxide replaced the cold trap; this apparatus had the disadvantage of allowing the processing of only small pieces 1 mm or less in thickness, but also the advantage of reducing the time required for adequate freeze-drying from several days to overnight (Hamberger, 1967). A recently developed new type of freeze-drier based on the Falck and Thieme models (Olsson and Ungerstedt, 1970) also uses phosphopentaoxide as adsorbent of the water removed from the specimen and, besides having the advantage of rapid freeze-drying and large capacity, permits the processing of large brain pieces with excellent results. The frozen tissue pieces, placed on a pre-cooled metal plate, are rapidly transferred to a stainless steel vacuum chamber which dips down into a cold bath containing ethanol. Two or three containers of granulated phosphopentaoxide are placed on top of each other just above the tissue pieces. The chamber is closed by a glass lid, and is connected to a two-stage gasballast vacuum pump giving a pressure of about 10^{-3} Torr. In this freeze-drier, even a large number (up to about 80) of thick brain specimens can be dried in 1—3 days. It is possible to determine when the freeze-drying can be safely terminated by reading the residual water pressure after closing the valve to the pump. The alcohol surrounding the stainless steel vacuum chamber is rapidly warmed to +35° C; and air is carefully admitted into the system about two hours later when the pecimens have reached room temperature. The metal-plate with the specimen s is ransferred as rapidly as possible to an exsiccator for subsequent formaldehyde gas reatment.

2. The critical factor in the formaldehyde gas exposure is the water content of the paraformaldehyde generating the formaldehyde gas. It is of critical importance to use water contents which will permit a strong reaction of the formaldehyde gas with the monoamines without causing any diffusion of the monoamines or their reaction products. To obtain paraformaldehyde of different water contents, samples of paraformaldehyde can be stored in exsiccators, each with a particular concentration of sulphuric acid placed in a Petri dish at the bottom (Hamberger, Malmfors and Sachs, 1965; Hamberger, 1967).

In our laboratory, optimal results have been obtained by using paraformaldehyde which has been stored in an environment having a relative humidity of 50 to 97.5%. In later experiments it was found that the most consistently good results are obtained by treating the specimens first with a relatively dry gas for 1 hour, and subsequently with a more humid formaldehyde gas for another hour (Fuxe and Jonsson, 1967). This procedure prevents diffusion of the fluorophors while still permitting the occurrence of the stronger reactions associated with the use of fairly humid formaldehyde gas. It is of particular value for the visualization of nerve terminals containing 5-HT.

The optimal humidity of the formaldehyde gas varies with the nature of the tissues studied. For example, drier gases are usually needed for embryonic tissues than would be optimal for adult material, and the same holds true for peripheral tissues in comparison with brain material. Owing to difficulties in standardizing the freeze-drying procedure, and to differences in quality of the paraformaldehyde from different batches and of different humidities, it is advisable to establish the best conditions for each laboratory in a preliminary test series, using a variety of formaldehyde gas humidities and different incubation times. Past experience has repeatedly emphasized the stringency of the requirements to be met, above all in the crucial steps of freeze-drying and formaldehyde gas treatment.

Generally the formaldehyde reaction is carried out conveniently in a well-closed one-liter jar containing 5—10 g of paraformaldehyde at a temperature of 80° C for 1 hour (Falck, 1962; Dahlström and Fuxe, 1964c; Hamberger, Malmfors and Sachs, 1965; Falck and Owman, 1965). It can, however, be of great advantage to have all specimens on the metal plate after completion of the freeze-drying, and place the plate in a prewarmed exsiccator containing 15—20 g paraformaldehyde in a Petri dish at the bottom (Ungerstedt, to be published). This procedure minimizes the loss of cracked specimens.

3. After the formaldehyde reaction, the tissue specimens are usually embedded in paraffin *in vacuo*. For this purpose a special glass equipment can be used (see Falck, 1962; Dahlström and Fuxe, 1964), designed so as to permit the specimens to fall from a shelf inside the vessel into the molten paraffin following air-evacuation of the vessel. Recently, this rather time-consuming procedure — which, moreover, often causes cracked tissue specimens to fall apart — has been replaced by a method (Ungerstedt, to be published) in which the tissue pieces, remaining on the metal plate after the formaldehyde gas treatment, are flooded with melted paraffin in an evacuated exsiccator heated in an oven at +60° C for 15—30 minutes. Sometimes it is necessary to use very short embedding times (Björklund, personal comm.) in order to avoid extraction of the fluorescent products.

In certain cases the specimens are advantageously embedded in Araldite (Hökfelt, 1965). This embedding medium permits thin sections to be made in which cytological details can be observed better than in paraffin sections. Araldite embedding has been useful, for example, in combined histochemical fluorescence and autoradiographic studies (Fuxe, Hökfelt, Ritzén and Ungerstedt, 1968) involving the intraventricular administration of labelled CA and 5-HT.

After embedding in paraffin each specimen is rapidly taken out to a small paper box in which it is oriented in melted paraffin, and the paraffin is allowed to set.

4. Before sectioning, the paper pieces attached to the tissue specimen must be removed. Usually the sections are made about 8 μ in thickness. The use of thicker sections leads to problems due to increased interference by background fluorescence. The sections are mounted on well-cleaned glass slides, and enclosed in Entellan (Merck p.a.) containing a small amount of xylene to dissolve the paraffin. The sections should be examined as soon as possible, since the neuronal fluorescence will decrease with time while the background fluorescence becomes intensified. However, stored in darkness the sections can be kept for at least two weeks without significant deterioration.

The use of cryostat sections in demonstrating biogenic monoamines has been advocated by several authors (Hamberger and Norberg, 1964; Csillik and Erulkar, 1964; Spriggs, Lever, Rees and Graham, 1966; Csillik, Kálmán and Knyihar, 1967; El-Badawi and Schenk, 1967; Laties, Lund and Jacobowitz, 1967; Nelson and Wakefield, 1968, Schiebler and Heene, 1968). Although this technique has the advantage of being less time-consuming and, moreover, permits cholinesterase staining of the same or adjacent sections, it carries a high risk of diffusion, especially in central nervous tissue. Recently, however, Placidi and Masuoka (1968) have published a technique which has produced good results in brain tissue. Some workers have suggested the combined use of catecholamine fluorescence technique and methylene-blue staining (Ehinger, Sporrong and Stenevi, 1967) on the basis of their belief that the latter technique preferentially stains cholinergic axons.

Various Modifications of the Formaldehyde Method

It has also been possible to visualize CA nerve terminals in brains that had been perfused with a formalin solution (Laties, Lund and Jacobowitz, 1967). Furthermore, adrenergic nerve fibers have been successfully demonstrated in iris incubated in a hypertonic salt solution with formalin (Eränkö and Räisänen, 1966). As in other departures from the standard procedure, it is difficult to avoid diffusion of the amines or their reaction products and obtain quantitative yields by these methods.

An improvement of considerable value in visualizing very fine monoamine nerve terminals in certain areas of the cerebral cortex has been the introduction of hypertonic salt perfusion of the living animal before dissection (Hökfelt, unpubl. observation). This improvement may be explained by the fact that diffusion is diminished under hypertonic conditions.

Stretch Preparations

Certain thin tissues and membranes, *e.g.* the iris and the mesentery, offer a unique opportunity to demonstrate adrenergic nerve terminals in their entire extent by means of stretch preparations (see Falck, 1962; Malmfors, 1965). The resected tissue is stretched out on a glass slide under the dissection microscope, and is then dried for at least one hour in an exsiccator containing phosphopentaoxide before being exposed to formaldehyde gas. After the formaldehyde treatment the object is ready for microscopic examination. No mounting is necessary, provided the diffraction of emitted light is not too high. By this simple procedure, it has been possible to make detailed studies of the adrenergic ground plexus around large vessels, in the bladder and in the plexus of Meissner and Auerbach (unpublished observations).

The DA, NA and 5-HT varicosities in the nerve-ending fraction of homogenized-tissue ultracentrifugates can be demonstrated by fluorescence histochemistry by spreading the fraction in a thin layer over glass, exsiccation with phosphopentaoxide, and exposure to formaldehyde gas (Fuxe, Grobecker, Hökfelt and Jonsson, 1967).

Examples of Pharmacological Means to Increase the Sensitivity of the Histochemical Fluorescence Method for CA and 5-HT

Following the *intra-vitam* administration of monoamine-oxidase (MAO) inhibitors such as nialamide, intraneuronal storage of 5-HT is increased considerably (Dahlström and Fuxe, 1964). In brain tissues taken from animals thus treated the visualization of the 5-HT nerve cell bodies, axons and terminals is greatly facilitated. This is of importance particularly because the fluorescence method is less sensitive for 5-HT than for primary CA.

Another important pharmacological aid in attempts to localize 5-HT is the recently introduced 6-HT (Jonsson and Sandler, 1969; Jonsson, Fuxe, Hamberger and Hökfelt, 1969), an indole derivative readily incorporated and accumulated in the 5-HT nerve terminals. Since 6-HT has a high affinity to 5-HT terminals and, moreover, its fluorescence yield is considerably higher and its photodecompostion rate lower than that of 5-HT, 5-HT nerve terminal systems not previously observed have been visualized recently by incubation of brain slices *in vitro* in a medium containing 6-HT. In higher concentrations 6-HT is also strongly incorporated by DA and NA nerve terminals, which consequently become likewise more strongly fluorescent than normal.

Significance of the Histochemical Fluorescence Method for CA and 5-HT

The development of the histochemical fluorescence technique opened up a new field of inquiry which may be called transmitter histochemistry. By virtue of its high sensitivity and specificity, the method has enabled investigators to localize the peripheral adrenergic transmitter at the cellular level and thus has contributed significantly to the knowledge of the morphology and function of the autonomic innervation apparatus (Falck, 1962; Norberg, 1965, 1967). Furthermore, the method has led to the disclosure in the central nervous systems of both vertebrates (see Fuxe, 1965; Fuxe, Hökfelt and Nilsson, 1965; Hillarp, Fuxe and Dahlström, 1966) and invertebrates (Dahl, Falck, Lindqwist and Mecklenburg, 1962; Dahl, Falck, Mecklenburg and Myhrberg, 1963; Clark, 1966; Rude, 1966; Myrhberg, 1967; Frontali, 1968) of neuron systems, each characterized by intraneuronal storage of one of the transmitter substances, DA, NA or 5-HT, mainly in the terminal parts of the axon (Fig. 8). For the first time a neuron system characterized by a known transmitter could be mapped out and changes in transmitter stores visualized at the cellular level. For some of these previously unsuspected chemically specified systems important functions have been identified. The amount of transmitter stored in the presynaptic structure (the so called varicosity) of the monoamine neurons is in the order of 4 to 5×10^{-3} pg. The high sensitivity of the fluorescence method is demonstrated by the finding that fluorescence, although very much weaker, can still be observed in varicosites which have suffered an 80—90% depletion of their transmitter store (Fuxe and Sedvall, 1964; Andén, Fuxe, Hamberger and Hökfelt, 1966; Jonsson, 1969). Since the varicosities are relatively small (0,4—1 μ) in diameter the concentrations (1,000—8,000 μg/g w. wt.) will still be very high. In the cell bodies (Figs. 9, 10), on the other hand, the concentrations are lower (10—100 μg/g w. weight) despite the fact that the total amount has been found to be in the order of no less than 0.5 pg in peripheral adrenergic perikarya. The large volume of the cell bodies accounts for this difference in concentration. Owing to these low concentrations, the lower limit of demonstrability lies at the level of at the most one-tenth of a pg of CA in most cell bodies.

A problem in the neuroanatomical studies with this fluorescence technique has been the fact that the non-terminal axons of the monoamine neurons may exhibit a very low fluorescence intensity or even none at all, due to the very low concentrations found in the axons. To solve this problem two possibilities exist: either monoamineoxidase inhibition (Fig. 11), which will increase markedly the amounts of amine in the entire *5-HT* neurons (Dahlström and Fuxe, 1964; Dahlström, Fuxe and Hillarp, 1965), or experimental interruption of the axons. In the latter case, strongly fluorescent material accumulates in the proximal stump of both CA and 5-HT axons (Fig. 12) (Dahlström and Fuxe, 1964a, b; Dahlström, 1966; Blümcke and Niedorf, 1965; Kapeller and Mayor ,1967; Mayor and Kapeller, 1967). This accumulation is

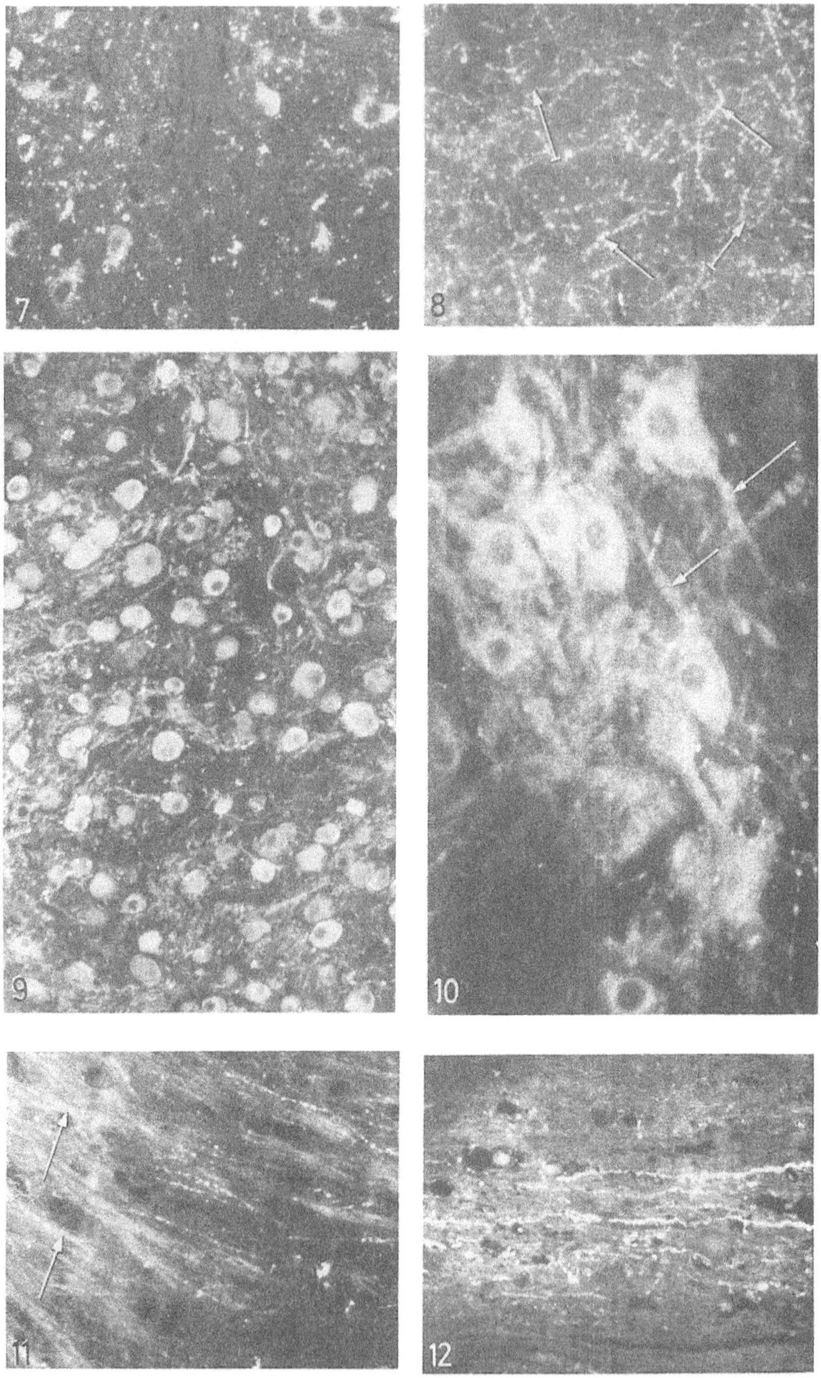

Fig. 7. Autofluorescent nerve cell bodies of the medulla oblongata of cat. The fluorescence exhibits a granular appearance and has a yellow-orange colour. The autofluorescent granules are usually located mainly on one side of the cell nucleus. ×190

probably caused by piling up of amine storage particles in transport from the cell-bodies to the nerve terminals (Dahlström and Fuxe, 1964a; Dahlström, 1966; Hökfelt, 1969).

In order to identify the axon system giving rise to a special terminal monoamine network, lesions of the monoamine cell bodies or non-terminal axons can be performed. After such lesions the monoamine terminals belonging to lesioned axons will undergo anterograde degeneration which is accompanied by a disappearance of fluorescence in these terminals due to degeneration of the storage mechanism for amines (Fig. 13). In the central DA neurons the capacity of axon terminals to store amines is partly lost already after 2 days (Hökfelt and Ungerstedt, 1969) whereas the terminals of central NA and 5-HT neurons exhibit the first signs of loss of amines stores after 3—4 days (Carlsson, Falck, Fuxe and Hillarp, 1964; Andén, Dahlström, Fuxe, Larsson, Olson and Ungerstedt, 1966).

Lesions as described above can also be used to determine which fibres correspond to which cell bodies, viz. by studying the retrograde cell changes of the body. These changes are revealed in the form of either an increased fluorescence intensity shortly after the lesion (Fig. 14) (Dahlström and Fuxe, 1965), or a decreased fluorescence intensity a longer time (e.g. 3—4 weeks) after the lesion (Andén, Dahlström, Fuxe and Larsson, 1965). In the latter cases, the affected cell bodies may also exhibit chromatolysis and cell shrinkage or swelling (Fig. 15).

Electrothermic lesions, even if very small, may involve neurons that form no part of the system studied. Recently Ungerstedt (1968) has avoided this problem by introducing 6-OH DA into the substantia nigra by intracerebral injection. 6-OH DA is taken up by the DA cell bodies which consequently undergo degeneration accompanied by the appearance of a yellowish fluorescence, not induced by formaldehyde, in the cell bodies. Degeneration of the terminal parts of NA neurons after 6-OH DA was reported earlier by a group of Swiss investigators (Tranzer and Thoenen, 1968) from electron-microscopic studies of peripheral andrenergic neurons. These earlier results were confirmed by fluorescence histochemical studies on the adrenergic nerve terminals of dilator muscle of the iris (Malmfors and Sachs, 1968; Jonsson and Sachs, 1970).

Fig. 8. Retrochiasmatic area of normal rat. Plexus of NA nerve terminals. Varicosities of various thickness are observed. Some are thick 1—2 μ (↗), some fine, 0.5—1 μ (↙ˣ). They probably establish mainly axodendritic contacts. ×120

Fig. 9. Superior cervical ganglion, monkey. NA-containing cell bodies fluorescent at varying intensity. Fluorescent fibres are seen among the cell bodies. ×120. (By courtesy of Dr. K.-A. Norberg)

Fig. 10. NA cell bodies of the locus coeruleus, rat. The cell bodies exhibit a strong fluorescence which is most prominent in a perinuclear zone. Note fluorescent processes (↗). (Dahlström and Fuxe, unpublished material.) ×200

Fig. 11. Horizontal section of the medial forebrain bundle in a nialamide-pretreated rat (100 mg/kg, 5 hours). Weakly fluorescent smooth bundles of CA axons (↗) among which a number of more strongly fluorescent varicose CA nerve terminals are present. ×190

Fig. 12. Proximal part of spinal cord in rat one day after transection. Sagittal section. There is an accumulation of green and yellow fluorescence in transsected NA and 5-HT axons respectively. (Dahlstrom and Fuxe, unpublished material.) ×120

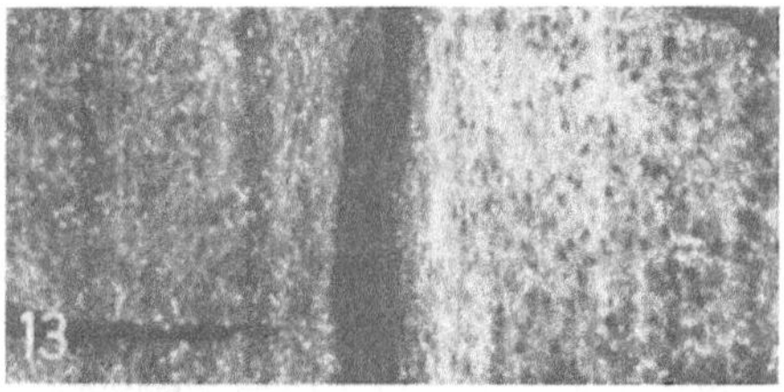

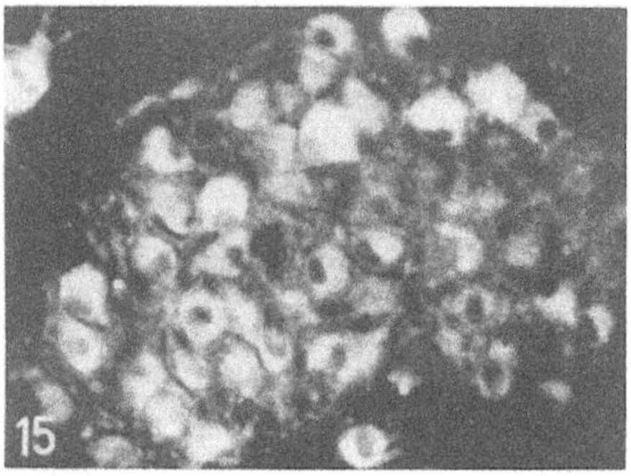

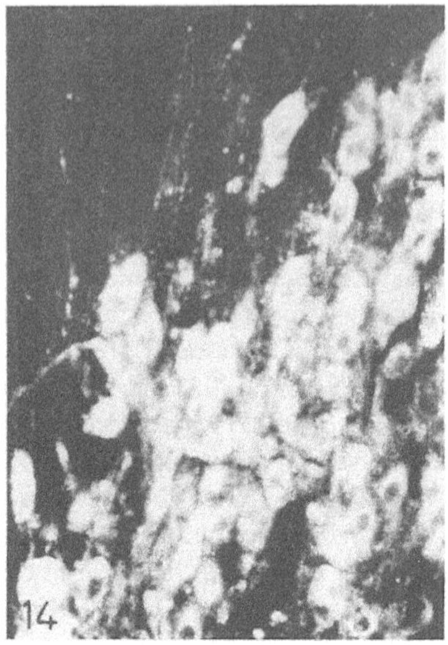

Fig. 13. NA terminals around the third ventricle. The considerable reduction in the number of terminals on the left side of the ventricle is caused by an electrothermal lesion of the ascending NA axons in the midbrain, 6 days before the animal was sacrificed. ×75

Fig. 14. Na cell bodies of the locus coeruleus one day after lesion of the ascending NA axons in the rostral midbrain. Increased fluorescence intensity in the cytoplasm of the cells. ×120

Fig. 15. Locus coeruleus, seven days after a lesion of the ascending NA axons in the caudal midbrain. Compare with Fig. 14. The NA cell bodies now exhibit clearcut retrograde changes: swelling of the cell body, and peripherally displaced nucleus; Nissl staining at this stage reveals chromatolysis. Some cells exhibit an increase in the fluorescence intensity of the cytoplasm, probably due to an accumulation of amine storage granules. ×190

The charting of central monoamine-containing pathways can also be facilitated by *in vitro* incubation of brain slices with monoamines. Since there is an uptake-concentration mechanism for amines at the cell membrane of the *entire* monoamine neuron, monoamine fibers will become strongly fluorescent and can thus be traced in the brain with greater ease, especially when displayed longitudinally (Fig. 16). The 5-HT fibres preferentially take up and accumulate 5-HT, whereas the NA and DA fibres preferentially take up and accumulate DA and NA. It has not yet been possible histochemically to distinguish between DA and NA fibres (see Fuxe and Ungerstedt, 1966; 1968a, b; Jonsson, Fuxe, Hamberger and Hökfelt, 1969).

Fig. 16. Sagittal slice of the capsula interna area of rat after incubation with α-methyl-NA (10^{-6} M). There is a marked accumulation of fluorescence in the ascending nigro-neostriatal DA fibres in the capsula interna. The fibres have a varicose-like appearance and are largely arranged in bundles. ×120

Fig. 17. Caudatus-putamen and globus pallidus of a reserpinenialamide pretreated rat killed 30 min after an intraneostriatal injection of NA (5 μg/5 μl). There is an accumulation of fluorescence in large parts of the DA nerve terminal network (right) and also a retrograde accumulation of fluorescence in the nigro-neostriatal DA nerve fibres (left). The mechanism underlying the latter phenomenon is as yet unknown. ×75

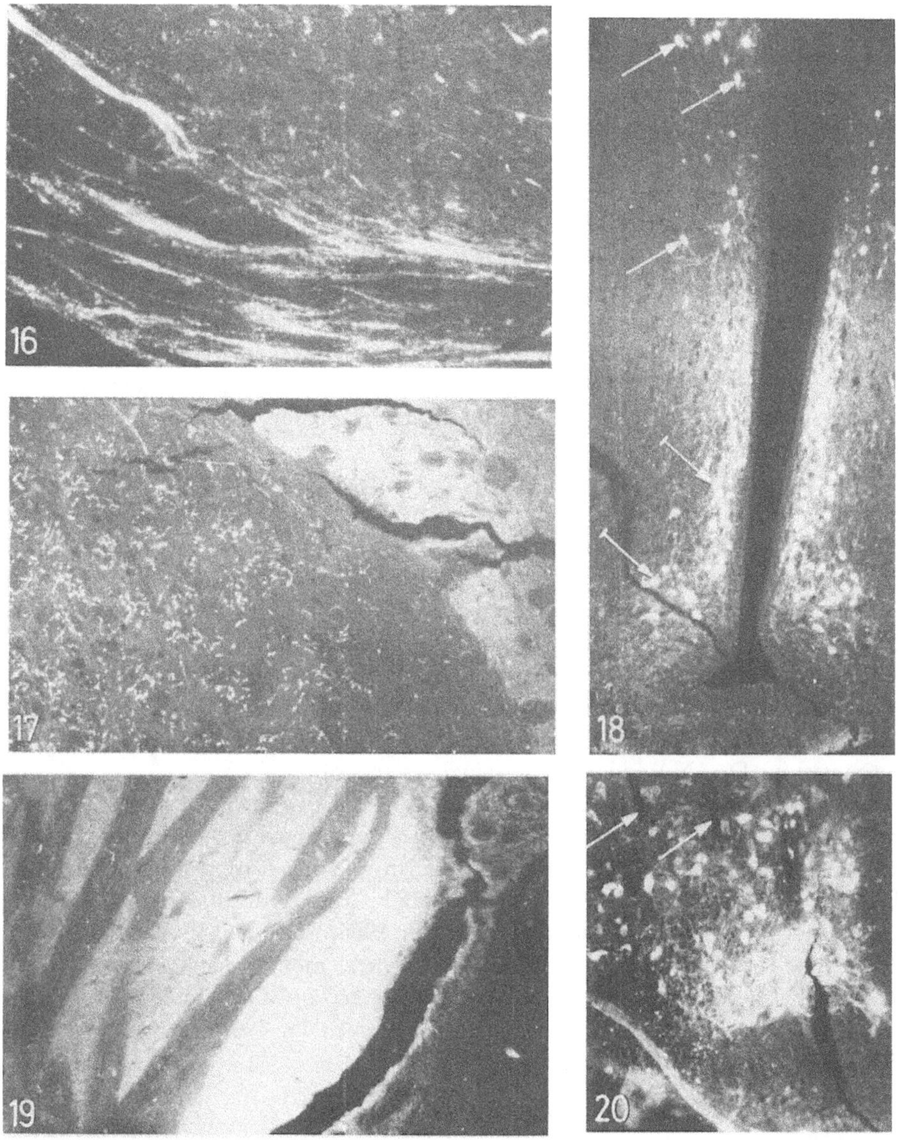

Fig. 18. The periventricular hypothalamic area of a rat killed 30 min after an injection of α-methyl-NA (5 μg/20 μl) into the lateral ventricle. Reserpine had been given 24 hours earlier There is a marked accumulation of fluorescence in periventricular CA nerve cells (↗), in CA nerve cells of the nuc. arcuatus (↗), in cells of the capillary walls and in periventricular CA nerve terminals. Outside the periventricular zone (about 300 μ wide) no accumulation of CA can be observed. ×75

Fig. 19. Reserpine-treated rat killed 30 min after an intraventricular injection of α-methyl-NA (5 μg/20 μl). There is a strong accumulation of amine in the dense DA terminals of the nucleus caudatus-putamen (left) whereas no fluorescence appears in the septal area (right). Note the strong concentration gradient extending from the lateral ventricle into the tissue. ×75

Fig. 20. The ventral midbrain of a reserpine-pretreated rat after an intraventricular injection of α-methyl-NA (5 μg/20 μl). There is a marked accumulation of fluorescence only in the CA cell bodies lying close to the subarachnoidal space (bottom left). The radices nervi III appear at upper left, (↗). ×75

Instead of by *in vitro* incubation, monoamines can also enhance the fluorescence of the corresponding neurons when administered by *in vivo* intracerebral injection, as discovered recently in studies on direct chemical stimulation of the neostriatum (Ungerstedt, Butcher, Butcher, Andén and Fuxe, 1969) After injection into the neostriatum of small volumes (5 μl) of saline solutions containing 5—10 μg of DA or NA a marked accumulation of these amines takes place in the dense plexus of DA axon terminals of the neostriatum, while a retrograde accumulation of CA will occur in the nigro-neostriatal DA fibres (Fio. 17) as far down as the rostral part of the substantia nigra. With the volumes used, the nigral axons are the only fibres outside the neostriatum which accumulate fluorescent material. However, when injected in large volumes such as 50—100 μl, the catecholamines diffuse out and are taken up by CA nerve terminals in large parts of the tel- and diencephalon and also to some degree in the mesencephalon.

Besides intracerebral administration, intraventricular injections of, and perfusion with, CA and 5-HT have been performed in attempts to by-pass the blood-brain barrier. It was found, however, that CA and 5-HT administered by intraventricular route are mainly taken up by those parts of the monoamine neurons lying close to ventricles (Fig. 18, 19) and the ventral part of the subarachnoidal space (Fig. 20) (Fuxe and Ungerstedt, 1966, 1968a, b). These results led to the conclusion that intraventricular injections, although of little use to the purpose of neuronal charting could be of value in pharmacological studies if the short distance of penetration into the brain is borne in mind, since the basic mechanisms probably are essentially the same in the various DA, NA and 5-HT terminal systems of the CNS.

By the combined use of all of the known principles and ancillary procedures of the histochemical fluorescence technique it has been possible to map out the main monoamine neuron systems in the mammalian and avian central nervous system (Dahlström and Fuxe, 1964, 1965; Fuxe 1965a b; Fuxe and Ljunggren, 1965; Fuxe and Andén, 1966; Hillarp, Fuxe and Dahlström, 1966; Fuxe, Hökfelt and Ungerstedt, 1968). The principal organization of these central monoamine systems has been confirmed at least in part in other laboratories (Barry and Leonardelli, 1967; Baumgarten and Braak, 1967; Braak, Baumgarten and Falck, 1968; Sharp and Follett, 1968).

Central DA Neurons

The Tubero-Infundibular DA Neurons (Fuxe, 1964; Fuxe and Hökfelt, 1966, 1967; Lichtensteiger and Langemann, 1966; Björklund, Enemar and Falck, 1968; Björklund, Falck and Ljunggren, 1968), a short-neuron system which innervates the median eminence (Fig. 21). Most of its cell bodies are probably localized in the nucleus arcuatus (Fig. 18). The system is probably involved in the control of the gonadotrophin secretion (Fuxe, Hökfelt and Nilsson, 1967, 1969; Fuxe and Hökfelt, 1967, 1969a, b).

Fig. 21. Median eminence of a nialamide-pretreated rat in sagittal section. The intense fluorescence in the outer layer of the median eminence is due to a high content of DA nerve terminals. Fluorescent DA cell bodies (↗) can be observed in the nuc. arcuatus. Weakly fluorescent strands, probably composed of DA fibres (↗) run to the median eminence from the nuc. arcuatus. ×75

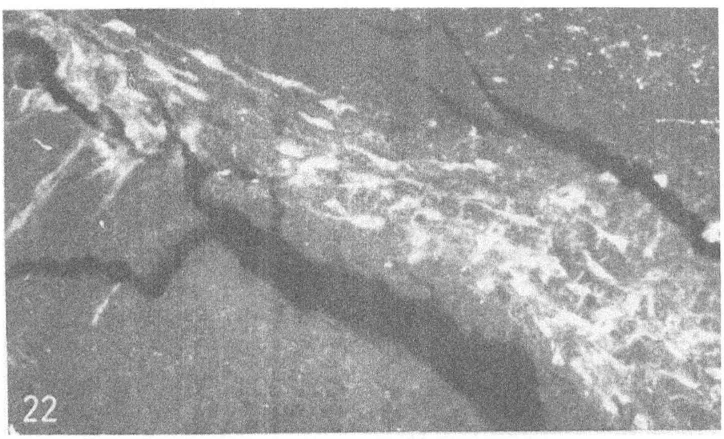

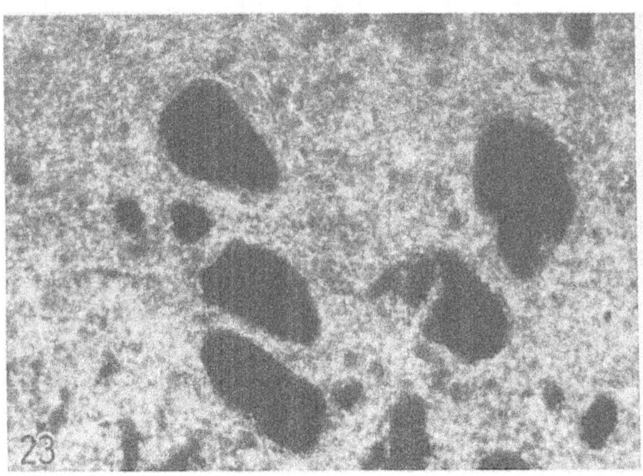

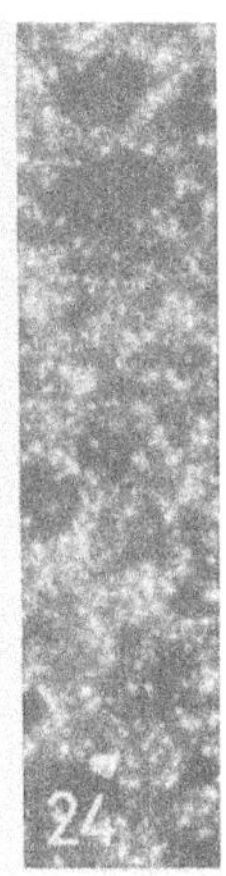

Fig. 22. Zona compacta of the substantia nigra of a normal rat. Fluorescence is observed in the DA cell bodies and their processes. ×120

Fig. 23. Caudatus-putamen of normal rat. The 10 μ-thick paraffin section reveals a very dense plexus of very fine, strongly fluorescent DA nerve terminals. ×190

Fig. 24. Caudatus-putamen of normal rat (Araldite section about 2 μ). The small varicosities of the densely packed DA nerve terminals are resolved in thin sections. ×480

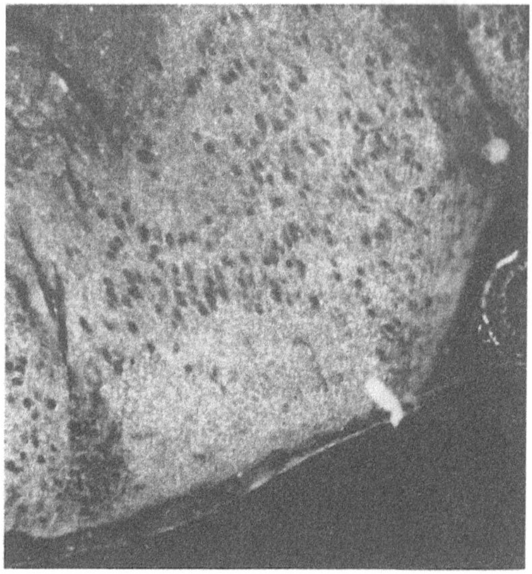

Fig. 25. Tuberculum olfactorium of normal rat. The very fine dots represent a dense plexus of DA nerve terminals of the same type as found in caudatus-putamen. × 120

The Nigro-Neostriatal DA Neurons. This large uncrossed system has been studied extensively both in fluorescencemicroscopic (Figs. 22, 23, 24) (Andén, Carlsson, Dahlström, Fuxe and Hillarp, 1964; Andén, Dahlström, Fuxe and Larsson, 1965, 1966; Ungerstedt, Butcher, Butcher, Andén and Fuxe, 1969; Ungerstedt, 1969) and electronmicroscopic studies (Hökfelt, 1968; Hökfelt and Ungerstedt, 1969). By the aid of the Fink-Heimer technique of silver impregnation the existence of this connection recently has been confirmed in more traditional neurohistological studies (Moore, personal comm.). The results of neurophysiological and neurochemical studies have indicated a particular importance of this system in certain motor mechanisms (Andén, Dahlström, Fuxe and Larsson, 1966; Sourkes and Poirier, 1966; Ungerstedt, 1968).

Fig. 26. Transverse section of reticular formation of a caudal midbrain immediately caudal to a lesion in the mesencephalic reticular formation. There is a marked accumulation of fluorescence in ascending NA axons which can be separated into a dorsal (top) and a ventral bundle. ×75

Fig. 27. Transverse section of the spinal cord about 5 mm cranial to a spinal cord transsection. The NA and 5-HT axons appear as small fluorescent dots, indicating small caliber of the axons. (Dahlström and Fuxe, unpublished material.) ×300

Fig. 28. Transverse section of the left half of the hypothalamus of a normal rat. A high-to-very high density of fluorescent nerve terminals is observed in the nucleus dorsomedialis hypothami (*DM*). Ventral to the fornix (*F*) very few terminals are observed. A low-to-medium density of mostly thick NA nerve terminals appears in the lateral hypothalamus (*LH*). A medium-to-high density of terminals is present in the most ventral parts (*V*). ×30

Fig. 29. Sagittal section of the normal rat spinal cord. There is a high density of strongly fluorescent NA and 5-HT nerve terminals in the lateral (sympathetic) column. ×120

Fig. 30. Lateral reticular formation of the medulla oblongata of normal rat, containing a plexus of NA nerve terminals, some of which appear to make axosomatic contacts. ×300

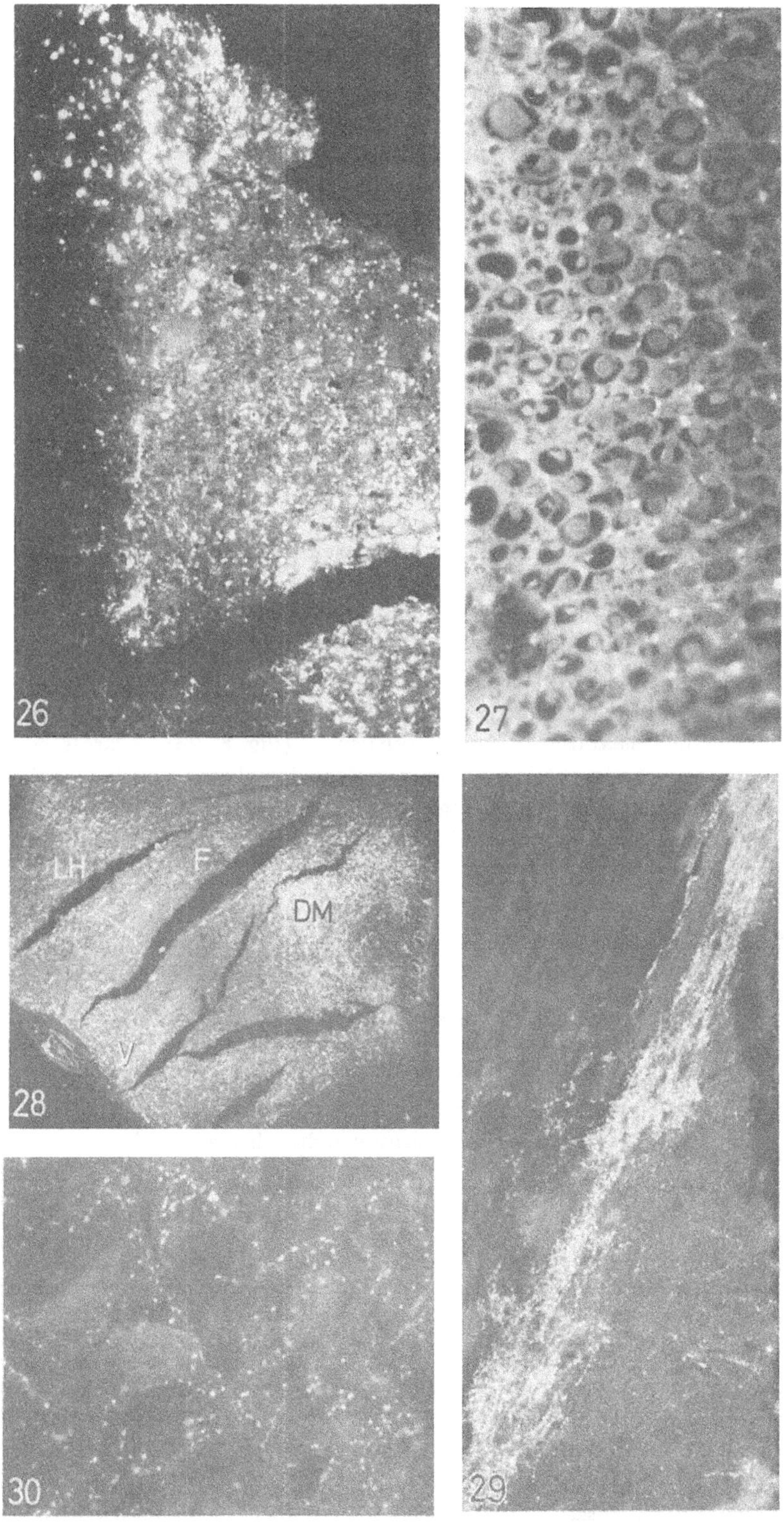

Figs. 26—30

The Mesencephalo-Limbic DA Neurons. The cell bodies of this system are mainly localized to the area dorsal to the cranial part of the nuc. interpeduncularis (Andén, Dahlström, Fuxe, Larsson, Olson and Ungerstedt, 1966; Dresse, 1966) and the ascending axons innervate mainly the nucleus accumbens and the tuberculum olfactorium (Fig. 25). The function of this system is at the present time unknown. An additional DA system has also been discovered in the retina (Malmfors, 1963; Häggendal and Malmfors, 1965). The cell bodies of these neurons are reminiscent of amacrine cells, and their axon terminals are located exclusively in the inner plexiform and nuclear layers (see also Ehinger, 1966). Their function is unknown.

Central NA Neurons

The central NA neurons are not as readily classified into various distinct systems as is the case with the DA neurons. The NA cell bodies (Fig. 10) are mainly localized to the reticular core of the medulla oblongata and pons (Dahlström and Fuxe, 1964; Andén, Dahlström, Fuxe, Olson and Ungerstedt, 1966) and many of the NA neurons seem to innervate large parts of the brain and spinal cord by way of long ascending (Fig. 26) and descending (Fig. 27) axons which in turn give rise to collaterals throughout the lower brain stem. Some cell bodies seem to emit mainly descending axons, whereas others have axons oriented predominantly in the rostral direction. Thus, the NA neurons, much like other reticular neurons, probably modulate transmission in a large number of areas in the CNS; *e.g.* in visceral autonomic mechanisms such as those of the hypothalamus (Fig. 8, 28), in areas containing preganglionic parasympathetic cell bodies (*e.g.* the dorsal motor nucleus of the vagus) or in preganglionic sympathetic regions such as the sympathetic lateral column (Fig. 29), in areas containing visceral afferent nuclei (Fig. 56), in parts of the reticular

Fig. 31. Septal area of normal rabbit, exhibiting locally high density of NA nerve terminals in central parts, and very low density close to the lateral ventricle (left). ×75

Fig. 32. Gyrus dentatus of normal rat, showing NA terminals in high density in the hilus fasciae dentatae (bottom left), low density in the molecular layer (upper right) and sparseness in the granular layer. Mainly axodendritic contacts are made. ×190

Fig. 33. Gyrus cinguli of normal rat. A medium density of Na nerve terminals is observed in the molecular layer (top). The density in the other layers is lower. ×75

Fig. 34. Second and third layers of neocortex in the normal rat, containing a plexus of fine, varicose NA nerve terminals. The dendrites appear as dark lines. ×190

Fig. 35. Longitudinal section of the substantia gelatinosa of normal rat. Longitudinally oriented varicose NA nerve terminals are observed, probably making axodendritic contacts. (Dahlström and Fuxe, unpublished material.) ×190

Fig. 36. Ventrolateral part of the ventral horn of the rat lumbar spinal cord, showing a plexus of fluorescent NA and 5-HT nerve terminalis. NA nerve terminals appear to establish axosomatic contacts (↗) with α-motoneurons. (Dahlström and Fuxe, unpublished material.) ×190

Fig. 37. Suprachiasmatic nucleus of the hypothalamus in a nialamide-treated rabbit (100 mg/kg i.p., 20 hours before sacrifice). A plexus of very fine yellow fluorescent 5-HT terminals is seen in the lower half of the figure. ×75

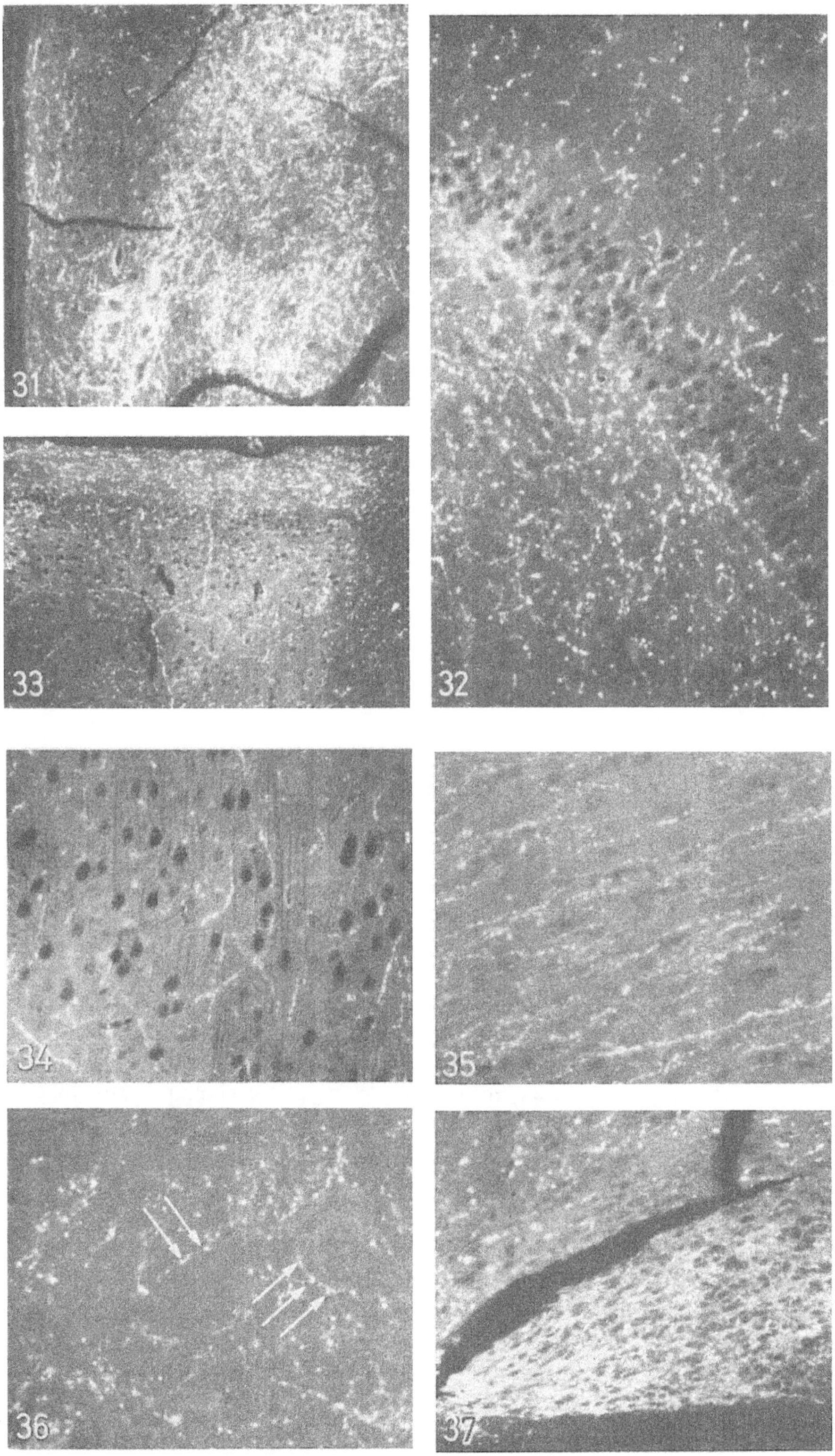

Figs. 31—37

formation (Fig. 30), in the limbic system (Figs. 31, 32, 33) and in the neo-cortex (Fig. 34) (see Fuxe, 1965; Fuxe, Hamberger and Hökfelt, 1968; Blackstad, Fuxe and Hökfelt, 1967). The NA nerve terminals probably can make both axo-dendritic (Fig. 35) and axo-somatic (Fig. 36) contacts. Many neuroanatomists (Moore, Heller and co-workers) have been reluctant to accept the fluorescence-microscopic evidence indicating that NA and 5-HT axons emanating from the lower brain stem extend directly into the neocortex and the limbic system (see Fuxe, 1965; Andén, Dahlström, Fuxe, Larsson, Olson and Ungerstedt, 1966), for the reason that such connections so far have not been demonstrated by traditional neurohistological techniques [although they are indicated in Golgi preparations by the Scheibels (1967)]. It would not be surprising, however, if the techniques of transmitter histochemistry, by virtue of their unique chemical and physical principles, would permit the visualization of neural pathways which have remained refractory to traditional experimental-neuro-anatomical demonstration. Moreover, it remains to be seen whether the existence of direct bulbo-cortical connections may perhaps be confirmed by the more recently developed Fink-Heimer (1967) technique in cases of lesion of the 5-HT cell bodies and the NA cell bodies.

The NA neurons seem to participate *e.g.* in producing sham rage (Fuxe and Gunne, 1964; Reis and Gunne, 1965; Reis and Fuxe, 1968a), aggressiveness and hyperactive intensive behaviour (Ungerstedt and Fuxe, unpublished data; Andén, Corrodi, Fuxe, Hökfelt, Hökfelt, Rydin and Svensson, 1970).

Central 5-HT Neurons

The 5-HT neuron systems originate from cell bodies with a localization very different from that of the NA and DA cell bodies, namely, the raphe nuclei and to some extent also the medial reticular core (Dahlström and Fuxe, 1964). The raphe nuclei in the pons and the medulla oblongata appear mainly to give rise to descending axons, whereas the mesencephalic raphe nuclei emit mostly ascending 5-HT axons (Fig. 51). Much like the NA nerve terminals, the 5-HT terminal systems have a wide distribution in spinal cord and brain (Fig. 37), so that most areas receive both NA and 5-HT nerve terminals. However, within each area, *e.g.* the hypothalamus (Fig. 28) or the hippocampal formation, the distribution patterns of NA and 5-HT nerve terminals differ markedly. The 5-HT neurons appear to be involved in thermo-regulation (Feldberg and Myers, 1963; Corrodi, Fuxe and Hökfelt, 1967, 1968), in the sleep-wakefulness mechanism (see Jouvet, 1968) and also in the neural mechanisms underlying "moods" (Carlsson, Fuxe and Ungerstedt, 1968; Fuxe and Ungerstedt, 1968; Carlsson, Corrodi, Fuxe and Hökfelt, 1969); they may mediate at least some of the psychic effects produced by hallucinogenic drugs of the indol alkylamine type (Andén, Corrodi, Fuxe and Hökfelt, 1968). It is important to emphasize the funda-mental neuroanatomical difference between the NA and 5-HT neurons: the NA neurons are spread widely through the reticular tegmentum, whereas the 5-HT neurons are largely confined to the raphe systems, an anatomical distinction suggest-ing that the functions of these two neurons systems may differ markedly from one another. — There probably exist monoaminergic mechanisms in the area postrema, for both CA and 5-HT cell bodies are found in this area (Fuxe and Owman, 1965).

Peripheral adrenergic neurons

The use of fluorescence histochemistry has led to important discoveries in the peripheral autonomic nervous system (see review by Norberg, 1967 and Malmfors, 1967). To begin with, fluorescence studies confirmed the existence of the adrenergic ground plexus postulated many years previously by Hillarp (1946). In view of the histochemical findings, the ground plexus can now be regarded as the true adrenergic innervation apparatus, for nerve impulses were found to release transmitter substance from all the varicosities in the plexus (Malmfors, 1964). Moreover, these studies produced evidence that the adrenergic ground plexus is composed of divergently and

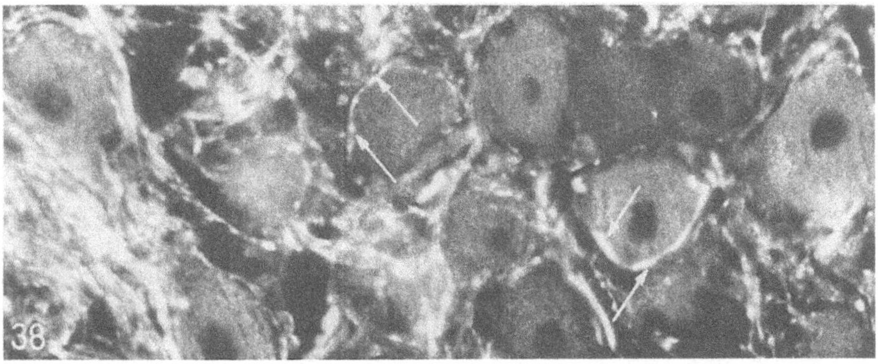

Fig. 38. Coeliac ganglion of cat, showing NA cell bodies fluorescent at varying intensities. Strongly fluorescent processes run between the cell bodies and some cell bodies are in close contact with varicose NA nerve terminals (↗). ×375. (Courtesy of Dr. K.-A. Norberg)

convergently distributed adrenergic axons originating from different NA cell bodies in the sympathetic ganglia, a morphological finding corroborating Cannon and Rosenblueth's view (1937) (see Hillarp, 1959; Malmfors and Sachs, 1965).

Fluorescence studies not only permitted the localization of NA cell bodies in the sympathetic ganglia (Fig. 9) (Norberg and Hamberger, 1964), but also disclosed a previously unknown adrenergic nerve terminal system in these ganglia (Fig. 38), especially in the abdominal ganglia (Hamberger and Norberg, 1963, 1964; Hamberger, Norberg and Sjöqvist, 1963a, b). These adrenergic terminals appear to form synapses with some of the ganglion cells, an observation suggesting that ganglionic transmission may somehow be modulated by adrenergic impulses affecting the ganglion cells (see Norberg and Sjöqvist, 1966). Denervation experiments have indicated that these terminals belong to interneurons or to collaterals arising from NA ganglion cells (Hamberger, Norberg and Ungerstedt, 1965; Jacobowitz and Woodward, 1968). Furthermore, cholinergic nerve cell bodies were found not to contain any NA (Hamberger, Norberg and Sjöqvist, 1965). Evidence was obtained that the NA storage granules are produced in the cell bodies and transported down to the terminals via the axons (Dahlström and Fuxe, 1964b; Dahlström, 1965).

Developmental studies of the sympathetic adrenergic neurons have demonstrated that the nerve growth factor causes an increase in number and intensity of the noradrenergic nerve terminals (Figs. 39, 40), as well as hypertrophy of the NA cell bodies (Olson, 1967). Adrenergic synaptic terminals likewise increased in number

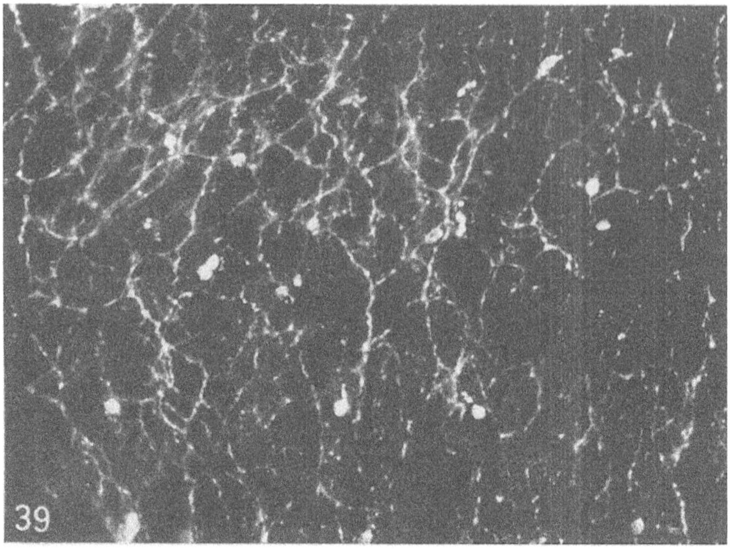

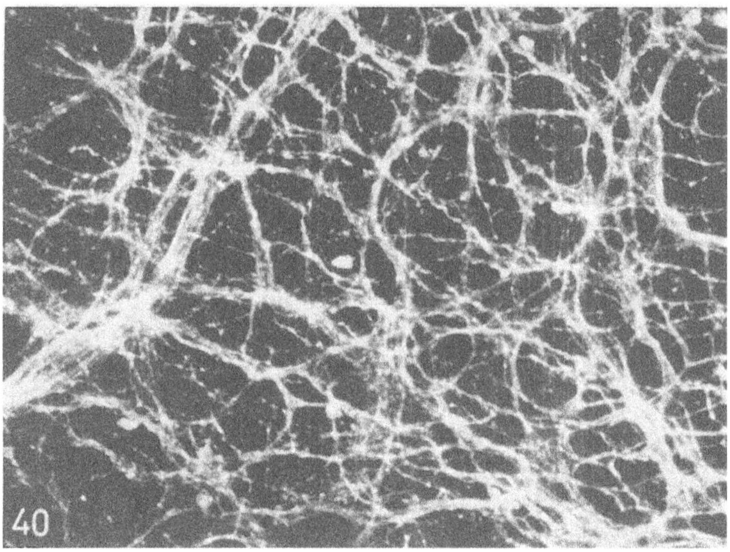

Fig. 39. Stretch preparations of iris from 14—16 day-old mouse. Untreated animal. The plexus of fluorescent nerve terminals over the dilator muscle is composed mainly of single strands. Varicosities are less distinct than in iris of adult mouse. Yellow fluorescent mast cells are scattered over the picture. ×210. [Reproduced from Olson, L., Z. Zellforsch. **81**, 155—173 (1967), by permission of the Publishers]

Fig. 40. Stretch preparation of iris of a 14—16 day-old mouse treated with nerve-growth-factor once daily since birth. Comparison with Fig. 39 shows a remarkable increase in the number of fluorescent fibers in many strands of the ground plexus. Some single varicose terminals are also seen. Yellow fluorescent mast cells are scattered throughout the picture. ×210. [Reproduced from Olson, L., Z. Zellforsch. **81**, 155—173 (1967), by permission of the Publishers] in the ganglia (Hammer, Norberg, Olson and Sjöqvist, 1967). In immunosympathectomized rats only some of the adrenergic neurons have been found to degenerate, whereas others, such as the short NA neurons innervating the vas deferens, are intact (Hamberger, Levi-Montalcini, Norberg and Sjöqvist, 1965). In recent transplantation experiments Malmfors and Olson (1968) have found that autologous or homologous iris grafts in the anterior eye chamber will become reinnervated from the adrenergic

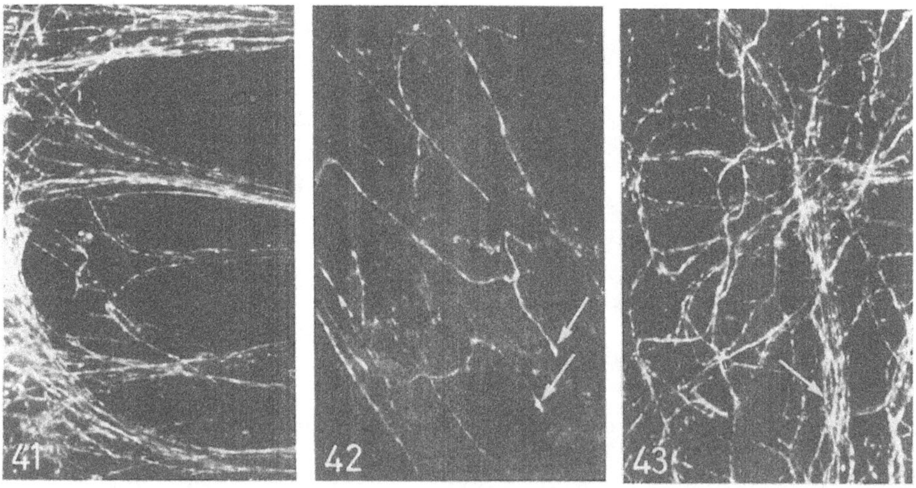

Figs. 41—43. Autologous iris transplant to the anterior chamber of the eye. Fluorescence microphotographs of stretch preparations. ×190. [Reproduced from Malmfors and Olson, Acta physiol. scand. **71**, 401—402 (1967) by permission of the Publishers]

Fig. 41. Bundles of nerve fibers bridging over from host iris (sphincter border at far left) towards the graft. Note bulgy appearance of nerves

Fig. 42. Graft, one week. Single nerve fibers arborizing over the dilator area. Enlarged growing tips (↗)

Fig. 43. Three-week graft, showing a well-developed adrenergic ground plexus. Arrow indicates large bundle of non-terminal axons

nerve terminals in the host iris (Figs. 41, 42, 43). By contrast, grafts from tissues which normally are not adrenergically innervated were found not to become so innervated. Similarly, NA cell bodies of autologous or homologous adrenergic ganglia transplanted into the anterior eye chamber will innervate the host iris (Olson and Malmfors, to be published). Studies on regenerating adrenergic axons have shown that the growth-cones and beads of such fibres are strongly fluorescent due to an accumulation of amine storage granules (Fig. 44) (Blümcke and Niedorf, 1965; Olson, 1969). Fluorescence histochemistry has also been used to study sympathetic NA nerve cells in tissue culture (Sano, Odake and Yonezawa, 1967).

In the cardio-vascular system there are a large number of noradrenergic afferents. The atria (Figs. 45, 46), and particularly the sinu-atrial and atrioventricular nodes, are heavily innervated by NA nerve terminals, whereas the ventricles receive a patchy NA innervation (Angelakos, Fuxe and Torchiana, 1963; Dahlström, Fuxe, Mya-Tu and Zetterström, 1965; Jacobowitz, Cooper and Barner, 1967; Jonsson and Sachs, 1969;

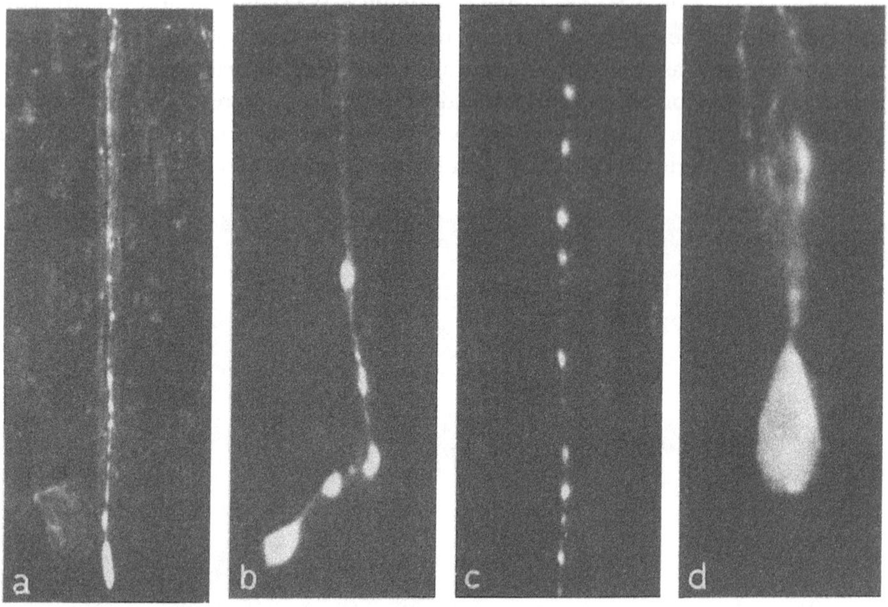

Fig. 44a—d. Different forms of regenerating adrenergic axons found distally to a crush lesion of the rat sciatic nerve. Thin, irregularly beaded fibers partly representing sprouting of the main axons (a—c), as well as typical growth cones (d) are seen. ×100 (a), ×160 (b, c) and ×250 (d). (Reproduced from Olson, Z. Zellforsch., in press, by permission of the Publishers)

Sachs, 1970). With the exception of the Purkinje cells, the cardiac conductile system receives a profuse NA innervation. Nielsen (1968) has reported evidence that these sympathetic NA neurons play a role in the occurrence of ventricular fibrillation during induced hypothermia (see also Nielsen and Owman, 1968). [In the frog heart, on the other hand, mainly adrenaline nerve terminals are found (Falck, Häggendal and Owman, 1963)]. — The NA innervation of the blood vessels is characterized by an adrener-

Fig. 45. Atrium of normal rat, showing a plexus of NA nerve terminals innervating the heart muscle cells (Dahlström and Fuxe, unpublished material.) ×120

Fig. 46. Same preparation as Fig. 45. Small arteries surrounded by an extremely dense plexus of fluorescent nerve terminals (↗). ×75

Fig. 47. Basilar artery of the normal rabbit. The artery is surrounded by a plexus of NA nerve terminals situated between tunica media and tunica adventitia. The autofluorescent lamina elastica interna is evident at the border between the tunica intima and the tunica media. ×75

Fig. 48. Pancreas of normal rat. An adrenergic groundplexus is shown surrounding the small intraglandular arteries. ×300

Fig. 49. Inferior concha of the normal rabbit. Note the dense adrenergic network surrounding the arteries and the sinusoids in the pseudo-cavernous tissue. (Dahlström and Fuxe, unpublished material.) ×75

Fig. 50. Submandibular gland of normal rat, containing an adrenergic ground plexus surrounding the serous acini of the gland. ×300

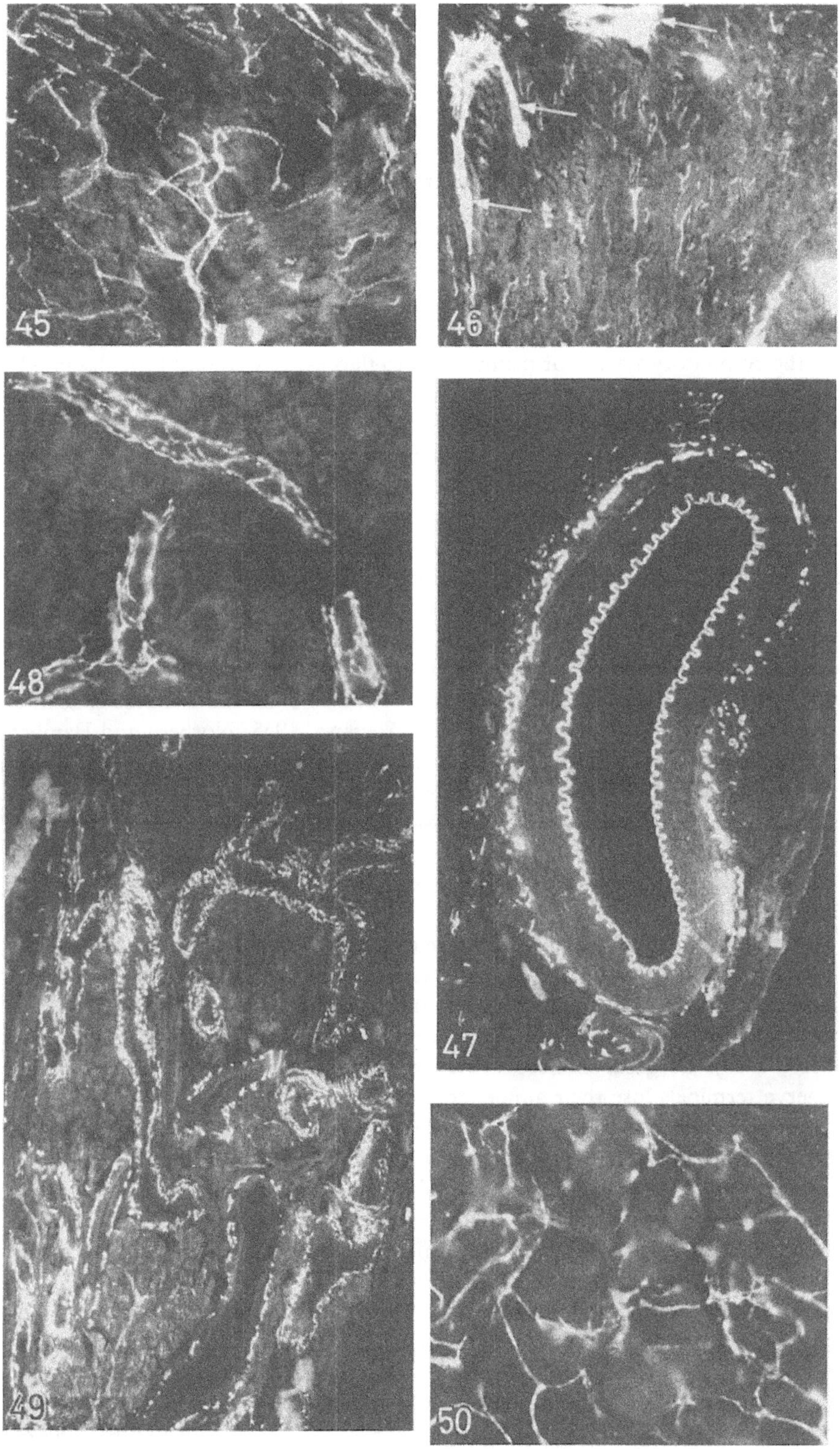

Figs. 45—50

gic ground plexus surrounding the outer layer of the media (Fig. 47); only rarely have adrenergic terminals been observed to penetrate into the muscular layer. In most tissues, *e.g.* skeletal muscle (Fuxe and Sedvall, 1965), the arteries are much more densely innervated than the veins (see also Lever, Spriggs and Graham, 1968). Furthermore, the large arteries usually are more sparsely innervated than the small arteries (Fig. 48) and arterioles (see Folkow, Fuxe and Sonnenschein, 1966). In pseudocavernous tissues the sinusoids (Fig. 49) are densely innervated (Dahlström and Fuxe, 1965). These morphological studies suggest that the influence of the sympathetic noradrenaline neurons may affect most or all levels of the vascular bed, but is particularly strong at the arteriolar level.

In the respiratory system of mammals histo-fluorescence studies have disclosed a dense adrenergic terminal network surrounding the bronchial muscle, an innervation which is likely to be involved in the mechanism regulating bronchial tonus (Dahlström, Fuxe, Hökfelt and Norberg, 1966). The adrenergic innervation of the lungs of the toad and the sleepy lizard has been described by McLean and Burnstock (1967a, b).

In the gastro-intestinal system fluorescence microscopy has produced evidence suggesting that the sympathetic inhibition of intestinal motility may be mediated in part at least by a noradrenergic innervation of the parasympathetic cell bodies of Auerbach's plexus, and to a somewhat lesser degree also of Meissner's plexus, for NA nerve terminals are heavily concentrated around the ganglion cells of these plexus (Norberg, 1964; Jacobowitz, 1965). In the submandibular (Fig. 50) and parotid glands a rich adrenergic ground plexus was found to surround the serous acini of these glands (Fujiwara, Tanaka, Honjo and Okegawa, 1965; Norberg and Hamberger, 1964; Norberg and Olson, 1965; Norberg, Hökfelt and Eneroth, 1969).

Studies by the aid of fluorescence histochemistry have greatly contributed to the discovery of the short noradrenaline neurons which appear to be a main source of innervation of the male accessory genital organs (Falck, Owman and Sjöstrand, 1965; Norberg, Risley and Ungerstedt, 1967; Baumgarten, Falck, Holstein and Owman, 1968).

In the female genital organs the isthmic muscular layer, which probably has a sphincter function (Brundin, 1965), receives a dense NA innervation part of which is derived from short NA neurons. These short neurons with their cell bodies close to the vaginal wall probably also innervate the uterus and the vagina (Owman and Sjöberg, 1966). Of great interest is Sjöberg's (1967) observation that the uterine NA nerve terminals lose their amine stores at the end of the pregnancy.

Studies on adipose tissue (Wirsén, 1964, 1965; Wirsén and Hamberger, 1967) have revealed that there is an adrenergic innervation of the fat cells in the brown adipose tissue, but probably not of those in the white adipose tissue, a finding which may explain some of the fundamental differences between these two types of fat tissue. In the white adipose tissue only an adrenergic vasomotor innervation has been found.

The adrenergic innervation of the iris has been studied extensively with the fluorescence-histochemical method by Malmfors (1965), mainly in stretch-preparations. These studies have contributed to the understanding of the fundamental mechanisms governing storage, release and especially uptake of NA by peripheral NA neurons (see also van Orden, Bloom, Barrnett and Giarman, 1966). A detailed account of the adrenergic innervation of the eye and the orbita has been published by Ehinger (1966). As to the adrenergic innervation of other sense organs, a network of sympathetic NA nerve terminals has been discovered in the lamina spiralis ossea and underneath the vestibular sensory epithelia of the labyrinth (Spoendelin and Lichtensteiger, 1965). Furthermore, certain mechanoreceptors may be directly influenced by NA released from adjacent NA nerve terminals (Fuxe and Nilsson, 1965). Of course, in many cases this influence of the NA nerve terminals in receptor areas may be mediated only indirectly, for example by way of an effect upon smooth muscle, as is probably the case in the sinus caroticus (Reis and Fuxe, 1968b).

The NA innervation of the pineal gland has been described in detail by Owman (1964) who demonstrated that these terminals in the rat store both NA and 5-HT. The latter are probably derived from the 5-HT containing pinealocytes.

As to other endocrine glands, the intermediate lobe of the pituitary gland is innervated by central CA nerve terminals (Dahlström and Fuxe, 1966; Björklund, 1968). The pancreatic islands of certain species likewise receive a NA innervation (Cegrell, 1968; Legg, 1968).

General properties of Monoamine Neurons

Monoamine transmitter histochemistry permits the study not only of the amines stored in the neurons, but also of the associated dynamic events such as the release (*e.g.* Dahlström, Fuxe, Kernell and Sedvall, 1965), uptake (*e.g.* Carlsson, Fuxe, Hamberger and Lindqvist, 1966) or synthesis of the transmitter in various parts of the monoamine neurons. Release of transmitter is revealed in the fluorescence microscope as a change in the distribution of fluorescence: the fluorescence intensity is decreased intraneuronally and increased extraneuronally in the form of a strong diffuse fluorescence immediately surrounding the monoamine neurons and their axon terminals. This phenomenon was first observed in regions rich in NA nerve terminals following treatment with the monoamine oxidase inhibitor MO 911 in a hot environment (Carlsson, Dahlström, Fuxe and Lindqvist, 1965). Similar observations with respect to 5-HT were subsequently made in rats treated with chlorimipramine, a 5-HT membrane pump blocking agent (Carlsson, Corrodi, Fuxe and Hökfelt, 1969) after pretreatment with reserpine which depletes monoamine stores — and nialamide, a monoamineoxidase inhibitor (Figs. 51, 52) (Carlsson, Fuxe, Jonasson and Lindqvist, 1969).

Uptake and accumulation of monoamines is readily observed in the fluorescence microscope if the monoamine stores previously have been depleted with reserpine, which acts by blocking the uptake — storage mechanism of the amine granules (Carlsson, 1966; Dahlström, Fuxe and Hillarp, 1965; Hökfelt, 1968), provided the breakdown of the amine by the monoamine-oxidase is prevented by injection of a monoamine oxidase inhibitor (Figs. 53, 54). Under these conditions, intravenous injection of even very small quantitites of NA will cause an appearance of fluorescence in the adrenergic nerve terminals, due to the very efficient uptake-concentration mechanism for amines localized at the level of the nerve cell membrane of the monoamine neurons. If the animal has been pretreated with a membrane pump blocker of the NA neurons such as desipramine, no fluorescence will accumulate (see Hillarp and Malmfors, 1964; Malmfors, 1965; van Orden, Bensch and Giarman, 1966). It is obvious that this model is suitable to test drugs for membrane pump blocking activity. Similar studies have been performed in the brain *in vitro* (Hamberger and

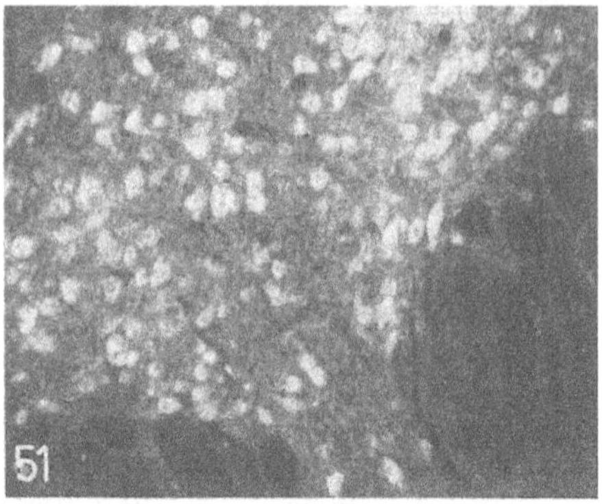

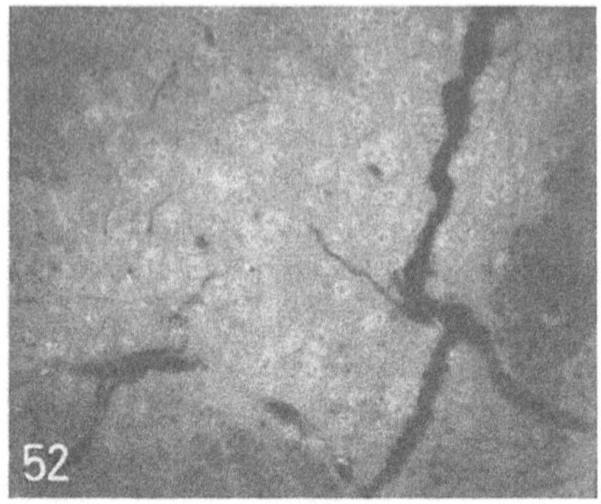

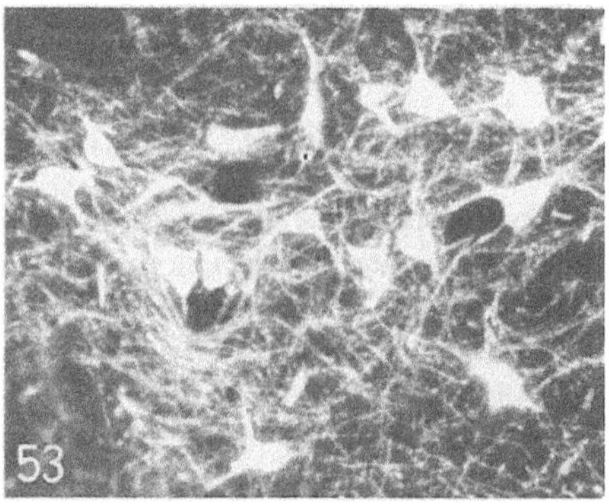

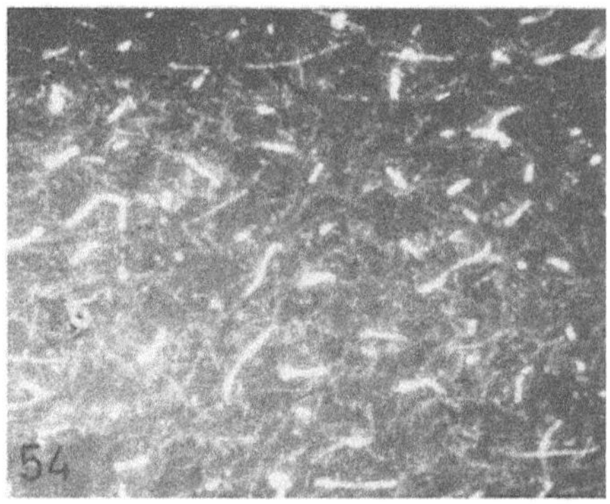

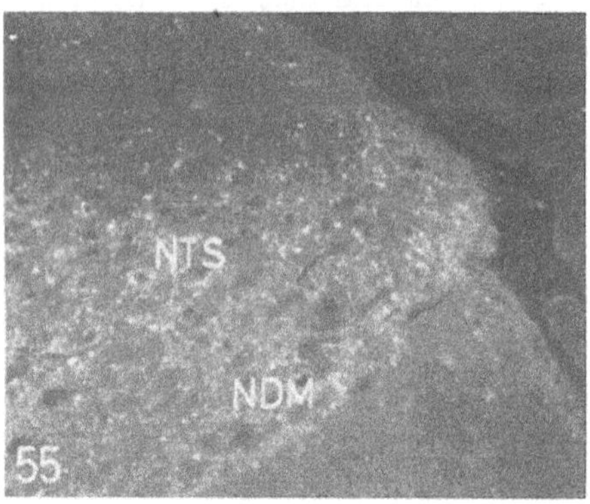

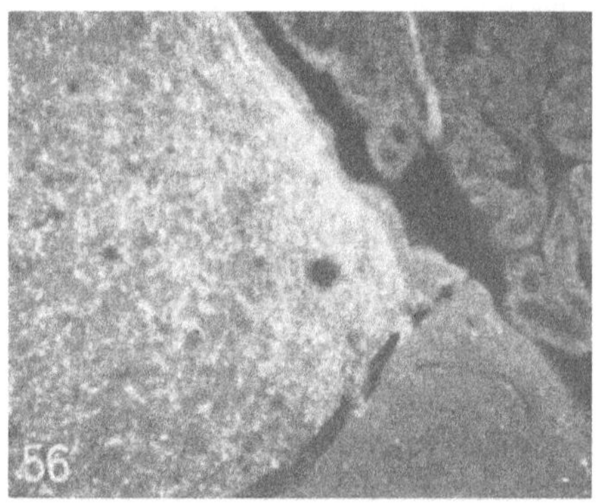

Fig. 51. Transverse section of the nucleus raphe dorsalis of a reserpine-nialamide-treated rat. 5-HT cell bodies are strongly fluorescent due to a marked accumulation of 5-HT. ×210

Fig. 52. Same area and treatment as in Fig. 51, except that chlorimipramine (25 mg/kg, i.p.) was given 30 min before sacrifice. An accumulation of extraneuronal fluorescence has appeared, with decreased fluorescence in the cell bodies, probably due to rapid outflow of 5-HT from the cell bodies. ×120

Fig. 53. Transverse slice of the substantia nigra area of a reserpine-pretreated rat after incubation with α-methyl-NA (10^{-6} M). There is a marked accumulation of fluorescent material in the DA cell bodies and their processes. The nuclei cannot be observed because of the strong cytoplasmic accumulation of α-methyl-NA. ×190

Masuoka, 1965; Hamberger, 1967) and *in vivo* (Fuxe and Ungerstedt, 1966, 1968a, b; Fuxe and Hillarp, 1964; Fuxe, Hamberger and Malmfors, 1967, Lichtensteiger and Langemann, 1966).

Injections of inhibitors of the monoamine synthesis, such as tyrosine and tryptophane hydroxylase inhibitors, cause a relatively rapid decrease in fluorescence intensity in the nerve terminals, non-terminal axons and cell bodies of CA and 5-HT neurons (Andén, Corrodi, Dahlström, Fuxe and Hökfelt, 1966; Andén, Fuxe and Hökfelt, 1966, 1967). This indicates that synthesis of monoamines occurs in the cell bodies as well as in proximal and terminal parts of the axon, a suggestion also arising from observations in experiments with nialamide and CA and 5-HT precursors (Dahlström and Fuxe, 1964; Fuxe, 1965; Andén, Fuxe and Hökfelt, 1967). The most important finding in these studies, performed in spinal rats, was that the degree of fluorescence disappearance after synthesis-inhibition is dependent on the neural impulse flow. The monoamine nerve terminals caudal to the spinal cord transection lack neuronal impulse flow, since all monoamine nerve terminals in the spinal cord arise from descending axons originating at supraspinal levels. Thus, only the terminals cranial to the spinal cord lesion receive nervous impulse flow, and only these showed depletion of fluorescence under conditions of synthesis-inhibition.

These results suggest that the degree of depletion obtained in the monoamine nerve terminal systems after synthesis-inhibition under various experimental and physiological conditions is related to the state of activity in the monoamine-neuron system, in the sense that high rates of fluorescence disappearance are associated with states of high neural activity. Thus, fluorescence microscopy, beside information concerning release and uptake of transmitter, can provide an indication of the prevailing state of activity of the monoamine neurons studied. By examining the effects of various conditions such studies, in combination with other approaches such as the experimental destruction of monoamine neuron systems, can help elucidate the functional role played by these systems. A few examples of such studies, performed during the past several years by the present authors in collaboration with Drs. Corrodi and Andén at the University of Gothenburg, will follow.

The stressful condition associated with physical restraint causes an increase in the rate at which fluorescence disappears from all NA terminals in the brain under conditions of synthesis inhibition (Corrodi, Fuxe and Hökfelt, 1968). High en-

Fig. 54. Tangential section of the gyrus hippocampi after an intraventricular α-methyl-NA injection (5 μg/20 μl, 30 minutes before sacrifice) in a reserpine-pretreated rat. Uptake of α-methyl-NA by the monoamine nerve terminals of this area and the cells of the capillary walls. ×120

Fig. 55. Dorsal motor nucleus of vagus (NDM) and nucleus tractus solitarii (NTS) of a rat injected with a tyrosine hydroxylas inhibitor (250 mg/kg, i.p., α-methyl-tyrosine-methylester) 4 hours before sacrifice. The treatment has caused a considerable reduction in the number and fluorescence intensity of visible NA verve terminals. ×120

Fig. 56. Same area and treatment as described in Fig. 55, except that cataprezan (1 mg/kg, i.p.) had been given two hours before the tyrosine hydroxylase inhibitor. There is now only a slight reduction in the number and fluorescence intensity of the visible NA nerve terminals. Strongly fluorescent NA terminals appear in relatively high numbers, due to a decreased activity of the NA neurons probably induced by the NA receptor stimulating action of cataprezan (Andén, Corrodi, Fuxe, Hökfelt and Hökfelt, 1969, unpublished.) ×120 vironmental temperature had similar effects, but in this case there was an increased disappearance of fluorescence also from the 5-HT neurons (Corrodi, Fuxe and Hökfelt, 1967). Furthermore, there is a markedly accelerated decrease of fluorescence in the DA nerve terminals of the median eminence after synthesis-inhibition during pregnancy and lactation (Fuxe, Hökfelt and Nilsson, 1967). Apomorphine, catapresan (Figs. 55, 56) and lysergic acid diethylamide have been found to reduce the rate of the fluorescence disappearance following synthesis inhibition in DA, NA and 5-HT nerve terminals, respectively (Andén, Fuxe, Hökfelt and Rubensson, 1967; Andén, Corrodi, Fuxe, Hökfelt, Hökfelt, Rydin and Svensson, 1969; Andén, Corrodi, Fuxe and Hökfelt, 1968). This is probably due to the fact that apomorphine, catapresan and lysergic acid diethylamide are specific stimulators of DA, NA and 5-HT receptors, respectively, involved in a compensatory neural feed-back to the presynaptic DA, NA and 5-HT cell bodies, resulting in a decreased activity in the neurons with a subsequent decrease in the turnover rate. All the studies using a semiquantitative evaluation of fluorescence intensity and number of fluorescent terminals were performed on coded preparation by at least two investigators independently of one another and of the biochemical results. In all of the 20—25 studies thus far completed, the histochemical and biochemical results have been in good agreement.

In recent studies on the adrenergic terminals of the iris, a satisfactory correlation between fluorescence intensity and uptake of labelled amines was established (Olson, Hamberger, Jonsson and Malmfors, 1968; Jonsson, 1969). A further convincing correlation has been found to exist between the number of NA terminals and the uptake of labelled amine after partial denervations of the iris (Olson and Malmfors, to be published).

In closing, it may be added that fluorescence microscopy has also permitted the localization of so-called "false monoaminergic transmitters" such as α-methyl-NA in the central NA neurons (Carlsson, Dahlström, Fuxe and Hillarp, 1965) and α-methyl-5-HT (Fuxe, unpublished data) and 6-HT (Jonsson, Fuxe, Hamberger and Hökfelt, 1969) in the 5-HT neurons. False adrenergic transmitters such as α-methyl-m-tyramine, *m*-hydroxy-norephedrine (metaraminol) are usually not observed in the fluorescence microscope, due to the fact that the corresponding 6-hydroxy-3,4-dihydroisoquinoline formed intraneuronally after formaldehyde gas treatment has only a blue emission peak (420 mμ). This fluorescence is not transmitted by the conventionally used secondary filter (cuts below 490 mμ), and is also concealed by the background fluorescence. By means of microspectrofluorometry, however, it has been possible to detect this emission peak (Jonsson and Ritzén, 1966) and thus, to demonstrate the intraneuronal localization of the substance.

Acknowledgements

The studies upon which this account is based were supported by grants from Karolinska Institutet (Reservationsanlaget), by research grants from the Swedish Medical Research Council (12X-715-01, 12X-715-02, 12X-715-03, 12X-715-04A, B12X-2295-01, 12X-2295-02, B 70-14X-2887-01), and by grants from "Stiftelsen G. och T. Svenssons Minne", "Stiftelsen Therese och Johan Anderssons Minne", "Magnus Bergvalls Stiftelse", "C.-H. Nathorsts Stiftelse", "Svenska Livförsäkringsbolags nämnd för medicinsk forskning" and "Ollie och Elof Ericssons Stiftelse".

Methods for the Counting of Neurons

Bruce W. Konigsmark[1]

General Survey

Only a short time after the neuron became identified as the essential unit of the nervous system, the first attempts were made to estimate the number of neurons in different parts of the nervous system. During the past century, a great number of methods have been used in making such estimates. Although the most widely used and accepted method is that of direct counting in the microscope, various other techniques, including photographic, projection, homogenate, automatic, and ocular methods have been designed. Brief descriptions of these techniques will be given in the following account.

Photographic Method

Neuronal populations may be estimated from photomicrographs, a method used by several investigators including Bouman (1928) and von Economo and Koskinas (1925). The latter authors photographed 25 μ thick sections from the cerebral cortex at 100 × and counted neurons from the plate, attempting to exclude glia and endothelial cells. Using these neuronal counts from a known area and section thickness, the number of neurons in a unit volume were calculated.

Several authors justifiably criticized this method (Agduhr, 1941; Rowland and Mettler, 1949). Agduhr pointed out that cells would be missed in a photographic count since cells in different planes of the section would be out of focus in the photograph. In addition, cells appearing at the surface of the section may be represented in more than one section; obviously such cells would be counted at least twice. Also, at the low magnification used, many glial cells could not be distinguished from small neurons. In view of these shortcomings it is understandable that the photographic method has not been extensively used since that time.

Projection Method

In this technique neurons are counted from a projected image of a histological section. This method has been used by Bianchi (1942), Dornfeld *et al.* (1942), Sullivan *et al.* (1958), and Escobar *et al.* (1968). Using a W and H (Wetzlar) microprojector with a magnification of 276 ×, Escobar *et al.* subdivided the area of the microscopic field into squares and counted the identifiable neurons in each unit. Estimates of the

[1] Supported by a grant from the John A. Hartford Foundation, Incorporated, by Research Grant No. NB808196, and by Career Fellowship Award No. NB5663, from the National Institute of Neurological Diseases and Stroke, Department of Health, Education and Welfare.

total number of neurons in the olivary nucleus were obtained by selecting a random sample of sections. At this magnification, nucleoli could be identified but were not counted.

Microphotometric Method

The use of a microphotometer in estimating cell numbers in the cerebral cortex was described by Ryzen (1956). Blocks of formalin-fixed cerebral cortex were embedded in paraffin, and sectioned at 10 μ; alternate slides were stained with cresyl violet-gallocyanin or by the Feulgen method. The slide was mounted on a mechanically driven microscope stage, and a narrow beam of light was directed through the slide onto the photocell of a microphotometer. As the section moved through the light, at a constant speed, the varying current in the photocell was recorded as a curve indicating optical density in terms of cortical depth. To translate density into cell numbers, the author obtained an α coefficient for an area of the cortex by dividing the average optical density by the counted cell numbers. The α coefficient was then used to calculate cell numbers from other densitometric recordings. Since the α coefficient varied with the stain intensity and section thickness, these variables were kept as nearly constant as possible. In cases of marked change, a new α coefficient was necessary.

Homogenate Method

A method of freezing nuclei from homogenates of brain tissue was developed by Heller and Elliott (1954). Nurnberger and Gordon (1957) and Nurnberger (1958) modified this technique, freeing cells from a measured sample of neural tissue to obtain an estimate of the number of neurons in the structure studied. In Heller and

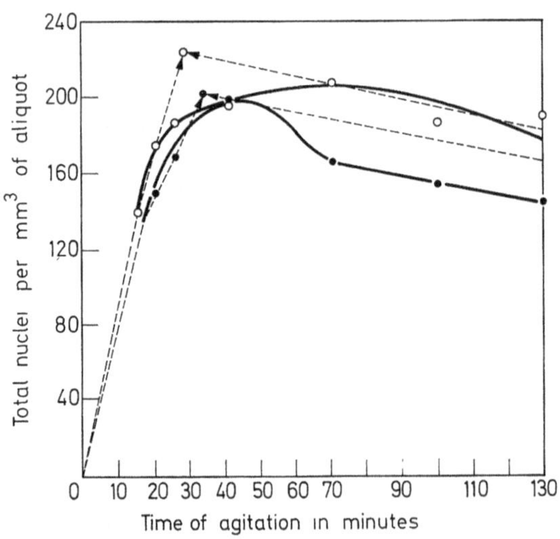

Fig. 1. The rates of liberation (rising lines) and eventual destruction (descending lines) of two rat brains disintegrated by agitation are plotted. Broken lines indicate the maximum rates of liberation and destruction. The points of intersection of the broken lines, taken as the total number of nuclei, are indicated by arrows. (From Nurnberger and Gordon, 1957); courtesy, P. B. Hoeber

Elliott's procedure, fresh whole brain or portions of brain were weighed in a glass bottle, suspended in a measured volume of 0.2 molar potassium chloride solution, and agitated with 6 mm glass beads. At about 15 min intervals, counts of the freed nuclei were made by taking aliquots of the suspension, diluting with 0.1% methylene blue, and counting in a hemocytometer chamber. The counts were plotted on a graph (Fig. 1). Since total counts increase for a period, then decrease as nuclei disintegrate from continued agitation, the theoretical maximal number of nuclei is extrapolated from the point of intersection of two lines, one indicating the rate of release of nuclei, taken from the rising part of the curve, and the other indicating the rate of disintegration of nuclei, taken from the descending portion of the curve. From this, the theoretical maximal number of nuclei in the sample is calculated.

In order to determine neuronal numbers, Nurnberger and Gordon (1957) and Nurnberger (1958) made differential counts of glia, neurons and other cell types in thin suspensions under a coverslip. This method permits reasonable estimates of the total number of cells — neurons and glia — per gram weight, but the differential counting is open to gross error in all cases in which neurons are difficult to distinguish from glia, such as in the study of young embryos or the cerebellum. The method has been found quite useful in estimating the total neuronal population in small brains.

Automation in Neuronal Counts

Since the counting of neurons is a laborious and slow process, many attempts have been made to automate the procedure. One such attempt has led to the development of the so-called flying spot microscope. This instrument, consisting of a microscope, a television scanning apparatus, and a computer, has been found useful for the detection, numbering, and sizing of particles. The microscope is used to focus a fine beam of light on the specimen. The beam, as in television scanning, sweeps the accessible field. The amount of illumination passing through each successive point of the field is recorded by a photocell and the resulting signal is amplified and fed into a cathode ray tube to produce a replica of the visual field. The flying-spot and its recording system can be programmed so as to measure the size of particles and to record the number of particles in each size group. Whenever the flying-spot passes over a denser area, the light to the photocell decreases. The length of time which it takes the spot to traverse the area is the measure of the size of the opaque region. Various programs have been designed to solve the problem of counting each particle only once. The system can be programmed to remember the last opaque area, and to count again only after a "clear" signal has been obtained. Causley and Young (1955) used a system involving scanning of the preparation with twin spots of light of different polarization. A count was registered when only one of the spots crossed the particle. When both crossed it, the pulses were made to cancel each other out. The authors made an estimate of neurons in the cortex of the rat by having the machine count separately all particles greater and smaller than 7.5 μ in diameter, and obtained results compatible with those obtained in a visual count.

Mansberg and Segarra (1962) encountered several problems in the use of the flying-spot scanner. Although the size discrimination circuit worked well for opaque round particles, it was not as accurate for irregularly shaped forms. Scanning at higher magnification had the result that some cells were counted several times. To facilitate the distinction between glial nuclei and neuronal nuclei, scanning tubes with absorption peaks close to that of the Nissl stain were selected and the scanning spot was de-focused to lower resolution, enough to eliminate glial cell detection. Despite these improvements, the problems of cell-type recognition and overlapping of cells, among others, thus far have not been solved to a degree warranting the use of this type of microscope for neuronal counts.

Ocular Micrometric Methods

Since the later part of the last century, the most frequently used method of estimating neuronal populations has been that of counting under direct microscopic observation. The structure under study is generally embedded in paraffin or celloidin, sectioned at 5 to 40 μ thickness, stained, and examined under the microscope. The labor of counting all cells of a given structure is usually reduced by the use of a sampling procedure: either cells are counted in a periodic selection of slides and the number thus obtained is multiplied by the period used, or the estimated number of neurons per unit volume is used to calculate the total population of the structure on a volumetric basis.

A number of problems have been encountered in the use of these methods, problems for which numerous solutions have been proposed. The various factors determining the accuracy of this category of counting techniques will be reviewed in the following account.

Histological Procedures

Fixation

The selection of a fixative is governed by two requirements: the fixative must produce a minimum of artifact, and it must allow for optimal affinity of the neurons for the stain to be used. The crush cell artifact or "dark" neuron can obscure nuclei and nucleoli and thus impair the accuracy of the counting process. Wherever possible, fixation by perfusion is therefore preferable to fixation by immersion (Cammermeyer, 1967a). Exsanguination by a saline solution or a mixture of saline and gum acacia (Koenig, Groat and Windle, 1945), followed by a fixative, such as 10% acrolein, 10% formalin in isotonic saline, or Heidenhain-Susa solution is usually quite effective. Human brains must be suspended in the fixative, usually 10% formalin in water or in isotonic saline.

Shrinkage

The volume of fresh tissue specimen changes drastically during the histological procedures leading to finished sections (Stowell, 1941). These changes can be determined by measuring the specimen, first after fixation, next after infiltration by the embedding medium, and finally after mounting of the sections.

Only a few authors have quantitated the volume lost during processing. Pakkenberg (1966), estimating neuronal numbers in the cerebral cortex, found a mean linear shrinkage of about 36% between the fresh unfixed cortex and the stained specimen. The linear shrinkage amounted to a volume reduction of 74%, only 26% of the original volume remaining.

Notching the Specimen

Although a histology technician can keep sides oriented in the serial processing of sections, it is impossible to differentiate the right from the left side after the block has been cut. Thus, since it is usually important to know one side from the other, particularly in a case with possible pathology, it is necessary to notch one side. This is done preferably before embedding is started, although it is possible to notch a specimen on the microtome. We routinely notch the right side of the specimen when it is trimmed for embedding.

Embedding Procedures

Features desired in an embedding medium for specimens in which neurons are to be counted are: an ability to accept any size of block, a consistency permitting uniformity of section thickness, easy storage of serial sections with possibility of recovering sections for staining at a later time, and reliability. The most frequently used embedding material for brain blocks used in neuronal counts are paraffin and celloidin.

Paraffin embedding is fast and allows thinner sections than does celloidin. Where serial sections are desired, paraffin material must be mounted on glass to be stored, requiring a large amount of storage space. Invariably, where serial sections are desired, even the best technician will lose some paraffin sections due to fragmentation or splitting, a loss the more likely to occur the larger the size of the block.

We prefer celloidin for the following reasons. Any size block can be embedded and sectioned. Sections are quite resistent to handling and need never be lost. The sections can be stored on numbered sheets of paper, remaining available for later staining. An excellent technique of celloidin embedding is described by the Manual of Histologic Staining Methods of the Armed Forces Institute of Pathology (1968). For blocks larger than 1 cm in thickness, the period of two weeks listed in the manual for both dehydration and celloidin embedding should be increased. The embedded block or the cut sections may be stored indefinitely in 80% ethanol.

Plane of Sectioning

Although any sectioning plane can be used, usually one is superior in demonstrating a specific structure of the brain. We have found that the availability of standard series of the human brain, sectioned in each of the three principal planes is extremely useful in helping define different cell groups and in deciding in which plane to section subsequent specimens.

Section Thickness

The following points need to be considered in selecting the section thickness. For several reasons it is generally agreed that the thicker the section the better. Fewer sections are necessary when a structure is serially cut, saving space, time and funds. Fewer cells, nuclei and nucleoli are split when fewer and thicker sections are taken, a circumstance decreasing the risk of counting such structures twice. However, sections can be too thick for efficient or accurate counting. If two or three neurons overlap, accurate counting is difficult. Thicknesses used range from 50 μ, in Chow's (1951) study of the monkey brain stem, to 10 μ in Ebbesson's (1968) study of the superior cervical ganglion. A thickness of 20 μ has been used by the majority of workers in this field (table). We have found this thickness quite convenient for our study of cell groups in the human brain stem.

Numbered Sheets for Serial Celloidin Sections

Intermediate quality paper cut to a size slightly larger than the largest celloidin section can be used for separating and numbering the serial celloidin sections. The paper can be numbered by hand or by a hand stamp automatic numbering device which advances one number with each stamp. Most recently we have been using a rapid and inexpensive method described by TAN CRETI (1968) in which a program for the IBM 1401 computer produces the numbering sequence on print-out paper. For rapid assembly the numbering is done in units of ten pages. When these are cut, the numbered slips fall into numerical order. A copy of the autocoder computer program is available from the present author.

Staining

The specificity and intensity of staining are important factors in neuronal counting procedures. Characteristics of a stain used in neuronal nucleolar counts should include: 1. intense staining of nucleoli, and 2. only enough staining of other cellular elements to allow neurons to be distinguished from other cell types. In one instance of counting cells in alternating series stained by, respectively, cresyl violet and Luxol Fast Blue-cresyl violet, cell counts made in sections stained by the combination of dyes were lower by 25% than those obtained in sections stained by cresyl violet alone (KONIGSMARK, KALYANARAMAN, COREY, and MURPHY, 1969). Most extensively used have been the Nissl stains including thionine, cresyl violet, methylene blue and toluidine blue. EBBESSON (1968) employed a hematoxylin-neutral red-safranin O stain for counts of sympathetic neurons, and a few authors have used other stains such as protargol impregnation (IRVING and HARRISSON, 1967). We have found the cresyl violet method described in the Manual of Histologic Staining Methods of the AFIP most satisfactory with respect to both clarity of background and intensity of staining of Nissl substance, nuclei and nucleoli.

Counting Procedures

Outlining the Cell Group

Although most gray matter structures in the nervous system are reasonably discrete, all have some areas where delimitation of the nucleus from adjacent cell groups is difficult. Thus, one of the major tasks confronting the investigator is that of defining the boundaries of the region within which neurons are to be counted. Obviously, no specific rules can be given that would be valid for all and any cell territory. In our laboratory, the nucleus is outlined as precisely as possible by a very fine ink line drawn on the cover slip. Higgins green ink has been found suitable for this purpose because of its high resistance to moderately vigorous cleaning of the cover slip. With the finest pen point avilable, the line can be placed precisely enough to separate adjacent cells. Since the counting itself can be delegated to a technician, the nucleus should be defined so that all neurons within its boundary can be counted as part of the nucleus.

Magnification in Neuronal Counts

Several factors should be considered in selecting a magnification for use in neuronal counts. These include the section thickness, neuronal size and population density, nucleolar size, and degree of artifact. In sparsely populated areas with large neurons, distinct nucleoli, and no artifact, lower magnifications can be used effectively. Where artifact is a problem, a higher magnification may be necessary to identify nucleoli in darkly staining nuclei. However, any increase in magnification necessitates more frequent field shifts, and thus increasing the possibility of errors of duplication or omission. Magnifications used in neuronal counts have varied from 50 (Hall, 1964; Chacko, 1948) to 800 and 1,000 (Irving and Harrison, 1967; Ebbesson, 1968). Most authors have used a magnification of 200 to 400 (table). In the case of neurons of intermediate size, such as those of the ventral cochlear nucleus, we have found that counts performed at 300× and 500×, respectively, usually fall within 3% of each other; within this range of powers at least, the factor of magnification appears to have no great importance. Cell bodies or nuclei can usually be counted with adequate accuracy at 125×, but reproducible counts of the much smaller nucleoli are difficult to obtain at this power. Microscopists in our laboratory find counting most convenient at about 300×, for cell density poses no great obstacles at this magnification and nucleoli are easily defined.

Units Counted

The neuronal cell body, its nucleus, or its nucleolus may be taken as the unit counted. The larger the unit that is used, and the thinner the section, the greater is the error due to duplicate counting of units split by the sectioning and thus appearing in several sections.

Many authors, including von Economo and Koskinas (1925) and Sullivan *et al.* (1958), have used the neuronal cell body as the unit counted, without using any correction factor for split neurons. In their study of the cerebral cortex, Bok and Kip (1940) counted cells in sections of varying thickness (15 and 30 μ) to correct for the overcount due to split cell bodies. In a quantitative study of the inferior olivary nucleus, Escobar *et al.* (1968) compared the numbers of cell bodies and nucleoli counted in twenty successive sections. They concluded that the neuronal cell body count divided by two corresponded to the nucleolar count, and used the neuronal body count for the remainder of their study.

The error caused by split cell bodies can be compounded in regions where cell bodies have a preferential axis of orientation, such as orientation of the neurons in the lateral geniculate body (Droogleever Fortuyn, 1963). In such cases the plane of sectioning could influence the count of neuronal cell bodies, for the number of sections in which a cell could appear and be counted is greatest in serial sectioning perpendicular to the long axis of the cell.

Neuronal nuclei have been used as the units counted by Guild *et al.* (1931), Chow (1951), and Irving and Harrison (1957). Only Guild *et al.* made a correction, estimating an overcount of about 10% due to split nuclei.

The method of counting neuronal nuclei or nucleoli is based on the assumption that each neuron has only one nucleus and one nucleolus. This appears to be valid for neuronal nucleoli, for neurons with more than one nucleolus per nucleus are extremely rare in the normal brain. On the other hand, binucleated neurons are seen, occasionally in normal brains and are common in some diseases, e.g., tuberous sclerosis. Moatamed (1966) in counting all neurons in serial sections through the inferior olivary complex found ten binucleated neurons, i.e. 0.0012% of a total population of 860,000 cells. In another series, he found one binucleated neuron among 88,000 neurons. We have found five binucleated neurons among 12,000 counted in the facial motor nucleus of one human brain, and none in over 200,000 neurons counted in the ventral cochlear nucleus of humans of all age groups. Judged by these figures, the problem of multiple nuclei or nucleoli appears insignificant.

Most workers have used the neuronal nucleolus as the unit counted. There are several advantages to this strategy, for the structures are quite distinct and can be recognized easily. Moreover, their small size, ranging from 1 to 5 μ diameter, reduces the frequency of double counting due to splitting. Even so, splitting of nucleoli occurs in sectioning and must be taken into account. Some authors have counted only distinct nucleoli and have made no correction factor for splitting (van Buskirk, 1945; Tomash and Malpass, 1958). Other authors have attempted to offset the effect of splitting and the resultant overcount by applying a variety of correction factors (Billingsley and Ranson, 1918; Agduhr, 1941), whereas yet others have included a correction factor for smaller portions of the split nucleoli not seen or recognized (Floderus, 1944; Schuknecht, 1960). These factors and formulas will be discussed in the next section.

Systematic Coverage of the Structure

The structure to be examined quantitatively can be covered systematically by the eyepiece-reticule method or by the eyepiece-line method. Although practically all workers have used the reticule method, Moatamed (1966), reviewing the irregularly-shaped inferior olivary complex, tallied cells as they passed by an eyepiece hairline as a result of moving the mechanical stage.

Another method of reviewing the structure was used by Tomash and Malpass (1958). These authors took photographs of the motor trigeminal nucleus and identified neurons with nucleoli in the corresponding sections at high magnification, circling their counterparts in the photograph. Finally a count of the circled cells was taken for the toal number of neurons. This method, used on serial sections, would seem to be quite accurate in obtaining a neuronal population estimate. For more routine purposes, however, it is more convenient to use one of the many types of reticules available. The reticule needs only to take a reasonable part of the optical field, breaking it into clear subdivisions. One we have found useful is the Howard disc[2] with a 10 mm square ruled into 36 squares. The eyepiece with the reticule is rotated so that its horizontal lines are parallel with the motion of the slide on the mechanical stage. Then, beginning in one corner of the area outlined for counting, all nucleoli in each of the squares are counted[3]. After finishing one field, the slide may be moved to the next, using any well defined landmark on the leading reticule line, and moving the slide carefully until the landmark falls on the contralateral reticule line. Neurons in successive fields are counted until the lateral margin of the outlined area is reached. Then a landmark is selected which falls on the lower line of the reticule, and the

[2] Bausch and Lomb Howard disc., 31-16-15.

[3] Useful is the Adams Hand Tally Counter or Adams Laboratory Counter.

slide moved vertically until the landmark falls on the upper reticule line. The entire structure under study can be covered in successive fields. It is important that the investigator keep some record of the last horizontal and vertical movement, for any distraction entails the risk of field reversals causing entire areas of the nucleus to be missed or recounted. Since some nucleoli will be bisected by the net reticule peripheral lines, the microscopist can average out this problem by counting bisected nucleoli on the upper and right-hand lines of the reticule, rejecting those bisected by the left-hand or bottom lines.

Sampling Methods and Correction Factors

Four different methods of using neuronal count data to estimate the neuronal population of a structure can be used. These include: 1. total count method, 2. systematic section method, 3. random section method, and 4. sample and volume method.

Method 1. Total Count Method

This technique, in which all neurons or neuronal nucleoli in serial sections through the structure studied are counted, has been used by Lassek (1940), Tomash and Malpass (1958), Radmanesh (1968), and Konigsmark *et al.* (1969). This technique, in which all neurons are counted, may be taken as the most accurate approximation of the total neuronal population of a structure. However, two kinds of error are possible in this estimate. These are: the *counting error*, in which units may be missed or counted twice by the microscopist, and the *split cell* error, in which the unit counted, whether it is the neuron cell body, nucleus or nucleolus, when near the section surface may be cut and appear for counting in two adjacent sections.

Split Cell Error

Although the neuronal cell body can be used as the unit counted, the larger the structure that is counted, the greater the problem of overcounting due to the appearance of the same cell in adjacent sections. Agduhr (1941) severely criticized von Economo and Koskinas' (1925) cortical neuronal counts, for they made no allowance for overcount due to split neuron bodies. Although some authors have not paid heed to this advice, others have developed different correction factors for this overcount. These formulas can be used for any spherical unit, whether it be the cell body, the nucleus or nucleolus, for the formulas are based on the diameter of the unit and the section thickness.

Agduhr (1941) developed several formulas, one based on the assumption that the unit diameter is less than the section thickness:

$$N_i = n_i \frac{2t-d}{2t} \tag{1}$$

in which N_i is the number of units that should be counted in section i, n_i is the number of units actually counted in section i, t is the section thickness in micra and d is the mean diameter of the unit. Thus, if the section thickness is 20 μ and the nucleolar diameter 5 μ, the correction factor is 0.875. This formula was derived from the assumption that as the unit moved through a distance AB $(t-d)$ in the section (Fig. 2), it would be counted whole in the section, while as it moved through the distance

BC (*d*) it would be represented in two sections and should be given only one-half count. Thus,

$$N_i = n_i \frac{t-d}{t} + \frac{1}{2}\frac{d}{t} = \frac{2t-2d+d}{2t} = \frac{2t-d}{2t}.$$

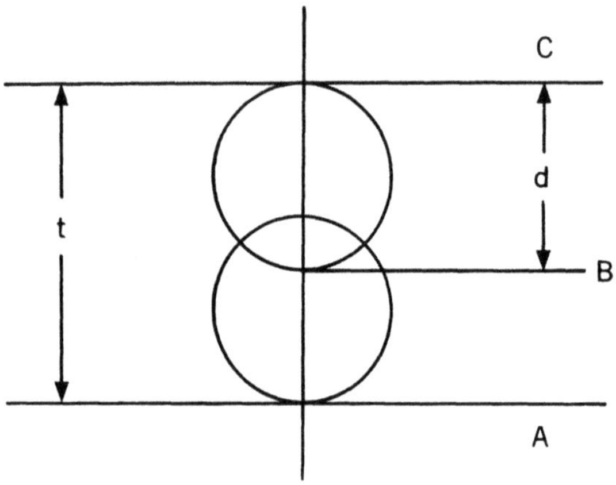

Fig. 2. Diagram illustrating the derivation of the split cell correction factor by Agduhr (1941) courtesy, Wistar Institute of Anat. and Biol.

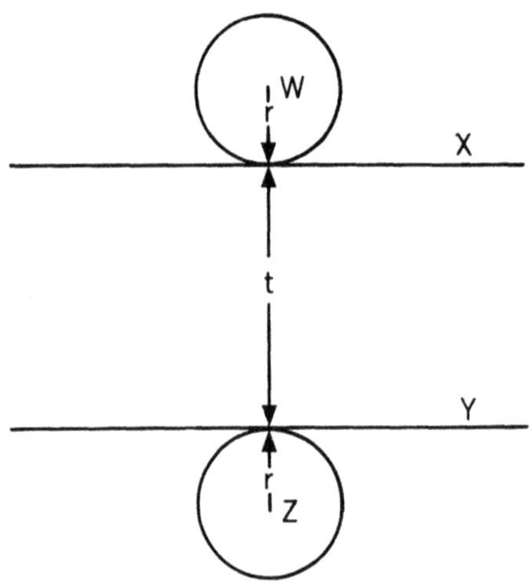

Fig. 3. Diagram modified from Abercrombie (1946) illustrating derivation of the split cell correction factor; courtesy, Wistar Institute of Anat. and Biol.

Abercrombie (1946) and Floderus (1944) independently reviewed and corrected Agduhr's formula. Abercrombie developed the formula:

$$N_i = n_i \frac{t}{t+d}. \tag{2}$$

As seen in Fig. 3, where $d=2r$, any nuclear or nucleolar center lying between *W* and *Z* will be represented in the section. If the distribution of units is random, then the proportion of units represented by the section is $\frac{WY}{WZ}$ or $\frac{t}{t+2r}$. With sections 20 μ thick and nucleoli 5 μ in diameter, the correction factor is 0.80.

Floderus (1944), studying the human hypophysis, emphasized another correction factor for unseen or uncounted nuclear or nucleolar fragments. His formula is

$$N_i = n_i \frac{t}{t+d-2b} \tag{3}$$

where b is the maximum height of the unseen or uncounted fragments (Fig. 4). However, as pointed out by Murphy (Konigsmark *et al.*, 1969), it is clear that b gives

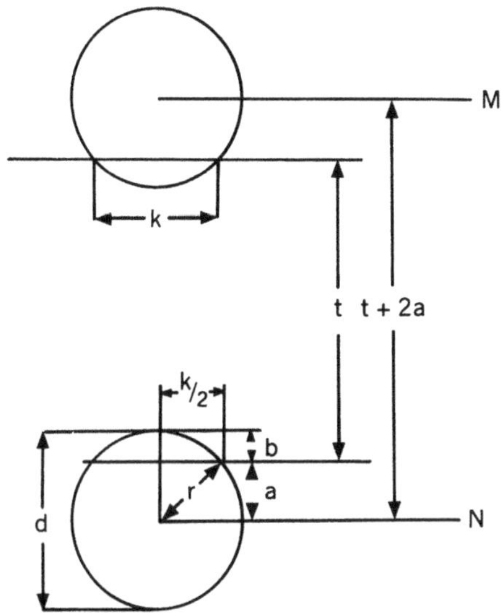

Fig. 4. Diagram illustrating correction factor for split cells including correction for uncounted fragments. Suppose nucleoli are distributed randomly throughout the section. All nucleoli whose centers lie between M and N will be included in the count. Thus, only a proportion $\frac{t}{t+2a}$ of the apparent nucleoli are representative of the section. (From Konigsmark, *et al.*, 1969)

only the smaller dimension of the fragment whose size, as seen by the microscopist, is more closely approximated by the diameter k of the fragment. All nucleoli whose centers lie between M and N, a distance of $t+2a$, will be included in the nucleolar count. Thus assuming a random distribution of nucleoli, the number of nucleoli representative of section i is estimated by:

$$\frac{N_i}{n_i} = \frac{t}{t+2a} \tag{4}$$

where N_i is the number of nucleoli representative of section i, and n_i is the number of nucleoli actually counted in section i. Since a is $\sqrt{r^2 - \left(\frac{k}{2}\right)^2}$ in sections 20 μ thick, with nucleoli 5 μ in diameter, and where k is arbitrarily set at 1 μ, the nucleolar correction factor is 0.803.

The size of unseen or uncounted fragments must be determined for the specific conditions of the experiment, for k will vary depending on the embedding material, stain used, optical system, and alertness of the observer (see Ebbesson and Tang, 1965). Floderus (1944), using a magnification of 600×, estimated k as 0.5 μ; he did not explain the technique used. Marrable (1962), although discussing theoretical aspects of k, did not define how he obtained his figure 1.45 μ. The value of "k" can be estimated pragmatically or theoretically. Since the limit of the resolving power of the light microscope is about $^1/_3$ of the wavelength of the illuminating light (Cosslett, 1966), objects smaller than about 0.16 μ could not be resolved, even when using the best microscope and a source of blue light with a wavelength of 0.475 μ. Taking σ as the resolving power, λ as the wavelength of light, and N.A. as numerical aperature: $\sigma = \frac{0.5\,\lambda}{\text{N.A.}}$. Thus, using a dry objective with a N.A. of 0.6, and with a properly aligned microscope, objects only greater than 0.5 μ in diameter could be seen clearly. Using an oil immersion objective with N.A. 1.0, the resolving power, with blue light, would be 0.24 μ. Thus k, at its theoretical limits, must be greater than 0.24 μ. Practically, the size of the smallest visible fragment can be measured by 1. focusing through the object and recording the microscope micrometer movement, 2. measuring the fragment by using the eyepiece micrometer reticule, or 3. recording the diameter of all or a proportion of the fragments counted, taking the smallest recorded fragments as k.

Another method of estimating the split cell error was described by Bok and Kip (1940). Taking neuronal counts from sections of cerebral cortex 15 and 30 μ thick, these authors estimated the number of neurons representative of a 15 μ thick section by determining the difference between the cell body counts obtained in 15 and 30 μ thick sections, respectively. Thus: $N_\alpha = n_\alpha - n_\beta$ where N_α is the number of neurons representative of a section α μ thick, and n_α and n_β are the total number of units counted per unit area in α and β μ thick sections.

Ebbesson and Tang (1965) improved the formula to the following form:

$$N_\alpha = \frac{t_\alpha}{t_\alpha - t_\beta}\,(n_\alpha - n_\beta)$$

in which t_α and t_β and sections α and β μ thick, and $t_\alpha > t_\beta$.

Two other studies should be mentioned in connection with the nucleolar correction factor, namely, those of Cammermeyer (1967b) and of Jones (1937a). To estimate the effect of split nucleoli on nucleolar counts, Jones took serial sections 6 to 12 μ in thickness from paraffin-embedded nodose ganglia of four cats. Theoretically, the use of 6 μ thick sections would result in splitting of twice the number of nucleoli that would be split if the tissue were cut at 12 μ, thus producing higher estimates for the neuronal population if no nucleolar correction factor were used. Nonetheless, Jones found very little difference between the total neuronal estimates based on sections 6, 8, or 12 μ in thickness, and therefore concluded that split nucleoli do not appreciably affect the nucleolar count. Jones next examined 300 cells in adjacent paraffin sections 6 to 12 μ thick, and found no more than seven with nucleoli in two sections, i.e. approximately 2.3%. He suggested that nucleoli are harder than the surrounding tissues and are pushed by the microtome knife into one or the other of the two sections, resulting in splitting of far fewer nucleoli than would be expected theoretically. Similar results were obtained by Foley and DuBois (1937) in their study of the paraffin-embedded nodose ganglion of the cat. Of 1,500 neurons in 12 μ thick sections, they found no more than 3% of the nucleoli had been divided by the knife. Work by Cammermeyer (1967b) supports the suggestion that nucleoli, in paraffin sections at least, are more resistant to sectioning than are other cellular constituents. The author studied nucleoli in brain sections from rabbit, cat, and chinchilla, fixed with a modified Heidenhain's-Susa solution and embedded in paraffin. Sections were cut at 10 or 20 μ thickness, stained with periodic acid-Schiff and gallocyanin-chrom alum, and studied under an oil immersion objective. The author found a number of nucleoli, some surrounded by nucleoplasm dislocated outside the nucleus, while other nucleoli, leaving an empty space in one nuclear hemisphere, were found intact in the adjacent section. The number of displaced nucleoli was counted in preparations representing different species, ages, methods of fixation, post-mortem intervals, and section thickness, and prepared by different technicians. In sections 10 μ thick 1.3 to 2.8% of nucleoli were displaced, while in sections 20 μ thick 0.1 to 2.3% were displaced. Cammermeyer concluded that the nucleolar displacement is an artifact, the incidence of which depends on cutting speed, cytoarchitectural pecularities, and section thickness. In all likelihood, the texture of the embedding medium is an additional determining factor. In Epon-embedded material, used for electron microscopy, no evidence of displacement of nucleoli has been found. It is likely that in celloidin-embedded material, nucleolar displacement would be even more common than Cammermeyer found it in paraffin sections but no data on this point are available.

All considered, it appears that the correction factor for nucleolar bisection cannot be determined on the basis of theoretical consideration alone. For the present it would seem most reasonable to report both the basic count and the figure resulting from the application of the nucleolar correction factor [Formula (4)].

Counting Error

Variable degrees of discrepancy, ranging from 2 to 10%, have been found when comparing repetitive counts of one and the same neuronal population (Gardner, 1940; Duncan and Keyser, 1936). Ebbesson (1963) in a quantitative study of the superior cervical ganglion in man, found the average counting error to be approximately 2%. This error is less than that found by many workers, probably because the repeat counts were done by the same observers, and because the neurons were large, distinct, of a single well-defined type and in a clearly demarcated area. In a review of the inferior olivary nucleus in man, Moatamed (1966) obtained a value of within 2% of the original value in repeat counts on the same slide. Altman (1966) recounted the neurons of the human facial motor nucleus in 64 sections, and arrived at a figure 2.4% lower than the original count on one side, and 5.2% less than the first count on the other side. In our laboratory, recounts by technicians are practically always within 3% and frequently within 1% of the original count. To keep check of counting errors, and to obtain accurate counts, we have found it useful to recount all sections used for counting.

Method II. Systematic Section Method

Many structures in the nervous system are so large that counting all cells in serial sections would be an almost impossible task. Also, the added accuracy in taking more than a certain sample may hardly warrant the added time and work involved in a total neuronal count. The types of samples of a neuronal population that can be taken include: 1. a systematic sample of periodic sections through the structure, 2. a random sample of sections through the structure, 3. systematic or random samples of neuron population per unit volume together with an estimate of the volume of the structure.

In many studies, including van Buskirk's (1945) review of the VII nerve nucleus, Escobar and colleagues' (1968) study of the inferior olivary nucleus, and Jones' (1937b) review of the nodose ganglion, sections were selected by a systematic sample technique (table). Random numbers may also be used to choose a number from 1 to P where P is the period of the systematic sample. If the number i is chosen, then slides i, $i+P$, $i+2P-$, will be studied. Periods of from 2 to 63 have been used by different authors (Irving and Harrison, 1967; Ebbesson, 1963) depending on the population structure. The period 10 appears to have been selected most commonly (table).

Formulas

Estimates of the total population, when the volume is known, can be obtained from the following formula:

$$\frac{\text{total count } N_t}{\text{sample count } n_s} = \frac{\text{total volume } V_t}{\text{sample volume } v_s} \tag{6}$$

and when the volume is not known, from:

$$N_t = n_s \frac{S_t}{s_s} \tag{7}$$

in which N_t is the estimated total number of units in the structure, n_s is the total number of units counted in the sample, S_t is the total number of sections through the structure, and s_s is the number of sections used in the sample.

A simpler formula, in which the product of the total number of nucleoli counted and the period gives the estimated neuronal population, can be used:

$$N_t = n_s \times P, \tag{8}$$

P being the period or frequency at which sections are sampled. This was derived by Corey using formula 6 (Konigsmark *et al.*, 1969). In the ratio estimator, the total volume V_t is not known but must be estimated. If we assume that any change in area from slide to slide is linear, a good approximation of V_t is given by:

$$V_t = l\,[A_f + A_{f+1} + \cdots + A_l]$$

in which l is the length in mm between sections in the sample, A_f is the area of the structure in the first section through the structure, A_{f+1} is the area of the next section studied, and A_l is the area of the last section through the structure.

Then:
$$V_t = l \times A_s = t_{\text{mm}}\, P A_s. \tag{9}$$

t_{mm} being the section thickness in mm, and A_s the sum of all the section areas through the structure. The estimator derived from Formula (6) then is

$$N_t = n_s \times \frac{V_t}{t_{mm} A_s} = n_s \frac{l A_s}{t_{mm} A_s} = n_s \frac{t_{mm} P A_s}{t_{mm} A_s} = n_s\, P, \qquad \text{(8, as above)}.$$

A comparison of these two Formulas, (7) and (8), was made by Konigsmark, Kalyanaraman, Corey and Murphy (1969) using a total count of neurons through the VIth nerve nucleus in man as the most accurate population estimate. Using a computer to calculate the population estimates, they found that the mean square error, that is the average of the squares of all deviations from the true population for all values in the period of 10, using Formula (7), was 4.176×10^6; using Formula (8) it was 3.074×10^6 and concluded that Formula (8) was the more accurate of the two methods.

Sample Size

The choice of sample size or period depends on the degree of accuracy required, and on the uniformity of the structure studied. Several authors, including Konigsmark *et al.* (1969), Radmanesh (1968), and Moatamed (1966) have compared results obtained by the use of different sample sizes. Moatamed, in a study of the inferior olivary complex of a 42-year-old male, counted all neurons in serial sections and arrived at the figures of 440,262 neurons on the right, and 424,642 neurons on the left side. These results were compared with estimated counts obtained by using different periods from 2 to 300. The counts of periodic sections up to every 50th were within 3% of the total count. A period of 100 was accurate to $\pm 5\%$, and the use of a period of 300 yielded a figure within 12% of the original value. In a quantitative study of the VIth nerve nucleus in a 43-year-old man we found the procedure of counting neurons in every tenth section to result in values within $\pm 5\%$ of the figures obtained by counting every section; counting with a period of 20 yielded figures differing from the total count by about 10% (Konigsmark *et al.*, 1969). Radmanesh (1968), in a study of the nucleus intercalatus in man, took neuronal counts on all serial sections in one case, and subsequently used different periods, from 2 to 10, for total population estimates. He found that the use of a period of 5 resulted in estimates within $\pm 5\%$ of the total count, whereas serial counting with a period of 10 yielded figures within $\pm 9\%$ of the population determined by counting in every section. The variation in cell counts from section to section in the nucleus intercalatus was not mentioned. It is likely that in this and other long thin structures showing considerable variation in cell number between adjacent sections, a higher percentage of slides must be counted than is necessary in the case of a large relatively equidimensional structure.

Method III. Random Section Method

A standard method of finding the number of cells in a structure consists in determining the number of cells per unit volume, and multiplying this figure by the total volume. In this method, the percentage of sections to be counted is first selected, most often 5—10%. Then random numbers are used to choose the specific sections from the total number of sections through the structure. As in Method II [Formula (7)], the estimated total population is the product of the sample mean and the total number of sections through the structure. Konigsmark *et al.* compared the results of this infrequently used technique with the figures obtained by counting in all serial sections through the VIth nerve nucleus. They found that the mean square error resulting from the use of Formula (7) was 2.7 times larger than that caused by using Formula (8), and therefore concluded that 8 is the better of the two formulas in quantitative studies of the central nervous system.

Method IV. Sample and Volume Method

This method differs from Methods II and III by incorporating a volume determination in the population estimate; it is similar to the two other methods in the sense that the ratio estimator formula (6) is used. Utilizing area and volume estimates of the structure, this method can applied in two ways: 1. by taking small samples of neuronal populations throughout the structure, or 2. by taking total neuronal counts in each section studied. First, however, an estimate of the volume must be obtained.

Area and Volume Determinations

Nervous system volumes change drastically during processing so that a determination of the volume of a structure from fresh or fixed brain cannot be taken as even an approximation of the final volume used in neuronal counts. Since cross-sectional area is a factor in volume calculation, an accurate determination of this value is essential. Three different methods can be used to determine the cross-sectional areas of a structure: 1. the eyepiece reticule method, 2. the planimeter method, 3. the graph paper method. CHOW (1951), in a study of the auditory system in two monkeys, determined the cross sectional areas by using an eyepiece reticule calibrated by means of a stage micrometer. This method includes a considerable subjective element, and leaves the accuracy open to question.

In using the planimeter or graph paper methods, enlargement of the section onto a flat surface is required. This can be done by photographing and enlarging the structure, by projecting and drawing the outline onto paper, or by drawing the outline from a camera lucida projection. Although each of these methods is effective, we have come to prefer the use of a camera lucida attachment to the Wild microscope[4] for both speed and accuracy. Once the structure has been outlined in ink, the outline can be transposed at any magnification onto paper. If the structure has not been outlined, neurons can be dotted onto a sheet of paper and the outline of the structure drawn by connecting the outermost dots by straight lines.

If the outline is made on graph paper, the number of squares within the outline can be counted. This method is somewhat tedious and subjective since many squares are cut by the figure. By using a planimeter[5] the area can be determined with speed and fair accuracy. The planimeter, giving area in square centimeters, can be calibrated to an accuracy of 0.3% in a 60 cm^2 area. Starting at one point on the outline, the tracer arm is moved on the outline through three complete rotations. The difference between the initial and final reading divided by three gives the area in cm^2 of the camera lucida drawing of the structure. To determine the true area, the magnification factor is determined. Using the camera lucida at the magnification used for making the outline, a stage micrometer is projected and the magnified line drawn. This length in millimeters, divided by the actual length will give the magnification factor. Thus $A_i = \frac{a_i}{m \times 100}$ where A_i is the area of section i in mm^2, and a_i is the apparent area in cm^2.

Two methods can be used to determine the total volume from the cross-sectional areas. DORNFELD, SALTER, and SCHEFFE (1942) plotted each of the section areas onto a graph with section numbers as the abcissa and area in cm^2 as the ordinate. The total area of the structure was obtained by taking the area subtended by the graph line, using a planimeter. The product of this area and the section thickness gives the volume of the structure. The volume can be determined more simply without resorting to a graph. Since the product of the area and thickness of the structure on each section will give its volume, the volume can be estimated from periodic sections by using Formula (9), taking the sum of the areas in mm^2 multiplied by the product of the section thickness and the period at which sections are taken. Obviously, the more sections are taken through a structure, the more accurate is the volume estimate.

[4] Wild MZO KGS with drawing tube, Traflenberg type.

[5] Keuffel and Esser Paragon 620005 planimeter.

Section Thickness

In the sample and volume method it is of essential importance to know the section thickness, for this thickness is the third dimension determining the sample volumes and the total volume as well as a factor determining the split cell error.

Several workers have demonstrated a variance between the measured section thickness and the microtome setting (Graf, 1948; Marengo, 1944). Marengo measured the thickness of paraffin sections of bull testes cut at a microtome setting of 3, 5, and 10 μ. By cutting re-embedded sections at right angles, he found an average thickness of, respectively, 3.31, 5.25, and 10.82 μ, representing an error of 5 to 10%. Lange and Engstrom (1954) discussed different methods of determining section thickness. These methods included: 1. surface focusing, 2. direct measurement of cross section, 3. light-profile method, 4. interferometric methods, and 5. stereoscopic techniques. Another method, employing a mechanical device ("microcator"), was described by Glimstedt and Hakansson (1951). Rahbek and Omar (1952) found the light-profile microscope accurate to within 2%. Not only is special equipment required for all but the first two methods, but appropriate procedures for use would have to be developed, since none but the first two methods have been put to practical use.

The surface-focusing method has the advantages of being applicable to the section studied, simple, and relatively accurate. The microscope must have negligible backlash and no other mechanical faults, and the objective must have high resolution. Using the best optical system commercially available, Lange and Engstrom (1954) studied the thickness of histological structures and concluded, from a sequence of 25 trials for each object, that the minimum error for thickness determination for a 5 μ thick object was about 10%, with greater accuracy for thicker objects. By using a Zeiss planapochromat objective (N.A. 1.30, 100×) and taking the difference in the fine adjustment micrometer readings when focusing on the upper and lower surfaces of the section, we have obtained reproducible readings with 1 μ of each other on a section measuring 19 μ in thickness, representing an accuracy of about 95%.

Total Neuronal Counts on Each Section Studied

This method includes an estimate of the volume so that the total number of cells is calculated from a known number of nucleoli in a known volume. The i-th slide may be considered a representative of one-half of the volume between the $i-1$ slide and the $i+1$ slide of the systematic sample. By assuming that any change in volume from slide to slide is linear, the volume v_{r_i} represented by section i is given by: $v_{r_i} = \frac{l(A_{i-l}+2A_i+A_{i+l})}{4}$ in which l is the distance in mm between sections in the sample. Hence the estimated total volume, determined using all sample sections, is the sum of the volumes represented by each section:

$$V_t = \sum_i \frac{l(A_{i-l}+2A_i+A_{i+l})}{4}.$$

Since the sample volume v_s and the sample count n_s are known, the estimated total number of neurons can be found using Formula (6) and solving for N_t:

$$N_t = \frac{n_s V_t}{v_s}. \qquad (10)$$

The results of using Formula (10) were compared with those obtained by the use of Formula (8) by Konigsmark *et al.* (1969). Matching the results against the figures from total counts on serial sections through the VIth nerve nucleus, it was found that the mean square error resulting from Formula (8) was 1.58 times the mean square error made by using Formula (10). This finding would suggest that Formula (10), used in conjunction with a volume estimate, can give slightly more accurate results.

Small Samples of Neuronal Populations Throughout the Structure

Many authors have used the packing density and a volume estimate as the factors in calculating neuronal populations (table). Although volume determination by planimetry is a relatively straight-forward procedure, methods of taking neuronal population samples yield greatly variable results. In the study of structures, such as the cerebral cortex, which are characterized by great variation in packing density from layer to layer and from area to area, samples are generally selected from specific areas and layers; the volume of each of these sections is determined and the neuronal population then calculated (Hammarberg, 1895; Thompson, 1899). Studying more homogeneous structures, other authors have taken only random counts through the structure (Chow, 1951). In a study of the superior cervical ganglion, however, Ebbesson and Tang (1967) found systematic sampling to be more accurate than random sampling. Tang and Ebbesson (1969) applied the systematic sampling method to a nerve trunk, using a serpentine pattern for the selection and numbering of grid squares. The percentage of squares in the reticule to be counted was first determined. A random number i between 1 and P, the Period, was then selected and the $i+P$, $i+2P$ . . . squares taken for counting. The same sampling principle might well be used to select periodic reticule fields in quantitative studies of brain structures.

Another technique of volume estimation was suggested by Chalkley (1943). In this method, later applied to the neuronal population of the cerebral cortex by Shariff (1953), the relative volumes of neurons and the remaining parenchyma are determined. The volume ratios are obtained by taking random samples of points, determined by four pointers in the eyepiece, throughout the structure. By determining the average volume of the neurons and the total volume of the cortex, it is possible to calculate the approximate total number of cortical neurons.

Suggested Protocol for Neuronal Population Estimates

To conclude this paper, a protocol is given which we have found useful for neuronal population estimates. Needless to say, it is subject to any modification that could be imposed by particular individual requirements.

1. Review all sections containing the nucleus to ascertain the number of slides and the degree of completeness with which they represent the structure in question.
2. Label sides A and B or L and R.
3. Outline the cell territory with ink on each section.
4. For cell count use a research microscope with a 25 × objective, and a Howard disc reticule in a 10 × or 12.5 × ocular.
5. Rotate the ocular until the orientation of the reticule is parallel with the movement of the slide on the mechanical stage.
6. Count all neuronal nucleoli within the area outlined by ink, beginning in the upper left margin of the nucleus. After completing count in one field, move the slide horizontally to the adjacent field, using any prominent landmark. At lateral margin move the slide down one field and proceed in the same manner until the entire field

has been counted. Include in the count intact nucleoli as well as nucleolar fragments larger than one-half the average nucleolar diameter. If a nucleolus is split by the outer lines of the reticule, include it only if it is on either the right or upper reticule lines.

7. Determine thickness of the section under oil immersion by focusing on the upper surface and then on the lower surface of the section, recording the difference in the fine micrometer scale readings to give the section thickness in micra. Take at least two readings on each section.

8. Calculate the neuronal population by using Formula (8) and also Formula (4) to correct for split nucleoli.

Key to Symbols Used in the Formulas

a	= a portion of the nucleolus equal to $\sqrt{r^2 - \left(\frac{k}{2}\right)^2}$.
A_f	= area of structure in the first section through the structure.
A_{f+1}	= area of next section in a periodic series of sections.
A_i	= true area of section i in mm^2.
a_i	= apparent area of section i in cm^2.
A_l	= area of the structure on last section through the structure.
A_s	= the sum of all of the section areas through the structure.
b	= the maximum height of uncounted fragments.
d	= mean diameter of the unit (cell, nucleus, or nucleolus).
i	= any random number or section from 1 to P.
$i-1$	= the counted section nearest i on one side.
$i+1$	= the nearest counted section on the other side of i.
k	= the maximum diameter of the unseen or uncounted units (nuclear or nucleolar fragment).
l	= length or distance in mm between section in sample; equal to $tP \div 1{,}000$.
m	= magnification factor: ratio of magnified line length divided by its actual length.
N_α, N_β	= the number of units per unit area representative of a section α or β μ thick.
n_α, n_β	= the number of units counted per unit area in α and β μ thick sections.
N_i	= the number of units representative of section i.
n_i	= the number of units actually counted in section i.
n_s	= the total number of units counted in the sample.
N_t	= the estimated total number of units in the entire structure.
P	= period or frequency at which sections are sampled.
r	= mean radius of the unit.
s_s	= the number of sections used in the sample.
S_t	= the total number of sections through the structure.
t	= measured section thickness in microns.
t_α, t_β	= sections of α and β μ thickness.
t_{mm}	= section thickness in mm.
v_{r_i}	= volume represented by section i in a periodic series of sections through the structure.
v_r	= volume represented by a sample section in a periodic series of sections through the structure.
v_s	= volume of the counted sample from the structure studied.
V_t	= estimated total volume of the structure studied.

Table. *Summary of the*

Author	Structure	Species	Age	Fixative	Embedding media	Section thick-ness	Stain	Method	Units counted
Nurnberger (1958)	CNS	Chicken	Embryo, newborn	—	—	—	Methylene blue	Homogenate	Neuron nuclei
Hammarberg (1895), Thompson (1899)	Cerebral cortex (one side)	Man	5 mo to. middle age	95% alcohol	Praffin	10 μ 20 μ	Methylene blue	Ocular M IV	Neuronal nuclei
von Economo and Koskinas (1925)	Cerebral cortex (one side)	Man	Adult	—	—	25 μ	—	Photo-graphic M IV	Neuron bodies
Shariff (1953)	Cerebral cortex (one side)	Man	1. Adult	4% formalin perfused	—	24 μ	Thionine	Ocular M IV	—
Pakkenberg (1966)	Cerebral cortex (one side)	Man (1)	18 yrs.	80% alcohol formalin perfusion	Paraffin	12 μ to 20 μ	Gallocyanin	Ocular M IV	Nucleoli (2.65 μ)
Lassek (1940)	Betz cells (one side)	Man (1)	22 yrs.	—	Proxylin	35 μ	Cresyl violet	Ocular M I	Neuron body
Tower (1954)	Cerebral cortex	1. Whale (3) 2. Elephant (1)	Adult	Formalin	Paraffin	20 μ	Cresyl violet	Photo-graphic	Visible nucleoli
Harman, Carpenter (1950)	1. Globus pallidus 2. Putamen 3. Caudate	Man	Adult	—	Celloidin	40 μ	Weigert, Pal	Projection, planimeter	—
Kreiner, Maksymowicz (1962)	1. Caudate and putamen 2. Globus pallidus 3. N. accum-bens 4. Amyg-doloid	Dog (5)	—	Formalin	Paraffin	20 μ	Kluver-Barrera, Nissl	Projection, wax model	—
Powell, Guillery, Cowen (1957)	1. Ant. thalamic nuclei 2. Medial mamillary nucleus	Man (2)	—	70% alcohol 2% acetic acid	Paraffin	15 μ to 25 μ	Thionine 1%	Volume-projection, planimeter. Ocular M IV	—
Chacko (1948)	Lateral geniculate body	Man (1)	—	—	Paraffin	10 μ or 15 μ	Methylene blue	Projection, planimeter. Ocular M IV	Neuron bodies
Sullivan *et al.* (1958)	Lateral geniculate body	Man (1)	28 yrs.	—	Celloidin	35 μ	Cresyl violet	Projection M IV	Neuron bodies
Chow (1951)	Inferior colliculus	Monkey (2)	Young	Formalin, 10% perfused	Celloidin	25 μ 50 μ	Thionine	Ocular M IV	Neuronal nuclei

techniques used in neuronal counts

Magnification	Volume	Packing density	% Counted	Actual count	Estimated total	Average	Comments
—	—	—	—	—	4×10^7	—	—
—	—	—	0.0017	1.596×10^5	4.6×10^9	—	Based on average number of cells/mm² of cortex × the area of cortex
50—100 ×	285 cm³ from thick-ness and area	24,500/mm³	—	—	7.0×10^9 (one side)	—	—
—	115.2 cm³ (one side)	Agranular 18,200/mm³ Conico-granular 97,200/mm³	—	—	3.45×10^9	—	Chalkley's method of relative volumes was used in this determination
400 ×	285 cm³ fresh	17,560/mm³	—	—	1.3×10^9	—	Fresh volume taken from previous workers. Determined % volume lost in processing (74%)
—	—	—	100%	R 34,182 L 34,562	Same	34,372	—
60 ×	—	1. 6,500 to 7,100 mm³ 2. 6,700 to 7,100 mm³	<1%	—	1. 6,800 mm³ 2. 6,900 mm³	—	No nucleoli correction factor
5—10 ×	1. 2,872 mm³ 2. 6,690 mm³ 3. 5,270 mm³	—	—	—	—	—	—
10 ×	1. 295 mm³ 2. 63.2 mm³ 3. 14.7 mm³ 4. 92.0 mm³	—	4%	—	—	—	Volume determined by weighing models therefore not real values
Oil emersion objective	1 76.67 mm³ 2. 24,840 mm³	8,500/mm³ 15,000/mm³	2—5%	—	1. 647,000 and 680,000 2. 385,000 and 392,000	1. 664,000 2. 389,000	Two counts were done, if 20% difference, another was done
50 μ areas counted	60.6 mm³	6,000 to 23,000/mm³		—	1,210,000	—	Ratio of small cell to large was 10:1
300 ×	—	—	8	80,191	1,018,190	—	—
3 mm field	—	43,000/mm³	1—2	—	361,000 430,000	392,000	Area was determined by estimate from eyepiece reticule

Table

Author	Structure	Species	Age	Fixative	Embedding media	Section thickness	Stain	Method	Units counted
Braitenberg, Atwood (1958)	1. Purkinje cells 2. Granule cells	Man	Adult	—	Paraffin	9 μ 12 μ	Cresyl violet	Ocular M IV	Cell bodies
Daron (1960)	Dentate	a) Man (4) b) Chimpanzee (1)	Newborn to adult	Formalin, 10%	Paraffin	15 μ	Luxol fast blue-cresyl violet	Photograph, planimeter	—
Tomash, Malpass (1958)	Motor V	Man (1)	42 yrs.	Formalin	Paraffin	20 μ	Toluidine blue	Ocular M I	Complete nucleoli 5 μ
Sheinin (1930)	Mesencephalic nucleus V	Dog (1)	—	Alcohol-acetic acid	Paraffin	10 μ	Toluidine blue 1%	Ocular M I	Nucleoli Type A 3.7 μ; Type B, C, D: 1.8 μ
Irving, Harrison (1967)	VI	a) Macaque b) Cat	Adult	Alcohol-acetic acid-formalin	Paraffin	16 μ	Protargol silver	Ocular M II	Nuclei (average 11 μ)
Konigsmark, Kalyanaraman, Corey, and Murphy (1969)	VI	Man (2)	—	Formalin 10%	Celloidin	20 μ	Cresyl violet	Ocular M I, II, III, IV	Nucleoli
van Buskirk (1945)	VII	a) Man (37) b) Dog (25) c) Cat (13)	a) Various ages b) and c) Adult	—	Paraffin	10 μ	Nissl	Ocular M II	Distinct nucleoli
Altman (1966)	VII	Man (1)	56 yrs.	Formalin, 10%	Celloidin	20 μ	Nissl	Ocular M II	Neuronal nucleoli
Hall (1962, 1964)	1. Dorsal cochlear N. 2. Ventrical Cochlear N.	Man (10)	Newborn	Formalin or Heidenhain Susa	Paraffin	15 μ	Thionine	Projection, planimeter Ocular M IV	Neuronal nucleoli
Konigsmark (1969)	Ventral cochlear N.	Man (8)	18 days 80 yrs.	Formalin 10%	Celloidin	20 μ	Cresyl violet	Ocular M II	Neuronal nucleoli
Moatamed (1966)	Inferior olivary N.	Man (4)	1 day to 56 yrs.	—	—	20 μ 25 μ	Nissl, Harggquist	Ocular I and II	Neuronal nucleolus (2—3 μ)
Escobar, Sampedro, Dow (1968)	Inferior olivary N.	a) Man b) Cat c) Bat	—	Formalin 10%	Paraffin	a) 20 μ b) and c) 15 μ	a) Kluver-Barrera b) Cresyl violet	Projection M II	Neuron bodies

(continued)

Magnification	Volume	Packing density	% Counted	Actual count	Estimated total	Average	Comments
0.3 mm field	—	1. 300 mm² 2. 3 to 7×10^6/mm³	—	—	1. 15,000,000 2. 1—10 $\times 10^{10}$	—	Assuming area of both hemispheres 150,000 mm²
—	a) Newborn to 71 mm³, Adult 124 to 230 mm³ b) 84.76 mm³	—	10	—	—	—	—
—	—	—	100	R 4,960 L 5,443	—	5,220	—
—	—	—	100	R 2,869 L 2,896	—	2,882	—
500 × 800 ×	—	—	50	—	a) 4,470 b) 1,040	—	No correction for split nuclei was made
310 ×	6.16 to 6.38 mm³	—	10 100	2,760 25,045	24,100 21,940	23,000	Nuclear correction factor was 0.876 where $t = 20\ \mu$, $r = 1.5\ \mu$, and $k = 1\ \mu$
16mm objective	—	—	10	—	a) 4,500 to 9,460 b) 6,800 to 11,510 c) 4,610 to 9,790	a) 6,811 b) 8,613 c) 7,734	—
—	—	—	50	R 3,229 L 3,534	R 6,458 L 7,068	6,763	—
50 ×	—	—	—	—	—	1. 25,100 2. 68,700	—
310 ×	—	—	10	—	60,580—71,540	66,580	Using nuclear correction factor of 0.865, average was 57,600
350 ×	—	—	100 (1) 10 (3)	—	421,655 to 472,434	441,339	—
276 ×	—	a) 5,000/mm³ b) 8,400/mm³ c) 65,406/mm³	10	—	a) 454,340 b) 70,260 c) 13,322	—	Correction factor of 50% for split cell bodies

Table

Author	Structure	Species	Age	Fixative	Embedding media	Section thickness	Stain	Method	Units counted
Lassek, Rasmussen (1938)	Spinal cord, gray matter	Man (12)	a) Newborn (6) b) Adult (6)	Formalin 10%	Not embedded	2 mm	Osmic acid	Camera lucida, drawing planimeter	—
Silver (1942)	Anterior horn neurons	Frog (1)	Adult	Formalin acetic acid alcohol	Double-embedded in celloidin paraffin	15 μ	Nissl or Protargol silver	Ocular M I	Neuron bodies
Guild, Crow, Bunch, Polvogt (1931)	Spiral ganglion	Man (3)	19 to 30 yrs.	Formalin	Celloidin	24 μ	—	Ocular M II	Neuronal nuclei
Jones (1937b)	Nodose ganglion	Cat (8)	—	—	—	10 μ	Azure-2-eosin	Ocular M II	Nucleoli
Foley, DuBois (1937)	Nodose ganglion	Cat (7)	—	Formalin-acetic acid	Paraffin	12 μ	Maximou's azur-eosin mixture	Ocular M II	Nucleoli
Ebbesson (1968)	Superior cervical ganglion	a) Man (1) b) Chimpanzee (2)	a) 92 yrs.	Formalin 10%	Paraffin	5 μ to 15 μ	Hematoxylin-neutral red-Safranin O	Ocular M I	Nucleoli
Gardner (1940)	Spinal ganglion T8, T9	Man (31)	a) 30 to 39 yrs. b) 80 to 89	Millers fluid	Paraffin	10 μ	Toluidine blue-eosin	Ocular M II	Nucleoli
Duncan, Keyser (1936)	Spinal ganglion thoracic	Cat (5)	0.75 to 4.5 kg	Formalin-acetic acid	—	10 μ	Toluidine blue	Ocular M I, II	Nucleoli
Hatai (1902)	Spinal ganglia	Rat (4)	10 to 167 gm	Osmic acid	Paraffin	12 μ		Ocular M I	Nucleoli
Cavanaugh (1951)	Spinal ganglia thoracic	Rat (8)	Adult		Paraffin		Toluidine blue	Ocular M I, ?	Distinct nucleoli

For a review of results in neuronal population studies see Blinkov and Glezer (1968). M I to IV refers

Summary

Neuronal counting methods have been reviewed including the photographic, projection, microphotometric, homogenate, automatic, and ocular-micrometric methods. The various factors contributing to errors in counting have been reviewed, including staining, delimination of the structure, magnification, choice of units to be counted, and counting error. An evaluation has been given of the various formulas introduced to correct for two further forms of inaccuracy, viz. the split-cell and the sampling error.

(continued)

Magnification	Volume	Packing density	% Counted	Actual count	Estimated total	Average	Comments
8.5 ×	Gray matter a) 1.00 cm^3 b) 5.01 cm^3	—	100	—	—	—	—
400 ×	—	—	100	30 g L 3,542 R 3,546	Same	3,544	Linear shrinkage was 3%. No split cell correction factor used
High-dry-objective	—	—	20	—	23,193—27,810	25,614	Estimated a 10% overcount due to split nuclei
—	—	—	20	—	25,800—36,200	29,600	No correction factor was used
260 ×	—	—	6—25	—	22,134—29,411	25,985	Corrected for an overcount of 3% due to split nucleoli
1,000 ×	—	—	—	—	a) L 1,003,384 R 760,370 b) 634,800 to 828,240	a) 881,877 b) 752,862	—
—	—	—	25	—	—	T8 a) 32,832 b) 24,367	—
—	—	—	100 (3) 25 (2)	—	5,363—8,561	6,811	No nucleolar correction factor used
8 mm objective	—	—	100	C_6: 9,793 to 12,200 T_4: 7,068 to 7,611 L_2: 8,200 to 9,514	Same	11,140 7,306 8,867	—
—	—	—	100 in some	—	4,757—7,286	5,793	No nucleolar correction factor used

to methods I to IV in text. In parenthesis after species is the number of specimens studied.

Discussion

Dr. Blackstad: I wonder if the interesting device described recently by Glaser and Van der Loos (1965) could be used to count neurons? They have mounted potentiometers to the microscope stage and the focussing mechanism, and can thus, by means of an X—Y plotter and computer, pinpoint a given cell in X, Y and Z coordinates. Van der Loos has used this apparatus especially to characterize dendrites in quantitative terms. If this device were used to record the positions of nerve cells in very thick sections and if the problem of identifying cells possibly recorded twice were handled by the computer, perhaps it would be possible to ignore the correction factors for cleaved particles.

Dr. Konigsmark: The device used by Van der Loos is quite good for studying dendritic patterns of a neuron, but it is adapted to studying only single cells, and would not be too helpful in enumerating numbers of cells. The flying spot microscope was developed by Causley and Young for neuronal counts. Other investigators, including Segarra, have found that the differentiation of cell types is a major problem in using a computer for neuronal population estimates, I am sure that, with advancing technology and interest in this problem, methods will be developed for using the computer in neuronal counts.

Dr. Blackstad: Surely the use of computerized identification would speed up procedures of this sort. But I can also easily imagine myself doing the identification first and thereafter feeding the *X-Y-Z* coordinates of each cell into a computer, going through the entire length, width and depth of the tissue section. I would undoubtedly collide with myself many times, but the idea would be to have the computer find out which cells had already been recorded.

Bibliography

Abercrombie, M.: Estimation of nuclear population from microtome sections. Anat. Rec. **94**, 239—247 (1946).

Adelstein, S. J., Lyman, C. P.: Pyrimidine nucleoside metabolism in mammalian cells: an *in vitro* comparison of two rodent species. Exp. Cell Res. **50**, 104—116 (1968).

— — O'Brien, R. C.: Variations in the incorporation of thymidine into the DNA of some rodent species. Comp. Biochem. Physiol. **12**, 223—231 (1964).

Adey, W. R.: An experimental study of the central olfactory connexions in the marsupial (*Trichosurus volpecula*). Brain **76**, 311—336 (1953).

— Meyer, M.: Hippocampal and hypothalamic connexions of the temporal lobe in the monkey. Brain **75**, 358—384 (1952).

— Rudolph, A. F., Hine, I. F., Harritt, N. J.: Glees staining of the monkey hypothalamus: A critical appraisal of normal and experimental material. J. Anat. (Lond.) **92**, 219—235 (1958).

Agduhr, E.: A contribution to the technique of determining the number of nerve cells per volume unit of tissue. Anat. Rec. **80**, 191—202 (1941).

Albrecht, M. H.: Mounting frozen sections with gelatin. Stain Technol. **29**, 89—90 (1954).

— A simple mechanical aid in staining frozen sections in quantity. Stain Technol **31**, 230—231 (1956).

— Fernstrom, R. C.: A modified Nauta-Gygax method for human brain and spinal cord. Stain Technol. **34**, 91—94 (1959).

Alksne, J. F., Blackstad, T. W., Walberg, F., White, L. E., Jr.: Electron microscopy of axon degeneration: A valuable tool in experimental neuroanatomy. Ergebn. Anat. Entwickl.-Gesch. **39**, 1—31 (1966).

Allison, A. C.: The morphology of the olfactory system in the vertebrates. Biol. Rev. **28**, 195—244 (1953a).

— The structure of the olfactory bulb and its relationship to the olfactory pathways in the rabbit and the rat. J. comp. Neurol. **98**, 309—354 (1953b).

— The secondary olfactory areas in the human brain. J. Anat. (Lond.) **88**, 481—488 (1954).

Altman, A. J.: Facial nucleus in Bell's Palsy. Bull. Johns Hopk. Hosp. **118**, 265—274 (1966b).

Altman, J.: The use of fine-resolution autoradiography in neurological and psychobiological research. In: Response of the Nervous System to Ionizing Radiation (Haley, T. J., and R. S. Snider, eds.), p. 336—359. Boston: Little, Brown & Co. 1964.

— Autoradiographic and histological studies of postnatal neurogenesis. II. A longitudinal investigation of the kinetics, migration and transformation of cells incorporating tritiated thymidine in neonate rats, with special reference to postnatal neurogenesis in some brain regions. J. comp. Neurol. **128**, 431—473 (1966a).

— Das, G. D.: Autoradiographic examination of the effects of enriched environment on the rate of glial multiplication in the adult rat brain. Nature (Lond.) **204**, 1161—1163 (1964).

Amano, M., Messier, B., Leblond, C. P.: Specificity of labelled thymidine as a deoxyribonucleic acid precursor in radioautography. J. Histochem. Cytochem. **7**, 153—155 (1959).

Amprino, R.: Correlazioni quantitative fra centri nervosi e territori d'innervazione periferica durante lo sviluppo. Arch. ital. Anat. Embriol. **49**, 261—300 (1943).

Andén, N.-E., Carlsson, A., Dahlström, A., Fuxe, K., Hillarp, N.-A.: Demonstration and mapping out of nigro-striatal dopamine neurons. Life Sci. **3**, 523—530 (1964).

—, Corrodi, H., Dahlström, A., Fuxe, K., Hökfelt, T.: Effects of tyrosine hydroxylase inhibition on the amine levels of central monoamine neurons. Life Sci. **5**, 561—568 (1966).

Andén, N.-E., Corrodi, H., Fuxe, K., Hökfelt, T.: Increased impulse flow in bulbospinal noradrenaline neurons by catecholamine receptor blocking agent. Europ. J. Pharmacol. **2**, 59—64 (1967).

— — — — Evidence for a central 5-hydroxytryptamine receptor stimulation by lysergic acid diethylamide. Brit. J. Pharmacol. **34**, 1—7 (1968).

— — — — Hökfelt, B., Rydin, C., Svensson, T.: Evidence for a central NA receptor stimulation by catapresan. (To be published.)

— Dahlström, A., Fuxe, K., Larsson, K.: Further evidence for the presence of nigro-neostriatal dopamine neurons. Amer. J. Anat. **116**, 329—333 (1965).

— — — — Functional role of the nigro-neostriatal dopamine neurons. Acta pharmacol. (Kbh.) **24**, 263—274 (1966).

— — — — Olson, L., Ungerstedt, U.: Ascending monoamine neurons to the telencephalon and diencephalon. Acta physiol. scand. **67**, 313—326 (1966).

— — — Olson, L., Ungerstedt, U.: Ascending noradrenaline neurons from the pons and the medulla oblongata. Experientia (Basel) **22**, 44 (1966).

— Fuxe, K., Hamberger, B., Hökfelt, T.: A quantitative study on the nigro-neostriatal dopamine neuron system in rat. Acta physiol. scand. **67**, 306—312 (1966d).

— — Hökfelt, T.: The importance of the nervous impulse flow for the depletion of the monoamines from central neurons by some drugs. J. Pharm. Pharmacol. **18**, 630—632 (1966).

— — — Effect of some drugs on central monoamine nerve terminals lacking nerve impulse flow. Europ. J. Pharmacol. **1**, 226—232 (1967)

— — — Rubensson, A.: Evidence for dopamine receptor stimulation by apomorphine. J. Pharm. Pharmacol. **19**, 627—629 (1967).

— — Larsson, K.: Effect of large mesencephalic-diencephalic lesions on the noradrenaline, dopamine and 5-hydroxytryptamine neurons of the central nervous system. Experientia (Basel) **22**, 842—847 (1966).

— — Ungerstedt, U.: Monamine pathways to the cerebellar and cerebral cortex. Experientia (Basel) **23**, 838—839 (1967).

Anderson, F. D.: Alcoholic-Thionin counterstaining following Golgi dichromate-silver impregnation. Stain Technol. **29**, 179—184 (1954).

Angelakos, E., Fuxe, K., Torchiana, M.: Chemical and histochemical evaluation of the distribution of catecholamines in the rabbit and guinea pig heart. Acta physiol. scand. **59**, 184—192 (1963).

Angelakos, E. T., King, M. P.: A new histochemical method for the use of specific demonstration of catecholamine (CA) — containing structures. Fed. Proc. **24**, 389 (1965).

— — Histochemical evaluation of "free" and "bound" catecholamines in nerve fibers. Fed. Proc. **25**, 383 (1966).

— — Demonstration of nerve terminals containing adrenaline by a new histochemical technique. Nature (Lond.) **213**, 391—392 (1967).

Angevine, J. B., Jr.: Time of neuron origin in the hippocampal region. An autoradiographic study in the mouse. Exp. Neurol., Suppl. **2**, 1—70 (1965).

— Autoradiographic study of neuronal proliferation gradients in the diencephalon of the mouse. Anat. Rec. **160**, 308 (1968).

— Sidman, R. L.: Autoradiographic study of cell migration during histogenesis of cerebral cortex in the mouse. Nature (Lond.) **192**, 766—768 (1961).

Apáthy, S.: Das leitende Element des Nervensystems und seine topographischen Beziehungen zu den Zellen. Mitt. Zool. Sta. Neapel. **12**, 495—748 (1897).

Apelgot, S., Ekart, B., Tisne, M. R.: Referred to by Cleaver, J. E., in: Thymidine Metabolism and Cell Kinetics, p. 21. Amsterdam and New York: North-Holland Publ. Co. 1967.

Arees, E. A., Mayer, J.: Anatomical connections between medial and lateral regions of the hypothalamus concerned with food intake. Science **157**, 1574—1575 (1967).

Armstrong, J. A.: An experimental study of the visual pathways in a reptile (*Lacerta vivipara*), vol. 84, part 2, p. 146—167. Cambridge: University Press 1950.

— An experimental study of the visual pathways in a snake (*Natrix natrix*), vol. 85, part 3, p. 275—288. Cambridge: University Press 1951.

Armstrong, J., Richardson, K. C., Young, J. Z.: Staining neural end-feet and mitochondria after postchroming and carbowax embedding. Stain Technol. **31**, 263—270 (1956).

Asbury, A. K.: Schwann cell proliferation in developing mouse sciatic nerve. A radioautographic study. J. Cell Biol. **34**, 735—743 (1967).

Ashley, C. A., Feder, N.: Glycol methacrylate in histopathology. Arch. Path. **81**, 391—397 (1966).

Atlas, M., Bond, V. P.: The cell generation cycle of the elevenday mouse embryo. J. Cell Biol. **26**, 19—24 (1965).

Auer, J.: Terminal degeneration in the diencephalon after ablation of frontal cortex in the cat. J. Anat. (Lond.) **90**, 30—41 (1956).

Avakian, O. V., Gillespie, J. S.: Uptake of noradrenaline by adrenergic nerves, smooth muscle and connective tissue in isolated perfused arteries and its correlation with the vasoconstrictor response. Brit. J. Pharmacol. Chermother **32**, 168—184 (1968).

Bachmann, L., Saltpeter, M. M.: Autoradiography with electron microscopy. A quantitative evaluation. Lab. Invest. **14**, 1041—1053 (1965).

Baginsky, B.: Über den Ursprung und den centralen Verlauf des Nervus acusticus des Kaninchens. Virchows Arch. path. Anat. **105**, 28—46 (1886).

Bahr, G. F.: Osmium tetroxide and ruthenium tetroxide and their reactions with biologically important substances. Electron stains III. Exp. Cell Res. **7**, 457—479 (1954).

Baker, J. R.: Principles of biological microtechnique. A study of fixation and dyeing. London: Methuen 1958.

Barnard, R. I.: Experimental changes in end-feet of Held-Auerbach in the spinal cord of the cat. J. comp. Neurol. **73**, 235—266 (1940).

Barr, M. L.: Some observations on the morphology of the synapse in the cat's spinal cord. J. Anat. (Lond.) **74**, 1—11 (1939).

— Axon reaction in motor neurons and its effect upon the end-bulbs of Held-Auerbach. Anat. Rec. **77**, 367—374 (1940).

Barron, D. H.: Structural changes in anterior horn cells following central lesions. Proc. Soc. exp. Biol. (N.Y.) **30**, 1327—1329 (1933).

Barry, J., Leonardelli, J.: Étude de la topographie des fibers et des neurones monoaminergiques au niveau de l'hypothalamus chez le Cobaye normal ou stéréotaxé. C. R. Acad. Sci. (Paris) **265**, 557—560 (1967).

Barter, R., Pearse, A. G. E.: Detection of 5-hydroxytryptamine in mammalian enterochromaffin cells. Nature (Lond.) **172**, 810 (1953).

— — Mammalian enterochromaffin cells as the source of serotonin (5-hydroxytryptamine). J. Path. Bact. **69**, 25—31 (1955).

Baserga, R., Kisielski, W. E.: Effects of histologic and histochemical procedures on the intensity of the label in radioautography. Lab. Invest. **12**, 648 (1963).

— Lisco, E.: Duration of DNA synthesis in Ehrlich ascites cells as estimated by double-labeling with C^{14}- and H^3-thymidine autoradiography. J. nat. Cancer Inst. **31**, 1559—1571 (1963).

— — Kisielski, W. E.: Further observations on induction of tumors in mice with radioactive thymidine. Proc. Soc. exp. Biol. (N.Y.) **110**, 687—690 (1962).

— — — Tumor induction in mice by radioactive thymidine. Radiat. Res. **29**, 583—596 (1966).

Bateman, A. J., Chandler, A. C.: Mutation induced in the mouse with tritiated thymidine. Nature (Lond.) **193**, 705—706. (1962).

Baumgarten, H. G., Braak, H.: Catecholamine im Hypothalamus vom Goldfisch (Carassius auratus). Z. Zellforsch. **80**, 246—263 (1967).

— Ehinger, B., Falck, B.: Microspectrofluorimetric differentiation between primary and secondary catecholamines in tissues. Life Sci. **6**, 2465—2468 (1967).

— Falck, B., Holstein, A.-F., Owman, Ch., Owman, T.: Adrenergic innervation of the human testis, epididymis, ductus deferens and prostate: A fluorescence microscopic and fluorimetric study. Z. Zellforsch. **90**, 81—95 (1968).

Beck, E.: The origin, course, and termination of the prefrontopontine tract in the human brain. Brain **73**, 368—391 (1950).

Becquerel, H.: Sur les radiations émises par phosphorescence. C. R. Acad. Sci. (Paris) **122**, 420—421 (1896).

Beech, R. H., Davenport, H. A.: The Bielschowsky staining technic. Stain Technol. **8**, 11—30 (1933).

Bélanger, L. F.: Opening address, in: The Use of Radioautography in Investigating Protein Synthesis (Leblond, C. P., and K. B. Warren, eds.), p. 1—6. New York: Academic Press 1965.

— Leblond, C. P.: A method for locating radioactive elements in tissues by covering histological sections with a photographic emulsion. Endocrinology **39**, 8—13 (1946).

Bennett, M. V. L., Nakajima, Y., Pappas, G. D.: Physiology and ultrastructure of electronic junctions. L. Supermedullary neurons. J. Neurophysiol. **30**, 161—179 (1967).

Beresfold, W. A.: A discussion of retrograde changes in nerve fibers. In: Progress in Brain Research. Degeneration Patterns in the Nervous System. (Singer, M., and J. P. Schadé, eds.), vol. 14, p. 33—56. New York: American Elsevier 1965.

— An evaluation of neuroanatomical methods and their relation to neurophysiology. Brain Res. **1**, 33—56 (1966).

Berkowitz, I. R., Fiorello, O., Kruger, L., Maxwell, D. S.: Selective staining of nervous tissue for light microscopy following preparation for electron microscopy. J. Histochem. Cytochem. **16**, 808—814 (1968).

Berry, M., Rogers, A. W.: The migration of neuroblasts in the developing cerebral cortex. J. Anat. (Lond.) **99**, 691—709 (1965).

— — Histogenesis of mammalian neocortex. In: Evolution of the Forebrain (Hassler, R., and H. Stephan, eds.), p. 197—205. New York: Plenum Press 1966.

Berthold, C.-H., Skoglund, S.: Postnatal development of feline paranodal myelin-sheath segments. I. Light Microscopy. Acta Soc. Med. upsalien. **73**, 113—126 (1968a).

— — Postnatal development of feline paranodal myelin-sheath segments. II. Electron Microscopy. Acta Soc. Med. upsalien. **73**, 127—144 (1968b).

Bertram, E. G., Ihrig, H. K.: Improvements of the Golgi method by pH control. Stain Technol. **32**, 87—94 (1957).

— — Zinc chromate solutions for impregnation of nervous tissue. Stain Technol. **33**, 187 (1958).

— — Staining formalin-fixed nerve tissue with mercuric nitrate. Stain Technol. **34**, 99—108 (1959).

— Sheppard, C.: A possible explanation for the Golgi impregnation of neurons. Anat. Rec. **148**, 413 (1964).

Bianchi, L.: Qualche considerazione sui metodi per il conteggio delle cellule nervose. Anat. Anz. **92**, 233—240 (1942).

Bielschowsky, M.: Die Silberimprägnation der Achsenzylinder. Neurol. Zbl. **21**, 579—584 (1902).

— Die Silberimprägnation der Neurofibrillen. J. Psychol. Neurol. (Lpz.) **3**, 169—188 (1904).

— Nervengewebe. In: Handbuch der mikroskopischen Anatomie des Menschen (W. von Mollendorf, ed.), vol. 4/1, p. 1—201. Berlin: Springer 1928.

Bignami, A., Ralston, H. J., III: The cellular reaction to Wallerian degeneration in the central nervous system of the cat. Brain Res. **3**, 444—461 (1969).

Billings, T. A.: A holder to facilitate the intravenous injection of mice. Lab. Animal Care **17**, 521—523 (1967).

Billingsley, P. R., Ranson, S. W.: On the number of nerve cells in the ganglion cervicale superius and of nerve fibers in the cephalic end of the truncus sympathicus in the cat and on the numerical relations of preganglionic and postganglionic neurons. J. comp. Neurol. **29**, 359—366 (1918).

Birks, R., Katz, B., Miledi, R.: Physiological and structural changes at the amphibian myoneural junction in the course of nerve degeneration. J. Physiol. (Lond.) **150**, 145—168 (1960).

Bishop, P. O.: The nature of the representation of the visual fields in the lateral geniculate nucleus. Proc. Aust. Ass. Neurol. **3**, 15—25 (1965).

Björklund, A.: Monoamine-containing fibres in the pituitary neurointermediate lobe of the pig and rat. Z. Zellforsch. **89**, 573—589 (1968).

— Ehinger, B., Falck, B.: A method for differentiating dopamine from noradrenaline in tissue sections by microspectrofluorometry. J. Histochem Cytochem. **16**, 263—270 (1968).

Björklund, A., Enemar, A., Falck, B.: Monoamines in the hypothalamo-hypophyseal system of the mouse with special references to the ontogenic aspects. Z. Zellforsch. **89**, 590—607 (1968).
— Falck, B., Håkansson, R.: Histochemical demonstration of tryptamine. Properties of the formaldehyde-induced fluorophores of tryptamine and related indole compounds in models. Acta physiol. scand. **74**, Suppl. 318, 1—31 (1968).
— — Ljunggren, L.: Monoamines in the bird median eminence. Failure of cocaine to block the accumulation of exogenous amines. Z. Zellforsch. **89**, 193—200 (1968).
Blackstad, T. W.: Commissural connections of the hippocampal region in the rat, with special references to their mode of termination. J. comp. Neurol. **105**, 417—537 (1956).
— On the termination of some afferents to the hippocampus and fascia dentata. Acta anat. (Basel) **36**, 202—214 (1958).
— Ultrastructural studies on the hippocampal region. In: The Rhinencephalon and Related Structures (W. Bargmann and J. P. Schadé, eds.), p. 122—148. Amsterdam: Elsevier 1963.
— Mapping of experimental axon degeneration by electron microscopy of Golgi preparations. Z. Zellforsch. **67**, 819—834 (1965).
— Studies on the hippocampus: methods in analysis. In: The Interneuron (M. Brazier, ed.). UCLA Forum in Medical Sciences, Los Angeles (in press) (1969).
— Brink, K., Hem, J., Jenne, B.: Distribution of hippocampal mossy fibers in the rat. An experimental study with silver impregnation methods. J. comp. Neurol., in press (1969).
— Fuxe, K., Hökfelt, T.: Noradrenaline nerve terminals in the hippocampal retion of the rat and the guinea-pig. Z. Zellforsch. **78**, 463—473. (1967).
— Kjaerheim, A.: Special axo-dendritic synapses in the hippocampal cortex. Electron and light microscopic studies on the layer of the mossy fibers. J. comp. Neurol. **117**, 133—159 (1961).
Blier, R.: Retrograde transsynaptic cellular degeneration in mammillary and ventral tegmental nuclei following limbic decortication in rabbits of various ages. Brain Res. **15**, 365—393 (1969).
— Bard, P., Woods, J. W.: Cytoarchitectonic appearance of the isolated hypothalamus of the cat. J. comp. Neurol. **128**, 255—312 (1966).
Blinkov, S. M., Glezer, I.: The human brain in figures and tables. (Translated by B. Haigh.) New York, N.Y.: Plenum Press 1968.
Blümcke, S., Niedorf, H. R.: Fluoreszenzmikroskopische und elektronenmikroskopische Untersuchungen an regenerierenden adrenergischen Nervenfasern. Z. Zellforsch. **68**, 724—732 (1965).
— — Rode, J.: Axoplasmic alterations in the proximal and distal stumps of transected nerves. Acta neuropath. **7**, 44—61 (1966).
Bodian, D.: A new method for staining nerve fibers and nerve endings in mounted paraffin sections. Anat. Rec. **65**, 89—97 (1936).
— The structure of the vertebrate synapse. J. comp. Neurol. **68**, 117—160 (1937).
— Taylor, N.: Synapse arising at central node of Ranvier and note on fixation of the central nervous system. Science **139**, 330—332 (1963).
Bok, S. T.: The branching of the dendrites in the cerebral cortex. Proc. Acad. Sci. (Amst.) **39**, 1209—1218 (1936).
— Histonomy of the cerebral cortex. Amsterdam: Elsevier. 1959.
— Kip, M. J.: The size of the body and the size and the number of the nerve cells in the cerebral cortex. Acta neerl. Morph. **3**, 1—22 (1940).
Bondareff, W.: Electron microscopic evidence for the existence of an intercellular substance in rat cerebral cortex. Z. Zellforsch. **72**, 487—495 (1966).
Bouman, K. H.: Die pathologische Anatomie des Zentralnervensystems bei Schizophrenie. Psychiat. neurol. Bl. (Amst.) **32**, 517—539 (1928).
Bowsher, D.: Trans-neuronal degeneration in the specific sensory nucleus of the thalamus. J. Anat. (Lond.) **91**, 611 (1957).
— Brodal, A., Walberg, F.: The relative values of the Marchi method and some silver impregnation techniques. Brain **83**, 150—160 (1960).

Boycott, B. B., Gray, E. G., Guillery, R. W.: A theory to account for the absence of boutons in silver preparations of the cerebral cortex, based on a study on axon terminals by light and electron microscopy. J. Physiol. (Lond.) **152**, 3—5P (1960).
— — — Synaptic structure and its alteration with environmental temperature: a study by light and electron microscopy of the central nervous system of lizards. Proc. roy. Soc. P **154**, 151—172 (1961).
Braak, H., Baumgarten, H. G., Falck, B.: 5-hydroxytryptamin im Gehirn der Eidechse (*Lacerta viridis* und *Lacerta muralis*). Z. Zellforsch. **90**, 161—185 (1968).
Braitenberg, V., Atwood, R. P.: Morphological observations on the cerebellar cortex. J. comp. Neurol. **109**, 1—27 (1958).
Brazier, M.: (editor) *The Interneuron*. UCLA Forum in Medical Sciences (in press) (1969).
Bregman, E.: Über experimentelle aufsteigende Degeneration motorischer und sensibler Hirnnerven. Arb. neurol. Inst. Univ. Wien **1**, 73—97 (1892).
Bremer, J. L.: Notes on staining methods. Anat. Rec. **4**, 265 (1910).
Bresciani, F.: DNA synthesis in alveolar cells of the mammary gland: acceleration by ovarian hormones. Science **146**, 653—655 (1964).
Brightman, M. W., Reese, T. S.: Astrocytic and ependymal junctions in the mouse brain. J. Cell Biol. **35**, 16A—17A (1967).
Brodal, A.: Experimentelle Untersuchungen über retrograde Zellveränderungen in der unteren Olive nach Läsionen des Kleinhirns. Z. ges. Neurol. Psychiat. **166**, 647—704 (1939).
— The reticular formation of the brainstem. Anatomical aspects and functional correlations. Edinburgh: Oliver & Boyd (1957).
— Grant, G.: Morphology and temporal course of degeneration in cerebellar mossy fibers following transection of spinocerebellar tracts in the cat. Exp. Neurol. **5**, 67—87 (1962).
— Hoivik, B.: Site and mode of termination of primary vestibulocerebellar fibers in the cat. An experimental study with silver impregnation methods. Arch. ital. Biol. **102**, 1—21 (1964).
— Pompeiano, O., Walberg, F.: The vestibular nuclei and their connections. Anatomy and functional correlations. Edinburgh: Oliver and Boyd (1962).
Brooks, C. McC., Eccles, J. C.: An electrical hypothesis of central inhibition. Nature (Lond.) **159**, 760—764 (1947).
Bryant, B. J.: Reutilization of leucocyte DNA by cells of regenerating liver. Exp. Cell Res. **27**, 70—79 (1962).
Bruesh, S. R., Arey, L. B.: The number of myelinated and unmyelinated fibers in the optic nerve of vertebrates. J. comp. Neurol. **77**, 631—665 (1942).
Brundin, J.: Distribution and function of adrenergic nerves in the human fallopian tube. Acta physiol. scand. **66**, 1—57 (Suppl.) (1965).
Buren, J. M. van, *see* Van Buren, J. M.
Buskirk, C. van, *see* Van Buskirk, C.
Cajal, S. Ramón y: Pequenas comunicaciones técnicas. Revista trimestral micrográfica (1900).
— Un sencillo método de coloración del reticulo protoplasmico. Trab. Lab. Invest. Biol. Univ. Madrid **2**, 129—222. (1903).
— Quelques méthodes de coloration des cylindresaxes, des neurofibrilles et des nids nerveux. Trav. Lab. Rech. biol. **3**, 1—7 (1904).
— Histologie du système nerveux, vol. II, p. 733—761. Paris: Maloine 1909.
— Histologie du Système Nerveux de l'Homme et des Vertébrés. (2 vols.). Paris: A. Maloine 1911.
— Quelques remarques sur les plaques motrices de la langue des mammifères. Trab. Lab. Invest. Biol. Univ. Madrid **23**, 245—254 (1925).
— Degeneration and Regeneration of the Nervous System. Translated and edited by R. M. May. New York: Hafner 1928.
— Elementos de Técnica Micrográfica del Sistema Nervioso, p. 103—128. Madrid: Tipografía Artística 1933.
— Histologie du système nerveux de l'homme et des vertébrés. Jl, p. 73. Consejo superior de investigaciónes cientificas, p. 986. Madrid: Institute Ramon y Cajal 1952.

Cajal, S. Ramón y, Castro, F. de: Elementos de técnica micrográfica del sistema nervioso. Madrid: Tipografía Artística. 1933.

Cammermeyer, J.: The post-mortem origin and mechanism of neuronal hyperchromatosis and nuclear pycnosis. Expl. Neurol. **2**, 379—405 (1960).

— An evaluation of the significance of the "dark" neuron. Ergebn. Anat. Entwickl.-Gesch. **36**, 1—61 (1962).

— Submerged heart method to prevent intracardial influx of air prior to perfusion fixation of the brain. Acta anat. (Basel) **67**, 321—337 (1967a).

— Artifactual displacement of neuronal nucleoli in paraffin sections. J. Hirnforsch. **9**, 209—224 (1967b).

Campbell, A. W.: Histological studies on the localisation of cerebral function. Cambridge: Cambridge University Press 1905.

Campbell, C. B. G., Ebbesson, S. O. E.: The visual system of a teleost; Holocentrus reexamined. In preparation (1969).

Campos-Ortega, J. A., Glees, P. : The termination of ipsilateral and contralateral optic fibers in the lateral geniculate body of *Galago crassicaudatus*. J. comp. Neurol. **129**, 279—284 (1967a).

— — The subcortical distribution of optic fibers in *Saimiri sciureus* (squirrel monkey). J. comp. Neurol. **131**, 131—142 (1967b).

Cannon, W. B., Rosenblueth, A.: Autonomic neuro-effector system. New York: MacMillan 1937.

Carlsson, A.: Drugs which block the storage of 5-hydroxytryptamine and related amines. In: Handbuch der experimentellen Pharmakologie (Eicher, O., and A. Farah, eds.), vol. 19, p. 529—592. Berlin-Heidelberg-New York: Springer 1966.

— Corrodi, H., Fuxe, K., Hökfelt, T.: Effect of some antidepressant drugs on the depletion of intraneuronal brain catecholamine stores caused by 4, a-dimethyl-meta-tyramine. Europ. J. Pharmacol., in press (1969a).

— Dahlström, A., Fuxe, K., Hillarp, N.-A.: Failure of reserpine to deplete noradrenaline neurons of a-methyl-noradrenaline formed from a-methyl-dopa. Acta pharmacol. (Kbh.) **22**, 270—276 (1965).

— — — Lindqvist, M.: Histochemical and biochemical detection of monoamine release from brain neurons. Life Sci. **4**, 809—816 (1965).

— Falck, B., Fuxe, K., Hillarp, N.-A.: Cellular localization of monoamines in the spinal cord. Acta physiol. scand. **60**, 112—119 (1964).

— — Hillarp, N.-A.: Cellular localization of brain monoamines. Acta physiol. scand. **56**, 1—27 (Suppl. 196) (1962).

— — — Thieme, G., Torp, A.: A new histochemical method for visualization of tissue catecholamines. Med. Exp. **4**, 123—125 (1961).

— Fuxe, K., Hamberger, B., Lindqvist, M.: Biochemical and histochemical studies on the effects of imipramine-like and (+)-amphetamine on central and peripheral catecholamine neurons. Acta physiol. scand. **67**, 481—497 (1966).

— — Jonasson, J., Lindqvist, M.: Demonstration of extraneuronal 5-hydroxytryptamine accumulation in brain following membrane-pump blockade by chlorimipramine. Brain Res., in press (1969).

— — Ungerstedt, U.: The effect of imipramine on central 5-hydroxytryptamine neurons. J. Pharm. Pharmacol. **20**, 150—151 (1968).

Carman, J. B., Cowan, W. M., Powell, T. P. S.: Cortical connexions of the thalamic reticular nucleus. J. Anat. (Lond.) **98**, 587—598 (1964a).

— — — The cortical projections upon the claustrum. J. Neurol. Neurosurg. Psychiat. **27**, 46—51 (1964b).

Carpenter, M.B., Strominger, N. L.: Efferent fiber projections of the subthalamic nucleus in the rhesus monkey. Anat. Rec. **157**, 225—226 (1967).

Caro, L. G.: High resolution autoradiography. II. The problem of resolution. J. Cell Biol. **15**, 189—199 (1962).

— High resolution autoradiography. In: Methods in Cell Physiology (Prescott, D. M., ed.), vol. 1, p. 327—363. New York and London: Academic Press 1964.

Caro, L. G.: Progress in high resolution autoradiography. Progr. Biophys. molec. Biol. **16**, 173—190 (1966).
Caspersson, T., Hillarp, N.-A., Ritzen, M.: Fluorescence microspectrophotometry of cellular catecholamines and 5-hydroxytryptamine. Exp. Cell Res. **42**, 415—428 (1966).
Castro, F. de: Evolución de los ganglios simpáticos vertebrales y prevertebrales. Conexiones y citoarquitectonia de algunos grupos de ganglies en el nino y hombre adulta. Trab. Lab. Invest. Biol. Univ. Madrid **20**, 113—208 (1923).
Catteneo, D.: I fenomeni degenerativi e regenerativi nelle vie visive in sequito a lesioni del nervo ottico. Riv. pat. nerv. ment. **28**, 61—118 (1923).
Caulfield, T. H., Robinson, J. S., Voneida, T. J.: Pupil response in cats after chiasm section. Psychon. Sci. **12**, 289—290 (1968).
Causley, D., Young, J. Z.: Counting and sizing of particles with the flying-spot microscope. Nature (Lond.) **176**, 453—454 (1955).
Cavanaugh, M. W.: Quantitative effects of the peripheral innervation area on nerves and spinal ganglion cells. J. comp. Neurol. **94**, 181—219 (1951).
Cegrell, L.: The occurrence of biogenic monoamines in the mammalian endocrine pancreas. Acta physiol. scand., Suppl. 314 (1968).
Chacko, L. W.: The laminar pattern of the lateral geniculate body in the primates. J. Neurol. Neurosurg. Psychiat. **71**, 211—224 (1948).
Chalkley, H. W.: Method for the quantitative morphologic analysis of tissues. J. nat. Cancer Inst. **4**, 47—54 (1943).
Chambers, W. W., Chung-Yu Liu, Chan-Nao Liu: A modification of the Nauta technique for staining of degenerating axons in the central nervous system. Anat. Rec. **124**, 391—392 (1956).
Champlin, A. K., McGill, T. E.: A simple restaining device for injecting mice. Lab. Animal Care **17**, 600—601 (1967).
Chase, H. B.: Studies on an anophthalmic strain of mice. V. Associated cranial nerves and brain centers. J. comp. Neurol. **83**, 121—140 (1945).
Chow, K. L.: Numerical estimates of the auditory central nervous system of the rhesus monkey. J. comp. Neurol. **95**, 159—175 (1951).
— Regional degeneration in the thalamic reticular nucleus following cortical ablations in monkeys. J. comp. Neurol. **97**, 37—60 (1952).
— Failure to demonstrate changes in the visual system of monkeys kept in darkness or in colored lights. J. comp. Neurol. **102**, 597—606 (1955).
— Dewson, J. H., III: Numerical estimates of neurons and glia in the lateral geniculate body during retrograde degeneration. J. comp. Neurol. **128**, 63—73 (1966).
Clark, M. E.: Histochemical localization of monoamines in the nervous system of the polychaete nephtys. Proc. roy. Soc. B **165**, 308—325 (1966).
Cleaver, J. E.: Thymidine Metabolism and Cell Kinetics, p. 259. Amsterdam and New York: North-Holland Publishing Co. 1967.
— Holford, R. M.: Investigations into the incorporation of (^{3}H) thymidine into DNA in L-strain cells and the formation of a pool of phosphorylated derivatives during pulse labelling. Biochem. biophys. Acta (Amst.) **103**, 654—671 (1965).
Cohn, R.: Laminar electrical responses in lateral geniculate body of cat. J. Neurophysiol. **19**, 317—324 (1956).
Coleman, P. D., Riesen, A. H.: Environmental effects on cortical dendritic fields. I. Rearing in the dark. J. Anat. (Lond.) **102**, 363—374 (1968).
Colonnier, M.: Experimental degeneration in the cerebral cortex. J. Anat. (Lond.) **98**, 47—53 (1964).
— Synapses in the cerebral cortex of the cat. Brain Res. **9**, 268—287 (1968a).
— Synaptic patterns on different cell types in the different laminae of the cat visual cortex. An electron microscope study. Brain Res. **9**, 268—287 (1968b).
— Gray, E. G.: Degeneration in cerebral cortex. In: Electron Microscopy. Fifth Internat. Congr. for Electron Microscopy (Breese, S. S., Jr., ed.), vol. 2, p. U3. New York: Academic Press (1962).
— Guillery, R. W.: Synaptic organization in the lateral geniculate nucleus of the monkey. Z. Zellforsch. **62**, 333—355 (1964).

Combs, C. M.: The distribution and temporal course of fiber degeneration after experimental lesions in the cat brain. J. comp. Neurol. **94**, 123—175 (1951).

Cook, W. H., Walker, J. H., Barr, M. L.: A cytological study of transneuronal atrophy in the cat and rabbit. J. comp. Neurol. **94**, 267—292 (1951).

Corrodi, H., Fuxe, K., Hökfelt, T.: Possible role played by central monoamine neurons in thermo-regulation. Acta physiol. scand. **71**, 224—232 (1967).

— — — Central serotonin neurons and thermoregulation. Advanc. Pharmacol. **6** B, 49—54 (1968a).

— — — The effect of immobilization stress on the activity of central monamine neurons. Life Sci. **7**, 107—112 (1968b).

— Hillarp, N.-A.: Fluoreszenzmethoden zur histochemischen Sichtbarmachung von Monoaminen. I. Identifizierung der fluoreszierenden Produkte aus Modellversuchen mit 6,7-Dimethoxyisochinolinderivaten und Formaldehyd. Helv. chim. Acta **46**, 2425—2430 (1963).

— — Fluoreszenzmethoden zur histochemischen Sichtbarmachung von Monoaminen. 2. Identifizierung des fluoreszierenden Produktes aus Dopamin und Formaldehyd. Helv. chim. Acta **47**, 911—918 (1964).

— — Jonsson, G.: Fluorescence methods for the histochemical demonstration of monoamines. 3. Sodium borohydride reduction of the fluorescent compounds as a specificity test. J. Histochem. Cytochem. **12**, 582—586 (1964).

— Jonsson, G.: Fluorescence methods for the histochemical demonstration of monoamines. 4. Histochemical differentiation between dopamine and noradrenaline in models. J. Histochem. Cytochem. **13**, 484—487 (1965a)

— — Fluoreszenzmethoden zur histochemischen Sichtbarmachung von Monaminen. 5. Identifizierung des fluoreszierenden Produktes aus Modellversuchen mit 5-Methoxytryptamin und Formaldehyd. Acta histochem. (Jena) **22**, 247—258 (1965b).

— — Fluoreszenzmethoden zur histochemischen Sichtbarmachung von Monaminen. 6. Identifizierung der fluoreszierenden Produkte aus m-Hydroxyphenyläthylaminen und Formaldehyd. Helv. chim. Acta **49**, 798—806 (1966).

— — The formaldehyde fluorescence method for the histochemical demonstration of biogenic monoamines. A review on the methodology. J. Histochem. Cytochem. **15**, 65—78 (1967).

— — Malmfors, T.: Factors affecting the quality and intensity of the fluorescence in the histochemical method for demonstration of catecholamines. Acta histochem. **25**, 367—370 (1966).

— Malfors, T.: The effect of nerve activity on the depletion of the adrenergic transmitters by inhibitors of noradrenaline synthesis. Acta physiol. scand. **67**, 352—357 (1966).

— — Sachs, Ch. Differences in the uptake of secondary catecholamines by the adrenergic nerves. Acta physiol. scand. **67**, 358—362 (1966).

Cosnettatos, G. F.: De la structure du centre visuel cérébral chez les anopthalmes congénitaux. Arch. Ophthal. (Paris) **48**, 282—289 (1931).

Cosslett, V. E.: Modern microscopy. Ithaca, New York: Cornell University Press 1966.

Coupland, R. E.: The natural history of the chromoffin cell. London: Longmans Green 1965.

— Hopwood, D.: Mechanism of a histochemical reaction differentiating between adrenaline- and noradrenaline-storing cells in the electron microscope. Nature (Lond.) **209** 590—591 (1966).

— Pypers, A. S., Hopwood, D.: A method for differentiating between noradrenaline-and adrenaline-storing cells in the light and electron microscope. Nature (Lond.) **201**, 1240—1242 (1964).

Cowan, W. M., Adamson, L., Powell, T. P. S.: An experimental study on the avian visual system. J. Anat. (Lond.) **95**, 545—563 (1961).

— Guillery, R. W., Powell, T. P. S.: The origin of the mamillary peduncle and other hypothalamic connexions from the midbrain. J. Anat. (Lond.) **98**, 345—363 (1964).

— Hart, J.: A study of retrograde transneuronal degeneration in the ventral tegmental nucleus of the rat. In preparation (1969).

Cowan, W. M., Martin, A. H., Wenger, E.: Mitotic patterns in the optic tectum of the chick during normal development and after early removal of the optic vesicle. J. exp. Zool. **169**, 71—92 (1968).
— Powell, T. P. S.: An experimental study of the relation between the medial mammillary nucleus and the cingulate cortex. Proc. roy. Soc. B **143**, 114—125 (1954).
— — The projection of the midline and intralaminar nuclei of the thalamus of the rabbit. J. Neurol. Psychiat. **18**, 266—279 (1955).
— — A note on terminal degeneration in the hypothalamus. J. Anat. (Lond.) **90**, 188—192 (1956).
— Wenger, E.: Degeneration in the nucleus of origin of the preganglionic fibers to the chick ciliary ganglion following early removal of the optic vesicle. J. exp. Zool. **168**, 105—124 (1968).
Cox, W.: Impregnation des centralen Nervensystems mit Quecksilbersalzen. Arch. mikr. Anat. **37**, 16—21 (1891).
Cox, W. H.: Impregnatie van het Centrale Zenuwstekel met Kwik-Zouten, Nederlandsch-Tijdschrift voor Genceshunde. 2. Reeks **26**, Afd. 2, No 15, 489 (1890).
Cragg, B. G.: Centrifugal fibers to the retina and olfactory bulb, and composition of the supraotpic commisures in the rabbit. Exp. Neurol. **5**, 406—427 (1962).
Creval, H. van, *see* Van Creval, H.
Cronkite, E. P.: In: The Kinetics of Cellular Proliferation (Stohlman, F., Jr., ed.). 456 pp. New York and London: Grune & Stratton 1959.
Crowter, D. G., Evans, E. A., Lambert, R. W.: Tritium labelled thymidine. Chem. Ind., pp. 899 (1960).
Csillik, B.: Histochemical model experiments on the effect of various drugs on the catecholamine content of adrenergic nerve terminals. J. Neurochem. **11**, 351—355 (1964).
— Erulkar, S. D.: Labile stores of monoamines in the central nervous system: a histochemical study. J. Pharmacol. exp. Ther. **146**, 186 (1964).
— Kalman, G., Knyihar, E.: Adrenergic nerve endings in the feline cervical superius ganglion. Experientia (Basel) **23**, 477—478 (1967).
Dahl, E., Falck, B., Lindqvist, M., Mecklenburg, C.: Monoamines in mollusc neurons. Kungl. Fysiol. Sällsk. Förh. Lund. **32**, 89—92 (1962).
— — Mecklenburg, C. V., Myhrberg, H.: An adrenergic nervous system in sea anemones. Quart. J. micr. Sci. **104**, 531—534 (1963).
Dahlström, A.: Observations on the accumulation of noradrenaline in the proximal and distal parts of peripheral adrenergic nerves after compression. J. Anat. (Lond.) **99**, 677—689 (1965).
— Studies on the intraneuronal distribution of noradrenaline and the transport and life-span of amine storage granules in the peripheral adrenergic neuron. M.D. Thesis, Stockholm (1966).
— Fuxe, K.: A method for demonstration of adrenergic nerve fibers in peripheral nerves. Z. Zellforsch. **62**, 606—607 (1964a).
— — A method for the demonstration of monoamine containing neurons in the central nervous system. Acta physiol. scand. **60**, 293—294 (1964b).
— — Evidence for the existence of monoamine containing neurons in the central nervous system. I. Demonstration of monoamines in the cell bodies of brain stem neurons. Acta physiol. scand. **62**, 1—55, Suppl. 232 (1964c).
— — The adrenergic innervation of the nasal mucosa of certain mammals. Acta oto-laryng. (Stockh.) **59**, 65—72 (1965a).
— — Evidence for the existence of monoamine containing neurons in the central nervous system. II. Experimentally induced changes in the intra-neuronal amine levels. Acta physiol. scand. **64**, 1—36, Suppl. 247 (1965b).
— — Monoamines and the pituitary gland. Acta endocr. (Kbh.) **41**, 301—314 (1966).
— — Hillarp, N.-A.: Site of action of reserpine. Acta pharmacol. (Kbh.) **22**, 277—292 (1965).
— — — Central monoamine neurons. In: Mechanisms on Release of Biogenic Amines (Euler, U. S. v., S. Rosell, and B. Uvnas, eds.), p. 31—37. London: Pergamon Press 1966a.

Dahlström, A., Fuxe, K., Hillarp, N.-A.: Demonstration and mapping out of central DA, NA, and 5-HT neuron systems and their reaction to psychopharmaca. Pharmacol. Rev. **18**, 727—741 (1966b).

— — Hökfelt, T., Norberg, K.-A.: Adrenergic innervation of the bronchial muscle of the cat. Acta physiol. scand. **66**, 507—508 (1966).

— — Kernell, D., Sedvall, G.: Reduction of the monoamine stores in the terminals of bulbospinal neurons following stimulation in the medulla oblongata. Life Sci. **4**, 1207—1212 (1965).

— — Larsson, O., Ungerstedt, U.: Ascending catecholamine system from the lower brain stem. Acta physiol. scand. **62**, 485—486 (1964).

— — Myatu, M., Zetterstrum, B.: Observations on the adrenergic innervation of the dog's heart. Amer. J. Physiol. **206**, 689—692 (1965).

— Häggendahl, J.: Some quantitative studies on the noradrenaline content in the cell bodies and terminals of a sympathetic adrenergic neuron system. Acta physiol. scand. **67**, 271—277 (1966).

— — Hökfelt, T.: The noradrenaline content of the varicosities of sympathetic adrenergic nerve terminals in the rat. Acta physiol. scand. **67**, 289—294 (1966d).

Dale, H. H.: Pharmacology and nerve endings. Proc. roy. Soc. Med. **28**, 319—332 (1935).

Daron, G. H.: Morphology of the cerebellar dentate nucleus in a chimpanzee. Anat. Rec. **138**, 81—92 (1960).

Davenport, H. A.: Acid versus neutral formalin solution as a neurological fixative. Stain Technol. **9**, 49—52 (1934).

— Combs, C. M.: Golgi's dichromate silver method. 3. Chromating fluids. Stain Technol. **29**, 165—173 (1954).

Davies, D. R., Wimber, D. E.: Studies of radiation-induced changes in cellular proliferation, using a double labelling autoradiographic technique. Nature (Lond.) **200**, 229—232 (1963).

Dawson, K. B., Field, E. O.: A procedure for computing cellular proliferation kinetics during "log phase" of growth. Exp. Cell Res. **34**, 507—511 (1964).

Delong, G. R., Sidman, R. L.: Effects of eye removal at birth on histogenesis of the mouse superior colliculus: An autoradiographic analysis with tritiated thymidine. J. comp. Neurol. **118**, 205—224 (1962).

De Robertis, E.: Submicroscopic morphology of the synapse. Exp. Cell Res. **5**, 347—369. (1958).

— Synaptic complexes and synaptic vesicles as structural and biochemical units of the central nervous system. In: Nerve as a Tissue (Rodahl, K., and B. Issekutz, eds.), p. 88—115. New York: Harper and Row, Publishers 1966.

Diderholm, H., Fichtelius, K. E., Linder, O.: Availability time of ^{3}H-label after administration of ^{3}H-thymidine *in vivo*. Exp. Cell Res. **27**, 431—435 (1962).

Doniach, I., Pelc, S. R.: Autoradiograph technique. Brit. J. Radiol. **23**, 184—192 (1950).

Dornfield, E. J., Slater, D. W., Scheffe, H.: A method for accurate determination of volume and cell numbers in small organs. Anat. Rec. **82**, 255—259 (1942).

Dowling, J. E., Cowan, W. M.: An electron microscope study of normal and degenerating centrifugal fiber terminals in the pigeon retina. Z. Zellforsch. **71**, 14—28 (1966).

Dresse, A.: Importance du systeme mesencephalo-telencephalique noradrenergique comme substratum anatomique du domportement d'autostimulation. Life Sci. **5**, 1003—1014 (1966).

Droogleever Fortuyn, J.: Geometrical properties of nervous structures. Internat. Congr. of Stereology, p. 1—5 (1963).

Duckworth, T., Cooper, E. R. A.: A study of anophthalmia in an adult. Acta anat. (Basel) **63**, 509—522 (1966).

Duncan, D., Keyser, L. L.: Some determinations of the ratio of nerve fibers to nerve cells in the thoracic dorsal roots and ganglia of the cat. J. comp. Neurol. **64**, 303—311 (1936).

Eager, R. P., Barrnett, R. J.: Morphological and chemical studies of Nauta-stained degenerating cerebellar and hypothalamic fibers. J. comp. Neurol. **126**, 487—510 (1966).

Eayrs, J. T., Goodhead, B.: Postnatal development of the cerebral cortex in the rat. J. Anat. (Lond.) **93**, 385—402 (1959).

Ebbesson, S. O. E.: A quantitative study of human superior cervical sympathetic ganglia. Anat. Rec. **146**, 353—356 (1963).
— Ascending fiber projections from the spinal cord in the Tegu lizard (*Tupinambis nigropunctatus*). Anat. Rec. **154**, 341—342 (1966).
— Ascending axon degeneration following hemisection of the spinal cord in the Tegu lizard (*Tupinambis nigropunctatus*). Brain Res. **5**, 178—206 (1967a).
— Retinal projections in two species of sharks (*Galeocerdo cuvier* and *Ginglymostoma cirratum*). Anat. Rec. **157**, 238 (1967b).
— — Quantitative studies of superior cervical sympathetic ganglia in a variety of primates including man. I. The ratio of preganglionic fibers to ganglionic neurons. J. Morph. **124**, 117—131 (1968a).
— Quantitative studies of superior cervical sympathetic ganglia in a variety of primates including man. II. Neuronal packing density. J. Morph. **124**, 181—185 (1968b).
— Retinal projections in two teleost fishes (*Opsanus tau* and *Gymothorax funebris*): An experimental study with silver impregnation methods. Brain Behav. Evol. **1**, 136—154 (1968c).
— A connection between the dorsal column nuclei and the dorsal accessory olive. Brain Res. **8**, 393—397 (1968d).
— Retinal projections in *Rana catesbeiana* with observations on the effect of various temperatures on the rate of degeneration. In preparation (1969a).
— Connections of the optic tectum in Holocentrus. In preparation. (1969b).
— Camunas, J.: Ascending spinal projections in a frog (*Rana catesbeiana*), in preparation (1969).
— Heimer, L.: Olfactory bulb projections in two species of sharks (*Galeocerdo cuvier* and *Ginglymostoma cirratum*). Abstract., Anat. Rec. **160**, 469 (1968).
— — Projections of the olfactory tract fibers in the nurse shark (*Ginglymostoma cirratum*). Brain Res. **17**; 47—57 (1970).
— Karten, H. J.: Tectal projections in the Tegu lizard (*Tupinambis nigropunctatus*). In preparation (1969).
— Ramsey, J. S.: The optic tracts in two species of sharks (*Galeocerdo cuvier* and *Ginglymostoma cirratum*). Brain Res. **8**, 36—53 (1968).
— Rubinson, K.: A simplified Nauta procedure. Physiol. Behav. **4**, 281—282 (1969).
— Tang, D.: A method for estimating the number of cells in histological sections. J. roy. micr. Soc. **84**, 449—464 (1965).
— — The sampling of structured cell population. Stereologia **6**, 15—16 (1967a).
— — A comparison of sampling procedures in a structured cell population. In: Proceedings of the Second Internat. Congr. for Sterology (Elias, H., ed.), p. 131—132. Berlin-Heidelberg-New York: Springer 1967b.
Eccles, J. C.: The Neurophysiological Basis of Mind. The principles of neurophysiology. 314 pp. Oxford: Clarendon Press 1953.
— The Physiology of Synapses. 316 pp. Berlin-Göttingen-Heidelberg-New York: Springer 1964.
— Fatt, P., Kotetsu, K.: Cholinergic and inhibitory synapses in a pathway from motor-axon collaterals to motoneurons. J. Physiol. (Lond.) **216**, 524—562 (1954).
Ehinger, B.: Adrenergic retinal neurons. Z. Zellforsch. **71**, 146—152 (1966a).
— Ocular and orbital vegetative nerves. Acta physiol. scand. **67**, 1—35, Suppl. 268 (1966b).
— Sporrong, B., Stenevi, U.: Combining the catecholamine fluorescence and methylene blue staining methods for demonstrating nerve fibers. Life Sci. **6**, 1973—1974 (1967).
— Thunberg, R.: Induction of fluorescence in histamine-containing cells. Exp. Cell Res. **47**, 116 (1967).
El-Badawi, A., Schenk, E. A.: Histochemical methods for separate, consecutive and simultaneous demonstration of acetylcholinesterase and norepinephrine in cryostat section. J. Histochem. Cytochem. **15**, 580—588 (1967).
— — Dual innervation of the mammalian urinary bladder. A histochemical study of the distribution of cholinergic and adrenergic nerves. Amer. J. Anat. **119**, 405—428 (1968).
Elzholz, A.: Zur Kenntnis der Veränderungen im centralen Stumpfe lädierter gemischter Nerven. J. Psychiat. Neurol. **17**, 323—359 (1898).

Eneström, S., Svalander, C.: Liquid formaldehyde in catecholamine studies. A new approach to the morphological localization of monoamines in the adrenal medulla. Histochemie **8**, 115—163 (1967).

Eränkö, O.: Histochemistry of noradrenaline in the adrenal medulla of rats and mice. Endocrinology **57**, 363—368 (1955).

— Histochemistry of nervous tissues; Catecholamines and cholinesterases. Ann. Rev. Pharmacol. **7**, 203—222 (1967a).

— The practical histochemical demonstration of catecholamines by formaldehyde-induced fluorescence. J. roy. Micr. **87**, 259—276 (1967b).

— Räisänen, L.: Demonstration of catecholamines in adrenergic nerve fibers by fixation in aqueous formaldehyde solution and fluorescence microscopy. J. Histochem. **14**, 690—691 (1966).

Erös, G.: Eine neue Darstellungsmethode der sogenannten "gelben" argentaffinen Zellen des Magendarmtraktes. Zbl. allg. Path. path. Anat. **54**, 385—391 (1932).

Escobar, A., Sampedro, E. D., Dow, R. S.: Quantitative data on the inferior olivary nucleus in man, cat, and vampire bat. J. comp. Neurol. **132**, 397—404 (1968).

Evans, D. H. L., Hamlyn, L. H.: A study of silver impregnation methods in the central nervous system. J. Anat. (Lond.) **90**, 193—202 (1956).

Falck, B.: Observations on the possibilities of the cellular localization of monoamines by a fluorescence method. Acta physiol. scand. **56**, 1—25, Suppl. 197 (1962).

— Häggendahl, J., Owman, Ch.: The localization of adrenaline in adrenergic nerves in the frog. Quart. J. exp. Physiol. **48**, 253—257 (1963).

— Hillarp, N.-A., Thieme, G., Torp, A.: Fluorescence of catecholamines and related compounds condensed with formaldehyde. J. Histochem. Cytochem. **10**, 348—354 (1962).

— Owman, Ch.: A detailed methodological description of the fluorescence method for the cellular demonstration of biogenic monoamines. Acta Univ. Lundens., Sect. II, **7**, 1—23 (1965).

— — Sjöstrand, N. O.: Peripherally located adrenergic neurons innervating the vas deferens and the seminal vesicle of the guinea-pig. Experientia (Basel) **21**, 98—100 (1965).

Fahimi, H. D., Drochmans, P.: Essais de standardisation de la fixation au glutaraldehyde. I. Purification et détermination de la concentration du glutaraldehyde. J. Microscopie **4**, 725—736 (1965a).

— — Essais de standardisation de la fixation au glutaraldehyde. II. Influence des concentration en aldehyde et de l'osmolalité. J. Microscopie **4**, 737—748 (1965b).

— — Purification of glutaraldehyde. Its significance for preservation of acid phosphatase activity. J. Histochem. Cytochem. **16**, 199—204 (1968).

Fajersztajn, J.: Ein neues Silberimprägnations-Verfahren als Mittel zur Färbung der Axencylinder. Neurol. Zbl. **20**, 98—106 (1901).

Farley, B. G.: The use of computer techniques in neural research. In Neural Theory and Modeling (Reiss, R. F., ed.), p. 72. Stanford, California: Stanford University Press 1964.

Farquhar, M. G., Palade, G. E.: Junctional complexes in various epithelia. J. Cell Biol. **17**, 375—412 (1963).

Feder, N.: A method for holding mice for intravenous injection. J. Lab. clin. Med. **51**, 662—663 (1958).

— Wolf, M. K.: Studies on nucleic acid metachromasy. II. Metachromatic and orthochromatic staining by toluidine blue of nucleic acids in tissue sections. J. Cell Biol. **27**, 327—336 (1965).

Feinendegen, L. E., Bond, V. P., Hughes, W. L.: ^{125}I-DU (5-iodo-2'-deoxyuridine) in autoradiographic studies of cell proliferation. Exp. Cell Res. **43**, 107—119 (1966).

Feldberg, W., Myers, R. D.: A new concept of temperature regulations by amines in the hypothalamus. Nature (Lond.) **200**, 1325 (1963).

Fernstrom, R. C.: A durable Nissl stain for frozen and paraffin sections. Stain Technol. **33**, 175—176 (1958).

Ficq, A.: Autoradiography. In: The Cell-Biochemistry, Physiology, Morphology (Brachet, J., ed.), vol. 1, p. 67—90. New York: Academic Press 1959.

Fink, R. P., Heimer, L.: Two methods for selective silver impregnation of degenerating axons and their synaptic endings in the central nervous system. Brain Res. **4**, 369—374 (1967).

Filogamo, G.: Consequence della demolizione dell'abbozzo dell'occhio sulla sviluppo del lobo ottico nell'embrione di pollo. Riv. Biol. **42**, 73—79 (1950).

Flatau, E.: Über die zweckmäßige Anwendung der Golgischen Sublimatmethode für die Untersuchung des Gehirns des erwachsenen Menschen. Arch. micr. Anat. **45**, 58—62 (**1895**).

Fliedner, T. M., Hass, R. J., Stehle, H., Adams, A.: Complete labeling of all cell nuclei in newborn rats with H^3-thymidine. A tool for the evaluation of rapidly and slowly proliferating cell systems. Lab. Invest. **16**, 249—259 (1968).

Floderus, S.: Untersuchungen über den Bau der menschlichen Hypophyse mit besonderer Berücksichtigung der quantitativen mikromorphologischen Verhältnisse. Kopenhagen: Munksgaard 1944.

Foerster, O., Gagel, O.: Die tigrolytische Reaktion der Ganglienzelle. Z. mikr.-anat. Forsch. **36**, 567—575 (1934).

— — Sheehan, D.: Veränderungen an den Endösen im Rückenmark des Affen nach Hinterwurzeldurchschneidung. Z. Anat. Entwickl.-Gesch. **101**, 553—565 (1933).

Foley, J. O., DuBois, F. S.: Quantitative studies of the vagus nerve in the cat. J. comp. Neurol. **67**, 49—64 (1937).

Folkow, B., Fuxe, K., Sonnenschein, R.: Responses of skeletal musculature and its vasculature during "diving" in the duck: Peculiarities of the adrenergic vasoconstrictor innervation. Acta physiol. scand. **67**, 327—342 (1966).

Forster, R., Taggart, J.: Use of isolated renal tubules for the examination of metabolic processes associated with active cellular transport. J. cell comp. Physiol. **36**, 251—270 (1950).

Fox, C. A., Hillman, D. E., Siegesmund, K. A., Dutta, C. R.: The primate cerebellar cortex: A Golgi and electron microscopic study: In: The Cerebellum (Fox, C. A., and R. S. Snider, eds.), p. 174—225. Amsterdam: Elsevier 1967.

— Ubeda-Purkiss, M., Ihrig, H. K., Biagoli, D.: Zinc chromate modification of the Golgi technic. Stain Technol. **26**, 109—114 (1951).

Fried, J.: Estimating the median generation time of proliferating cell systems in steady state. Biophys. J. **8**, 710—729 (1968).

Friedkin, M., Tilson, D., Roberts, D.: Studies of deoxyribonucleic acid biosynthesis in embryonic tissues with thymidine-C^{14}. J. biol. Chem. **220**, 627—637 (1956).

Fromman, C.: Über die Färbung der Binde- und Nervensubstanz des Rückenmarkes durch Argentum nitricum und über die Struktur der Nervenzellen. Virchows Arch. path Anat. **31**, 129—153 (1864).

Frontali, N.: Histochemical localization of catecholamines in the brain of normal and drug-treated cockroaches. J. Insect. Physiol. **14**, 881—886 (1968).

Frontera, J. G.: Improved Golgi-type impregnation of nerve cells. Anat. Rec. **148**, 371—372 (1964) (abstract).

Fujita, S.: Kinetics of cellular proliferation. Exp. Cell Res. **28**, 52—60 (1962).

— The matrix cell and cytogenesis in the developing central nervous system. J. comp. Neurol. **120**, 37—42 (1963).

— Analysis of neuron differentiation in the central nervous system by tritiated thymidine autoradiography. J. comp. Neurol. **122**, 311—328 (1964).

— Applications of light and electron microscopic autoradiography to the study of cytogenesis of the forebrain. In: Evolution of the Forebrain, Phylogenesis and Ontogenesis of the Forebrain (Hassler, R., and H. Stephen, eds.), p. 180—196. Stuttgart: Thieme 1966.

— Quantitative analysis of cell proliferation and differentiation in the cortex of the postnatal mouse cerebellum. J. Cell Biol. **32**, 277—287 (1967).

— Horii, M.: Analysis of cytogenesis in chick retina by tritiated thymidine autoradiography. Arch. Histol. Japon. **23**, 359—366 (1963).

— Shimada, M., Nakamura, T.: H^3-thymidine autoradiographic studies on the cell proliferation and differentiation in the external and internal granular layers of the mouse cerebellum. J. comp. Neurol. **128**, 191—208 (1966).

Fujiwara, M., Tanaka, Ch., Honjo, T., Okegawa, T.: Histochemical demonstration of noradrenaline in rat salivary glands. Jap. J. Pharmacol. **15**, 369—377 (1965).

Fuxe, K.: Cellular localization of monoamines in the median eminence and infundibular stem of some animals. Z. Zellforsch. **61**, 710—724 (1964).

— Evidence for the existence of monoamine neurons in the central nervous system. M.D. Thesis, Stockholm 1965a.

— Evidence for the existence of monoamine containing neurons in the central nervous system. III. Presence of monoamine containing terminals in the lower brain stem. Z. Zellforsch. **64**, 573—596 (1965b).

— Evidence for the existence of monoamine containing neurons in the central nervous system. IV. The distribution of monoamine terminals in the central nervous system. Acta physiol. scand. **64**, 39—85 (Suppl. 247) (1965c).

— Andén, N.-E.: Studies on central monoamine neurons with special reference to the nigro-neostriatal dopamine neuron system. In: Biochemistry and Pharmacology of the Basal Ganglia (Costa, E., L. J. Cote, and M. D. Yahr, eds.), p. 123—130. New York: Raven Press: Hewlett 1966.

— Grobecker, H., Hökfelt, T., Jonsson, G.: Identification of dopamine, noradrenaline and 5-hydroxytryptamine varicosities in a fraction containing nerve particles. Brain Res. **6**, 475—480 (1967).

— Gunne, L.-M.: Depletion of the amine stores in brain catecholamine terminals on anygdaloid stimulation. Acta physiol. scand. **62**, 493—494 (1964).

— Hamberger, B., Hökfelt, T.: Distribution of noradrenaline nerve terminals in cortical areas of the rat. Brain Res. **8**, 125—131 (1968).

— — Malmfors, T.: The effect of drugs on accumulation of amines in tubero-infundibular dopamine neurons. Europ. J. Pharmacol. **1**, 334—341 (1967).

— Hillarp, N.-A.: Uptake of L-dopa and noradrenaline by central adrenergic neurons. Life Sci. **13**, 1403—1406 (1964).

— Hökfelt, T.: Further evidence for the existence of tuberoinfundibular DA neurons. Acta physiol. scand. **66**, 245—246 (1966).

— — The influence of central catecholamine neurons on the hormone secretion from the anterior and posterior pituitary. In: Neurosecretion (Stutinsky, F., ed.). Berlin-Heidelberg-New York: Springer 1967.

— — Catecholamines in the hypothalamus. In: Frontiers in Neuroendocrinology (Ganong, F., and L. Martini, eds.). New York: Oxford Press 1969a.

— — Monoaminergic afferent input to the hypothalamus and the dopamine afferent input to the median eminence. 3. Int. Congr. Endocrinol., Mexico, **30**, 6—57 (1969b).

— — Nilsson, O.: A fluorescence and electron-microscopic study on certain brain regions in monoamine terminals. Amer. J. Anat. **117**, 33—46 (1965).

— — — Activity changes in the tuberoinfundibular dopamine neurons of the cat during various stages of the reproductive cycle. Life Sci. **6**, 2057—2061 (1967).

— — — Castration, sex hormones and tubero-infundibular dopamine neurons. Neuroendocrinology, in press (1969).

— — Ritzén, M., Ungerstedt, U.: Studies on uptake of intraventricularly administered tritiated noradrenaline and 5-hydroxytryptamine by combined fluorescence histochemical and autoradiographic techniques. Histochemie **16**, 186—194 (1968).

— — Ungerstedt, U.: Localization of indolealkylamines in CNS. Advanc. Pharmacol. **6** A, 235—251 (1968).

— Jonsson, G.: A modification of the histochemical fluorescence method for the improved localization of 5-hydroxytryptamine. Histochemie **11**, 161—166 (1967).

— Ljunggren, L.: The cellular localization of monoamines in the upper brain stem of the pigeon. J. comp. Neurol. **125**, 255—382 (1965).

— Nilsson, B. Y.: Mechanoreceptors and adrenergic nerve terminals. Experientia (Basel) **21**, 641—642 (1965).

— Owman, Ch.: The cellular localization of monoamines in the area postrema of certain mammals. J. comp. Neurol. **125**, 337—354 (1965).

— Sedvall, G.: A histochemical and biochemical observation of the effect of reserpine on the noradrenaline storage in vasoconstrictor nerves. Acta physiol. scand. **61**, 121—129 (1964).

Fuxe, K., Sedvall, G.: The distribution of adrenergic nerve fibers to the blood vessels in skeletal muscle. Acta physiol. scand. **64** 75—86 (1965).

— Ungerstedt, U.: Localization of catecholamine uptake in rat brain after intraventricular injection. Life Sci. **5**, 1817—1824 (1966).

— — Histochemical studies on the effect of (+)-amphetamine and drugs of the imipramine-group on central catecholamine and 5-hydroxytryptamine neurons after intraventricular injection of catecholamines and 5-hydroxytryptamine. Europ. J. Pharmacol. **4**, 135—144 (1968a).

— — Histochemical studies on the distribution of catecholamines and 5-hydroxytryptamine after intraventricular injection. Histochemie **13**, 16—28 (1968b).

Gallasi, L.: Delayed and direct labeling after a system injection of thymidine-^{3}H. J. Histochem. Cytochem. **15**, 565—572 (1967).

Gamble, H. J.: An experimental study of the secondary olfactory connexions in *Lacerta viridis*. J. Anat. (Lond.) **86**, 180—196 (1952).

— Goldby, F., Smith, G. M. R.: Effect of temperature on the degeneration of nerve fibers. Nature (Lond.) **179**, 527 (1957).

Ganser, S.: Über die periphere und zentrale Anordnung der Sehnervenfasern und über das Corpus bigeminum anterius. Arch. Psychiat. Nervenkr. **13**, 341—381 (1882).

Gardner, E.: Decrease in human neurons with age. Anat. Rec. **77**, 529—536 (1940).

Garey, J. J., Jones, E. G., Powell, T. P. S.: Interrelationships of striate and extrastriate cortex with the primary relay sites of the visual pathway. J. Neurol. Neurosurg. Psychiat. **31**, 135—157 (1968).

Geffen, L. B., Livett, B. G., Rush, R. A.: Immunohistochemical localization of protein components of catecholamine storage vesicles. J. Physiol (Lond.) **204**, 593—605 (1969).

Gehuchten, A. van, *see* Van Gehuchten, A.

Gibson, W. C.: Degeneration of the boutons terminaux in the spinal cord. Arch. Neurol. Psychiat. (Chic.) **38**, 1145—1157 (1937).

— Degeneration and regeneration of sympathetic synapses J. Neurophysiol. **3**, 237—247 (1940).

Gillespie, J. S., Hamilton, D. N. H.: Binding of noradrenaline to smooth muscle cells in the spleen. Nature (Lond.) **212**, 252—254 (1967).

Gillis, C. N., Schneider, F. H., Orden, L. S. van, Giarman, N. J.: Biochemical and micro-fluorometric studies of norepinephrine redistribution accompanying sympathetic nerve stimulation. J. Pharmacol, exp. Ther. **151**, 46—54 (1966).

Glaser, E. M., Van der Loos, H.: A semi-automatic computer-microscope for the analysis of neuronal morphology. JEEE Trans. Bio-med. Eng. BME **12**, 22—31 (1965).

Glees, P.: The termination of optic fibers in the lateral geniculate body of the cat. J. Anat. (Lond.) **75**, 434—440 (1941).

— The terminations of optic fibers in the lateral geniculate body of the rabbit. J. Anat. (Lond.) **76**, 313—318 (1942).

— Terminal degeneration within the central nervous system as studied by a new silver method. J. Neuropath. exp. Neurol. **5**, 54—59 (1946).

— The time factor in central nerve fiber degeneration. Acta anat. (Lond.) **6**, 447—450 (1948).

— Terminal degeneration and transynaptic atrophy in the lateral geniculate body of the monkey. In: Neurophysiologic und Psychophysik des visuellen Systems (Jung, R., and Kornhuber, eds.), p. 104—110. Berlin-Göttingen-Heidelberg: Springer (1961).

— Hasan, M.: The signs of synaptic degeneration — a critical appraisal. Acta anat. (Basel) **69**, 153—167 (1968).

— Le Gros Clark, W. E.: The termination of optic fibers in the lateral geniculate body of the monkey. J. Anat. (Lond.) **75**, 295—308 (1941).

— Livingston, R. B., Soler, J.: Der intraspinale Verlauf und die Endigungen der sensorischen Wurzeln in den Nuclei gracilis und cuneatus. Arch. Psychiat. Nervenkr. **187**, 190 (1951).

— Meller, K., Eschner., J.: Terminal degeneration in the lateral geniculate body of the monkey; an electron microscope study. Z. Zellforsch. **71**, 29—40 (1966).

Glees, P., Meyer, A., Meyer, M.: Terminal degeneration in the frontal cortex of the rabbit following the interruption of afferent fibers. J. Anat. (Lond.) **80**, 101—105 (1946).
— Nauta, W. J. H.: A critical review of studies on axonal and terminal degeneration. Mschr. Psychiat. Neurol. **129**, 74—91 (1955).
— Soler, J., Bailey, R. A.: Retrograde axonal changes of the de-afferentated nucleus gracilis following mid-brain tractotomy. J. Neurol. Neurosurg. Psychiat. **14**, 281—286 (1951).
— Zander, E.: Der Tractus tegmento-olivris des Menschen. Mschr. Psychiat. Neurol. **120**, 21—30 (1950).
Glickstein, M.: Laminar structure of the dorsal lateral geniculate nucleus in the tree shrew (*Tupaia glis*). J. comp. Neurol. **131**, 93—102 (1967).
Glimstedt, G., Hakansson, R.: Measurement of thickness in various parts of histological sections. Nature (Lond.) **167**, 397—398 (1951).
Globus, A., Scheibel, A. B.: Loss of dendrite spines as an index of pre-synaptic terminal patterns: An experimental application of the Golgi method. Nature (Lond.) **212**, 463—465 (1966).
— — Synaptic loci on visual cortical neurons of the rabbit. The specific afferent radiation. Exp. Neurol. **18**, 116—131 (1967a).
— — The effect of visual deprivation on cortical neurons: A Golgi study. Exp. Neurol. **19**, 331—345 (1967b).
— — Synaptic loci on parietal cortical neurons: Terminations of corpus callosum fibers. Science **156**, 1127—1129 (1967c).
Glücksmann, A.: Cell death in normal vertebrate ontogeny. Biol. Rev. **26**, 59—86 (1951).
Goldberg, J. M., Brown, P. B.: Functional organization of the dog superior olivary complex: An anatomical and electrophysiological study. J. Neurophysiol. **31**, 639—656 (1968).
Goldby, F.: A note on transneuronal atrophy in the human lateral geniculate body. J. Neurol. Neurosurg. Psychiat. **20**, 202—207 (1957).
— Robinson, L. R.: The central connections of dorsal spinal nerve roots and the ascending tracts in the spinal cord of *Lacerta viridis*. J. Anat. (Lond.) **96**, 153—170 (1962).
Golgi, C.: Sulla struttura della sostanza grigia dell cervello. Gazz. med. ital. lombarda **33**, 244—246 (1873).
— Un nuovo processo di tecnica microscopia. R. C. Inst. Lombardo Sci., 2nd ser. **12**, 5 (1878).
— Di una nuova reasione apparentemente nera delle cellule nervose cerebrali ottenuta col bichloruro di mercurio. Archs. Sci. med. **3**, 1—7 (1879).
— Sulla fina anatomia delgi organi centrali del sistema nervosa. Pavia 1886.
— Modificazione del metodo di colorazione deli elementi nervosi col bicloruro di mercurio. Riv. med. Napoles **7**, 193—194 (1891).
— Opera omnia, Istologia normale, iv Sulla Sostanza grigia del cervello (Gazzetta Medica Italiana, 1873). Vol. I, p. 91—98. Milano: Ulrico Hoepli 1903.
Goldberg, J. M., Brown, P. B.: Functional organization of the dog superior olivery complex: an anatomical and electrophysiological study. J. Neurophysiol. **31**, 639—656 (1968).
Goodman, L.: Effect of total absence of function on the optic system of rabbits. Amer. J. Physiol. **100**, 46—63 (1932).
Gracheva, Nina D.: Autoradiography of nuclei acid and protein synthesis in the nervous system [in Russian]. 230 p. Leningrad: Nauka Press (1967).
Graf, W.: The microtome as an error-producing factor in quantitative histological investigations. Hist. Invest. **6**, 141—144 (1948).
Grant, G.: Projections of the external cuneate nucleus onto the cerebellum in the cat: An experimental study using silver methods. Exp. Neurol. **5**, 179—195 (1962).
— Degenerative changes in dendrites following axonal transection. Experientia (Basal) **21**, 722 (1965).
— Silver impregnation of degenerating dendrites, cells and axons central to axonal transection. II. A Nauta study on spinal motor neurons in kittens. Exp. Brain Res. **6**, 284—293 (1968).
— Aldskogius, H.: Silver impregnation of degenerating dendrites, cells and axons central to axonal transection. I. A Nauta study on the hypoglossal nerve of kittens. Exp. Brain Res. **3**, 150—162 (1967).

Grant, G., Ekvall, L., Westman, J.: Transganglionic degeneration in the vestibular nerve. In press. (1969).
— Westman, J.: Degenerative changes in dendrites central to axonal transection. Electron microscopical observations. Experientia (Basel) **24**, 169—170 (1968a).
— — The lateral cervical nucleus in the cat. IV. A light and electron microscopical study after midbrain lesions with demonstration of indirect Wallerian degeneration at the ultrastructural level. Exp. Brain Res. **7**, 51—67 (1968b).
Gray, E. G.: Axo-somatic and axo-dendritic synapses of the cerebral cortex; an electron microscope study. J. Anat. (Lond.) **93**, 420—433 (1959).
— A morphological basis for pre-synaptic inhibition? Nature (Lond.) **193**, 82—83 (1962).
— Guillery, R. W.: A basis of silver staining of synapses of the mammalian spinal cord: A light and electron microscope study. J. Physiol. (Lond.) **157**, 581—588 (1961).
— — Synaptic morphology in the normal and degenerating nervous system. Int. Rev. Cytol. **19**, 111—182 (1966).
— Hamlyn, L. H.: Electron microscopy of experimental degeneration in the avian optic tectum. J. Anat. (Lond.) **96**, 309—316 (1962).
— Whittaker, V. P.: The isolation of synaptic vesicles from the central nervous system. J. Physiol. (Lond.) **153**, 35—37 (1960).
Greenfield, J. G.: General pathology of the nerve cell and neuroglia. In: Neuropathology (Greenfield, J. G., W. Blackwood, W. H. McMenemy, A. Meyer, and R. M., Norman, eds.), p. 1—66. London: Edward Arnold 1958.
Greenfield, W., McMenemy, W. H., Meyer, A., Norman, R. M., Russell, D. S.: In: Greenfield's Neuropathology. 679 p. Baltimore: William & Wilkings, Co. 1967.
Grimley, P. M.: Selection for electron microscopy of specific areas in large epoxy tissue sections. Stain Technol. **40**, 259—263 (1965).
Grisham, J. W.: Inhibitory effect of tritiated thymidine on regeneration of the liver in the young rat. Proc. Soc. exp. Biol. (N.Y.) **105**, 555—558 (1960).
Gudden, B.: Experimentaluntersuchungen über das peripherische und centrale Nervensystem. Arch. Psychiat. Nervenkr. **2**, 693—723 (1870).
— Beitrag zur Kenntniss des Corpus mammillare und der sogenannten Schenkel des Fornix. Arch. Psychiat. Nervenkr. **11**, 428—452 (1880).
— Ueber das Corpus mammillare und die sogenannten Schenkel des Fornix. Verslg dtsch. Naturforsch. **57**, 126 (1884).
Guild, S. R., Crowe, S. J., Bunch, C. C., Polvogt, L. M : Correlations of differences in the density of innervation of organ of Corti with differences in the acuity of hearing, including evidence as to the location in the human cochlea of the receptors for certain tones. Acta oto-laryng. (Stockh.) **15**, 269—308 (1931).
Guillery, R. W.: Degeneration in the post-commissural fornix and the mamillary peduncle of the rat. J. Anat. (Lond.) **90**, 350—370 (1956).
— Afferent fibers to the dorso-medial thalamic nucleus in the cat. J. Anat. (Lond.) **93**, 403—419 (1959).
— Fibre degeneration in the efferent mamillary tracts of the cat. In: Cytology of Nervous Tissue, p. 64—67. London: Taylor and Francis 1961.
— Some electron microscopical observations of degenerative changes in central nervous synapses. In Degenerative Patterns in the Nervous System (Singer, M., and J. P. Schadé, eds.). Amsterdam: Elsevier, 11. Progr. Brain Res. **14**, 57—76 (1965).
— A light and electron microscopical study of neurofibrils and neurofilaments at neuroneuronal junctions in the dorsal lateral geniculate nucleus of the cat. Amer. J. Anat. **120**, 583—603 (1967a).
— Patterns of fiber degeneration in the dorsal lateral geniculate nucleus of the cat following lesions in the visual cortex. J. comp. Neurol. **130**, 197—222 (1967b).
— The organization of synaptic interconnections in the laminae of the dorsal lateral geniculate nucleus of the cat. Z. Zellforsch., **96**, 1—38 (1969).
— Ralston, H. J., III: Nerve fibers and terminals: electron microscopy after Nauta staining. Science **143**, 1331—1332 (1964).
— Shirra, B., Webster, K. E.: Differential impregnation of degenerating nerve fibers in paraffin-embedded material. Stain Technol. **36**, 9—13 (1961).

Gustavson, K. H.: The chemistry of tanning processes. New York: Academic Press 1956.
Guth, L.: "Trophic" influences of nerve on muscle. Physiol. Rev. **48**, 587—645 (1968).
Ha, H., Liu, C.-N.: Synaptology of spinal afferents in the lateral cervical nucleus of the cat. Exp. Neurol. **8**, 318—327 (1963).
Haddock, J. N., Berlin, L.: Transsynaptic degeneration in the visual system. Arch. Neurol. Psychiat. (Chic.) **64**, 66—73 (1950).
Häggendahl, J., Malmfors, T.: Evidence of dopamine-containing neurons in the retina of the rabbit. Acta physiol. scand. **59**, 295—296 (1963).
— — Identification and cellular localization of the catecholamines in the retina and choroid of the rabbit. Acta physiol. scand. **64**, 58—66 (1965).
Haggar, R. A., Barr, M. L.: Quantitative data on the size of the synaptic end bulbs in the cat's spinal cord, with a note on the preparation of cell models. J. comp. Neurol. **93**, 17—36 (1950).
Hakansson, R., Owman, Ch.: Concomitant histochemical demonstration histamine and catecholamines in enterochromaffin-like cells of gastric mucosa. Life Sci. **6**, 759 (1967).
Hall, J. G.: A histological investigation of the auditory pathways in neonatal asphyxia. Acta otolaryng. **54**, 369—375 (1962).
— On the neuropathological changes in the central nervous system following neonatal asphyxia with special reference to the auditory system in man. Acta oto-laryng. (Stockh.) **56**, 331—339, Suppl. 188 (1964).
Hamberger, B.: Reserpine-resistant uptake of catecholamines in isolated tissues of the rat. A histochemical study. Acta physiol. scand., Suppl. 295 (1967).
— Levi-Montalcini, R., Norberg, L.-A., Sjöqvist, F.: Monoamines in immunosympathectomized rats. Int. J. Neuropharmacol. **4**, 91—95 (1965).
— Malmfors, T., Sachs, Ch.: Standardization of paraformaldehyde and of certain procedures for the histochemical demonstration of catecholamines. J. Histochem. Cytochem. **13**, 147 (1965).
— Masuoka, D.: Localization of catecholamine uptake in rat brain slices. Acta pharmacol. (Kbh.) **22**, 363—368 (1965).
— Norberg, K.-A.: Monoamines in sympathetic ganglia studied with fluorescence microscopy. Experientia (Basel) **19**, 580—581 (1963).
— — Histochemical demonstration of catecholamines in fresh frozen section. J. Histochem. Cytochem. **12**, 48—49 (1964).
— — Sjöqvist, F.: Cellular localization of monoamines in sympathetic ganglia of the cat. Life Sci. **2**, 659—661 (1963a).
— — — Evidence for adrenergic nerve terminals and synapses in sympathetic ganglia. Int. J. Neuropharmacol. **2**, 279—282 (1963b).
— — — Correlated studies of monoamines and AChE in sympathetic ganglia, illustrating the distribution of adrenergic and cholinergic neurons. In: Pharmacology of Cholinergic and Adrenergic Transmission (Koelle, J. B., W.. W. Douglas, and A. Carlsson, eds.). Proc. Soc. int. pharmacol., Meeting, Prague, vol. 3, p. 41—54. Oxford: Pergamon 1963.
— — Ungerstedt, U.: Adrenergic synaptic terminals in autonomic ganglia. Acta physiol. scand. **64**, 285—286 (1965).
Hamburger, V., Levi-Montalcini, R.: Proliferation, differentation and degeneration in the spinal ganglia of the chick embryo under normal and experimental conditions. J. exp. Zool. **95**, 223—242 (1949).
— — Some aspects of neuroembryology. In: Genetic Neurology (Weiss, P., ed.). Chicago, Illinois: Chicago University Press 1950.
Hamlyn, L. H.: The effect of preganglionic section of the neurons of the superior cervical ganglion in rabbits. J. Anat. (Lond.) **88**, 184—191 (1954).
— Electron microscopy of mossy fibre endings in Ammon's horn. Nature (Lond.) **190**, 645—646 (1961).
— The fine structure of the mossy fibre endings in the hippocampus of the rabbit. J. Anat. (Lond.) **96**, 112—120 (1962).
Hammarberg, C.: Studien über Klinik und Pathologie der Idiotie. Upsala 1895.
Hammarström, L., Ritzen, M., Ullberg, S.: Combined autoradiography and fluorescence microscopy. Localization of labelled 5-hydroxytryptophan. Experientia (Basel) 22, 213 1966.

Hammer, W., Norberg, K.-A., Olson, L., Sjöqvist, F : Excessively growing monoamine-containing nerves in mice during treatment with a nerve growth factor (NGF). Acta physiol. scand. **69**, 127—128 (1967).

Hammond, W. S., Yntema, C L.: Origin of ciliary ganglia in the chick. J. comp. Neurol. **110**, 367—389 (1958).

Hare, W. K., Hinsey, J. C.: Reactions of dorsal root ganglion cells to section of peripheral and central processes. J. comp. Neurol. **73**, 489—502 (1940).

Harman, P. J., Carpenter, M. B.: Volumetric comparisons of the basal ganglia of various primates including man. J. comp. Neurol. **93**, 125—137 (1950).

Harreveld, A. van, *see* Van Harreveld, A.

Hatai, S.: Number and size of the spinal ganglion cells and dorsal root fibers in the white rat at different ages. J. comp. Neurol. **12**, 107—124 (1902).

Hayashi, Y., Sumitomo, I., Iwana, K.: Activation of lateral geniculate neurons by electrical stimulation of superior colliculus in cats. Jap. J. Physiol. **17**, 638—651 (1967).

Hayhow, W. R.: Experimental degeneration of optic axons in the lateral geniculate body of the cat. Acta anat. (Basel) **37**, 281—298 (1959).

Heene, R.: Histochemischer Nachweis von Katecholaminen und 5-hydroxytryptamin am Kryostatschnitt. Histochemie **14**, 324—327 (1968).

Heimer, L.: Silver impregnation of terminal degeneration in some forebrain fiber systems: A comparative evaluation of current methods. Brain Res. **5**, 86—108 (1967).

— The tracing of pathways in the central nervous system. Proc. IEEE **56**, 950—959 (1968a).

— Synaptic distribution of centripetal and centrifugal nerve fibers in the olfactory system of the rat. An experimental anatomical study. J. Anat. (Lond.) **103**, 413—432 (1968b).

— Silver impregnation of degenerating axons and their terminals on Epon-Araldite sections. Brain Res. **12**, 246—249 (1969).

— Ebner, F. F., Nauta, W. J. H.: A note on the termination of commissural fibers in the neocortex. Brain Res. **5**, 171—177 (1967).

— Ekholm, R.: Neuronal argyrophilia in early degenerative states: A light and electron microscopic study of Glees and Nauta techniques. Experientia (Basel) **23**, 237—239 (1967).

— Nauta, W. J. H.: The hypothalamic distribution of the stria terminalis in the rat. Brain Res. **13**, 284—297 (1969).

— Wall, P. D.: The dorsal root distribution to the substantia gelatinosa of the rat with a note on the distribution in the cat. Exp. Brain Res. **6**, 89—99 (1968).

Held, H.: Die centrale Gehörleitung. Arch. Anat. Entwickl. Lpz., Abt., 201—248 (1893).

Heller, A., Moore, R. Y.: Control of brain serotonin and norepinephrone by specific neural systems. Advanc. Pharmacol. **6**A, 191—206 (1968).

Heller, I. H., Elliott, K. A. C.: Deoxyribonucleic acid content and cell density in brain and human brain tumors. Canad. J. Biochem. Physiol. **32**, 584—592 (1954).

Henle, J.: Über das Gewebe der Nebenniere und der Hypophyse. Z. rationelle Med. **24**, 143 (1965).

Herndon, R. M.: Thiophen induced granule cell necrosis in the rat cerebellum. An electron microscopic study. Exp. Brain Res. **6**, 49—68 (1968).

Herman, C. J., Lapham, L. W.: DNA content of neurons in the cat hipocampus. Science **160**, 537 (1968).

Hess, A.: The experimental embryology of the foetal nervous system. Biol. Rev. **32**, 231—260 (1957a).

— Optic centers and pathways after eye removal in fetal guinea pigs. J. comp. Neurol. **109**, 91—116 (1957b).

Hicks, S. P., D'Amato, C. J.: Cell migrations to the isocortex in the rat. Anat. Rec. **160**, 619—627 (1968).

Hill, D. K.: Resolving power with tritium autoradiographs. Nature (Lond.) **194**, 831—832 (1962).

Hill, M.: Intercellular passage of tritium-thymidine-labeled DNA from donor lymphocytes to recipient bone marrow cells. Exp. Cell Res. **28**, 21—26 (1962).

Hillarp, N.-A.: Structure of the synapse and the peripheral innervation apparatus of the autonomic nervous system. M. D. Thesis. Acta anat. (Basel) **2**, 1—153, Suppl. 4 (1946).

Hillarp, N.-A.: The construction and functional organization of the autonomic innervation apparatus. Acta physiol. scand. **46**, 1—38, Suppl. 157 (1959).
— Fuxe, K., Dahlström, A.: Central monamine neurons. In: Mechanisms of Release of Biogenic Amines (Euler, U. S. von, S. Rosell, and B. Uvnäs, eds.), p. 31—57. London: Pergamon 1966.
— Hökfelt, B.: Evidence of adrenaline and noradrenaline in separate adrenal medullary cells. Acta physiol. scand. **30**, 65—68 (1953).
— — Histochemical demonstration of noradrenaline and adrenaline in the adrenal medulla. J. Histochem. Cytochem. **3**, 1—5 (1955).
— Malmfors, T.: Reserpine and cocaine blocking of the uptake and storage mechanisms in adrenergic nerves. Life Sci. **3**, 703—708 (1964).
Hilscher, W., Maurer, W.: Autoradiographische Bestimmung der Dauer der DNS-Verdopplung und ihres zeitlichen Verlaufs bei Spermatogenien der Ratte durch Doppelmarkierung mit C-14-and-H-3-Thymidin. Naturwissenschaften **49**, 352 (1962).
Hinds, J. W.: Autoradiographic study of histogenesis in the mouse olfactory bulb. I. Time of origin of neurons and neuroglia. J. comp. Neurol., in press (1969a).
— Autoradiographic study of histogenesis in the mouse olfactory bulb. II. Cell proliferation and migration. J. comp. Neurol., in press (1969b).
Hirano, A., Zimmerman, H. M., Levine, S.: Myelin in the central nervous system as observed in experimentally induced edema in the rat. J. Cell Biol. **31**, 397—411 (1966).
Hökfelt, T.: A modification of the histochemical fluorescence method for the demonstration of catecholamines and 5-hydroxytryptamine, using araldite as embedding medium. J. Histochem. Cytochem. **13**, 518—519 (1965).
— Distribution of noradrenaline storing particles in peripheral adrenergic neurons as revealed by electron microscopy. Acta physiol. scand., in press (1969).
— *In vitro* studies on central and peripheral monoamine neurons at the ultrastructural level. Z. Zellforsch. **91**, 1—74 (1968).
— Ungerstedt, U.: Electron and fluorescence microscopical studies on the nucleus caudatus putamen of the rat after unilateral lesions of ascending nigro-neostriatal dopamine neurons. Acta physiol. scand., in press (1969).
Hoff, E. C.: Central nerve terminals in the mammalian spinal cord and their examination by experimental degeneration. Proc. roy. Soc. B **3**, 175—188 (1932a).
— The distribution of spinal terminals (boutons) of the pyramidal tract, determined by experimental degeneration. Proc. roy. Soc. B **3**, 226—237 (1932b).
Holländer, H.: Personal communication (1969).
— Vaaland, J. L.: A reliable staining method for semi-thin sections in experimental neuroanatomy. Brain Res. **10**, 120—126 (1968).
Hollyfield, J. G.: Differential addition of cells to the retina in *Rana pipiens* tadpoles. Develop. Biol. **18**, 163—179 (1968).
Holmes, W.: Silver staining of nerve axons in paraffin sections. Anat. Rec. **86**, 157—185 (1943).
Holt, E. J., Hicks, R. M.: Studies on formalin fixation for electron microscopy and cytochemical staining purpose. J. biophys. biochem. Cytol. **11**, 31—45 (1961).
Hommes, O. R., Leblond, C. P.: Mitotic division of neuroglia in the normal adult rat. J. comp. Neurol. **129**, 269—278 (1967).
Howard, A., Pelc, S. R.: Synthesis of deoxyribonucleic acid in normal and irradiated cells and its relation to chromosome breakage. Heredity, Suppl. **6**, 261—273, Biol. Abstr., **27**, 1383—1384 (1953).
Hoyle, G.: Potassium ions and insect nerve muscle. J. exp. Biol. **30**, 121—135 (1953).
Hubel, D. H., Wiesel, T. N.: An anatomical demonstration of columns in the monkey striate cortex. Nature (Lond.) **221**, 747—750 (1969).
Hughes, A. F. W.: Cell degeneration in the larval horn of *Xenopus laevis* (Daudin). J. Embryol. exp. Morph. **9**, 269—284 (1961).
— Aspects of neural ontogeny, p. 249. New York and London: Academic Press 1968.
Hughes, W. L., Bond, V. P., Brecher, G., Cronkite, E. P., Painter, R. B., Quastler, H., Sherman, F. G.: Cellular proliferation in the mouse as revealed by autoradiography with tritiated thymidine. Proc. nat. Acad. Sci. (Wash.) **44**, 476—483 (1958).

Hunt, C. C., Nelson, P. G.: Structural and functional changes in the frog sympathetic ganglion following cutting of the presynaptic nerve fibers. J. Physiol. (Lond.) **177**, 1—20 (1965).

Irving, R., Harrison, J. M.: The superior olivary complex and audition: A comparative study. J. comp. Neurol. **130**, 77—86 (1967).

Jacobowitz, D.: Histochemical studies of the autonomic innervation of the gut. J. Pharmacol. exp. Ther. **149**, 358—364 (1965).

— Cooper, T., Barner, H. B.: Histochemical and chemical studies of the localization of adrenergic and cholinergic nerves in normal and denervated cat hearts. Circulat. Res. **20**, 289—298 (1967).

— Woodward, J. K.: Adrenergic neurons in the cat superior cervical ganglion and cervical sympathetic nerve trunk: A histological study. J. Pharmacol. exp. Ther. **162**, 213—226 (1968).

Jacobson, S.: Adaptation of the Nauta method for use on degenerating fibers in rat cortex and thalamus. Stain Technol. **38**, 275—279 (1963).

Jacobson, W., Sidman, R. L., Diamond, L. K.: The effect of testosterone on the uptake of tritiated thymidine by the bone marrow of children. Ann. N.Y. Acad. Sci. **149**, 389—405 (1968).

Jakob, A.: Über die feinere Histologie der sekundären Faserdegeneration in der weißen Substanz des Rückenmarks (mit besonderer Berücksichtigung der Abbauvorgänge). In: Histologische und Histopathologische Arbeiten über die Großhirnrinde (Nissl, F., and A. Alzheimer, eds.), vol. 5, 1. Jena: Fischer (1912).

Johnson, H. A., Cronkite, E. P.: The effect of tritiated thymidine on mouse spermatogonia. Radiat. Res. **11**, 825—831 (1959).

— — The effect of tritiated thymidine on mortality and tumor incidence in mice. Radiat. Res. **30**, 488—495 (1967).

Johnston, G., Bowsher, D.: A new method for the selective impregnation of degenerating axon terminals. Brain Res. **12**, 47—53 (1969).

Jones, R. L.: Split nucleoli as a source of error in nerve cell counts. Stain Technol. **12**, 91—95 (1937a).

— Cell hyphen fiber ratios in the vagus nerve. J. comp. Neurol. **67**, 469—482 (1937b).

Jones, W. H., Thomas, D. B.: Trans-synaptic atrophy in the cerebral cortex. Nature (Lond.) **178**, 47—48 (1956).

— — Changes in dendritic organization of neurons in the cerebral cortex following deafferentation. J. Anat. (Lond.) **96**, 375—381 (1962).

Jonsson, G.: Fluorescence studies on some 6, 7-substituted 3, 4-dihydroisoquinolines formed from 3-hydroxytyramine (dopamine) and formaldehyde. Acta. chem. scand. **20**, 2755—2762 (1966).

— Binding of catecholamines and related compounds to proteins after formaldehyde gas treatment. Histochemie **8**, 122—130 (1967a).

— Further studies on the specificity of the histochemical fluorescence method for the demonstration of catecholamines. Acta histochem. (Jena) **26**, 1—11 (1967b).

— The formaldehyde fluorescence method for the histochemical demonstration of biogenic monoamines: A methodological study. M. D. Thesis, Stockholm (1967c).

— Microfluorimetric studies on the formaldehyde-induced fluorescence of noradrenaline in adrenergic nerves of rat iris. In preparation (1969).

— Fuxe, K., Hamberger, B., Hökfelt, T.: 6-Hydroxytryptamine — a new tool in fluorescence microscopy. Brain Res., **1**^{3}, 190—195 (1969).

— Ritzén, M.: Microspectrofluorometric identification of metaraminol in sympathetic adrenergic neurons. Acta physiol. scand. **67**, 505—512 (1966).

Sachs, Ch.: Subcellular distribution of ^{3}H-noradrenaline in adrenergic nerves of mouse atrium — effect of reserpine, monoamine oxidase and tyrosine hydroxylase inhibition. Acta physiol. scand. **77**, 344—357 (1969).

— — Effects of 6-hydroxydopamine on the uptake and storage of noradrenaline in sympathetic adrenergic neuron. Europ. J. Pharmacol. **9** (1970) (in press).

— Sandler, M.: Fluorescence of indolylethylamines condensed with formaldehyde. Histochemie, in press (1969).

Joseph, B., Whitlock, D.: The spatio-temporal course of Wallerian degeneration: A Nauta study in the toad (*Bufo marinus*). Anat. Rec. **160**, 372 (1968).

Jouvet, M.: Insomia and decrease of cerebral 5-hydroxytryptamine after destruction of the raphe system in the cat. Advanc. Pharmacol. **6** B, 265—279 (1968).

Juhlin, L., Shelley, W. B.: Detection of histamine by a new fluorescent *o*-phthalaldehyde stain. J. Histochem. Cytochem. **14**, 525—528 (1966).

Källén, B.: Degeneration and regeneration in the vertebrate central nervous system during embryogenesis. Progr. Brain Res. **14**, 77—96 (1965).

Kallius, E.: Enzyklopädie der mikroskopischen Technik, Bd. I. Golyische Methode (Ehrlich, P., R. Krause, M. Mosse, H. Rosin, and K. Weigert, eds.), 2nd ed., p. 551—582. Berlin and Wien: Urban & Schwarzenberg 1910.

Kapeller, K., Mayor, D.: The accumulation of noradrenaline in constructed sympathetic nerves as studied by fluorescence and electron microscopy. Proc. roy. Soc. B **167**, 282—292 (1967).

Karlsson, U.: Three-dimensional studies of neurons in the lateral geniculate nucleus of the rat. I. Organelle organization in the perikaryon and its proximal branches. J. Ultrastruct. Res. **16**, 429—481 (1966a).

— Three-dimensional studies of neurons in the lateral geniculate nucleus of the rat. II. Environment of perikarya and proximal parts of their branches. J. Ultrastruct. Res. **16**, 482—504 (1966b).

— Three-dimensional studies of neurons in the lateral geniculate nucleus of the rat. III. Specialized neuronal contacts in the neuropil. J. Ultrastruct. Res. **17**, 137—157 (1967).

— Schultz, R. L.: Fixation of the central nervous system for electron microscopy by aldehyde fixation. I. Preservation with aldehyde perfusates versus direct perfusion with osmium tetroxide with special references to membrane and the extra cellular space. J. Ultrastruct. Res. **12**, 160—186 (1965).

Karnovsky, M. J.: A formaldehyde-glutaraldehyde fixative of high osmolality for use in electron microscopy. J. Cell Biol. **27**, 137 A (1965) (abstract).

Karten, H. J.: Projections of the optic tectum of the pigeon (*Columba livia*). Anat. Rec. **151**, 369 (1965).

— The ascending auditory pathway in the pigeon (*Columba livia*). II. Telencephalic projections of the nucleus ovoidalis thalami. Brain Res. **2**, 134—153 (1968a).

— Unpublished observations (1968b).

— Ebbesson, S. O. E.: Retinal projections in the Tegu lizard. Manuscript in preparation (1969).

— Nauta, W. J. H.: Organization of the retino-thalamic projections in the pigeon and owl. Anat. Rec. **160**, 373 (1968).

Kauffman, S. L.: Lengthening of the generation cycle during embryonic differentiation of the mouse neural tube. Exp. Cell Res. **49**, 420—424 (1968).

Key, A., Retzius, G.: Studien in der Anatomie des Nervensystems. Arch. mikr. Anat. **9** (1873).

— — Studien in der Anatomie des Nervensystems und des Bindegewebes. II. I. Stockholm: P. A. Norstedt and Söner 1876.

Kisieleski, W. E., Samuels, L. D., Hiley, P. C.: Dose-effect measurements of radiation following administration of tritiated thymidine. Nature (Lond.) **202**, 458—459 (1964).

Klüver, H.: Certain effects of lesions of the occipital lobes in macaques. J. Psychol. **4**, 383—401 (1937).

Koburg, E., Maurer, W.: Autoradiographische Untersuchungen mit ^{3}H-Thymidin über die Dauer der Deoxyribonukleinsäure-Synthese und ihren zeitlichen Verlauf bei den Darmepithelien und anderen Zelltypen der Maus. Biochim. biophys. Acta (Amst.) **61**, 229—242 (1962).

Koenig, H.: The autofluorescence of lysomones. Its valuc for the identification of lysomal constituents. J. Histochem. Cytochem. **11**, 556—557 (1963).

— Groat, R. A., Windle, W. F.: A physiological approach to perfusion-fixation of tissues with formalin. Stain Technol. **20**, 13—21 (1945).

Kollros, J.: The development of the optic lobes in the frog. J. exp. Zool. **123**, 153—187 (1953).

Konigsmark, B. W.: Neuronal population of the ventral cochlear nucleus in man. Presented at the meeting of the Amer. Assoc. of Anatomists, Boston, Mass., April 2, 1969.
— Kalyanaraman, U. P., Corey, P., Murphy, E. A.: An evaluation of techniques in neuronal population estimates. The VIth nerve nucleus. Johns Hopk. med. J. **125**, 146—158 (1969).
— Sidman, R. L.: Origin of brain macrophages in the mouse. J. Neuropath. exp. Neurol. **22**, 643—676 (1963).
Kopriwa, B. M., Leblond, C. P.: Improvements in the coating techniques of radioautography. J. Histochem. Cytochem. **10**, 269—284 (1962).
Kovacs, Ö., Fodor, G.: Beitrag zur Synthese der Tetrahydroisochinolin-Alkaloide unter physiologischen Bedingungen. Ber. dtsch. chem. Ges. **84**, 795—801 (1951).
Krause, C.: Haut. In: Rudolf Wagner's Handwörterbuch der Physiologie, p. 108—186. Braunschweig: Vieweg & Sohn 1844.
Kreiner, J., Maksymowicz, K.: A three-dimensional model of the striatal nuclei in the dog's brain. Acta Biol. exp. (Warszawa) **22**, 69—79 (1962).
Kupfer, C.: The distribution of cell size in the lateral geniculate nucleus of man following transneuronal cell atrophy. J. Neuropath. exp. Neurol. **24**, 653—661 (1965).
— Downer, J. L.: Ribonucleic acid content and metabolic activity of lateral geniculate nucleus in monkey following afferent denervation. J. Neurochem. **14**, 257—263 (1967).
— Palmer, P.: Lateral geniculate nucleus histological and cytochemical changes following afferent denervation and visual deprivation. Exp. Neurol. **9**, 400—409 (1964).
Laatsch, R. H., Cowan, W. M.: Electron microscopic studies of the dentate gyrus of the rat. II. Degeneration of commissural afferents. J. comp. Neurol. **130**, 241—262 (1967).
Lagunoff, D., Phillips, M., Benditt, E. P.: The histochemical demonstration of histamine in mast cells. J. Histochem. Cytochem. **9**, 534—541 (1961).
Lala, P. K.: Measurement of S period in growing cell populations by a graphic analysis of double labeling with ^{3}H-and-^{14}C-thymidine. Exp. Cell Res. **50**, 459—463 (1968).
Lamerton, L. F., Fry, R. J. M.: In: Cell Proliferation, p. 241. Philadelphia: F. A. Davis, Co. 1963.
Lampert, P. W.: A comparative electron microscopic study of reactive degenerating, regenerating and dystrophic axons. J. Neuropath. exp. Neurol. **26**, 345—368 (1967).
— Cressman, M.: Axonal degeneration in the dorsal columns of the spinal cord of adult rats. An electron microscopic study. Lab. Invest. **13**, 825—839 (1964).
Lange, P. W., Engstrom, A.: Determination of thickness of microscopic objects. Lab. Invest. **3**, 116—131 (1954).
Lapham, L.: Personal communication (1968).
Larsell, O.: The effect of experimental excision of one eye on the development of the optic lobe and opticus layer in larvae of the treefrog (*Hyla regilla*). II. The effect on cell size and differentiation of cell processes. J. exp. Zool. **58**, 1—20 (1931).
Lassek, A. M.: The human pyramidal tract. Arch. Neurol. Psychiat. (Chic.) **44**, 718—724 (1940).
— Rasmussen, G. L.: A quantitative study of the newborn and adult spinal cords of man. J. comp. Neurol. **69**, 371—379 (1938).
Laties, A. M., Lund, R., Jacobowitz, D.: A simplified method for the histochemical localization of cardiac catecholamine-containing nerve fibers. J. Histochem. Cytochem. **15**, 535—541 (1967).
La Velle, A., La Velle, F. W.: The nucleolar apparatus and neuronal reactivity to injury during development. J. exp. Zool. **137**, 285—315 (1958).
Leblond, C. P., Messier, B., Kopriwa, B.: Thymidine-H^3 as a tool for the investigation of the renewal of cell populations. Lab. Invest. **8**, 296—308 (1959).
Legg, P. G.: Fluorescence studies on neural structures and endocrine cells in the pancreas of the cat. Z. Zellforsch. **88**, 187—495 (1968).
Leghissa, A.: A proposito dello svillupo del tetto ottico nei Teleostei (*Salmo faris*). Bol. Zool. **18**, 355—365 (1951).
Leghissa, S.: Studio sperimentale sul differenziamento dei neuroni del tetto ottico di Gallus. I. Esperienze di asportazione della vescicola ottica. Atti Accad. Sci. Bologna **6**, 56—75 (1959).

Le Gros Clark, W. E.: A morphological study of the lateral geniculate body. Brit. J. Ophthal. **16**, 264—284 (1932).
— The anatomy of cortical vision. Trans. ophthal. Soc. **62**, 229—245 (1943).
— Inquiries into the anatomical basis of olfactory discrimination. Proc. roy. Soc. B **146**, 299—319 (1957).
— Meyer, M.: The terminal connections of the olfactory tract of the rabbit. Brain **70**, 304—328 (1947).
— Penman, G. G.: The projection of the retina in the lateral geniculate body. Proc. roy. Soc. B **114**, 292—313 (1934).
— Warwick, R. T. T.: The pattern of olfactory innervation. J. Neurol. Neurosurg. Psychiat. **9**, 101—111 (1946).
Lever, J. D., Spriggs, T. L. B., Graham, J. D. P.: A formol-fluorescence, fine-structural and autoradiographic study of the adrenergic innervation of the vascular tree in the intact and sympathectomized pancreas of the cat. J. Anat. (Lond.) **103**, 15—34 (1968).
Levi-Montalcini, R.: Regressione secondaria del ganglio ciliare dopo asportazione della vesicola mesencefalica in embrione di pollo. R. Acad. Naz. Lincei. **3**, 144—416 (1947).
— The development of the acoustico-vestibular centers in the chick embryo in the absence of the afferent root fibers and of descending fiber tracts. J. comp. Neurol. **91**, 209—242 (1949).
— Amprino, R.: Recherches expérimentales sur l'origine du ganglion ciliare dans l'embryon de Poulet. Arch. Biol. (Liège) **58**, 265—288 (1947).
Lichtensteiger, W., Langemann, H.: Uptake of exogenous catecholamines by monoamine containing neurons of the central nervous system: Uptake of catecholamines by arcuato-infundibular neurons. J. Pharmacol. exp. Ther. **151**, 469—492 (1966).
Liesegang, R. E.: Untersuchungen über die Golgi-Färbung. J. Psychol. Neurol. (Lpz.) **17**, 1—18 (1910—1911).
Lison, L.: Histochemie et cytochemie animales. Principes et méthodes. Paris: Gauthier-Villars 1960.
Lohman, A. H. M.: The anterior olfactory lobe of the guinea pig. Doctoral thesis. Nigmegen, The Netherlands 1963.
— Mentink, G. M.: The lateral olfactory tract: The anterior commissure and the cells of the olfactory bulb. Brain Res., in press (1969).
Loos, H. van der, *see* Van der Loos, H.
Lorente de Nó, R.: Studies on the structure of the cerebral cortex. II. Continuation of the study of the ammonic system. J. Psychol. Neurol. (Lpz.) **46**, 113—177 (1934).
— Cerebral cortex: architecture, intracortical connections, motor projections. In: Physiology of the Nervous System. (Fulton, J. F., ed.), 3rd edit., p. 288—330. London: Oxford University Press 1949.
Luft, J. H.: Improvements in epoxy resin embedding methods. J. biophys. biochem. Cytol. **9**, 409—414 (1961).
Lubińska, L.: Axoplasmic streaming in regenerating and normal fibers. In: Mechanisms of Neural Regeneration (Singer, M., and J. P. Shade, eds.). Brain Res. **13**, 1—71. (1964).
Lund, R. D.: Synaptic patterns of the superficial layers of the superior colliculus of the rat. J. comp. Neurol., **135**, 179—208 (1969).
— Westrum, L. E.: Neurofibrils and the Nauta method. Science **151**, 1397—1399 (1966).
Lundborg, P.: Studies on the uptake and subcellular distribution of catecholamines and their a-methylated analogues. Acta physiol. scand., Suppl. 302 (1967).
Malmfors, T.: Evidence of adrenergic neurons with synaptic terminals in the retina of rats demonstrated with fluorescence and electron microscopy. Acta physiol. scand. **58**, 99—100 (1963).
— Release and depletion of the transmitter in adrenergic terminals produced by nerve impulse after the inhibition of noradrenaline synthesis or reabsorption. Life Sci. **3**, 1397—1402 (1964).
— Studies on adrenergic nerves. The use of rat and mouse iris for direct observations on their physiology and pharmacology at cellular and subcellular levels. Acta physiol. scand. **64**, 1—93, Suppl. 248 (1965a).

Malmfors, T.: The adrenergic innervation of the eye as demonstrated by fluorescence microscopy. Acta physiol. scand. **65**, 259—267 (1965b).
— Fluorescent histochemical studies on the uptake, storage, and release of catecholamines. Circulat. Res. **20** and **21**, Suppl. III, 25—42 (1967).
— Sachs, Ch.: Degeneration of adrenergic nerves produced by 6-hydroxydopamine. Europ. J. Pharmacol. **3**, 89—92 (1968).
— — Direct demonstration of the systems of terminals belonging to an individual adrenergic neuron and their distribution in the rat iris. Acta physiol. scand. **64**, 337—382 (1965).
Mann, G.: Physiological histology, methods and theory. Oxford: H. Hart 1902.
Mansberg, H. H., Segarra, J. M.: Counting on neurons by flying spot microscope. Ann. N.Y. Acad. Sci. **99**, 309—322 (1962).
Manual of Histologic Staining Methods of the Armed Forces Institute of Pathology: (Luna, Lee G., ed.), 3rd edit. New York: McGraw Hill 1968.
Marchi, V., Algeri, G.: Sulle degenerazioni discendenti consecutive a lesioni sperimentale in diverse zone della corteccia cerebrale. Riv. sper. Freniat. **11**, 492—494 (1885).
Marengo, N. P.: Paraffin section thickness — a direct method of measurement. Stain Technol. **19**, 1—10 (1944).
Marrable, A. W.: The counting of cells and nuclei in microtome sections. Quart. J. micr. Sci. **103**, 331—347 (1962).
Marshall, Wade H.: An application of the frozen sectioning technic for cutting serial sections thru the brain. Stain Technol. **15**, 133—138 (1940).
Marsland, T. A., Glees, P., Erikson, L. B.: Modification of the Glees silver impregnation for paraffin sections. J. Neuropath. exp. Neurol. **13**, 587—591 (1954).
Martin, A., Langman, J.: The development of the spinal cord examined by autoradiography. J. Embryol. exp. Morph. **14**, 25—35 (1965).
Masuda, N.: Über das Brückengrau des Menschen (*Griseum pontis*) und dessen nähere Beziehungen zum Kleinhirn und Großhirn. Arb. Hirnanat. Inst. Zürich **9**, 1—249 (1914).
Matthews, M. R.: Further observations on transneuronal degeneration in the lateral geniculate nucleus of the macaque monkey. J. Anat. (Lond.) **98**, 255—263 (1964).
— Cowan, W. M., Powell, T. P. S.: Transneuronal cell degeneration in the lateral geniculate nucleus of the macaque monkey. J. Anat. (Lond.) **94**, 145—169 (1960).
— Powell, T. P. S.: Some observations on transneuronal cell degeneration in the olfactory bulb of the rabbit. J. Anat. (Lond.) **96**, 89—102 (1962).
Maturana, H. R.: Number of fibres in the optic nerve and the number of ganglion cells in the retina of anurans. Nature (Lond.) **183**, 1406—1407 (1959).
Maunsbach, A. B.: The influence of different fixatives and fixation methods on the ultrastructure of rat kidney proximal tubule. II. Effects of varying osmolality, ionic strength, buffer system and fixative concentration of glutaraldehyde solutions. J. Ultrastruct. Res. **15**, 283—309 (1966).
Maurer, W., Primbsch, E.: Größe der Selbstabsorption bei der ^{3}H-Autoradiographie. Exp. Cell Res. **33**, 8—18 (1964).
Mayor, D., Kapeller, K.: Fluorescence microscopy and electron microscopy of adrenergic nerves after construction at two points. J. roy. micr. Soc. **76**, 277—294 (1967).
McGee-Russel, S. M., Gosztonyi, G.: Assembly of Semliki Forest virus in brain. Nature (Lond.) **214**, 1204—1206 (1967).
McLean, J. R., Burnstock, G.: Innervation of the lungs of the sleepy lizard (*Trachysaurus Rugosus*). I. Fluorescent histochemistry of catecholamines. Comp. Biochem. Physiol. **22**, 809—813 (1967a).
— — Innervation of the lungs of the toad (*Bufo Marinus*). II. Fluorescent histochemistry of catecholamines. Comp. Biochem. Physiol. **22**, 767—773 (1967b).
McMahan, U. J.: Fine structure of synapses in the dorsal nucleus of the lateral geniculate body of normal and blinded rats. Z. Zellforsch. **76**, 116—146 (1967).
McMurray, V. M.: The development of the optic lobes in *Xenopus laevis*. The effect of repeated crushing of the optic nerve. J. exp. Zool. **125**, 247—264 (1954).
Mehler, W. R.: Further notes on the centre median nucleus of Luys. In: The Thalamus (Purpura, D. P., and M. D. Yohr, eds.), p. 109—127. New York: Columbia University Press 1966.

Meikle, T. H., Sprague, J. M.: The neural organization of the visual pathways in the cat. Int. Rev. Neurobiol. **6**, 149—189 (1964).

Mendelsohn, M. L.: Autoradiographic analysis of cell proliferation in spontaneous breast cancer of C3H mouse. II. Growth and survival of cells labeled with tritiated thymidine. J. nat. Cancer Inst. **25**, 485—500 (1960).

Meyer, M., Allison, A. C.: An experimental investigation of the connections of the olfactory tract in the monkey. J. Neurol. Neurosurg. Psychiat. **12**, 274—286 (1949).

Miale, I., Sidman, R. L.: An autoradiographic analysis of histogenesis in the mouse cerebellum. Exp. Neurol. **4**, 277—296 (1961).

Millonig, G.: Advantages of a phosphate buffer for OsO_4-solutions in fixation. J. appl. Phys. **32**, 1637 (1961a).

— A modified procedure for lead staining of thin sections. J. biophys. biochem. Cytol. **11**, 736—739 (1961b).

Minckler, J.: The morphology of the nerve terminals of the human spinal cord as seen in black silver preparations, with estimates of the total number per cell. Anat. Rec. **77**, 9—25 (1940).

Minkowski, M.: Über den Verlauf, die Endigung und Zentrale Repräsentation von gekreuzten und ungekreuzten Sehnervenfasern bei einigen Säugetieren und beim Menschen. Schweiz. Arch. Neurol. Psychiat. **6**, 201—252; **7**, 268—303 (1920).

Mitra, N. L.: Quantitative analysis of cell types in mammalian neocortex. J. Anat. (Lond.) **89**, 467—483 (1955).

Moatamed, F.: Cell frequencies in the human inferior olivary nuclear complex. J. comp. Neurol. **128**, 109—116 (1966).

Moffat, G. H., Pelc, S. R.: Delay after plucking of hairs between the appearance of ^{3}H-thymidine in cells and its incorporation into DNA. Exp. Cell Res. **42**, 460—466 (1966).

Monakow, C. von, *see* Von Monakow, C.

Mondino, C.: Sull'uso del bicloruro di mercurio nello studio delgi organi centrali del sistema nervoso. Z. wiss. Mikr. **2**, 157—163 (1885).

Montero, V. M., Guillery, R. W.: Degeneration in the dorsal lateral geniculate nucleus of the rat following interruption of the retinal or cortical connections. J. comp. Neurol., in press (1968).

Morest, D. K., Morest, R. R.: Perfusion-fixation of the brain with chrome-osmium solutions for the rapid Golgi method. Amer. J. Anat. **118**, 811—832 (1966).

Moscowitz, N., Noback, C. R.: The human lateral geniculate body of normal development and congenital unilateral anophthalamia. J. Neuropath. exp. Neurol. **21**, 377 (1962).

Moses, M. J.: Application of autoradiography of electron microscopy. J. Histochem. Cytochem. **12**, 115—130 (1964).

Münzer, E., Wiener, H.: Das Zwischen- und Mittelhirn des Kaninchens und die Beziehungen dieser Teile zum übrigen Centralnervensystem, mit besonderer Berücksichtigung der Pyramidenbahn und Schleife. Mschr. Psychiat. Neurol. **12**, 241—279 (1902).

Mugnaini, E.: "Dark Cells" in electron micrographs from the central nervous system of vertebrates. J. Ultrastruct. Res. **12**, 235—236 (1965).

— Walberg, F.: An experimental electron microscopical study on the mode of termination of cerebellar corticovestibular fibers in the cat lateral vestibular nucleus (Deiters' nucleus). Exp. Brain Res. **4**, 212—236 (1967).

— — Brodal, A.: Mode of termination of primary vestibular fibers in the lateral vestibular nucleus: An experimental electron microscopical study in the cat. Exp. Brain Res. **4**, 187—211 (1967).

Mukherji, M., Ray, A. K., and P. B. Sen.: Histochemical identification of noradrenaline in fluorescence microscopy by borohydride-periodic acid sequence. J. Histochem. Cytochem. **14**, 479—482 (1966).

Murray, M.: Degeneration of some intralaminar thalamic nuclei after cortical removals in the cat. J. comp. Neurol. **127**, 341—368 (1966).

Myhrberg, H. E.: Monaminergic mechanisms in the nervous system of *Lumbricus terrestris* (L.). Z. Zellforsch. **81**, 311—343 (1967).

Mylrea, K. C., Albrecht, P. H.: An apparatus for tail vein injection in mice. Lab. Animal Care **17**, 602—603 (1967).

Nafstad, P. H. J.: An electron microscope study of the termination of the perforant path fibres in the hippocampus and the fuscia dentata. Z. Zellforsch. **76**, 532—542 (1967).
Nakamura, K.: A modified Nauta method. Kobe J. med. Sci. **9**, 135—138 (1963).
Narkiewicz, O.: Degenerations in the claustrum after regional neocortical ablations in the cat. J. comp. Neurol. **123**, 335—356 (1964).
Nashold, B. S., Hanbery, J., Olszewski, J.: Observations on the diffuse thalamic projections. Electronenceph. clin. Neurophysiol. **7**, 609—620 (1955).
Nauta, W. J. H.: Über die sogenannte terminale Degeneration im Zentralnervensystem und ihre Darstellung durch Silberimprägnation. Arch. Neurol. Psychiat. **66**, 353—376 (1950).
— Silver impregnation of degenerating axons. In: New Research Techniques of Neuroanatomy (W. F. Windle, ed.), p. 17—26. Springfield, Illinois: Thomas 1957.
— Gygax, P. A.: Silver impregnation of degenerating axon terminals in the central nervous system: (1) Technic, (2) Chemical notes. Stain Technol. **26**, 5—11 (1951).
— — Silver impregnation of degenerating axons in the central nervous system: A modified technique. Stain Technol. **29**, 91—93 (1954).
— Ryan, L. F.: Selective silver impregnation of degenerating axons in the central nervous system. Stain Technol. **27**, 175—179 (1952).
Nelson, J. S., Wakefield, P. L.: The cellular localization of catecholamines in frozen-dried cryostat sections of the brain and autonomic nervous system. J. Neuropath. exp. Neurol. **27**, 221—233 (1968).
Nichterlein, O. E., Goldby, F.: An experimental study of optic connexions in the sheep. J. Anat. (Lond.) **78**, 59—67 (1944).
Nielsen, K. C.: Experimental studies on the involvement of adrenergic mechanism in the development of ventricular fibrillation during induced hypothermia. M. D. Thesis 1968.
— Owman, Ch.: Differences in cardiac adrenergic innervation between hibernators and non-hibernating mammals. Acta physiol. scand., Suppl. 316 (1968).
Nissl, F.: Die Großhirnanteile des Kaninchens. Arch. Psychiat. Nervenkr. **52**, 867—953 (1913).
Nonidez, J. F.: Studies on the innervation of the heart. Amer. J. Anat. **65**, 361—413 (1939).
Norberg, K.-A.: The sympathetic adrenergic neuron and certain adrenergic mechanisms. M. D. Thesis, Stockholm (1965a).
— Drug-induced changes in monoamine levels in the sympathetic adrenergic ganglion cells and terminals: A histochemical study. Acta physiol. scand. **65**, 221—234 (1965b).
— Transmitter histochemistry of the sympathetic adrenergic nervous system. Brain Res. **5**, 125—170 (1967).
— Hamberger, B.: The sympathetic adrenergic neuron. Some characteristics revealed by histochemical studies on the intraneuronal distribution of the transmitter. Acta physiol. scand. **63**, 1—42, Suppl. 238 (1964).
— Hökfelt, T., Eneroth, C.-M.: The autonomic innervation of human submandibular and parotid glands. J. Neuro-Visceral Relat. **31**, 280—290 (1969).
— Olson, L.: Adrenergic innervation of the salivary glands in the rat. Z. Zellforsch. **68**, 183—189 (1965).
— Risley, P. L., Ungerstedt, U.: Adrenergic innervation of the male reproductive ducts in some mammals. I. The distribution of adrenergic nerves. Z. Zellforsch. **76**, 278—286 (1967).
— Ritzen, M., Ungerstedt, U.: Histochemical studies on a special catecholamine-containing cell type in sympathetic ganglia. Acta physiol. scand. **76**, 260—270 (1966).
— Sjökvist, F.: New possibilities for adrenergic modulation of ganglionic transmission. Pharmacol. Rev. **18**, 743—751 (1966).
Nordlander, R. H., Edwards, J. S.: Morphological cell death in the post-embryonic development of the insect optic lobes. Nature (Lond.) **218**, 780—781 (1968).
Numberger, J. I.: Direct enumeration of cells of the brain. In: Biology of Neuroglia (Windle, W. F., ed.), p. 193—211. Springfield, Illinois: Ch. C. Thomas 1958.
— Gordon, M. W.: The cell density of neural tissues. Direct counting method and possible applications as a biologic referent. In: Ultrastructure and Cellular Chemistry of Neural Tissue (Waelsch, H., ed.), p. 100—138. New York, N.Y.: Hoeber-Harper 1957.

Nyberg-Hansen, R.: Sites and mode of termination of reticulo-spinal fibers in the cat: An experimental study with silver impregnation methods. J. comp. Neurol. **124**, 71—100 (1965).

Olson, L: Outgrowth of sympathetic adrenergic neurons in mice treated with a nerve-growth factor (NGF). Z. Zellforsch. **81**, 155—173 (1967).

— Hamberger, B., Jonsson, G., Malmfors, T.: Combined fluorescence histochemistry and ^{3}H-noradrenaline measurements of adrenergic nerves. Histochemie **15**, 38—45 (1968).

Orden, L. S. van, *see* Van Orden, L. S.

Owman, Ch.: New aspects of the mammalian pineal gland. Acta. physiol. scand. **63**, 1—40, Suppl. 240 (1964).

— Sjöberg, N.-O.: Short adrenergic neurons and catecholamine containing cells in vas deferens and accessory male genital glands of different mammals. Z. Zellforsch. **66**, 300—320 (1965).

— — Adrenergic nerves in the female genital tract of the rabbit. With remarks on cholinesterase-containing structures. Z. Zellforsch. **74**, 182—197 (1966).

Packer, A. D.: An experimental investigation of the visual system in the phalanger, Trichosurus vulpecula. J. Anat. (Lond.) **75**, 309—329 (1941).

Pakkenberg, H.: The number of nerve cells in the cerebral cortex of man. J. comp. Neurol. **128**, 17—20 (1966).

Pal, J.: Ein Beitrag zur Nervenfärbetechnik. Z. wiss. Mikr. **4**, 92—96 (1887).

Palay, S. L.: The morphology of synapses in the central nervous system. Exp. Cell Res. **5**, 275—293, Suppl. (1958).

— The role of neuroglia in the organization of the central nervous system. In: Nerve as a Tissue (Rodahl, K., and B. Issekutz, eds.), p. 3—10. New York: Hoeber-Harper 1966.

— Principles of cellular organization in the nervous system. In: The Neurosciences (Quarton, G. C., T. Melnechuck, and F. O. Schmitt, eds.), p. 24—31. New York: The Rockefeller University Press 1967.

— McGee-Russell, S., Gordon, S., Grillo, M. A.: Fixation of neural tissues for electron microscopy by perfusion with solutions of osmium tetroxide. J. Cell Biol. **12**, 385—410 (1962).

— Sotelo, C., Peters, A., Orkand, P. M.: The axon hillock and the initial segment. J. Cell. Biol **38**, 193—201 (1968).

Pappas, G. D.: Electron microscopy of neuronal junctions involved in transmission in the central nervous system. In: Nerve as a Tissue (Rodahl, K., B. Issekutz, eds.), p. 49—87. New York: Hoeber-Harper 1966.

— Bennett, M. V. L.: Specialized junctions involved in electrical transmission between neurons. Ann. N.Y. Acad Sci. **137**, 495—503 (1966).

— Cohen, E. B., Purpura, D. P.: The structure of synaptic and non-synaptic neuronal relations in the thalamus of the cat. In: The Thalamus (Purpura, D. P., and M. D. Yahr, eds.), p. 47—75. New York: Columbia University Press 1966.

Peacock, J. H., Combs, C. M.: Retrograde cell degeneration in diencephalic and other structures after hemidecortication of rhesus monkeys. Exp. Neurol. **11**, 367—399 (1965).

Pearse, A. G. E.: Histochemistry. Theoretical and applied, 2nd ed., reprinted. Boston: Little Brown & Co. 1961.

Pease, D. C.: Buffered formaldehyde as a killing agent and primary fixative for electron microscopy. Anat. Rec. **142**, 342 (1962).

— Histological Techniques for Electron Microscopy, 2nd ed., reprinted. New York and London: Academic Press 1964.

Pelc, S. R.: Nuclear uptake of labelled adenine in the seminal vesicle of the mouse. Exp. Cell Res. **14**, 301—315 (1958).

— On the question of renewal of differentiated cells. Exp. Cell Res. **29**, 194—198 (1963a).

— Incorporation of labelled precursors of DNA in nondividing cells. In: Cell Proliferation (Lamerton, L. F., and R. J. M. Fry, eds.), p. 94—109. Philadelphia: F. A. Davis Co. 1963b.

— Labelling of DNA and cell division in so called nondividing tissues. J. Cell Biol. **22**, 21—28 (1964).

Pelc, S. R., Appleton, T. C.: Distribution of tritiated thymidine in various tissues. Nature (Lond.) **205**, 1287—1289 (1965).
— — Welton, M. E.: State of light autoradiography. In: The Use of Radioautography in Investigating Protein Synthesis (Leblond, C. P., and K. C. Warren, eds.). New York: Academic Press 1965.
— Welton, M. G. E.: Quantitative evaluation of tritium in autoradiography and biochemistry. Nature (Lond.) **216**, 925—927 (1967).
Penman, J., Smith, M. C.: Degeneration of the primary and secondary sensory neurones after trigeminal injection. J. Neurol. Neurosurg. Psychiat. **13**, 36—46 (1950).
Perkin-Elmer Corporation: Demonstration photograph of evaporated gold specimen produced by Hitachi Perkin-Elmer HS-8. Commercially presented in Science **159**, 144 (1968).
Peters, A.: The node of Ranvier in the central nervous system. Quart. J. exp. Physiol. **55**, 229—236 (1966).
— Palay, S. L.: The morphology of laminae A and A_1 of the dorsal nucleus of the lateral geniculate body of the cat. J. Anat. (Lond.) **100**, 451—486 (1966).
— Proskauer, C. C., Abramof, I. R. K.: The small pyramidal neuron of the cat cerebral cortex. The axon hillock and initial segment. J. Cell Biol. **39**, 604—619 (1968).
Peters, T., Jr., Ashley, C. A.: An artefact in radioautography due to binding of free amino acids to tissues by fixatives. J. Cell Biol. **33**, 53—60 (1967).
Petras, J. M.: Some fiber connections of the precentral cortex (areas 4 and 6) with the diencephalon in the monkey (*Macaca mulatta*). Anat. Rec. **148**, 322 (1964).
— The substantia gelatinosa of Rolando. Experientia (Basel) **24**, 1045—1047 (1968).
Pflugfelder, O.: Weitere volumetrische Untersuchungen über die Wirkung der Augenexstirpation und der Dunkelhaltung auf das Mesencephalon und die Pseudobranchien von Fischen. Arch. Entwickl.-Mech. Org. **145**, 549—560 (1952).
Phalen, G. S., Davenport, H. A.: Pericellular end bulbs in the central nervous system of vertebrates. J. comp. Neurol. **63**, 67—82 (1937).
Pierce, E. T.: An autoradiographic study of mouse brain stem histogenesis. Anat. Rec. **151**, 400 (1965).
— Histogenesis of the nuclei griseum pontis, corporis pontobulbaris and reticularis tegmenti pontis (Bechterew) in the mouse. An autoradiographic study. J. comp. Neurol. **126**, 219—240 (1966).
— Histogenesis of the dorsal and ventral cochlear nuclei in the mouse: An autoradiographic study. J. comp. Neurol. **131**, 27—38 (1967).
— Sweet, E.: Histogenesis of deep cerebellar nuclei studies autoradiographically with thymidine-H^3 in the mouse. Anat. Rec. **157**, 301 (1967).
Pilgrim, C. H., Maurer, W.: Autoradiographische Untersuchung über die Konstanz der DNS-Verdopplungs-Dauer bei Zellarten von Maus und Ratte durch Doppelmarkierung mit ^{3}/H- und ^{14} C-Thymidin. Exp. Cell Res. **37**, 183—199 (1965).
Placidi, G.-F., Masuoka, D. T.: Histochemical demonstration of fluorescent catecholamine terminals in cryostat sections of brain tissue. J. Histochem. Cytochem. **16**, 491—492 (1968).
Polyak, S.: The vertebrate visual system. Chicago, Illinois: Chicago Univ. Press 1957.
Porter, R. W., Davenport, H. A.: Golgi's dichromate-silver method 1. Effects of embedding, 2. Experiments with modifications. Stain Technol. **24**, 117—126 (1949).
Post, J., Hoffman, J.: Some effects of tritiated thymidine as a deoxyribonucleic acid label in the rat liver. Radiat. Res. **14**, 713—720 (1961).
Powell, T. P. S.: Transneuronal cell degeneration in the auditory relay nuclei. In: Cytology of Nervous Tissue (Taylor and Francis, eds.), p. 68—69. London 1961.
— Transneuronal cell degeneration in the olfactory bulb shown by the Golgi method. Nature (Lond.) **215**, 425—426 (1967).
— The fine structure of neurons in the dorsal horn of the cat spinal cord. J. comp. Neurol. **132**, 275—301 (1968a).
— Dorsal root projections to dorsal horn neurons in the cat spinal cord. J. comp. Neurol. **132**, 303—329 (1968b).
— The synaptic organization of lemniscal projections to the ventro-basal thalamus of the cat. Brain Res., in press (1969).

Powell, T. P. S., Cowan, W. M.: An experimental study of the projection of the cochlea. J. Anat. (Lond.) **96**, 269—284 (1962).
— — A note on retrograde fiber degeneration. J. Anat. (Lond.) **98**, 579—585 (1964).
— — The interpretation of the degenerative changes in the intralaminar nuclei of the thalamus. J. Neurol. Neurosurg. Psychiat. **30**, 140—153 (1967).
— Eruklar, S. D.: Transneuronal cell degeneration in the auditory relay nuclei of the cat. J. Anat. (Lond.) **96**, 249—268 (1962).
— Guillery, R. W., Cowan, W. M.: A quantitative study of the fornix-mamillo-thalamic system. J. Anat. (Lond.) **91**, 419—435 (1957).
— Kruger, L.: The thalamic projections upon the telencephalon in *Lacerta viridis*. J. Anat. (Lond.) **94**, 528—542 (1960).
Prestige, M. C.: Cell turnover in the spinal ganglia of *Xenopus laevis* tadpoles. J. Embryol. exp. Morph. **13**, 63—72 (1965).
Price, J. L.: The termination of centrifugal fibres in the olfactory bulb. Brain Res. **7**, 483—486 (1968).
Puck, T. T.: Studies on the life cycle of mammalian cells. In: *Human Genetics*. Cold Spring Harbor Symposis on Quantitative Biology. Cold Spr. Harb. Lab. quant. Biol. **29**, 167—176 (1964).
Quastler, H., Sherman, F. G.: Cell population kinetics in the intestinal epithelium of the mouse. Exp. Cell Res. **17**, 420—438 (1959).
Radmanesh, H.: Cell sizes and frequencies in the human nucleu sintercalatus. Anat. Rec. **160**, 729—734 (1968).
Rahbek, H., Omar, M.: Measurement of the thickness of transparent films with the light-profile microscope. Nature (Lond.) **169**, 1008—1009 (1952).
Raisman, G., Cowan, W. M., Powell, T. P. S.: The extrinsic afferent, commissural and association fibres of the hippocampus. Brain Res. **88**, 963—996 (1965).
— — — An experimental analysis of the efferent projection of the hippocampus. Brain Res. **89**, 83—108 (1966).
Rakic, P., Sidman, R. L.: Supravital DNA synthesis in the developing human and mouse brain. J. Neuropath. exp. Neurol. **27**, 246—276 (1968a).
— — Subcommissural organ and adjacent ependyma: Autoradiographic study of their origin in the mouse brain. Amer. J. Anat. **122**, 317—335 (1968b).
— — Telencephalic origin of pulvinar neurons in the fetal human brain. In press (1969).
Rall, W.: Theory of physiological properties of dendrites. Ann. N.Y. Acad. Sci. **96**, 1071—1092 (1962).
— Burke, R. E., Smith, T. G., Nelson, P. G., Frank, K.: Dendritic location of synapses and possible mechanisms for the monosynaptic EPSP in motor neurons. J. Physiol. (Lond.) **30**, 1168—1193 (1967).
— Shepard, G. M., Reese, T. S., Brightman, M. W.: Dendrodentritic pathway for inhibition in the olfactory bulb. Exp. Neurol. **14**, 44—56 (1966).
Ralston, H. J., III: The organization of the substantia gelatinosa Rolandi in the cat lumbosacral spinal cord. Z. Zellforsch. **67**, 1—23 (1965).
— The fine structure of neurons in the dorsal horn of the cat spinal cord. J. comp. Neurol. **132**, 275—301 (1968a).
— Dorsal root projections to dorsal horn neurons in the cat spinal cord. J. comp. Neurol. **132**, 303—329 (1968b)
— The synaptic organization of lemniscal projections to the ventro-basal thalamus of the cat. Brain Res., **14**, 99—116 (1969).
Ramón y Cajal, S., *see* Cajal, S. Ramon y.
Ramón-Moliner, E.: A chlorate-formaldehyde modification of the Golgi method. Stain Technol. **32**, 105—116 (1957).
— A tungstate modification of the Golgi-Cox method. Stain Technol. **33**, 19—29 (1958).
— The histology of the postcruciate gyrus in the cat. II. A statistical analysis of the dentritic distribution. J. comp. Neurol. **117**, 63—76 (1961a).
— The histology of the post-cruciate gyrus in the cat. III. Further observations. J. comp. Neurol. **117**, 229—249 (1961b).

Ramón-Moliner, E.: A source of error in the study of Golgi-stained material. Arch. ital. Biol. **105**, 139—148 (1967).
— Vane, M. A., Fletcher, G. V.: Basic dye counterstaining of sections impregnated by the Golgi-Cox method. Stain Technol. **39**, 65—70 (1964).
Rasmussen, G. L.: Selective silver impregnation of synaptic endings. In: New Research Techniques of Neuroanatomy (Windle, W. F., ed.), p. 27—39. Springfield: Thomas 1957.
Rauch, H. C., Raffel, S.: Immuno fluorescent localization of encephalitogenic protein in myelin. J. Immunol. **92**, 452—455 (1964).
Recklinghausen, V.: Eine Methode, mikroskopisch hohle und solide Gebilde voneinander zu unterscheiden. Virchows Arch. path. Anat. **19**, 451—459 (1860).
Reese, T. S., Brightman, M. W.: Electron microscope studies on the rat olfactory bulb. Anat. Rec. **151**, 492 (1965).
— — Personal communication (1968).
— Karnovsky, M. J.: Fine structural localization of a blood-brain barrier to exogenous peroxidase. J. Cell Biol. **34**, 207—217 (1967).
Reichard, P., Estborn, B.: Utilization of desoxyribosides in the synthesis of polynucleotides. J. biol. Chem. **188**, 839—846 (1951).
Reichel, W.: Lipofuscin pigment accumulation and distribution in five rat organs as a function of ages. J. Geront. **23**, 145—153 (1968).
Reimer, L.: Elektronenmikroskopische Untersuchungs- und Präparationsmethoden, 2. Aufl., S. 598. Berlin-Heidelberg-New York: Springer 1967.
Reis, D., Fuxe, K.: Depletion of noradrenaline in brain stem neurons during sham rage behavior produced by acute brain stem transection in cat. Brain Res. **7**, 448—451 (1968a).
— — Adrenergic innervation of the carotid sinus. Amer. J. Physiol. **215**, 1054—1057 (1968).
— Gunne, L.-M.: Brain catecholamines: Relation to the defense reaction evoked by amygdaloid stimulation in the cat. Science **149**, 450—451 (1965).
Reiss, F.: The editor's introduction. In: Neural Theory and Modeling (Reiss, F., ed.), p. 427. Stanford: University Press 1964.
Retzius, G.: Biografiska Anteckningar och Minnen (Biographical notes and recollections), vol. II, p. 145 (footnote). Stockholm: Almkvist & Wiksells Boktryckeri 1933.
Richardson, K. C.: Electron microscopic observations on Auerbach's plexus in the rabbit, with special reference to the problem of smooth muscle innervation. Amer. J. Anat. **103**, 99—136 (1958).
— The fine structure of the albino rabbit iris with special reference to the identification of adrenergic and cholinergic nerves and nerve endings in its intrinsic muscles. Amer. J. Anat. **114**, 173—205 (1964).
— Electron microscopic identification of autonomic nerve endings. Nature (Lond.) **210**, 756 (1966).
Riss, W., Knapp, H., Scalia, F.: Optic pathways in *Cryptobranchus alleghеniensis* as revealed by the Nauta technique. J. comp. Neurol. **121**, 31—43 (1963).
Ritzen, M.: Quantitative fluorescence microspectrophotometry of catecholamine-formaldehyde products. Model experiments. Exp. Cell Res. **44**, 505—520 (1966).
— Quantitative fluorescence microspectrophotometry of 5-hydroxytryptamine-formaldehyde products in models and in mast cells. Exp. Cell Res. **45**, 178—194 (1967a).
— Cytochemical identification and quantitation of biogenic monoamines: A microspectrofluorometric and autoradiographic study. M. D. Thesis, Stockholm (1967b).
— Hammarstrom, L., Ullberg, S.: Autoradiographic distribution of 5-hydroxytryptamine and 5-hydroxytryptophan in the mouse. Biochem. Pharmacol. **14**, 313—321 (1965).
Robertis de, E., *see* De Robertis, E.
Robertson, J. D.: The occurrence of a subunit pattern in the unit membranes of club endings in Mauthner cell synapses in goldfish brains. J. Cell Biol. **19**, 201—221 (1964).
— Current problems of unit membrane structure and contact relationships. In: Nerve as a Tissue (Rodahl, K., and B. Issekutz, eds.), p. 11—48. New York: Harper & Rowe, Publishers 1966.

Robertson, J. D., Bodenheimer, T. S., Stage, D. E.: The ultrastructure of Mauthner cell synapses and nodes in goldfish brains. J. Cell Biol. **19**, 159—199 (1963).

Robertson, J. S., Bond, V. P., Cronkite, E. P.: Resolution and image spread in autoradiographs of tritium-labeled cells. Int. J. appl. Radiat. **7**, 33—37 (1959).

Robertson, W. F.: A textbook of pathology in relation to mental diseases (Clay, W. F., ed.). Edinburgh 1900.

Robinson, S. H., Breacher, G.: Delayed incorporation of tritiated thymidine into DNA. Science **142**, 393 (1963).

Rogers, A. W.: Techniques of autoradiography, p. 335. New York: American Elseview Publ. Co. 1967.

Romanes, G. J.: The motor cell columns of the lumbo-sacral spinal cord of the cat. J. comp. Neurol. **94**, 313—363 (1951).

Romer, A.: The development of atomic physics. Canton, New York: Publication of the author (1955).

Ronai, P. E.: The use of sodium sulphide to facilitate tail vein injections in mice and rats. Transplantation **4**, 208—209 (1966).

Rose, J. E.: The cortical connections of the reticular complex of the thalamus. Res. Publ. Ass. nerv. ment. Dis. **30**, 454—479 (1952).

— Woolsey, C. N.: A study of the thalamocortical relations in the rabbit. Bull. Johns Hopk. Hosp. **73**, 65—128 (1943).

— — Structure and relations of limbic cortex and anterior thalamic nuclei in rabbit and cat. J. comp. Neurol. **89**, 279—348 (1948).

Roth, R. L., Gomez, D., Voneida, T. J., Edwards, J. S.: A modification of the Nauta technique for fresh water teleosts and insects. Stain Technol., in press (1969).

Routtenberg, A., Sladek, J., Bondareff, W.: Histochemical fluorescence after application of neurochemicals to caudate nucleus and septal area *in vivo*. Science **161**, 272—274 (1968).

Rowland, L. P., Mettler, F. A.: Cell concentration and laminar thickness in the frontal cortex of psychotic patients; studies on cortex removed at operation. J. comp. Neurol. **90**, 255—280 (1949).

Ruben, R. J.: Development of the inner ear of the mouse. A radioautographic study of terminal mitoses. Acta oto-Laryng., Suppl. **220**, 1—44 (1967).

Rubinson, K.: Projections of the optic tectum of the frog. Brain Behav. Evol., in press (1969).

Rude, S.: Monoamine-containing neurons in the nerve cord and body wall of *Lumbricus terrestris*. J. comp. Neurol. **128**, 397—412 (1966).

Ruiz-Marcos, A., Valverde, F.: The temporal evolution of the distribution of dendritic spines in the visual cortex of normal and dark raised mice. Exp. Brain Res. **8**, 284—294 (1969).

Ryzen, M.: A microphotometric method of cell enumeration within the cerebral cortex of man. J. comp. Neurol. **104**, 233—245 (1956).

Sabatini, D. D., Bensch, K., Barrnett, R. J.: Cytochemistry and electron microscopy. The preservation of cellular ultrastructure and enzymatic activity by aldehyde fixation. J. Cell Biol. **17**, 19—58 (1963).

Sachs, Ch.: Demonstration of metaraminol in nervous and non-nervous tissues by a histochemical fluorescence reaction. Biochem. Pharmacol. **14**, 1773—1778 (1966).

— Noradrenaline uptake mechanism in mouse atrium. Acta physiol. scand. (1970) Suppl. (in press).

Sainte-Marie, G.: The distribution of the autofluorescent cells and of the yellow autofluorescent granules in the rat tissues. Anat. Rec. **153**, 71—83 (1965).

Saint-Victor, M. N. de: Sur une nouvelle action de la lumiere. C. R. Acad. Sci. (Paris) **65**, 505—507 (1867).

Sakai, S.: Some observations on the cortico-thalamic fiber connections in the monkey. Proc. Jap. Acad. **43**, 822—826 (1967).

Saltpeter, M. M.: General area of autoradiography at the electron microscope level. In: Methods in Cell Physiology (Prescott, D. M., ed.), p. 229—253. New York and London: Academic Press 1966.

Sano, Y., Odakes, G., Yonezawa, T.: Fluorescence microscopic observations of catecholamines in cultures of the sympathetic chains. Z. Zellforsch. **80**, 345—352 (1967).

Sauer, M. E.: Radioautographic study of the location of newly synthesized deoxyribonucleic acid in the neural tube of the chick embryo: Evidence for intermitotic migration of nuclei. Anat. Rec. **133**, 456 (1959).

— Walker, B. E.: Radioautographic study of interkinetic nuclear migration in the neural tube. Proc. Soc. exp. Biol. (N.Y.) **101**, 557—560 (1959).

Saunders, J. W., Jr., Fallon, J. F.: Cell death in morphogenesis. In: Major Problems in Developmental Biology (Locke, M., ed.), p. 289—314. New York and London: Academic Press 1966.

Scalia, F., Halpern, M., Knapp, H., Riss, W.: The efferent connections of the olfactory bulb of the frog: A study of degenerating unmyelinated fibers. J. Anat. (Lond.) **103**, 245—262 (1968).

Schade, J. P., Backer, H. van, Colon, E.: Quantitative analysis of neuronal parameters in the maturing cerebral cortex. In: Progress in Brain Research, vol. 4, Growth and Maturation of the Brain. Amsterdam: Elsevier 1964.

Schadewald, M.: Effects of cutting the trochlear and abducens nerves on the end-bulbs about the cells of the corresponding nuclei. J. comp. Neurol. **74**, 239—246 (1941).

— Transynaptic effect of neonatal axon section on bouton appearance about somatic motor cells. J. comp. Neurol. **77**, 739—746 (1942).

Scheibel, M. E., Scheibel, A. B.: Some structural and functional substrates of development in young cats. In: The Developing Brain (Himwich, W., and H. Himwich, eds.), p. 6—25. Elsevier Press, Amsterdam 1964.

— — The organization of the nucleus reticularis thalami: A Golgi study. Brain Res. **1**, 43—62 (1966a).

— — Spinal motoneurons, interneurons and Renshaw cells. A Golgi study. Arch. ital. Biol. **104**, 328—353 (1966b).

— — Anatomical basis of attention mechanisms in vertebrate brains. In: The Neurosciences (Quarton, G. C., T. Melnechuk, and F. O. Schmitt, eds.), p. 577—607. New York: The Rockfeller University Press 1967.

— — A structural analysis of spinal interneurons and Renshaw cells. In: The Interneuron (Brazier, M. A. B., ed.). Berkeley: University of California Press 1969a (in press).

— — Terminal patterns in cat spinal cord. III. Primary afferent collaterals. Brain Res., in press (1969b).

Schiebler, T. H., Heene, R.: Nachweis von Katecholaminen im Rattenherzen während der Entwicklung. Histochemie **14**, 328—334 (1968).

Schimert, J.: Die Endigungsweise des Tractus vestibulospinalis. Z. Ges. Anat. I. Z. Anat. Entwickl.-Gesch. **108**, 761—767 (1938).

— Das Verhalten der Hinterwurzelkollateralen im Rückenmark. Z. Anat. Entwickl.-Gesch. **109**, 665—687 (1939).

Schneider, G. E.: Retinal projections characterized by differential rate of degeneration revealed by silver impregnation. Anat. Rec. **160**, 423 (1968).

Schuknecht, J. F.: Neuroanatomical correlates of auditory sensitivity and pitch discrimination in the cat. In: Neural Mechanisms of the Auditory and Vestibular Systems (Rasmussen, G. L., and W. Windle, eds.), p. 76—90. Springfield, Ill.: Ch. C. Thomas 1960.

Schultz, R. L., Case, N. M.: Microtubule loss with acrolein and bicarbonate-containing fixatives. J. Cell Biol. **38**, 633—637 (1968).

— Karlsson, U.: Fixation of the central nervous system for electron microscopy by aldehyde perfusion. II. Effect of osmolarity, pH of perfusate, and fixative concentration. J. Ultrastruct. Res. **12**, 187—206 (1965).

— Pease, D. C.: Cicatriz formation in rat cerebral cortex as revealed by electron microscopy. Amer. J. Path. **35**, 1017—1042 (1959).

Schultze, M.: Allgemeines über die Strukturenelemente des Nervensystems. In: Handbuch der Lehre von den Geweben des Menschen und der Thiere (Stricker, S., ed.), vol. 1, p. 108—136. Leipzig: Engelmann 1871.

Sefton, A. J.: The innervation of the lateral geniculate nucleus and anterior colliculus in the rat. Vision Res. **8**, 867—881 (1968).

Shariff, G. A.: Cell counts in the primate cerebral cortex. J. comp. Neurol. **98**, 381—400 (1953).
Sharp, P. J., Pollett, B. K.: The distribution of monoamines in the hypothalamus of the Japanese quail (*Corturnix coturnix japonica*). Z. Zellforsch. **90**, 262 (1968).
Sheinin, J. J.: Typing of the cells of the mesencephalic nucleus of the trigeminal nerve in the dog, based on Nissl-granule arrangement. J. comp. Neurol. **50**, 109—131 (1930).
Sholl, D. A.: The quantitative investigation of the vertebrate brain and the applicability of allonetic formulae to its study. Proc. roy. Soc. **135**, 243—258 (1948).
— Dendritic organization in the neurons of the visual and motor cortices of the cat. J. Anat. (Lond.) **87**, 387—406 (1953).
— The organization of the cerebral cortex. London: Methuen 1956.
— A comparative study of the neuronal packing density in the cerebral cortex. J. Anat. (Lond.) **93**, 143—158 (1959).
Sidman, R. L.: The structure and concentration of solids in photoreceptor cell studied by refractometry and interference microscopy. J. biophys. biochem. Cytol. **3**, 15—30 (1957).
— Histogenesis of mouse retina studied with thymidine-H^3. In: The Structure of the Eye (Smelser, G. K., ed.), p. 487—506. New York: Academic Press 1961.
— Organ culture analysis of inherited retinal degeneration in rodents. J. nat. Cancer Inst., monogr. No 11, 227—246 (1963).
— Development of interneuronal connections in brains of mutant mice. In: Physiological and Biochemical Aspects of Nervous Integration (Carlson, F. D., ed.), p. 163—193 (Soc. of Gen. Physiologists monograph). New Jersey: Prentice-Hall 1968.
— Miale, I.: Histogenesis of the mouse cerebellum studies by autoradiography with tritiated thymidine. Anat. Rec. **133**, 429—430 (1959).
— Feder, N.: Cell-proliferation in the primitive ependymal zone: An autoradiographic study of histogenesis in the nervous system. Exp. Neurol. **1**, 322—333 (1959a).
— — Cell migrations in the immature neuroepithelium of mouse embryos studies by autoradiography with tritiated thymidine. Anat. Rec. **133**, 430 (1959b).
— Mottla, P. A., Feder, N.: Improved polyester wax embedding method for histology. Stain Technol. **36**, 279—283 (1961).
— Rakic, P.: Cerebral origin of pulvinal thalamic neurons in the human fetal brain. (Abstr.). Anat. Rec. **160**, 428—429 (1968).
Silver, M. L.: The motorneurons of the spinal cord of the frog. J. comp. Neurol. **77**, 1—33 (1942).
Sisken, J. E.: Methods for measuring the length of the mitotic cycle and the timing of DNA synthesis for mammalian cells in culture. In: Methods in Cell Physiology (Prescott, D. M., ed.), p. 387—401. New York and London: Academic Press 1964.
Sjöberg, N.-O.: The adrenergic transmitter of the female reproductive tract: Distribution and functional changes. Acta physiol. scand., Suppl. 305 (1967).
Sjöstrand, F.: Über die Eigenfluoreszenz tierischer Gewebe mit besonderer Berücksichtigung der Säugetierniere. M. D. Thesis, Stockholm (1944).
Skougaard, M. R., Stewart, P. A.: Comparative effectiveness of intraperitoneal and intramuscular ^{3}/H-TDR injection routes in mice. Exp. Cell Res. **45**, 158—166 (1967).
Smart, I.: The subependymal layer of the mouse brain and its cell production as shown by radioautography after thymidine-H^3/injection. J. comp. Neurol. **116**, 325—347 (1961).
— Leblond, C. P.: Evidence for division and transformations of neuroglia cells in the mouse brain, as derived from radioautography after injection of thymidine-H^3/. J. comp. Neurol. **116**, 349—367 (1961).
Smith, C. A., Rasmussen, G. L.: Degeneration in the efferent nerve endings in the cochlea after axonal section. J. Cell Biol. **26**, 63—77 (1965).
Smith, J. M., O'Leary, J. L., Harris, A. B., Gay, A. J.: Ultrastructural features of the lateral geniculate nucleus of the cat. J. comp. Neurol. **123**, 357—378 (1964).
Smith, K. R., Hudgens, R. W., O'Leary, J. L.: An electron microscopic study of degenerative changes in the cat cerebellum after extrinsic and intrinsic lesion. J. comp. Neurol. **126**, 15—36 (1966).
Smith, W. W., Breacher, G., Stohlman, F., Jr., Cornfield, J.: Toxicity of tritiated thymidine to bone marrow transplants. Radiat. Res. **16**, 201—207 (1962).

Snodgress, A. B., Dorsey, C. H.: Egg albumin embedding: A procedure compatible with neurological staining techniques. Stain Technol. **38**, 149—155 (1963).

Song, C. W., Tabachnick, J.: Epidermal cell population kinetics in albino guinea-pig skin. Nature (Lond.) **217**, 650—651 (1968).

Sotelo, C.: Permanence of postsynaptic specializations in the frog sympathetic ganglion cells after denervation. Exp. Brain Res. **6**, 294—305 (1968).

— Palay, S. L.: The fine structure of the lateral vestibular nucleus in the rat. I. Neurons and neuroglial cells. J. Cell Biol. **36**, 151—180 (1968).

Sourkes, T. L., Poirier, L. J.: Neurochemical bases of tremor and other disorders of movement. Canad. med. Ass. J. **94**, 53—60 (1966).

Spoendlin, H., Lichtensteiger, W.: The adrenergic innervation of the labyrinth. Acta oto-laryng. (Stockh.) **61**, 423—434 (1965).

Sprague, J. M.: The distribution of dorsal root fibres on motor cells in the lumbosacral spinal cord of the cat, and the site of excitatory and inhibitory terminals in monosynaptic pathways. Proc. roy. Soc. B **149**, 534—556 (1958).

— Meyer, M.: An experimental study of the fornix in the rabbit. J. Anat. (Lond.) **84**, 354—368 (1950).

Spriggs, T. L. B., Lever, J. D., Rees, P. M., Graham, J. D. P.: Controlled formaldehyde-catecholamine condensation in cryostat sections to show adrenergic nerves by fluorescence. Stain Technol. **41**, 323—327 (1966).

Staroscik, R. N., Jenkins, W. H., Mendelsohn, M. L.: Availability of tritiated thymidine after intravenous administration. Nature (Lond.) **202**, 456—458 (1964).

Steel, G. G.: Delayed uptake by tumors of tritium from thymidine. Nature (Lond.) **210**, 806—810 (1966).

Stell, W. K.: Correlated light and electron microscope observations on Golgi preparations of goldfish retina. J. Cell Biol. **23**, 89 A (1964).

— Thesis (for Doctor of Philosophy in Anatomy), University of Chicago (1965a).

— Correlation of retinal cytoarchitecture and ultrastructure in Golgi preparations. Anat. Rec. **153**, 389—397 (1965b).

— The structure and relationships of horizontal cells and photoreceptor-bipolar synaptic complexes in goldfish retina. Amer. J. Anat. **121**, 401—424 (1967).

Sterling, P., Kuypers, H. G. J. M.: Simultaneous demonstration of normal boutons and degenerating nerve fibers and their terminals in the spinal cord. J. Anat. (Lond.) **100**, 723—732 (1966).

Sternschlein, E.: Das Ganglion cervicale supremun nach prae- und postganglionärer Durchschneidung. Arb. neurol. Inst. Wien **23**, 155—176 (1920).

Stevens, A. R.: High resolution autoradiography. In: Methods in Cell Physiology (Prescott, D. M., ed.), p. 255—310. New York and London: Academic Press 1966.

Stohlman, F., Jr. (ed.): In: The Kinetics of Cellular Proliferation, p. 456. New York and London: Grune and Stratton 1959.

Stotler, W. A.: An experimental study of the cells and connections of the superior olivary complex of the cat. J. comp. Neurol. **98**, 401—423 (1953).

Stowell, R. E.: Effect on tissue volume of various methods of fixation, dehydration, and embedding. Stain Technol. **16**, 67—83 (1941).

Stretton, A. O. W., Kravitz, E. A.: Neuronal geometry: Determination with a technique of intracellular dye injection. Science **162**, 132—134 (1968).

Sullivan, P. R., Kuten, J., Atkinson, M. S., Angevine, J. B., Yakovlev, P. I.: Cell count in the lateral geniculate nucleus of man. Neurol. **8**, 566—567 (1958).

Swank, R. L., Davenport, H. A.: Marchi's staining method: Studies of some of the underlying mechanisms involved. Stain Technol. **9**, 11—19 (1934).

Szentágothai, J.: The anatomical basis of synaptic transmission of excitation and inhibition in motorneurons. Acta morph. Acad. Sci. hung. **8**, 287—309 (1958).

— The structure of the synapse in the lateral geniculate body. Acta anat. (Basel) **55**, 166—185 (1963).

— Flerkó, B., Mess, B., Halasz, B.: Hypothalamic control of the anterior pituitary. Budapest: Akademiai Kiado 1968.

Szentágothai, J., Hámori, J., Tömböl, T.: Degeneration and electron microscope analysis of the synaptic glomeruli in the lateral genicualate body. Exp. Brain Res. **2**, 283—307 (1966).

Tal, I.: Modificazione al metodo di Golgi nella preparazione delle cellule gangliari nel sistema nervoso centrale. Z. wiss. Mikr. **4**, 497 (1886).

Tan Creti, D. M.: Production by an IBM 1401 computer of numbered paper insets for serial celloidin sections. Stain Technol. **43**, 44—45 (1968).

Tang, D. B., Ebbesson, S. O. E.: A comparison of the sampling errors of a systematic sampling method with complete random sampling in the estimation of total number of nerve fibers. (Submitted for publication) (1969).

Taxi, J.: Étude au microscope electronique de synapses ganglionnaires chez quelques vertébrés. In: IVth Internat. Congress of Neuropathology (Jacob, H., ed.), vol. 2, p. 197—203. Stuttgart: Thieme 1962.

— Contribution a l'étude des connexions des neurones moteurs du système nerveux autonome. Ann. Sci. Nat. Zool. Biol. Anim. **7**, 143—674 (1965).

Tello, F.: Terminaciones en los musculos estriados. Trab. Lab. Invest. Biol. Univ. Madr. **4**, 105—114 (1906).

— Degeneration et regeneration des plaques motrices apes la section des nerfs. Trab. Lab. Invest. Biol. Univ. Madr. **5**, 117—149 (1907).

Terry, R. J., Gordon, J.: The effects of unilateral and bilateral enucleation on optic lobe development and pigmentation of the skin in *Rana catesbiana* larvae. J. exp. Zool. **143**, 245—257 (1960).

— Roland, A. L., Race, J.: Effect of eye enucleation and eyelid closure upon the brain and associated visual structures in the mouse. I. A report of degenerative changes. J. exp. Zool. **150**, 165—183 (1962).

Thieme, G.: A versatile device for microscopic spectrofluorometry. Acta physiol. scand. **67**, 514—520 (1966).

Thompson, H. B.: The total number of functional cells in the cerebral cortex of man, and the percentage of the total volume of the cortex composed of nerve cell bodies, calculated from Karl Hammarberg's data; together with a comparison of the number of giant cells with the number of pyramidal fibers. J. comp. Neurol. **9**, 113—140 (1899).

Thrasher, J. D.: Analysis of renewing epithelial cell populations. In: Methods in Cell Physiology (Prescott, D. M., ed.), p. 323—357. New York and London: Academic Press 1966.

Tomasch, J., Malpass, A. J.: The human motor trigeminal nucleus. Anat. Rec. **130**, 91—102 (1958).

Torvik, A.: Afferent connections to the sensory trigeminal nuclei, the nucleus of the solitary tract and adjacent structures: An experimental study in the rat. J. comp. Neurol. **106**, 51—141 (1956a).

— Transneuronal changes in the inferior olive and pontine nuclei in kittens. J. Neuropath. exp. Neurol. **15**, 119—145 (1956b).

Tower, D. B.: Structural and functional organization of mammalian cerebral cortex: The correlation of neurone density with brain size. J. comp. Neurol. **101**, 19—51 (1954).

Tramezzani, J. H., Chiocchio, S., Wassermann, G. F.: A technique for light and electron microscopic identification of adrenalin-and-noradrenalin-storing cells. J. Histochem. Cytochem. **12**, 890—899 (1964).

Tranzer, J. P., Thoenen, H.: An electron microscopic study of selective, acute degeneration of sympathetic nerve terminals after administration of 6-hydroxydopamine. Experientia (Basel) **24**, 155 (1968).

Tsang, Y.: Visual centers in blinded rats. J. comp. Neurol. **66**, 211—261 (1937).

Turcotte, A., Ramón-Moliner, E.: Counterstaining solution for sections stained with the Golgi-Cox method. Stain Technol. **40**, 310 (1965).

Ungerstedt, U.: 6-hydroxydopamine induced degeneration of central monoamine neurons. Europ. J. Pharmacol., in press (1968).

— Butcher, L. L., Butcher, S. G., Andén, N.-E., Fuxe, K.: Direct chemical stimulation of dopaminergic mechanisms in the neostriatum of the rat. Brain Res., in press (1969).

Valverde, F.: Studies on the Piriform Lobe. Cambridge: Harvard University Press 1965.
— The pyramidal tract in rodents. A study of its relations with the posterior column nuclei, dorsolateral reticular formation of the medulla oblongata end cervical spinal cord (Golgi and electron microscopic observations). Z. Zellforsch. **71**, 297—363 (1966).
— Apical dendritic spines of the visual cortex and light deprivation in the mouse. Exp. Brain Res. **3**, 337—352 (1967).
— Structural changes in the area striata of the mouse after enucleation. Exp. Brain Res. **5**, 274—292 (1968).
— Esteban, M. E.: Peristriate cortex of mouse: Location and the effects of enucleation on the number of dendritic spines. Brain Res. **9**, 145—148 (1968).
— Ruiz-Marcos, A.: Dendritic spines in the visual cortex of the mouse. Introduction to a mathematical model. Exp. Brain Res. **8**, 269—283 (1969).
Van Buren, J. M.: The retinal ganglion cell layer. Springfield, Illinois: Ch. C. Thomas 1963a.
— Trans-synaptic retrograde degeneration in the visual system of primates. J. Neurol. Neurosurg. Psychiat. **26**, 402—409 (1963b).
Van Buskirk, C.: The seventh nerve complex. J. comp. Neurol. **82**, 303—333 (1945).
Van Crevel, H., Verhaart, W. J. C.: The rate of secondary degeneration in the central nervous system. I. The pyramidal tract of the cat. J. Anat. (Lond.) **97**, 429—449 (1963a).
— — The rate of secondary degeneration in the central nervous system. II. The optic nerve of the cat. J. Anat. (Lond.) **97**, 451—464 (1963b).
Van der Loos, H.: Une combinaison de deux vieilles methodes histologiques pour le système nerveux central. Mschr. Psychiat. Neurol. **132**, 330—334 (1956).
— Similarities and dissimilarities in submicroscopic morphology of interneuronal contact sites of presumably different functional character. In: Topics of Basic Neurology. Progress in Brain Research, vol. 6 (Bargmann, W., and J. P. Scháde, eds.), p. 43—58. Amsterdam: Elsevier 1964.
Van Gehuchten, A.: La dégénérescence dite rétrograde ou dégénérescence Wallérienne indirecte. Névraxe **5**, 1—107 (1903).
— Anatomie du Système Nerveux de l'Homme, ed. 2 (1906).
— Les lois de la dégénérescence Wallérienne directe. Névraxe **11**, 73—130 (1910).
— Molhant, M.: Les lois de la dégénérescence Wallérienne directe. Névraxe **11**, 73—130 (1910).
Van Harreveld, A., Crowell, J., Malhotra, S.: A study of extracellular space in central nervous tissue by freeze substitution. J. Cell Biol. **25**, 117—137 (1965).
Van Orden, L. S., Bensch, K. G., Giarman, N. J.: Histochemical and functional relationships of catecholamines in adrenergic nerve endings. II. Extravesicular norepinephrine. J. Pharmacol. exp. Ther. **155**, 428—439 (1966a).
— — Langer, S. Z., Trendelenburg, U.: Histochemical and fine structural aspects of the onset of denervation supersensitivity in the nicitating membrane of the spinal cat. J. Pharmacol. exp. Ther. **157**, 274—283 (1967).
— Bloom, F. E., Barrnett, R. J., Giarman, N. J.: Histochemical and functional relationships of catecholamines in adrenergic nerve endings. I. Participation of granular vesicles. J. Pharmacol. exp. Ther. **154**, 185—199 (1966b).
— Wugman, I., Giarman, N. J.: 5-hydroxytryptamine in single neuroplastic mast cells: A microscopic spectrofluorometric study. Science **148**, 642—644 (1965).
Vaughn, J. E., Peters, A.: Aldehyde fixation of nerve fibers. J. Anat. (Lond.) **100**, 687 (Abstract) (1966).
Virolainen, M.: Intravenous injections in small laboratory rodents. Transportation **5**, 1530—1532 (1968).
Von Economo, C. F., Koskinas, G. N.: Die Cytoarchitektonik der Hirnrinde des erwachsenen Menschen. Wien u. Berlin: Springer 1925.
Von Monakow, C.: Über einige durch Exstirpation circumscripter Hirnrindenregionen bedingte Entwicklungshemmungen des Kaninchengehirns. Arch. Psychiat. Nervenkr. **12**, 141—156 (1882).
— Experimentelle und pathologisch-anatomische Untersuchungen über die optischen Centren und Bahnen. Arch. Psychiat. Nervenkr. **20**, 714—789 (1889).

Voneida, T. J., Ebbesson, S. O. E.: Pallial commissural systems in the Tegu lizard (*Tupinambis nigropunctatus*). (Abstract.) Anat. Rec. **160**, 443 (1968).
— — Pallial commissures in the Tegu lizard (*Tubinampis nigropunctatus*). In preparation (1969).
Walberg, F.: Descending connections to the inferior olive. An experimental study in the cat. J. comp. Neurol. **104**, 77—173 (1956).
— Role of normal dendrites in removal of degenerating terminal boutons. Exp. Neurol. **8**, 112—124 (1963).
— The early changes in degenerating boutons and the problem of argyrophilia. Light and electron microscopical observations. J. comp. Neurol. **122**, 113—137 (1964).
— An electron microscopic study of terminal degeneration in the inferior olive of the cat. J. comp. Neurol. **125**, 205—222 (1965).
— The fine structure of the cuneate nucleus in normal cats following interruption of afferent fibers: An electron microscope study with particular references to findings made in Glees and Nauta sections. Exp. Brain Res. **2**, 107—128 (1966).
— Bowsher, D., Brodal, A.: The termination of primary vestibular fibers in the vestibular nuclei in the cat. An experimental study with silver methods. J. comp. Neurol. **110**, 391—419 (1958).
— Mugnaini, E.: Distinction of degenerating fibres and boutons of cerebellar and peripheral origin in the Deiter's nucleus of the same animal. Brain Res., in press (1969).
Walker, A. E.: The primate thalamus. Chicago, Illinois: Chicago University Press 1938.
Wall, P. D., Glees, P., Fulton, J. F.: Corticofugal connections of posterior orbital surface in rhesus monkey. Brain **74**, 66—71 (1951).
Waller, A.: Experiments on the section of the glossopharyngeal and hyperglossal nerves of the frog, and observations of the alterations produced thereby in the structure of their primitive fibers. Phil. Trans. **140**, 423—469 (1850).
Walsh, F. B.: Clinical neuro-ophthalmology. Baltimore, Maryland: Williams & Wilkins 1947.
Warrington, W. B.: On the structural alterations observed in nerve cells. J. Physiol. (Lond.) **23**, 112—129 (1898, 1899).
Webster, K. E.: The cortico-striatal projection in the cat. J. Anat. (Lond.) **99**, 328—329 (1965).
Weisschedel, E., Jung, R.: Die anatomische Auswertung und das Studium der sekundären Faserdegeneration nach lokalisierter subcorticaler Ausschaltung durch Elektrokoagulation. Z. Anat. Entwickl.-Gesch. **109**, 374—395 (1939).
Weston, J. A.: A radioautographic analysis of the migration and localization of trunk neural crest cells in the chick. Develop. Biol. **6**, 279—310 (1963).
Westrum, L. E.: Electron microscopy of degeneration in the prepyriform cortex. J. Anat. (Lond.) **100**, 683—685 (1966).
— Lund, R. D.: Formalin perfusion for correlative light and electron microscopical studies of the nervous system. J. Cell Sci. **1**, 229—238 (1966).
— White, L. E., Jr., Ward, A. A.: Morphology of the experimental epileptic focus. J. Neurosurg. **21**, 1033—1046 (1964).
Whaley, W. M., Govindarchari, T. R.: Organic Reactions, vol. 6. Coll. New York, N.Y.: John Wiley & Sons, Inc. 1951.
White, L. E., Jr.: Ipsilateral afferents to the hippocampal formation in the albino rat. I. Cingulum projections. J. comp. Neurol. **113**, 1—42 (1959).
— Olfactory bulb projections of the rat. Anat. Rec. **152**, 465—480 (1965).
— Westrum, L. E.: Dendritic spine changes in prepyriform cortex following olfactory bulb lesions — rat: Golgi method. Anat. Rec. **148**, 410—411 (1964).
Wiesel, T. N., Hubel, D. H.: Effects of visual deprivation on morphology and physiology of cells in the cat's lateral geniculate body. J. Neurophysiol. **26**, 978—993 (1963).
— — Extent of recovery from the effects of visual deprivation in kittens. J. Neurophysiol. **28**, 1060—1072 (1965).
Wigglesworth, V. B.: The use of osmium in the fixation and staining of tissues. Proc. roy. Soc. B **147**, 185—199 (1957).

Wimber, D. E.: Methods for study of cell proliferation with emphasis on DNA labels. In: Cell Proliferation (Lamerton, L. F., and R. J. M. Fry, eds.), p. 1—17. Philadelphia: F. A. Davis Co. 1963.
— Lamerton, L. F.: Cell population kinetics in the intestine of continuously irradiated mice, using double-labelling autoradiography. Radiat. Res. **28**, 694—700 (1966).
— Quastler, H.: A^{14}/C- and ^{3}/H-thymidine double labeling technique in the study of cell proliferation in Tradescantia root tips. Exp. Cell Res. **30**, 8—22 (1963).
Winkler, C. A.: The olfactory tract in the rabbit. In: Opera Omnia, vol. 5, p. 397—413. Haarlem: F. Bohn 1918.
Wirsén, C.: Adrenergic innervations of adipose tissue examined by fluorescence microscopy. Nature (Lond.) **202**, 913 (1964).
— Distribution of adrenergic nerve fibers in brown and white adipose tissue. In: Handbook of Physiology (Renold, A. E., and G. F. Cahill, Jr., eds.). Amer. Physiol. Soc., Sect. 5. Washington 1965.
— Hamberger, B.: Catecholamines in brown fat. Nature (Lond.) **214**, 625 (1967).
Wolman, M.: The reaction of osmium tetroxide with tissue components. Exp. Cell Res. **12**, 231—240 (1957).
Wood, J. G.: Electron microscopic localization of 5-hydroxytryptamine (5-HT). Tex. Rep. Biol. Med. **23**, 828—837 (1965).
— Barrnett, R. J.: Histochemical demonstration of norepinephrine at a fine structural level. J. Histochem. Cytochem. **12**, 197—209 (1964).
Wuerker, R. B., Palay, S. L.: Neurofilaments and microtubules in ventral horn cells of the rat. In: Tissue and Cell; in press (1968).

Subject Index

Universitätsdruckerei H. Stürtz AG, Würzburg

www.ingramcontent.com/pod-product-compliance
Lightning Source LLC
Chambersburg PA
CBHW081124180926
51611CB00004B/47
9783642859885